Advances in Parasitology

Volume I

Advances in Parasitology
Volume I

Edited by **Cherilyn Jose**

R<small>EFERENCE</small> **C**<small>ALLISTO</small>

New York

Published by Callisto Reference,
106 Park Avenue, Suite 200,
New York, NY 10016, USA
www.callistoreference.com

Advances in Parasitology: Volume I
Edited by Cherilyn Jose

International Standard Book Number: 978-1-63239-050-9 (Hardback)

Contents

Preface

The study of parasites, which focuses on parasitic interaction with the host is called parasitology. Parasites can be in living form as an intracellular/intercellular parasite, or as an ectoparasite, which do not let any disease or discomfort happen to host. There are two types of parasites: Non-pathogenic parasites, which never cause any diseases and Pathogenic parasites, which causes diseases to the host. Pathogenic parasites are responsible for various diseases spread amongst domesticated animals across the globe. The numerous threats imposed by pathogenic parasites attract the parasitologists and experts in biology and medicine.

This book presents the impact of parasitology in medicinal studies and public issues of health. Intense researches on genetically evolved parasites have been discussed in this book to help in the development of chemotherapeutic strategies. The reason for the growth in parasite invasion in organisms, no matter they are human, or animals, or aquatic creatures, can be attributed to excessive world population and its impact on agriculture and aquaculture.

Instead of organizing the book into a pre-formatted table of contents with chapters, sections and then asking the authors to submit their respective chapters based on this frame, the authors were encouraged by the publisher to submit their contributions based on their area of expertise. The editor was then commissioned to examine the reading material and put it together as a book.

I thank all the authors for allocating much of their scarce time to this project. Not only do I appreciate their participation, but also their adherence as a group to the time parameters set for this publication. I hope that this book proves to be a resourceful guide for both basic and advanced concepts in parasitology research. Finally, I hope that this book will achieve success, serving as a support to the readers and prove to be very useful to the scientific community.

Editor

Association between Micronutrients (Vitamin A, D, Iron) and Schistosome-Specific Cytokine Responses in Zimbabweans Exposed to *Schistosoma haematobium*

Liam Reilly,[1,2] **Norman Nausch,**[1] **Nicholas Midzi,**[3]
Takafira Mduluza,[4,5] **and Francisca Mutapi**[1]

[1] *Ashworth Laboratories, Institute for Immunology and Infection Research, School of Biological Sciences, The University of Edinburgh, King's Buildings, Edinburgh EH9 3JT, UK*
[2] *Institute of Tropical Medicine Antwerp, Nationalestraat 155, 2000 Antwerp, Belgium*
[3] *Schistosomiasis Section, National Institute of Health Research, Box CY 570, Causeway, Harare, Zimbabwe*
[4] *Biochemistry Department, University of Zimbabwe, P.O. Box MP 167, Mount Pleasant, Harare, Zimbabwe*
[5] *Harvard School of Public Health, Botswana Havard Aids Institute, P. Bag 320, Gaborone, Botswana*

Correspondence should be addressed to Liam Reilly, l.j.reilly@doctors.org.uk

Academic Editor: Sungano Mharakurwa

Micronutrients play an important role in the development of effective immune responses. This study characterised a populations exposed to schistosome infections in terms of the relationship between micronutrients and immune responses. Levels of retinol binding protein (RBP; vitamin A marker), vitamin D, ferritin and soluble transferrin receptor (sTfR), and C reactive protein (CRP) were related to levels of schistosome specific cytokines (IFN-γ, IL-4/5/10) in 40 Zimbabweans (7–54 years) exposed to *Schistosoma haematobium* infection. 67.2% of the participants were deficient in vitamin D. RBP levels were within normal ranges but declined with age. The two indicators of iron levels suggested that although levels of stored iron were within normal levels (normal ferritin levels), levels of functional iron (sTfR levels) were reduced in 28.6% of the population. Schistosome infection alone was not associated with levels of any of the micronutrients, but altered the relationship between parasite-specific IL-4 and IL-5 and levels of ferritin and sTfR.

1. Introduction

Micronutrients are known to play an important role in health and the development of an effective immune system. In tropical and subtropical regions there is an overlapping distribution of helminth infections and micronutrient deficiencies. [1–3]. Schistosomiasis is a global health burden with over 200 million people infected by one of five *Schistosoma* trematode species [1, 4, 5]. *Schistosoma haematobium* is the causative agent of urogenital schistosomiasis and is widely distributed in Africa [1]. Infection is linked to significant morbidity and functional disability [6]. Simultaneously, according to the Global Progress Report on vitamin and Mineral Deficiency, more than half of Africa's population lack critical vitamins and minerals. Deficiencies in iron and Vitamin A each

rank among the top 10 leading causes of death in developing countries through disease. A recent study in Nigeria showed that infection with *S. haematobium* affected growth and nutritional status of children [7]. It is clear that micronutrient supplementation though programmes such as Expanded Programme of Immunisation (EPI) and Child Health Days can help reduce under 5 mortality, which is the stated aim of millennium development goal 5. With growing calls for integrated approaches to improving human health, it is important to characterise the interaction between micronutrient deficiencies and the immune response to schistosomiasis so that public health programs can plan their interventions accordingly.

Acquired immunity to schistosomiasis develops slowly and only provides partial protection [8]. Schistosomes can survive in human hosts for up to 40 years [9]. Helminth infection including infection by schistosomes, modulates the host immune response, manifesting as diminished allergic responses, amelioration of autoimmune disease, and chronic parasitic infection [9–11]. Immunomodulation is mediated by regulatory T cells (T_{REG}) through direct contact stimulation and IL-10 production [12, 13]. While the switch to T_H2 which occurs during helminth infection is an effective antiparasitic response, it is unclear whether superimposition of regulatory responses primarily benefits the worms or the host. Downregulation of the inflammatory response would reduce host mediated immunopathology but also reduce protection [9, 14]. These effects are seen as a diminished allergic response, amelioration of autoimmune disease and chronic parasitic infection.

Traditionally vitamin A has been known for its role in vision, with deficiency resulting in xerophthalmia, which is the leading cause of preventable childhood blindness. However, it has a wide range of physiological functions and is essential for haematopoiesis and prevention of anaemia, as well as immune function. It is acquired from foods such as liver, milk, cheese, eggs, green leaves, carrots and ripe mangos. Infants acquire vitamin A through breast feeding [2]. Vitamin A has now been implicated in the development of T_H2, T_H17 and T_{REG} responses through the activation of retinoid receptors. Retinoic acid activates the FoxP3 transcription factor, which stimulates the development of naïve T cells into T_{REG} [15–17]. Vitamin supplementation studies suggest that adequate Vitamin A is required for normal antihelminthic responses [18]. Hypovitaminosis A is an immunodeficient state linked to decreased antibody production, typically diminished T_H2 antibodies IgE, IgG1, and IgA [19].

Vitamin D is historically known for its role in calcium and bone homeostasis. It is produced in the skin when 7-dehydrocholesterol reacts with UVB radiation to form vitamin D3, which modified in the liver to form 25(OH) vitamin D3, and converted to its active metabolite 1,25(OH)$_2$ vitamin D3 in the kidney [20]. Vitamin D2 and D3 can also be acquired from dietary sources. They are then metabolised by the liver in the same manner as cutaneously derived vitamin D3 [21]. A role has been suggested for vitamin D in diseases with an immunological aetiology such as psoriasis, multiple sclerosis and diabetes mellitus. It may also have a role in blood pressure homeostasis [21]. The immuno-regulatory functions of vitamin D are being increasingly understood. It suppresses the T_H1 cytokines IFN-γ and IL-2, and upregulates IL-4 to create a T_H2 polarisation. Vitamin D can stimulate T_{REG} through production of TGFβ-1 and CD25 expression by CD4$^+$ T cells [22–24]. It also diminishes expression of dendritic cell (DC) costimulatory markers CD40, CD80, and CD86, again linked to T_{REG} induction [14, 23].

Anaemia affects 1.62 billion people worldwide [25], and around 500 million of those people have iron deficiency anaemia. A causal relationship between infection with *S. japonicum* and iron deficiency anaemia has been established [26]. It is linked to increased infectious mortality and morbidity, and can itself be caused by chronic infection [27, 28]. Its relationship with infection is complex as both pathogen and host use body iron stores. It has been shown that iron supplementation during active infection can increase the infectious load of some pathogens [27, 29]. Experimental studies on mice have found that those with high iron indices had a significantly increased fibrosis around egg granulomata [26]. Iron deficiency is associated with IgG1, IgE, and T_{REG} responses whereas iron supplementation has been linked to T_H1 responses and decreased IL-10 [30, 31]. The soluble transferrin receptor (sTfR) is a diagnostic tool for differentiating between iron deficiency anemia (IDA) and anemia of chronic disease [32] since ferritin levels reflect amounts of stored iron while the sTfR reflects the functional iron compartment.

A few studies have shown a recent review of data collected in Zimbabwe between 1980 and 2006 showed that a significant proportion of preschool children, school children, and adult women (lactating or pregnant) experienced malnutrition with significant proportions of these groups suffering from vitamin A and iron deficiencies [33].

The aim of this study was to determine the relationship between the micronutrients vitamin A, D, and iron as well as a measure of inflammatory responses C-reactive protein (CRP) and schistosome-specific cytokine levels in Zimbabweans exposed to *S. haematobium* infection.

2. Methods

2.1. Ethical Statement. The study received ethical and institutional approval from the Medical Research Council of Zimbabwe and the University of Zimbabwe, respectively. Permission to conduct the work in this province was obtained from the Provincial Medical Director. Informed consent/assent was obtained from all participants or their parents/guardians prior to enrolment into the study. Project aims and procedures were explained to the community, school children, and their teachers prior the study, and survey was conducted amongst all compliant participants. After sample collection, all participants were offered treatment with the standard dose of 40 mg/Kg body weight of the antihelminthic drug Praziquantel.

2.2. Study Area and Population. The study was conducted in two rural villages in the Mashonaland East Province of Zimbabwe (31°30′E; 17°45′S) where *S. haematobium* is endemic. Participants were part of a larger immunoepidemiology study which was carried out between 2002 and 2005, and the study area is described in detail elsewhere [34]. The main activity in these villages is subsistence farming mainly of maize and vegetables. Drinking water is collected from open wells while bathing and washing is conducted in two main rivers in the villages. Most families maintain a garden located near the river where water is collected for watering the crops and the schools surveyed were all in close proximity to rivers.

All samples used in this study were obtained at baseline in 2002 were selected using following criteria: (1) participants should be life-long residents in this area (assessed by questionnaire), (2) should not have received antihelminthic treatment prior this study, (3) should have provided at least two urine and 2 stool samples on consecutive days to allow

Association between Micronutrients (Vitamin A, D, Iron) and Schistosome-Specific Cytokine Responses in Zimbabweans Exposed to Schistosoma haematobium

3

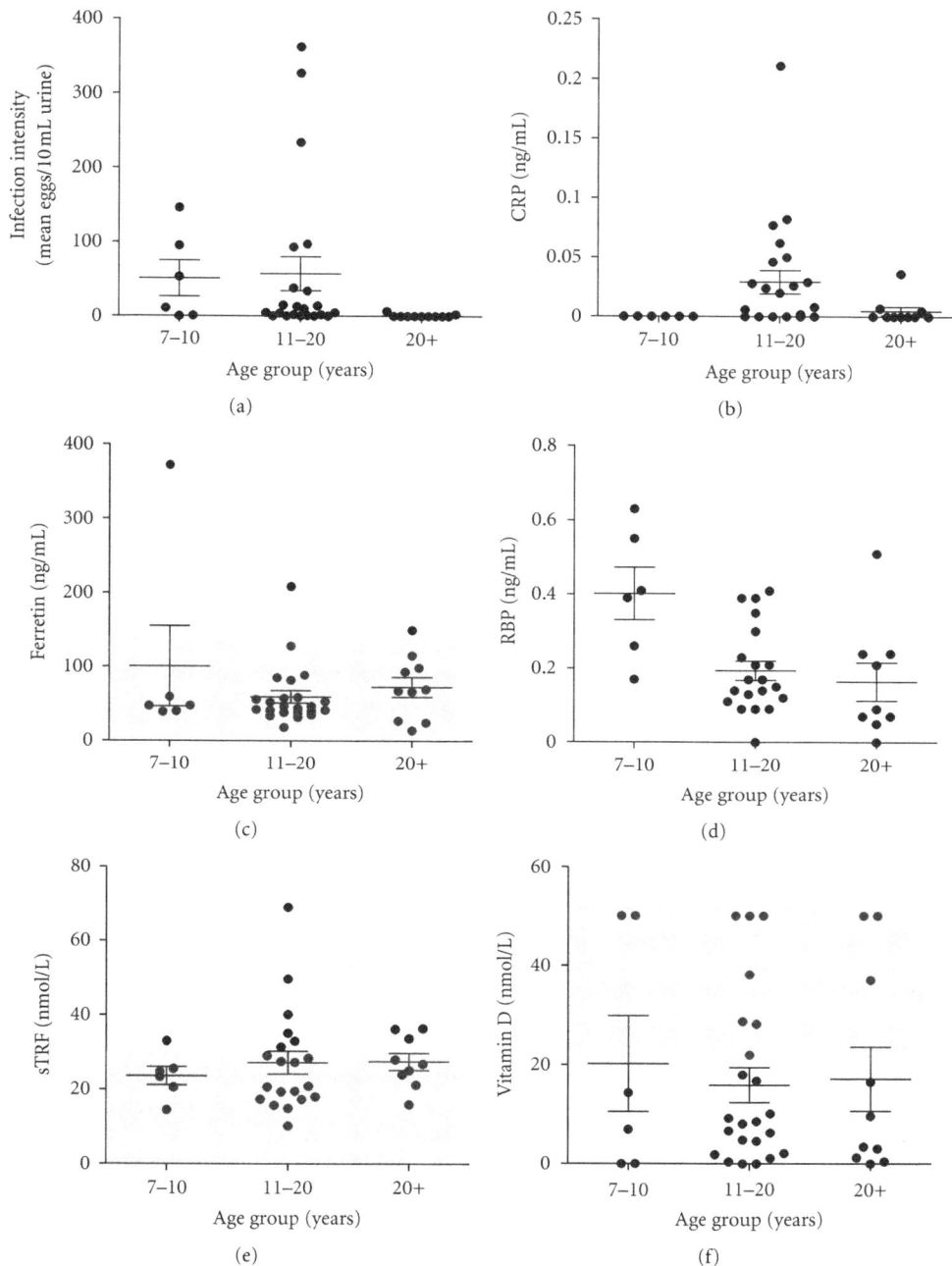

FIGURE 1: Age profiles of the host population infection and micronutrient levels. Samples for each age group are $n = 6$ for ≤10 years, $n = 23$ for 11–20 years, and $n = 11$ for 21+ years. Bars represent means and standard error of the mean. Shaded regions represent normal ranges of micronutrients. (a) Infection intensity, (b) C-reactive protein (CRP) levels, (c) ferritin levels (measure of stored iron levels), (d) retinol binding protein (RBP) levels (a measure of vitamin A levels), (e) soluble transferrin receptor (sTfR) levels (measure of functional iron levels), and (f) vitamin D levels.

parasitological diagnosis, (4) should have been test negative for soil transmitted helminth and *S. mansoni* as well as negative for HIV and *Plasmodium falciparum*, (5) should have provided a blood sample to obtain sera. Furthermore, only sera samples were used for these analyses, which have not been used previously and therefore were defrosted for the first time. Following these criteria samples from 40 people aged 7–54 years (13 male, 27 female) were included in this

study. Data were subsequently separated into 3 age groups: 7–10 years ($N = 6$), 11–20 years ($N = 23$), 21+ years ($N = 11$), which represent a typical age-infection profile for *S. haematobium* as shown in Figure 1(a).

2.3. Sample Collection. Parasitology samples (at least 2 urine and 2 stool samples collected on 3 three consecutive days) and 20 mL of venous blood were collected from each participant. Stool samples were processed following the Kato-Katz procedure [35] to detect *S. mansoni* eggs and other intestinal helminths, while the urine filtration method [36] was used to detect *S. haematobium* eggs in urine samples. Serum samples obtained from 20 mL of venous blood from each participant were frozen and stored in duplicate at $-20°C$ in the field and transferred to a $-80°C$ freezer in the laboratory. One complete set of the samples was subsequently transported frozen from Zimbabwe to the UK, stored at $-80°C$ and defrosted for the first time for use in this study. Small aliquots of blood were used to prepare thick and thin smears for the microscopic detection of *Plasmodium* parasites.

2.4. Immunoassays. The parasite-specific cytokines IFN-γ, (marker for T_H1 responses) IL-4, IL-5 (markers of T_H2 responses), and IL-10 (marker for regulatory responses) were measured by enzyme linked immunosorbent assays (ELISA) in supernatants obtained after stimulation of whole blood samples using cercarial, egg, and adult schistosome antigens following published methods [37]. Spontaneous cytokine production was determined in unstimulated controls containing media alone while the mitogen Concanavalin A (ConA) was used as a positive control for the restimulations. Values of cytokines obtained from the media alone incubations were subtracted from those of the antigen-specific restimulations to remove the effects of background cytokine production in the statistical analyses.

2.5. Micronutrient Assays. Micronutrients and C reactive Protein (CRP) were measured using enzyme linked immunosorbent assay (ELISA) kits according to manufacturers' instructions. Serum transferrin receptor (sTFR) is a marker of iron deficiency and is required for lymphocyte activation and proliferation. It was assayed using an ELISA kits from R&D Systems (Cat. #DTFR1). Ferritin is a marker of iron status, but rises with inflammation [27, 38] and this was measured by an ELISA kit from BioQuant (Cat. #BQ065T). CRP is an inflammatory marker [39] and was measured by an ELISA kit from Anogen (Cat. #EL 10022). Retinol Binding Protein a measure of vitamin A status [39] was assayed using an ELISA kit from Phoenix Pharmaceuticals (Cat. #EK-028-28), and 25(OH) vitamin D was used to assess the inactive vitamin D status [40] although through a kit from Immunodiagnostik (Cat. #K2110).

2.6. Statistical Analyses. Statistical analyses were performed using the software PASW 17 (formerly SPSS). Vitamin D status was described using previously published ranges (replete ≥ 50.00 nmol/L, mild deficiency 25.00–49.99 nmol/ L, moderate deficiency 12.50–24.99 nmol/L, severe deficiency ≤ 12.49 nmol/L) [41]. The World Health Organisation reference range for ferritin was used (female normal range 15.0–150.0 μg/L, male normal range 15.0–200.0 μg/L) [42]. R&D Systems provided a 2.5–97.5 percentile range (8.7– 28.1 nmol/l) for sTFR from a survey of 225 ethnically diverse participants of both sexes. Their mean value for Afro-Carribeans was significantly higher than for other ethnic groups. There is no peer-reviewed reference range

for sTFR [42]. There is no published reference range for RBP, although the World Health Organisation has produced retinol reference ranges for use in public health [43, 44]. The ratio of sTfR/log Ferritin (sTfR-F index) has been suggested as an alternative estimate of body iron, so this was also calculated in this study and used in the statistical analyses.

For the statistical analyses, host infection intensity was recorded into infection status, that is, infected and uninfected, cytokine absorbencies were square root transformed, and levels of all micronutrients were log transformed to satisfy the assumptions of parametric tests. In order to determine if the relationship between micronutrients and immune responses differed between schistosome infected versus uninfected people, a multivariate analysis of variance (MANOVA) was conducted. The dependent variables were the transformed micronutrient data and the independent variables were cytokine levels, infection status (infected/uninfected) age (categorical (7–10 years, 11–20 years, 21+ years)), sex (categorical male/female). The effects of interactions between infection status and micronutrients were also included in the MANOVA model. Sequential sums of squares were used to calculate the test statistics so that the potentially confounding effects of all other variables could be allowed for testing for the effects of infection status which was entered last in the single effects list. *P* values ≤ 0.05 were taken as significant.

3. Results

3.1. Population Characteristics. Schistosome infection prevalence in the study population was 60% (95% CI: 43–75%) and the mean infection intensity was 39.3 eggs/10 mL urine (SEM = 13.5) with a range of 0–362 eggs/10 mL urine. Infection intensity followed the typical schistosome age-infection pattern, rising with age to a peak in childhood and declining thereafter (Figure 1(a)). The age profiles of the micronutrients are given in Figures 1(b)–1(f). There is no reference range for RBP [45]. The study population had a mean RBP of 0.23 ng/mL with a range of 0–0.63 ng/mL. Most values for ferritin were within published ranges. 25(OH) vitamin D titres in this population were low when compared to published values with 32.8% ($n = 12$) of the population being classified as vitamin D replete (≥ 50.00 nmol/L); 17.9% ($n = 7$) were mildly deficient (25–49.90 nmol/L), 10.3% ($n = 4$) were moderately deficient (12.50–24.90 nmol/L), and 38.5% ($n = 15$) were severely deficient (≤ 12.49 nmol/L). Levels of CRP were within the normal range while 28.6% ($n = 10$) of the participants had elevated sTfR based on the 95 percentile data provided with the assay as detailed in the methods section.

The statistical analyses showed that sex affected only levels of ferritin, which was significantly lower in females and did not have a significant effect on levels of any of the other micronutrients (Table 1). Age significantly affected levels of RBP, with RBP levels falling with age ($r = -0.315, P = 0.033$) as shown in Figure 1, but did not affect levels of any of the other micronutrients of CRP. Although Figure 1(b) shows differences in the age profile of CRP levels, the statistical analyses show that after allowing for other variables such as

Association between Micronutrients (Vitamin A, D, Iron) and Schistosome-Specific Cytokine Responses in Zimbabweans Exposed to Schistosoma haematobium

5

TABLE 1: List of factors whose association with micronutrient levels was tested with ANOVA. F and P values are given for each factor.

	Sex F value (P value)	Age group F value (P value)	Schistosome infection status F value (P value)
Vitamin D	3.24 (0.083)	1.97 (0.160)	0.004 (0.953)
RBP	0.500 (0.485)	**5.39 (0.010)**	0.195 (0.663)
sTfR	1.28 (0.268)	0.639 (0.536)	0.482 (0.493)
Ferritin	**4.146 (0.050)** ($M > F$)	0.673 (0.517)	1.506 (0.229)
sTfR/ferritin ratio	0.255 (0.618)	0.388 (0.682)	1.294 (0.265)
CRP	1.710 (0.200)	2.670 (0.085)	0.652 (0.425)

The effects of the factors sex, age was allowed for first before testing for the effects of infection status on the micronutrient levels using sequential sums of squares to calculate the F value. Significant P values are highlighted in bold.

sex and for example, age, there are no significant differences in CRP levels between the 2 age groups.

3.2. Association between Parasite-Specific Cytokines and Levels of Micronutrients. Overall, there was a significant positive association between RBP and levels of parasite-specific IL-10 ($P = 0.049$, $\beta = 0.314$) as well as between ferritin and parasite-specific IL-4 ($P = 0.035$, $\beta = 0.317$). In some cases, the relationship between the cytokine levels and micronutrients varied with schistosome infection status as shown in Table 1. Thus, levels of vitamin D showed a significant negative correlation with IL-4 in egg positive children but no association in egg negative children (Figure 2(a)). Levels of parasite-specific IFN-γ showed a significant positive correlation with sTFR in egg negative people but a negative but nonsignificant association in egg positive people (Figure 2(b)). In egg positive people levels of parasite-specific IL-5 went down with ferritin levels but went up in egg positive people although this later association was not significant (Figure 2(c)). When considering the ratio of sTfR, levels of both IFN-γ and IL-4 went down with the sTfR-F index in egg positive people and up in egg negative people as shown in Figures 2(d) and 2(e).

4. Discussion

This study describes the micronutrient status of a rural black Zimbabwean population and then characterises the relationships between micronutrients and immune responses to schistosomiasis. While this study showed that there was vitamin D deficiency in the population, levels of all other micronutrients and markers of inflammation were within normal ranges. The global micronutrient report in 2001 has classified Zimbabwe as having a vitamin A deficiency prevalence of 10–15%. The study population had easy access to good dietary sources of micronutrients, including fortified foods (margarine and some vegetable oils during the study period were fortified with VitA) as well as from home-grown vegetables. Vegetables are amongst the prominent cash crops for commercial and small-scale farmers [46]. This may explain why the population was predominantly micronutrient replete. Iron supplementation for pregnant women at ante-natal clinics and targeted vitamin A supplementation

were not commenced in Zimbabwe until 2 years after this current study was conducted [33].

In this study serum retinol levels declined with age which is contradictory to reports from primary aged school children in Zimbabwe and Kenya [47, 48] which show retinol levels increasing with age. Work on RBP levels in exercise programs in South Korean women revealed a larger decrease in older women than younger women after a structured exercise regime [49]. This is consistent with our finding that RBP decreased with age, since our study captures a wider age range than the 2 previous studies in primary school children. However, the major occupation amongst our population is subsistence farming and so they are likely to be more physically active, therefore it is not clear whether our observations represent a normal decline in RBP with age, or whether there is an interaction between physical activity, age, and RBP level.

Friis et al. found no association between *S. haematobium* infections with serum retinol levels in Zimbabwe, similar to observations in this current study. Interestingly, Friis et al. found, a strong negative association between *S. mansoni* infection and serum retinol levels in both Zimbabwe and Kenya, which suggests that the intestinal niche of *S. mansoni* infection may interfere with vitamin A absorption [47, 48]. However, experimental studies show that vitamin A deficiency leads to reduced schistosome-specific antibody responses [50], which may suggest that vitamin A deficiecy leads to susceptibility to *S. mansoni* infections. However, all participants of our study were negative for *S. mansoni* and therefore it was excluded as confounding factor.

It has also been shown that all trans retinoic acid (ATRA) binds retinoic acid receptors, which induce FoxP3 expression polarizing immune responses towards a regulatory phenotype [16]. Our finding that RBP is correlated with IL-10 suggests that vitamin A may be important in augmenting schistosome-specific regulatory responses.

Vitamin D produced the most surprising data, with 38.5% of subjects being severely deficient. There is a paucity of Vitamin D surveys in Africa compared to those conducted in Western countries. Since no clinical examination were conducted in this study, it is impossible to say whether the deficiencies observed in this study results are associated with pathology or remained asymptomatic. Production of pre-vitamin D_3 occurs in the skin under the influence of

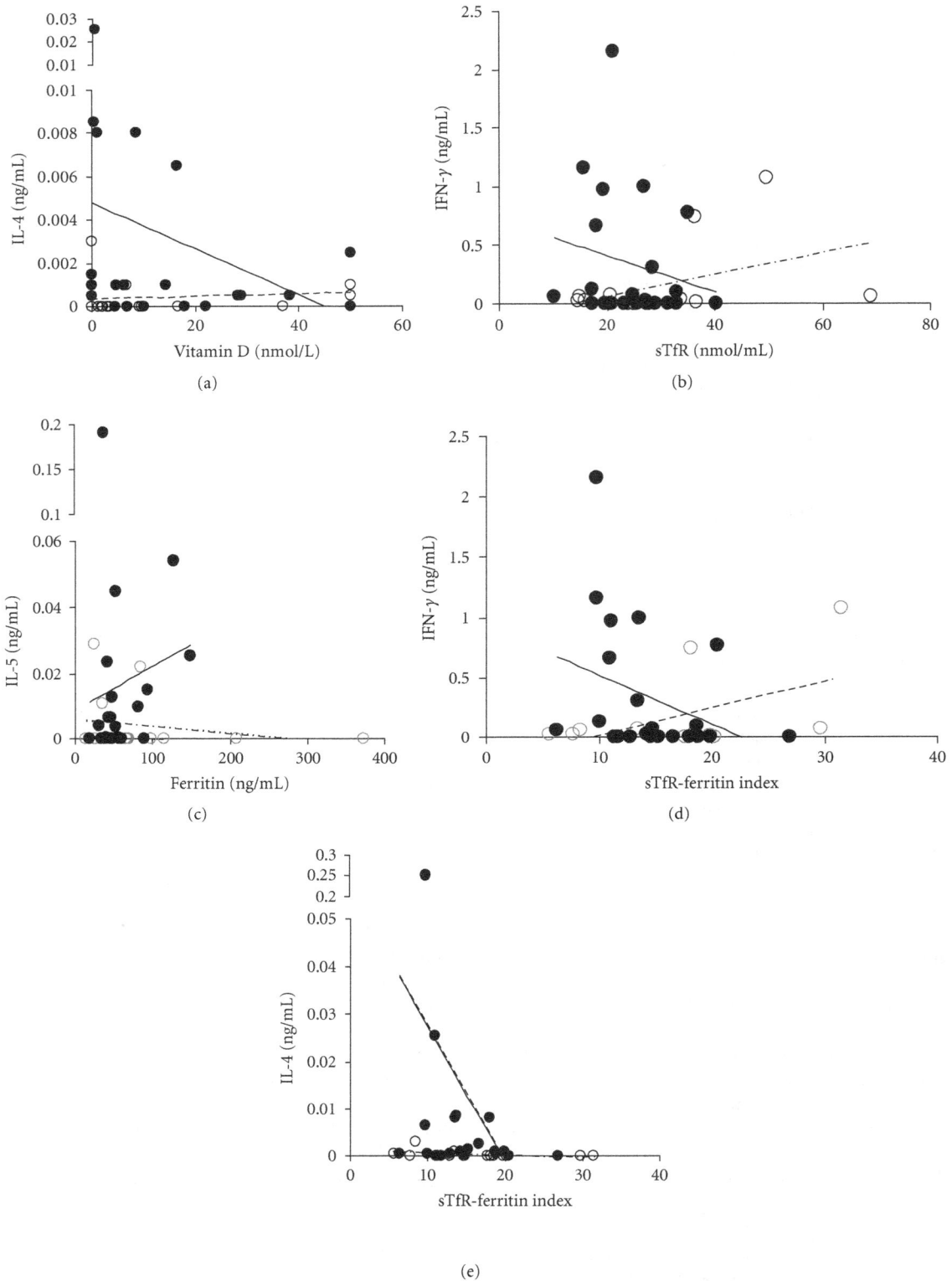

FIGURE 2: Relationship between micronutrients and cytokines showing associations significant that are significant from the ANOVA analyses (Table 2). Solid symbols and lines indicate egg positive people, open symbols and dashed lines represent egg negative people. (a) IL-4 level versus vitamin D, (b) IFN-γ versus soluble transferrin receptor (sTfR), (c) IL-5 versus ferritin levels (measure of stored iron levels), and (d) IFN-γ versus sTfR-F index (ratio soluble transferrin receptor/log Ferritin), a measure of stored and functional iron levels. (e) IL-4 versus sTfR-F index (ratio soluble transferrin receptor/log ferritin), a measure of stored and functional iron levels.

Association between Micronutrients (Vitamin A, D, Iron) and Schistosome-Specific Cytokine Responses in Zimbabweans
Exposed to Schistosoma haematobium

7

TABLE 2: F and P values obtained from ANOVA determining the association between cytokine levels and micronutrient levels.

| | Vit D | RBP | sTfR | Ferritin | sTfR/ferritin ratio | CRP |
	F value (P value)	F value (P value)	F value (P value)	F value (P value)	F value (P value)	F value (P value)
IFN-γ	0.008 (0.931)	0.153 (0.701)	0.017 (0.898)	1.741 (0.206)	0.081 (0.779)	0.256 (0.616)
IFN-γ* infection status	0.047 (0.790)	0.806 (0.383)	**4.631 (0.047)**	1.312 (0.269)	**7.516 (0.011)**	0.009 (0.926)
IL-4	**9.662 (0.004)**	0.105 (0.751)	2.649 (0.123)	0.288 (0.599)	2.218 (0.140)	0.543 (0.467)
IL-4* infection status	**10.487 (0.003)**	0.894 (0.358)	4.412 (0.052)	0.126 (0.727)	**7.702 (0.010)**	0.465 (0.500)
IL-5	0.311 (0.560)	0.001 (0.975)	0.293 (0.596)	1.005 (0.331)	0.236 (0.631)	1.793 (0.190)
IL-5* infection status	0.003 (0.960)	0.742 (0.402)	0.122 (0.732)	**10.706 (0.005)**	0.080 (0.780)	0.006 (0.937)
IL-10	1.509 (0.237)	**5.786 (0.023)**	0.001 (0.970)	0.875 (0.364)	0.042 (0.838)	0.520 (0.476)
IL-10* infection status	0.372 (0.550)	2.831 (0.104)	1.336 (0.265)	0.237 (0.633)	0.038 (0.846)	2.243 (0.144)

The effects of the potential confounders sex, age was allowed for first before testing for the effects of the cytokine and the interaction between cytokine and infection status using sequential sums of squares to calculate the F value. Significant P values are highlighted in bold.

ultraviolet light. Most studies of Vitamin D levels have been in Caucasian populations with reference to osteoporosis. It is possible that our findings may be explained by ethnic differences in skin pigmentation and skin UV penetration [41, 51]. Given that the reference ranges come from studies on osteoporosis, they may not be applicable in Zimbabwean population. Nonetheless, they remain an important starting point for analysis and suggest that further work is required to examine the biological relevance of these categories to immunology [52]. In this study, vitamin D levels in egg negative children showed a significant positive association with IL-4 levels, consistent with the role of vitamin D in upregulating IL-4 to polarize responses towards a T_H2 phenotype [23].

Iron deficiency is one of the most prevalent micronutrient deficiencies in the world affecting at least half of all pregnant women and young children in developing countries. In a survey conducted by the Ministry of Health and Child Welfare in 1997, 9% of the surveyed population (pregnant women, lactating women, preschool children, and adult males) had depleted iron stores that is, ferritin. At the time of the study, pregnant and postpartum women were not offered iron supplementation by local healthcare providers, thus pregnancy and childbirth-related iron and blood loss may explain why male participants have significantly higher levels of ferritin. In this study, while ferritin levels were within normal ranges, sTfR levels were elevated in 28.6% of the population. Ferritin is an indicator of stored iron reserves in the body while sTfR indicates the functional iron component of the body and becomes elevated soon after the onset of iron deficiency. Ferritin is often decreased in iron deficiency anaemia, but can be raised in inflammatory conditions [27, 39]. However, we observed normal CRP levels, which excluded excess inflammation in the participants. Similarly the lack of association between schistosome infection intensity/status and levels of sTfR implies that schistosome infection does not explain the elevated levels of sTfR. In this population the measures of body iron (sTfR-F index) showed a negative association with IFN-γ and IL-4 in egg positive people, while IL-5 levels showed a positive association with ferritin in the same people. Iron replete people use iron in mounting inflammatory immune responses [26]. Iron supplementation has been shown

to increase dendritic cell stimulation and promote T_H1 responses [30], but also an increased burden of immunopathology in those already infected [26]. However, increased IFN-γ is seen in iron deficiency, where it has a role in preserving iron stores [27, 28]. Thus in this population the inverse association between measures of body iron and the cytokines IFN-γ and IL-4 may be adaptive to preserving iron stores during schistosome infection. However, in the absence of mechanistic studies, this remains speculative.

In conclusion the study showed that while levels of vitamin A and iron where within normal ranges, there was a deficiency of vitamin D in 67.2% of the study population as well as elevated levels of sTfR in 28.6% of the participants. Thus, the 2 indicators of iron levels suggested that although levels of stored iron were within normal levels (normal ferritin levels), levels of functional iron (measure by sTfR) may have been reduced in some participants. Schistosome infection intensity or status was not associated with levels of any of the micronutrients, but altered the relationship between parasite-specific IL-4 and IL-5 and the measures of iron levels (ferritin and sTfR). Cohort studies following a larger group of people through a cycle of antihelminthic treatment will clarify the effects of helminth infection on micronutrient levels and their subsequent effect on immune responses.

Acknowledgments

The investigation received financial support from the Medical Research Council, UK, The Wellcome Trust, UK (Grant no. WT082028MA), and the Carnegie Trust for the Universities of Scotland. F. Mutapi acknowledges support from the MRC, UK (Grant no. G81/538). The authors are grateful for the co-operation of the Ministry of Health and Child Welfare in Zimbabwe, the Provincial Medical Director of Mashonaland East, the Environmental Health Workers, residents, teachers and school children in Mutoko and Rusike. We also thank Members of the National Institutes for Health Research (Zimbabwe) for technical support.

References

[1] P. Steinmann, J. Keiser, R. Bos, M. Tanner, and J. Utzinger, "Schistosomiasis and water resources development: systematic

review, meta-analysis, and estimates of people at risk," *Lancet Infectious Diseases*, vol. 6, no. 7, pp. 411–425, 2006.

[2] World Health Organization, Global prevalence of vitamin A deficiency in population at risk: 1995–2005 In: *WHO report*, Geneva, Switzerland, 2009, http://www.who.int/vmnis/vitamina/prevalence/report/en/.

[3] A. C. D. Tiong, M. S. Patel, J. Gardiner et al., "Health issues in newly arrived Afican refugees attending general practice clinics in Melbourne," *Medical Journal of Australia*, vol. 185, no. 11-12, pp. 602–606, 2006.

[4] A. O. Laosebikan, S. R. Thomson, and N. M. Naidoo, "Schistosomal portal hyptertension," *Journal of the American College of Surgeons*, vol. 200, no. 5, pp. 795–806, 2005.

[5] D. Engels, L. Chitsulo, A. Montresor, and L. Savioli, "The global epidemiological situation of schistosomiasis and new approaches to control and research," *Acta Tropica*, vol. 82, no. 2, pp. 139–146, 2002.

[6] B. Gryseels, K. Polman, J. Clerinx, and L. Kestens, "Human schistosomiasis," *Lancet*, vol. 368, no. 9541, pp. 1106–1118, 2006.

[7] C. J. Uneke and M. U. Egede, "Impact of urinary schistosomiasis on nutritional status of school children in south-eastern Nigeria," *The Internet Journal of Health*, vol. 9, no. 1, 2009.

[8] M. E. J. Woolhouse and P. Hagan, "Seeking the ghost of worms past," *Nature Medicine*, vol. 5, no. 11, pp. 1225–1227, 1999.

[9] D. W. Dunne and A. Cooke, "A worm's eye view of the immune system: consequences for evolution of human autoimmune disease," *Nature Reviews Immunology*, vol. 5, no. 5, pp. 420–426, 2005.

[10] C. F. Anderson, M. Oukka, V. J. Kuchroo, and D. Sacks, "CD4+CD25-Foxp3- Th1 cells are the source of IL-10-mediated immune suppression in chronic cutaneous leishmaniasis," *Journal of Experimental Medicine*, vol. 204, no. 2, pp. 285–297, 2007.

[11] P. Smith, N. E. Mangan, C. M. Walsh et al., "Infection with a helminth parasite prevents experimental colitis via a macrophage-mediated mechanism," *Journal of Immunology*, vol. 178, no. 7, pp. 4557–4566, 2007.

[12] S. G. Kang, H. W. Lim, O. M. Andrisani, H. E. Broxmeyer, and C. H. Kim, "Vitamin A metabolites induce gut-homing FoxP3+ regulatory T cells," *Journal of Immunology*, vol. 179, no. 6, pp. 3724–3733, 2007.

[13] C. Mottet and D. Golshayan, "CD4+CD25+Foxp3+ regulatory T cells: from basic research to potential therapeutic use," *Swiss Medical Weekly*, vol. 137, no. 45-46, pp. 625–634, 2007.

[14] E. van Riet, F. C. Hartgers, and M. Yazdanbakhsh, "Chronic helminth infections induce immunomodulation: consequences and mechanisms," *Immunobiology*, vol. 212, no. 6, pp. 475–490, 2007.

[15] K. M. Elias, A. Laurence, T. S. Davidson et al., "Retinoic acid inhibits Th17 polarization and enhances FoxP3 expression through a Stat-3/Stat-5 independent signaling pathway," *Blood*, vol. 111, no. 3, pp. 1013–1020, 2008.

[16] F. Schambach, M. Schupp, M. A. Lazar, and S. L. Reiner, "Activation of retinoic acid receptor-α favours regulatory T cell induction at the expense of IL-17-secreting T helper cell differentiation," *European Journal of Immunology*, vol. 37, no. 9, pp. 2396–2399, 2007.

[17] C. B. Stephensen, R. Rasooly, X. Jiang et al., "Vitamin A enhances in vitro Th2 development via retinoid X receptor pathway," *Journal of Immunology*, vol. 168, no. 9, pp. 4495–4503, 2002.

[18] K. Z. Long, T. Estrada-Garcia, J. L. Rosado et al., "The effect of vitamin A supplementation on the intestinal immune response in Mexican children is modified by pathogen infections and diarrhea," *Journal of Nutrition*, vol. 136, no. 5, pp. 1365–1370, 2006.

[19] C. B. Stephensen, X. Jiang, and T. Freytag, "Vitamin A deficiency increases the in vivo development of IL-10-positive Th2 cells and decreases development of Th1 cells in mice," *Journal of Nutrition*, vol. 134, no. 10, pp. 2660–2666, 2004.

[20] W. G. Tsiaras and M. A. Weinstock, "Factors influencing vitamin d status," *Acta Dermato-Venereologica*, vol. 91, no. 2, pp. 115–124, 2011.

[21] A. Mithal, D. A. Wahl, J. P. Bonjour et al., "Global vitamin D status and determinants of hypovitaminosis D," *Osteoporosis International*, vol. 20, no. 11, pp. 1807–1820, 2009.

[22] M. T. Cantorna and B. D. Mahon, "Mounting evidence for vitamin D as an environmental factor affecting autoimmune disease prevalence," *Experimental Biology and Medicine*, vol. 229, no. 11, pp. 1136–1142, 2004.

[23] E. Van Etten and C. Mathieu, "Immunoregulation by 1,25-dihydroxyvitamin D3: basic concepts," *Journal of Steroid Biochemistry and Molecular Biology*, vol. 97, no. 1-2, pp. 93–101, 2005.

[24] M. D. Griffin, N. Xing, and R. Kumar, "Vitamin D and its analogs as regulators of immune activation and antigen presentation," *Annual Review of Nutrition*, vol. 23, pp. 117–145, 2003.

[25] World Health Organisation, Worldwide prevalence of anaemia 1993–2005. In: *WHO report*. Geneva, Switzerland, 2008, http://whqlibdoc.who.int/publications/2008/9789241596657_eng.pdf.

[26] C. J. McDonald, M. K. Jones, D. F. Wallace, L. Summerville, S. Nawaratna, and V. N. Subramaniam, "Increased iron stores correlate with worse disease outcomes in a mouse model of schistosomiasis infection," *PLoS ONE*, vol. 5, no. 3, Article ID e9594, 2010.

[27] J. Jason, L. K. Archibald, O. C. Nwanyanwu et al., "The effects of iron deficiency on lymphocyte cytokine production and activation: preservation of hepatic iron but not at all cost," *Clinical and Experimental Immunology*, vol. 126, no. 3, pp. 466–473, 2001.

[28] G. Weiss, "Pathogenesis and treatment of anaemia of chronic disease," *Blood Reviews*, vol. 16, no. 2, pp. 87–96, 2002.

[29] S. M. Grant, J. A. Wiesinger, J. L. Beard, and M. T. Cantorna, "Iron-deficient mice fail to develop autoimmune encephalomyelitis," *Journal of Nutrition*, vol. 133, no. 8, pp. 2635–2638, 2003.

[30] S. Bisti, G. Konidou, F. Papageorgiou, G. Milon, J. R. Boelaert, and K. Soteriadou, "The outcome of Leishmania major experimental infection in BALB/c mice can be modulated by exogenously delivered iron," *European Journal of Immunology*, vol. 30, no. 12, pp. 3732–3740, 2000.

[31] S. Bisti and K. Soteriadou, "Is the reactive oxygen species-dependent-NF-κB activation observed in iron-loaded BALB/c mice a key process preventing growth of Leishmania major progeny and tissue-damage?" *Microbes and Infection*, vol. 8, no. 6, pp. 1473–1482, 2006.

[32] B. S. Skikne, C. H. Flowers, and J. D. Cook, "Serum transferrin receptor: a quantitative measure of tissue iron deficiency," *Blood*, vol. 75, no. 9, pp. 1870–1876, 1990.

[33] T. H. Gadaga, R. Madzima, and N. Nembaware, "Status of micronutrient nutrition in Zimbabwe: a review," *African Journal of Food, Agriculture, Nutrition and Development*, vol. 9, pp. 502–522, 2009.

[34] F. Mutapi, R. Burchmore, T. Mduluza et al., "Praziquantel treatment of individuals exposed to *Schistosoma haematobium*

Association between Micronutrients (Vitamin A, D, Iron) and Schistosome-Specific Cytokine Responses in Zimbabweans
Exposed to Schistosoma haematobium

9

enhances serological recognition of defined parasite antigens," *Journal of Infectious Diseases*, vol. 192, no. 6, pp. 1108–1118, 2005.

[35] N. Katz, A. Chaves, and J. Pellegrino, "A simple device for quantitative stool thick-smear technique in Schistosomiasis mansoni," *Revista do Instituto de Medicina Tropical de Sao Paulo*, vol. 14, no. 6, pp. 397–400, 1972.

[36] K. E. Mott, "A reusable polyamide filter for diagnosis of *S. haematobium* infection by urine filtration," *Bulletin de la Societe de Pathologie Exotique et de ses Filiales*, vol. 76, no. 1, pp. 101–104, 1983.

[37] F. Mutapi, G. Winborn, N. Midzi, M. Taylor, T. Mduluza, and R. M. Maizels, "Cytokine responses to *Schistosoma haematobium* in a Zimbabwean population: contrasting profiles for IFN-γ, IL-4, IL-5 and IL-10 with age," *BMC Infectious Diseases*, vol. 7, article 139, 2007.

[38] G. Porto and M. De Sousa, "Iron overload and immunity," *World Journal of Gastroenterology*, vol. 13, no. 35, pp. 4707–4715, 2007.

[39] J. G. Erhardt, J. E. Estes, C. M. Pfeiffer, H. K. Biesalski, and N. E. Craft, "Combined measurement of ferritin, soluble transferrin receptor, retinol binding protein, and C-reactive protein by an inexpensive, sensitive and simple sandwich enzyme-linked immunosorbent assay technique," *Journal of Nutrition*, vol. 134, no. 11, pp. 3127–3132, 2004.

[40] M. F. Holick, "High prevalence of vitamin D inadequacy and implications for health," *Mayo Clinic Proceedings*, vol. 81, no. 3, pp. 353–373, 2006.

[41] P. Lips, "Which circulating level of 25-hydroxyvitamin D is appropriate?" *Journal of Steroid Biochemistry and Molecular Biology*, vol. 89-90, pp. 611–614, 2004.

[42] World Health Organisation, Iron deficiency anaemia: assessment, prevention and control. A guide for programme managers. In: *WHO guidelines*. Geneva, Switzerland, 2001, http://www.who.int/nutrition/publications/en/ida_assessment_prevention_control.pdf.

[43] World Health Organisation, Serum retinol concentrations for determing the prevalence of vitamin A deficiency in populations. In: *WHO guidelines*. Geneva, Switzerland, 2011, http://www.who.int/vmnis/indicators/retinol.pdf.

[44] World Health Organisation, Vitamin and Mineral Requirements in Human Nutrition. In: *WHO report*. Geneva, Switzerland, 1998, http://whqlibdoc.who.int/publications/2004/9241546123.pdf.

[45] World Health Organisation, Guidelines for the evaluation of soil-transmitted helminthiasis and schistosomiasis at community level. In: *WHO guidelines*. Geneva, Switzerland, World health Organisation; 1998, http://whqlibdoc.who.int/hq/1998/WHO_CTD_SIP_98.1.pdf.

[46] P. Konje, "Market Brief Focus on Mashonaland East," 2011, http://www.zimtrade.co.zw/pdf/market%20briefs/MashonalandEast.pdf.

[47] H. Friis, D. Mwaniki, B. Omondi et al., "Serum retinol concentrations and Schistosoma mansoni, intestinal helminths, and malarial parasitemia: a cross-sectional study in Kenyan preschool and primary school children," *American Journal of Clinical Nutrition*, vol. 66, no. 3, pp. 665–671, 1997.

[48] H. Friis, P. Ndhlovu, K. Kaondera et al., "Serum concentration of micronutrients in relation to schistosomiasis and indicators of infection: a cross-sectional study among rural Zimbabwean schoolchildren," *European Journal of Clinical Nutrition*, vol. 50, no. 6, pp. 386–391, 1996.

[49] S. Lim, H. C. Sung, I. K. Jeong et al., "Insulin-sensitizing effects of exercise on adiponectin and retinol-binding protein-4 concentrations in young and middle-aged women," *Journal of Clinical Endocrinology and Metabolism*, vol. 93, no. 6, pp. 2263–2268, 2008.

[50] G. Parent, R. Rousseaux-Prevost, Y. Carlier, and A. Capron, "Influence of vitamin A on the immune response of Schistosoma mansoni-infected rats," *Transactions of the Royal Society of Tropical Medicine and Hygiene*, vol. 78, no. 3, pp. 380–383, 1984.

[51] N. G. Jablonski and G. Chaplin, "The evolution of human skin coloration," *Journal of Human Evolution*, vol. 39, no. 1, pp. 57–106, 2000.

[52] J. F. Aloia, "African Americans, 25-hydroxyvitamin D, and osteoporosis: a paradox," *American Journal of Clinical Nutrition*, vol. 88, no. 2, pp. 545S–550S, 2008.

The Role of Vitamin D and Vitamin D Receptor in Immunity to *Leishmania major* Infection

James P. Whitcomb,[1] Mary DeAgostino,[1] Mark Ballentine,[1] Jun Fu,[1] Martin Tenniswood,[2] JoEllen Welsh,[2] Margherita Cantorna,[3] and Mary Ann McDowell[1]

[1] *Eck Institute for Global Health, Department of Biological Sciences, University of Notre Dame, Notre Dame, IN 46556, USA*
[2] *Cancer Research Center, University at Albany, 1 Discovery Drive, Rensselaer, NY 12144, USA*
[3] *Department of Veterinary and Biomedical Sciences, Pennsylvania State University, University Park, PA 16802, USA*

Correspondence should be addressed to Mary Ann McDowell, mcdowell.11@nd.edu

Academic Editor: Marcela F. Lopes

Vitamin D signaling modulates a variety of immune responses. Here, we assessed the role of vitamin D in immunity to experimental leishmaniasis infection in vitamin D receptor-deficient mice (VDRKO). We observed that VDRKO mice on a genetically resistant background have decreased *Leishmania major*-induced lesion development compared to wild-type (WT) mice; additionally, parasite loads in infected dermis were significantly lower at the height of infection. Enzymatic depletion of the active form of vitamin D mimics the ablation of VDR resulting in an increased resistance to *L. major*. Conversely, VDRKO or vitamin D-deficient mice on the susceptible Th2-biased background had no change in susceptibility. These studies indicate vitamin D deficiency, either through the ablation of VDR or elimination of its ligand, 1,25D3, leads to an increase resistance to *L. major* infection but only in a host that is predisposed for Th-1 immune responses.

1. Introduction

The initial metabolism of vitamin D_3 occurs in the skin where 7-dehydrocholesterol is converted to previtamin D_3 after exposure to UVB radiation. After isomerization, vitamin D_3 is metabolized by hepatic 25-hydroxylase (CYP2R1) to form 25-hydroxyvitamin D_3 (25OHD3), which is the major circulating form of vitamin D_3. 25OHD$_3$ is metabolized by 1α-hydroxylase (CYP27B1) that is present in the kidney and many other target tissues including the skin. The active form of vitamin D_3, 1α,25 dihydroxyvitamin D3 (1,25D3) translocates to the nucleus of target cells, where it binds to the vitamin D receptor (VDR), a member of the nuclear receptor supergene family. The VDR heterodimerizes with the retinoid X receptor and this complex recognizes vitamin D response elements in the promoters of many genes to modulate their transcription.

1,25D3 is well characterized for its function in maintaining appropriate serum calcium concentrations as well as its critical requirement for proper bone formation. In addition to these roles, 1,25D3 is important in the regulation of immune responses. Vitamin D has important roles in leukocyte differentiation, dendritic cell maturation, and modulation of the T-helper cell dichotomy [1–5]. Symptoms of Th-1-mediated autoimmune diseases can be reduced or eliminated by treatment with 1,25D3 [6–8]. In addition to its role in autoimmune disease, recent studies have explored a role for vitamin D-mediated signaling in infectious disease resistance [9]. For example, *Mycobacterium tuberculosis* infection induces both VDR and 1-α hydroxylase, which are necessary for production of cathelicidin as part of an antimicrobial peptide response against the *M. tuberculosis* infection in vitro [10]. Here, we investigated the role of VDR in immunity to murine experimental cutaneous leishmaniasis (CL), the prototypical *in vivo* model of the Th-1/Th-2 dichotomy. Parasites from the genus *Leishmania* are responsible for a spectrum of diseases ranging from self-healing cutaneous disease to the potentially fatal visceral disease. Immunity to *L. major*, a parasite responsible for cutaneous leishmaniasis, is highly dependent on a strong

IFNγ-mediated T-helper 1 (Th-1) response. This Th-1 response induces production of nitric oxide (NO) which is critical in elimination of parasites.

We report that VDRKO mice exhibit decreased lesion development as well as decreased parasite loads at the height of *L. major* infection. Additionally, we demonstrate that resistant mice that are rendered vitamin D-deficient by ablation of the 1-α hydroxylase enzyme in CYP27B1 KO mice mimic VDRKO mice with regards to decreased lesion development, suggesting that the effect is dependent on the presence of the ligand. Surprisingly, deficiency of vitamin D signaling in genetically susceptible mice has no effect on *L. major* infection.

2. Materials and Methods

2.1. Mice. VDRKO [11], CYP27B1KO [12], and WT mice, on both BALB/c and C57BL/6 backgrounds, were maintained at Friemann Life Sciences Center at the University of Notre Dame (Notre Dame, Ind, USA). To generate the BALB/c VDRKO strain, female C57BL/6 VDRKO mice were bred with WT male BALB/c mice. The female offspring from this mating were genotyped by PCR [11] and the mice identified as heterozygous for the VDRKO allele were backcrossed to WT male BALB/c mice. This process was repeated 7 times resulting in mice with a greater than 99.2% BALB/c background. Male and female heterozygous mice were bred to each other to generate offspring that were homozygous for either the VDRKO allele or for the WT allele. These homozygous mice were then used to establish a breeding colony of both VDRKO and WT mice on a BALB/c background. VDRKO and WT mice were maintained from weaning on a high-calcium, high-lactose rescue diet (TD96348; Teklad, Madison, Wis, USA) to prevent hypocalcemia associated with VDR deficiency [13]. All animal protocols were approved and reviewed by the University of Notre Dame Institutional Animal Care and Use Committee (IACUC). All mice used in the experiments described below were females between 1 and 4 months of age. WT female mice were age-matched for all experiments.

2.2. Parasites and Infection. *L. major* NIH Friedlin V1 strain (MHOM/IL/80/FN) was used in this study. The parasites were cultivated and infective stage, metacyclic promastigotes were generated as previously described [14]. For infections, 1×10^5 metacyclic promastigotes in $20\,\mu L$ of PBS were injected intradermally into the outside surface of the ear. Lesion diameter and thickness and ulcer diameter was measured weekly with a digital vernier caliper. A lesion "score" was calculated by adding the values obtained for lesion diameter, ulcer diameter, and lesion thickness. Infected ear tissue was homogenized to determine relative parasite load by performing a limiting dilution assay as previously described [15].

2.3. Histology. Segments of ear tissue were embedded in Tissue-Tek OCT freezing media (Fisher Scientific, Pittsburgh, PA), sectioned and stained with Mayer's hematoxylin and eosin to distinguish cytoplasm and nuclei, respectively. Images were obtained using a Nikon E400 light microscope (Nikon, Melville, NY, USA).

2.4. IgG Subclass Determination. Collected sera was used in a soluble *Leishmania* antigen (SLA) specific ELISA. IgG subclass specific secondary antibodies (Southern Biotech, Birmingham, Ala, USA) were used to determine IgG subclass levels. Absorbances were recorded using a SpectraMax M2 plate reader (Molecular Devices, Sunnyvale, Calif, USA). Sera from mice 12 weeks after post-infection were pooled and run as a normalizing sample on each plate. For each isotype, the absorbance for each normalizing sample was used to generate a mean absorbance (ABS). Plate to plate variation was eliminated by using the equation (mean ABS/plate ABS) × (plate ABS) = normalized ABS.

2.5. Quantitative RT-PCR Analysis. Total RNA was isolated from pulverized ears using an RNeasy Mini kit (Qiagen, Valencia, Calif, USA) according to the manufacturer's instructions and contaminating DNA was removed via DNase I treatment (Invitrogen, Carlsbad, Calif, USA). Reverse transcription was performed using $1\,\mu g$ of DNA-free total RNA, 250 ng random primers (Invitrogen), and SuperScript III kit (Invitrogen). Real-time PCR was performed on an ABI 7900HT sequence detection analyzer (Applied Biosystems, Foster City, Calif, USA) using the 2x SYBR Green Kit (Applied Biosystems). The primers (300 nM; IDT, Coralville, Iolua, USA) used were HPRT, IFNγ, IL-4, IL-12p40 [16], Arg1 [17], iNOS [18], IL-10: 5'-CAC AAA GCA GCC TTG CAG AA-3'/ 5'-CTG GCC CCT GCT GAT CCT-3', TNFα: 5'-GAA ACA CAA GAT GCT GGG ACA GT-3'/ 5'-CAT TCG AGG CTC CAG TGA ATT C-3'. For FoxP3, gene expression was determined using a pre-made gene expression assay (Applied Biosystems) according to manufacturer's instructions. Relative copy number was determined using the comparative CT method [19].

2.6. Restimulation Assay. Lymph nodes from infected mice were disrupted using a syringe plunger and a cell strainer (BD Biosciences, San Jose, Calif, USA) and cells from 4 mice per condition were pooled and stimulated with $50\,\mu g$ of SLA. Cell supernatant was harvested at 120 hr after treatment and was analyzed on a Luminex 200 instrument using a multiplex biomarker immunoassay (Millipore, Billerica, Mass, USA). Two biological replicates were assessed, the first in quadruplicate on two separate Luminex plates and the second in duplicate on one plate.

2.7. Flow Cytometry. FcReceptors were blocked with 10% normal mouse serum and stained using 1% BSA in PBS. Cell surface labeling was performed using the following anti-mouse antibodies: α-CD45, α-CD11b, α-CD11c, α-LY-6C/G, α-CD4, α-CD8, α-TCR (BD Biosciences), α-F4/80 (Invitrogen), and α-FoxP3 (eBioscience, San Diego, Calif, USA). Appropriate isotype controls were used as negative controls. Flow cytometry was performed using an FC-500 flow

(a)

(b)

(c)

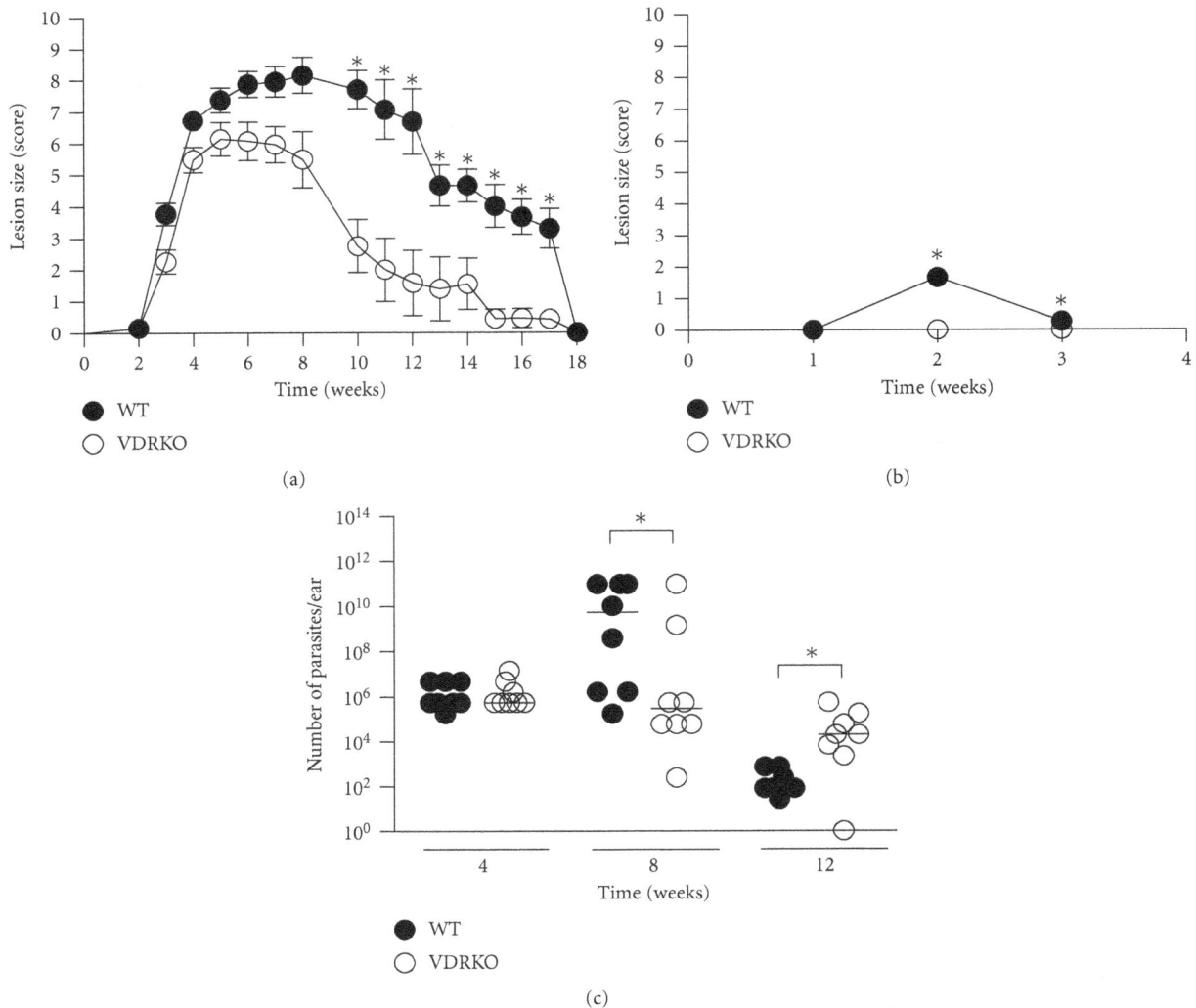

FIGURE 1: VDRKO mice develop smaller lesions when infected with *L. major*. C57BL/6 WT (closed circles) and VDRKO (open circles) mice were infected in the ear with 10^5 *L. major* parasites. (a) Cumulative measurement of lesion size (score) was measured over the course of infection. One representative of 5 independent experiments is presented; *$P \leq 0.05$. (b) Mice were reinfected upon complete resolution of the primary infection (wk 18) and lesions were measured until completely healed. Mean lesion score $\pm$ SE is presented; $n = (11–21$ ears). One representative of 2 independent experiments is presented. (c) Parasite burden of infected ears from VDRKO and WT mice was determined at 4, 8, and 12 weeks after infection by performing a limiting dilution assay. Each point indicates the parasite load in a single ear. One representative of 2 independent experiments is presented. *$P \leq 0.05$.

cytometer (Beckman Coulter, Fullerton, CA) and analyzed with MXP analysis software (Beckman Coulter).

2.8. Diet Studies.

For vitamin D deficiency experiments, mice were fed TD.04179, a vitamin D-deficient diet (Harlan Teklad) from birth prior to mating with male mice from the same background. Once weaned, the offspring from these mice were fed solely using the vitamin D-deficient diet throughout the study. Mice were infected in the ears and lesions were measured as previously described. Vitamin D analysis of serum was performed using both a 25(OH)D3 ELISA and a 1,25D3 ELISA (IDS, Fountain Hills, AZ, USA) following the manufacturer's instructions.

2.9. Generation of Bone Marrow Derived Macrophages and Bone Marrow-Derived Dendritic Cells (DC).

Bone marrow was flushed from femurs with RPMI-C. Red blood cells were lysed using ice-cold sterile ACK lysis buffer (0.15 M Nh_4Cl, 10 mM $KHCO_3$, 0.1 mM Na_2EDTA). Macrophages were generated as previously described [14]. For the generation of DC, progenitor cells were counted and resuspended at 2×10^5/mL in RPMI-C containing 20 ng/mL granulocyte macrophage colony-stimulating factor (GM-CSF) (Pepro-tech, Rocky Hill, NJ, USA). The cells were supplemented with fresh RPMI-C containing 20 ng/mL GM-CSF on days 3, 6, 8, and 10. On day 12, the cells were transferred to 24 well plates for the duration of the experiment.

(a)

(b)

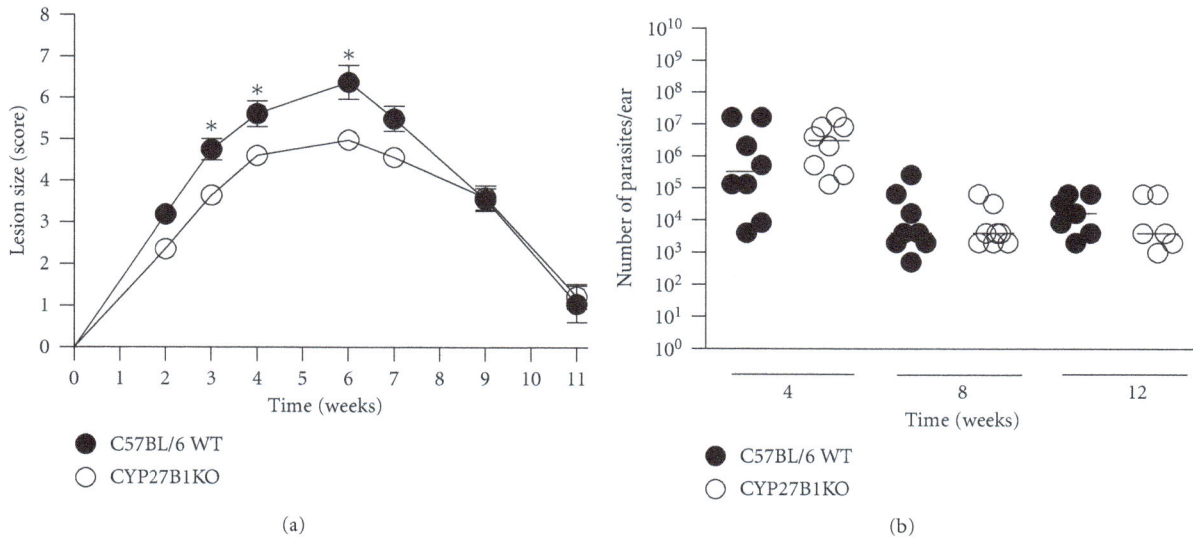

FIGURE 2: Lesion size is reduced in CYP27B1KO mice. C57BL/6 WT (closed circles) and CYP27B1KO (open circles) mice were infected with 10^5 *L. major* parasites and the resulting lesions were measured at the indicated time points. Mean $\pm$ SE is presented; $n = (8–17)$. (b) Infected ears were harvested at 4, 8, and 12 weeks after infection, and parasite load was determined by limiting dilution assay. Each point indicates parasite burden in one ear. One representative of two independent experiments is presented. $^*P \le 0.05$.

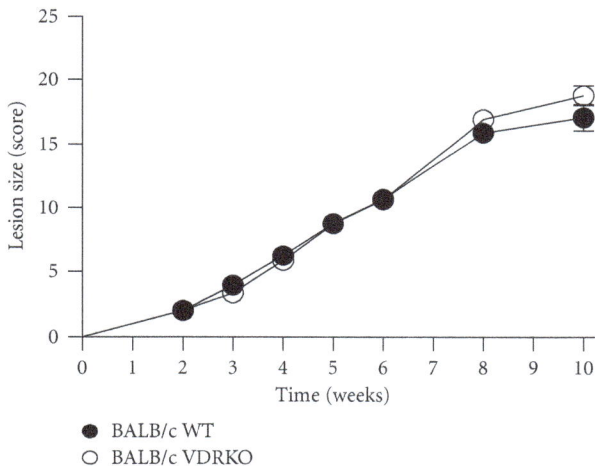

FIGURE 3: Deficiency of vitamin D signaling does not modulate resistance in a susceptible mouse strain. VDRKO mice on a susceptible BALB/c background (open circles) or WT BALB/c (closed circles) were infected with 10^5 *L. major* parasites and the resulting lesions were measured at the indicated time points. Mean $\pm$ SE is presented; $n = 14$. One representative of two independent experiments is presented.

2.10. Phagocytosis Assays.

Macrophages or DC were treated with IFN-γ (500 U/mL) and/or 1,25D3 (Biomol, Plymouth Meeting, Pa, USA) (40 nM) for 24 hr or left untreated. Cells were infected at a ratio of 5 : 1 with V1 strain *L. major* metacyclic promastigotes. Parasites were opsonized with 5% normal mouse serum for 30 minutes, washed, then coincubated with the macrophages or dendritic cells for 1, 2, 4, and 24 hr. Coverslips were stained with Diff-Quick (Fisher Scientific).

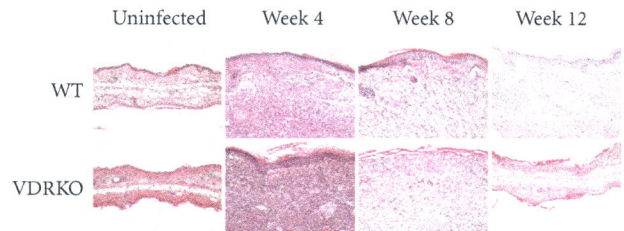

FIGURE 4: Inflammation is decreased in the infected ears of VDRKO mice. C57BL/6 WT and VDRKO were infected in the ear with 10^5 *L. major* parasites. Histological cross sections were H&E stained at the indicated time points and inflammation of the dermis was assessed visually by microscopy. One representative of 2 independent experiments is presented.

2.11. IL-12 ELISA/Nitric Oxide (NO) Assays.

Levels of IL-12 were analyzed using anti-IL-12p40 ELISA (Pierce, Rockford, Ill, USA). NO production was assessed by Griess reaction according to the manufacturer's instructions (Promega, Madison, Wis, USA).

2.12. Statistical Analyses.

For statistical analysis of lesion size, analysis of variance (ANOVA) between WT and KO strains was utilized with a subsequent Bonferonni's post-test to determine at what time points the strains were different. Paired t-tests were performed to determine differences between WT and KO populations for parasite burdens, IgG levels, cytokine regulation, and all *in vitro* analyses. All statistical analyses were performed using GraphPad Prism 4.0 software. In all cases, a P value ≤ 0.05 was considered statistically significant.

3. Results

3.1. VDRKO Mice on a C57BL/6 Background Exhibit Increased Resistance to L. major Infection.

To investigate the role of VDR in immunity to *L. major* infection, C57BL/6 VDRKO and WT mice were infected in the ear dermis with 10^5 *L. major* metacyclic promastigotes. Both groups of mice developed lesions that eventually healed; however, VDRKO mice developed lesions that were significantly smaller and healed 3 weeks faster than their WT counterparts (Figure 1(a)). Upon reinfection, VDRKO mice did not develop lesions at the site of secondary infection whereas the WT mice developed small lesions that resolved within 3 weeks (Figure 1(b)). These results indicate that VDRKO mice have an increased resistance to *L. major* infection compared to WT mice as demonstrated by decreased lesion development. Additionally, VDRKO mice have an intact and possibly heightened memory response to secondary infection.

Because VDRKO mice had decreased lesion development, we compared the parasite load at the site of infection between VDRKO and WT mice using a limiting dilution assay. At 4 weeks after infection, both groups of mice had similar levels of parasites at the site of infection and at week 8, the VDRKO mice had significantly fewer parasites compared to the WT mice (Figure 1(c)). However, by 12 weeks after infection, VDRKO mice harbored significantly elevated parasite loads, suggesting that wound healing and parasite killing may be occurring via different mechanisms [20].

3.2. 1-α Hydroxylase-Deficient Mice Exhibit Heightened Resistance to L. major Infection.

CYP27B1KO mice, which lack the 1-α hydroxylase enzyme required to produce 1,25D3, displayed prototypical lesion development that is observed in resistant (C57BL/6) strains of mice; that is, lesions develop after a few weeks and eventually resolve. Similar to VDRKO mice, CYP27B1KO mice developed significantly smaller lesions compared to their WT counterparts throughout the first 6 weeks of infection (Figure 2(a)). CYP27B1KO mice resolved their infections at a similar rate as WT mice and both groups were completely healed by week 11 after infection. Parasite quantification of infected ears from these groups of mice demonstrated that that CYP27B1KO and WT mice possess similar parasite burdens throughout the infection (Figure 2(b)).

3.3. VDR Ablation Does Not Affect L. major Infection in BALB/c Mice.

C57BL/6 mice have the propensity to completely heal after *Leishmania* infection, whereas BALB/c mice are unable to heal or resolve *L. major* induced lesions. Unlike C57BL/6 VDRKO mice, VDRKO mice on a susceptible BALB/c background did not develop smaller lesions compared to their WT counterparts (Figure 3).

3.4. VDRKO Mice Exhibit Decreased Inflammation at the Site of Infection.

To investigate the causes of reduced lesions and parasite burden in C57BL/6 VDRKO mice, we performed further analyses on lesions and immune responses to *L. major* infection. VDRKO mice exhibit alopecia, develop dermal cysts, and do not recycle epidermal layers properly [21, 22]. To assess if these skin differences were involved in the differential lesion development, we performed a histological study to explore lesion architecture. Uninfected C57BL/6 VDRKO and WT mice do not have any differences in their skin architecture and displayed no characteristics indicative of inflammation (Figure 4). At 4 weeks after infection, both VDRKO and WT mice exhibit similar levels of inflammation at the site of infection. Although numbers of infiltrating cells were similar in the two mouse strains through 8 weeks of infection (data not shown), VDRKO mice appeared to display decreased inflammation by 8 weeks after infection (Figure 4). Inflammation continued to decrease in both groups of mice, and by 12 weeks after infection lesions from VDRKO mice exhibited little signs of inflammation. Inflammation was still observed in the WT mice at 12 weeks after infection although greatly reduced.

There were no significant differences between WT and VDRKO mice in terms of macrophage (CD11b+/F480+), neutrophil (Ly-6C/G+/F480-), T cell (CD4+ or CD8+), or dendritic cell (CD11c+) cell infiltration (data not shown).

WT mice produced more IFNγ and IL-10 mRNA at the infection site 2 weeks after *L. major* infection (Figure 5). In addition, IL-4 and IL-12 mRNA was elevated in WT mice compared to VDRKO mice. Inducible nitric oxide synthase (iNOS) was upregulated during *L. major* infection, however no differences between WT and VDRKO were detected. As previously reported [17], WT mice express more arginase mRNA than VDRKO mice (Figure 5).

3.5. Systemic Immune Responses in VDR KO and WT Mice.

Flow cytometry analysis of the draining lymph nodes indicated that T-helper cells (CD4+), cytotoxic T-cells (CD8+), and FoxP3+ Treg (CD4+/CD25+) cells are elevated in VDRKO mice at most time points relative to WT animals (Suppl. Figure 1 which is available online at doi:10.1155/2012/134645). The number of each of these cell types increased as the lesions progressed and then decreased during healing. No obvious differences in other cell types such as macrophages (CD11b+/F480+), dendritic cells (CD11c+), or neutrophils (LY-6C/G+/F480-) were observed in the draining lymph nodes (data not shown).

VDRKO lymph node cells produced significantly lower amounts of inflammatory cytokines IFNγ, TNF-α, GM-CSF, and MIP-1α at the height of infection than the cells from WT mice (Suppl. Figure 2). Conversely, more MCP-1 was produced by lymph node cells from VDRKO mice compared to the cells from WT mice. Early after infection (2 weeks), cells from both mouse strains up regulated IL-4, IL-5, and IL-13, production that decreased by 12 wks post-infection (Suppl. Figure 2). Furthermore, VDRKO and WT lymph nodes generated equivalent amounts of IL-1β, IL-6, IL-10, Rantes, and KC at all time points.

Total IgG serum levels increased in both groups of mice as lesion development progressed and remained elevated even after the mice had healed (Figure 6). Significantly elevated levels of IgG2a/c were detected in VDRKO mice beginning at week 4 after infection and remained significantly elevated

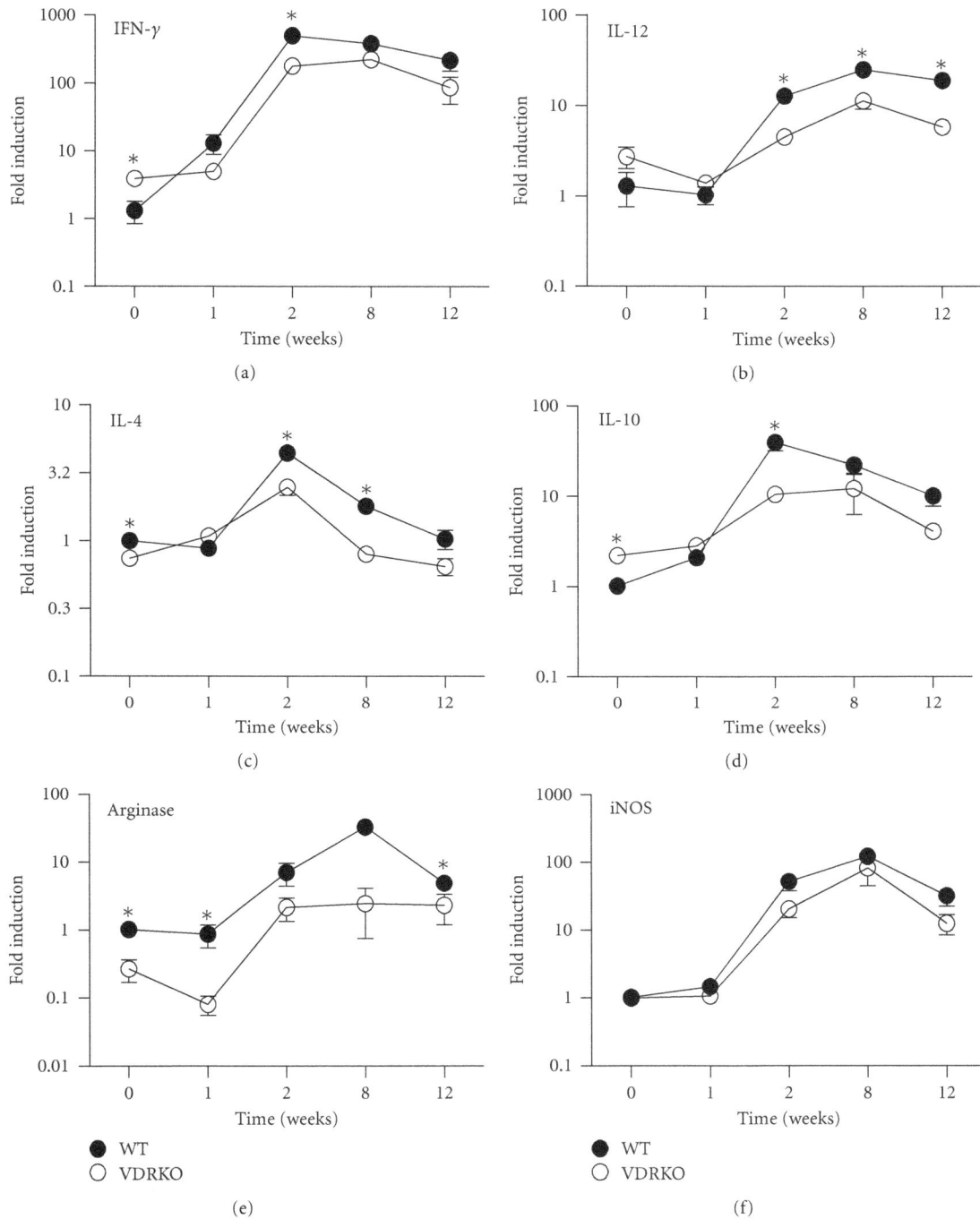

FIGURE 5: Quantitative RT-PCR analysis of cytokine production at the site of infection. C57BL/6 WT (closed circles) and VDRKO (open circles) mice were infected with 10^5 L. major parasites and their ears were harvested at the indicated time points. Quantitative RT-PCR was performed to determine the fold increase over uninfected WT ears for each cytokine. One representative of two independent experiments is presented. $*P \leq 0.05$. $n = 4$ mice/time point.

throughout infection (Figure 6). No differences in IgG1 titers were observed between the VDRKO and WT mice at any time point (data not shown).

3.6. NO and IL-12 Production Are Increased in DC from VDRKO C57BL/6 Mice.
NO production is a potent mechanism used by antigen presenting cells to eliminate L.

major parasites. DC from both groups of mice were either left untreated or preincubated with combinations of IFN-γ and/or 1,25D3 prior to L. major infection and 48 hours after infection, cell culture supernatant was analyzed for levels of NO production. DC from VDRKO generate significantly more NO upon stimulation with IFNγ or IFNγ and 1,25D3 (Figure 7(a)). Preincubation of IFNγ or a combination of IFNγ and 1,25D3 resulted in significantly more NO production as compared to untreated cells in both cell types. We did not observe an inhibition of NO production by the DC upon treatment with 1,25D3 in either the VDRKO or WT derived cells. In addition, there is no effect of either the VDR or 1,25D3 on NO production in macrophages (Suppl. Figure 3(a)).

Regardless of the IFNγ and/or 1,25D3 treatment, IL-12p40 production by VDRKO DC is significantly increased compared to WT DC, suggesting that VDR may contribute to regulation of IL-12p40 production by DC (Figure 7(b)).

4. Discussion

Leishmania are obligate intracellular parasites that are eliminated by a strong Th-1 host response. Vitamin D treatment reduces inappropriate Th-1 responses thus decreasing or eliminating symptoms of autoimmune diseases [6–8, 23]. As vitamin D exerts these effects through the VDR, we hypothesized that ablation of the VDR would tilt the Th-1/Th-2 balance towards a Th-1 bias and lead to enhanced resistance to *L. major* infection. Using a mouse model of cutaneous leishmaniasis, in which the parasites were injected into the ear dermis, we observed that VDRKO mice developed significantly smaller lesions and have fewer parasites at the site of infection during the height of infection than their WT counterparts. These results are similar to those observed by previous studies using a foot pad model of cutaneous leishmaniasis [17]. As VDRKO mice healed earlier than WT mice, we anticipated that VDRKO mice would exhibit decreased parasite loads compared to WT mice, because VDRKO mice resolve their lesions more quickly than WT mice. However, at 12 weeks after infection, VDRKO mice had elevated levels of parasites at the infection site even though they have decreased lesion size. This disparity may indicate that different processes are contributing to wound healing and parasite killing. The ability of VDRKO mice to produce increased dermal depositions of collagen may contribute to the increased healing phenotype we observed in VDRKO mice, as research indicates that orderly collagen fiber deposition in the skin is one factor that may contribute to increased healing in *L. major* infected mice [20].

Successful clearance of a *L. major* infection depends on initiation of a robust Th-1 response that leads to parasite killing via production of NO. VDRKO mice produce significantly lower levels of inflammatory cytokines locally at the infection site and following restimulation in vitro suggesting that VDRKO mice generate a decreased Th-1 response to *L. major* infection. We expected to detect upregulated transcription of iNOS in VDRKO mice as this enzyme leads to production of the NO necessary for killing of *L. major*; however, iNOS is not upregulated at any point post *L. major* infection in VDRKO compared to WT mice. Rather, arginase transcript levels are higher in WT compared to VDRKO mice, supporting the suggestion by others that upregulation of arginase antagonizes the metabolism of NO by competing with iNOS for a common substrate, L-arginine [17]. The lack of such competition would allow VDRKO mice to generate more NO than WT mice, leading to increased parasite

killing. This hypothesis is further supported by studies demonstrating that inhibition of arginase by N-hydroxy-L-arginine, a precursor of NO, results in increased macrophage killing of *L. major* parasites *in vitro*, while induction of arginase contributes to the growth of *L. major* by providing the parasites with the polyamines required for replication [24, 25].

Toll like receptor activation initiates upregulation of VDR and CYP27B1 leading to an antimicrobial peptide response against *Mycobacteria* [10, 26], which contributes to killing of *M. tuberculosis* in human macrophages [27]. This pathway is unlikely to play a role in resistance to *L. major* as antimicrobial peptides have little effect in killing of *Leishmania* [28, 29]. VDR expression, treatment with IFNγ and/or vitamin D does not affect the ability of macrophage and DC to phagocytose *L. major* (data not shown and [17]). Furthermore, 1,25D3 treatment alone had no effect on IFNγ-induced NO production in infected macrophage and DC in vitro. This contrasts with other published data suggesting that NO production by *L. major* infected macrophages is inhibited by 1,25D3 [17]. These authors suggest that enhanced production of arginase competes with iNOS for a common substrate, ultimately resulting in decreased production of NO in macrophage. Stimulation of VDRKO macrophage with IFNγ and LPS resulted in significantly less NO than similarly treated WT macrophage (data not shown) suggesting that VDR may play a role in modulation of NO production in macrophage. Conversely, IFNγ-induced NO production by infected VDRKO DC was significantly higher than WT DC implying that VDR ablation increases rather than decreases NO production in DC. Our data suggests that vitamin D and the VDR differentially regulate NO production in macrophage versus DC.

In addition to overproducing NO, VDRKO DC overproduce IL-12p40 compared to WT. Other studies demonstrate that 1,25D3 inhibits IFNγ signaling in both T cells, macrophages and DC [17, 30–33]. Indirect inhibition of DC produced NO then could result via VDR inhibition of IFNγ activated STAT1 signaling. The ablation of the VDR results in elimination of this inhibition resulting in upregulation of NO and IL-12p40 production. As a similar increase of IL-12p40 and NO production was not observed in VDRKO macrophage this implies that macrophage differ from DC in their IFNγ induced production of IL-12p40.

VDRKO mice generate significantly more antigen specific IgG2a/c antibodies, that serve as an indicator of Th-1 biased immune responses [34, 35] than WT mice. The elevated IgG2a/c production in VDRKO mice is not surprising as treatment with 1,25D3 inhibits IgG2a production in mice [36], in cattle [37], and in pigs [38]. These data suggest that elimination of inhibitory effects of 1,25D3 skews antibody response towards production of IgG2a/c. The higher levels of IgG2a/c possibly explain the increased resistance to secondary infection with *L. major* observed in the VDRKO mice (Figure 1(b)). Indeed, elevated antigen specific and nonspecific induction of IgG2a/c has been observed in older VDRKO mice infected with *Listeria monocytogenes* [39]. However, the exact relationship between IgG2a/c production and vitamin D signaling remains to be elucidated.

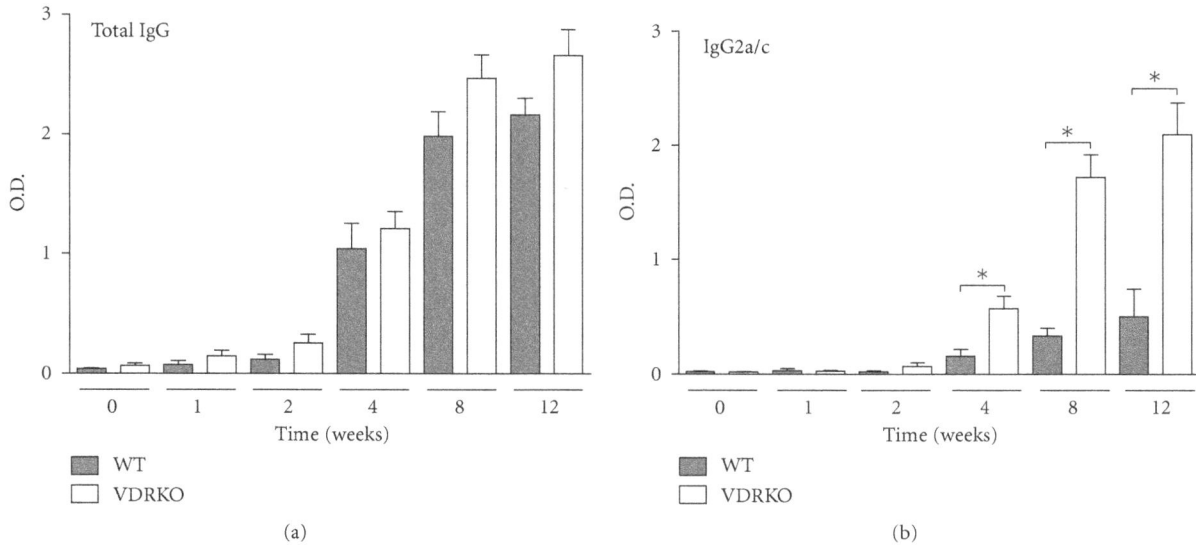

FIGURE 6: VDRKO mice exhibit elevated IgG2a/c production in response to *L. major* infection. C57BL/6 WT (closed bars) and VDRKO (open bars) mice were infected with 10^5 *L. major* parasites and serum samples were collected at the indicated time points. Isotype specific ELISA was performed to determine levels of total IgG and IgG2a/c in the serum. Mean $\pm$ SE is presented; $n = (8-10)$. $^*P \leq 0.05$. One representative of two independent experiments is presented.

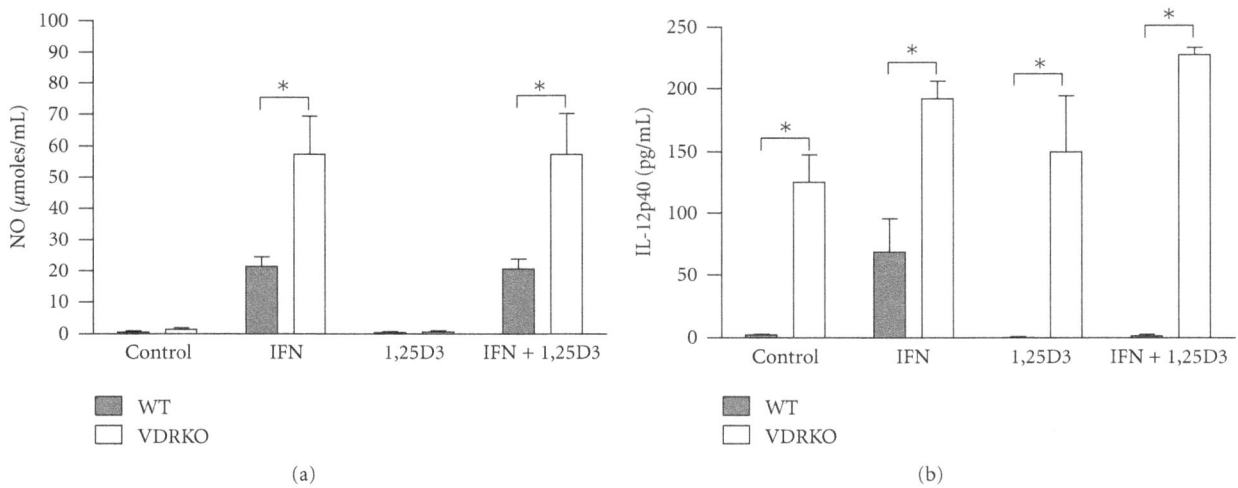

FIGURE 7: NO and IL-12p40 production by host cells. C57BL/6 WT (closed bars) and VDRKO (open bars) mice. DC were treated with IFNγ and/or 1,25D3 for 24 hr prior to being infected at a ratio of 5 : 1 with *L. major*. The cells were infected for 4 hr and washed to remove extracellular parasites. Supernatants were harvested 48 hr after infection and analyzed for production of NO (a) and IL-12p40 (b). Mean $\pm$ standard error is presented; $(n = 4)$. $^*P \leq 0.05$.

We demonstrate that neither genetic removal of the VDR or depletion of vitamin D from the diet (data not shown) of mice on the BALB/c background does not alter susceptibility to *L. major* infection. Susceptibility in WT BALB/c mice is attributed to their inability to mount a Th-1 response against infection. Our results indicate that the effect of the VDR may depend on the nature of the host immune response. Additional mechanisms, such as wound healing and genetic background, have also been shown to be important in resistance to *L. major* infection [20, 40]. Our

data clearly show that the absence of vitamin D signaling cannot overcome the susceptibility traits of BALB/c mice.

Contrary to the results observed in VDRKO or CYP27B1KO mice, dietary deficiency of vitamin D in resistant C57BL/6 mice did not reduce the severity of *L. major* lesions (data not shown). A similar disparity has been observed in studies investigating the role of VDR deletion versus dietary vitamin D deficiency in the development of diabetes [23, 41] and MS [8, 42]. Our data may indicate that some of the effects of VDR on the course of *L. major* infection are ligand independent, as has been shown in the case of alopecia [43], or that sufficient 1,25D3 persists in mice maintained on the deficient diet to activate the receptor.

In summary, we have demonstrated that VDRKO mice on the C57BL/6 background develop smaller lesions than WT mice upon infection with *L. major*, but this phenotype is not observed on the BALB/c genetic background. Enzymatic depletion of 1,25D3 also enhances resistance to *L. major* infection in C57BL/6 mice. The data further suggest that increased IL-12p40 and NO production by VDRKO dendritic cells may contribute to increased resistance to *L. major*. These studies indicate that ablation of VDR or elimination of its ligand, 1,25D3, is able to increase resistance to *L. major* infection, but only in a host that is predisposed for Th-1 immune responses. Our data confirm an important role for vitamin D for regulating immune responses that depend on Th-1 cells.

Acknowledgments

The authors thank Freimann Life Science Center at the University of Notre Dame for excellent animal care. This work was supported in part by NIH Grants no. R01AI056242 (M. A. McDowell) and American Heart Association no. 0435333Z (M. A. McDowell).

References

[1] M. D. Griffin and R. Kumar, "Effects of 1α,25(OH)2D3 and its analogs on dendritic cell function," *Journal of Cellular Biochemistry*, vol. 88, no. 2, pp. 323–326, 2003.

[2] H. P. Koeffler, T. Amatruda, and N. Ikekawa, "Induction of macrophage differentiation of human normal and leukemic myeloid stem cells by 1,25-dihydroxyvitamin D3 and its fluorinated analogues," *Cancer Research*, vol. 44, no. 12 I, pp. 5624–5628, 1984.

[3] A. Berer, J. Stöckl, O. Majdic et al., "1,25-Dihydroxyvitamin D3 inhibits dendritic cell differentiation and maturation in vitro," *Experimental Hematology*, vol. 28, no. 5, pp. 575–583, 2000.

[4] M. O. Canning, K. Grotenhuis, H. de Wit, C. Ruwhof, and H. A. Drexhage, "1-α,25-dihydroxyvitamin D3 (1,25(OH)2D3) hampers the maturation of fully active immature dendritic cells from monocytes," *European Journal of Endocrinology*, vol. 145, no. 3, pp. 351–357, 2001.

[5] G. Penna and L. Adorini, "1α,25-dihydroxyvitamin D3 inhibits differentiation, maturation, activation, and survival of dendritic cells leading to impaired alloreactive T cell activation," *Journal of Immunology*, vol. 164, no. 5, pp. 2405–2411, 2000.

[6] M. T. Cantorna, C. E. Hayes, and H. F. Deluca, "1,25-dihydroxyvitamin D3 reversibly blocks the progression of relapsing encephalomyelitis, a model of multiple sclerosis," *Proceedings of the National Academy of Sciences of the United States of America*, vol. 93, no. 15, pp. 7861–7864, 1996.

[7] M. T. Cantorna, C. E. Hayes, and H. F. DeLuca, "1,25-Dihydroxycholecalciferol inhibits the progression of arthritis in murine models of human arthritis," *Journal of Nutrition*, vol. 128, no. 1, pp. 68–72, 1998.

[8] M. T. Cantorna, C. Munsick, C. Bemiss, and B. D. Mahon, "1,25-Dihydroxycholecalciferol prevents and ameliorates symptoms of experimental murine inflammatory bowel disease," *Journal of Nutrition*, vol. 130, no. 11, pp. 2648–2652, 2000.

[9] D. Bruce, J. H. Ooi, S. Yu, and M. T. Cantorna, "Vitamin D and host resistance to infection? Putting the cart in front of the horse," *Experimental Biology and Medicine*, vol. 235, no. 8, pp. 921–927, 2010.

[10] P. T. Liu, S. Stenger, D. H. Tang, and R. L. Modlin, "Cutting edge: vitamin D-mediated human antimicrobial activity against Mycobacterium tuberculosis is dependent on the induction of cathelicidin," *Journal of Immunology*, vol. 179, no. 4, pp. 2060–2063, 2007.

[11] Y. C. Li, A. E. Pirro, M. Amling et al., "Targeted ablation of the vitamin D receptor: an animal model of vitamin D-dependent rickets type II with alopecia," *Proceedings of the National Academy of Sciences of the United States of America*, vol. 94, no. 18, pp. 9831–9835, 1997.

[12] O. Dardenne, J. Prud'homme, A. Arabian, F. H. Glorieux, and R. St-Arnaud, "Targeted inactivation of the 25-hydroxyvitamin D3-1α-hydroxylase gene (CYP27B1) creates an animal model of pseudovitamin D-deficiency rickets," *Endocrinology*, vol. 142, no. 7, pp. 3135–3141, 2001.

[13] Y. C. Li, M. Amling, A. E. Pirro et al., "Normalization of mineral ion homeostasis by dietary means prevents hyperparathyroidism, rickets, and osteomalacia, but not alopecia in vitamin D receptor-ablated mice," *Endocrinology*, vol. 139, no. 10, pp. 4391–4396, 1998.

[14] O. E. Akilov, R. E. Kasuboski, C. R. Carter, and M. A. McDowell, "The role of mannose receptor during experimental leishmaniasis," *Journal of Leukocyte Biology*, vol. 81, no. 5, pp. 1188–1196, 2007.

[15] C. R. Carter, J. P. Whitcomb, J. A. Campbell, R. M. Mukbel, and M. A. McDowell, "Complement receptor 3 deficiency influences lesion progression during Leishmania major infection in BALB/c Mice," *Infection and Immunity*, vol. 77, no. 12, pp. 5668–5675, 2009.

[16] M. J. Donovan, A. S. Messmore, D. A. Scrafford, D. L. Sacks, S. Kamhawi, and M. A. McDowell, "Uninfected mosquito bites confer protection against infection with malaria parasites," *Infection and Immunity*, vol. 75, no. 5, pp. 2523–2530, 2007.

[17] J. Ehrchen, L. Helming, G. Varga et al., "Vitamin D receptor signaling contributes to susceptibility to infection with Leishmania major," *FASEB Journal*, vol. 21, no. 12, pp. 3208–3218, 2007.

[18] L. Poluektova, S. Gorantla, J. Faraci, K. Birusingh, H. Dou, and H. E. Gendelman, "Neuroregulatory events follow adaptive immune-mediated elimination of HIV-1-infected macrophages: studies in a murine model of viral encephalitis," *Journal of Immunology*, vol. 172, no. 12, pp. 7610–7617, 2004.

[19] K. J. Livak and T. D. Schmittgen, "Analysis of relative gene expression data using real-time quantitative PCR and the 2-ΔΔCT method," *Methods*, vol. 25, no. 4, pp. 402–408, 2001.

[20] T. Baldwin, A. Sakthianandeswaren, J. M. Curtis et al., "Wound healing response is a major contributor to the severity of cutaneous leishmaniasis in the ear model of infection," *Parasite Immunology*, vol. 29, no. 10, pp. 501–513, 2007.

[21] Z. Xie, L. Komuves, Q. C. Yu et al., "Lack of the vitamin D receptor is associated with reduced epidermal differentiation and hair follicle growth," *Journal of Investigative Dermatology*, vol. 118, no. 1, pp. 11–16, 2002.

[22] S. Meindl, A. Rot, W. Hoetzenecker, S. Kato, H. S. Cross, and A. Elbe-Bürger, "Vitamin D receptor ablation alters skin architecture and homeostasis of dendritic epidermal T cells," *British Journal of Dermatology*, vol. 152, no. 2, pp. 231–241, 2005.

[23] J. B. Zella, L. C. McCary, and H. F. DeLuca, "Oral administration of 1,25-dihydroxyvitamin D3 completely protects NOD mice from insulin-dependent diabetes mellitus," *Archives of Biochemistry and Biophysics*, vol. 417, no. 1, pp. 77–80, 2003.

[24] V. Iniesta, L. C. Gómez-Nieto, and I. Corraliza, "The inhibition of arginase by Nω-hydroxy-L-arginine controls the growth of Leishmania inside macrophages," *Journal of Experimental Medicine*, vol. 193, no. 6, pp. 777–783, 2001.

[25] V. Iniesta, L. C. Gómez-Nieto, I. Molano et al., "Arginase I induction in macrophages, triggered by Th2-type cytokines, supports the growth of intracellular Leishmania parasites," *Parasite Immunology*, vol. 24, no. 3, pp. 113–118, 2002.

[26] P. T. Liu, S. Stenger, H. Li et al., "Toll-like receptor triggering of a vitamin D-mediated human antimicrobial response," *Science*, vol. 311, no. 5768, pp. 1770–1773, 2006.

[27] S. Thoma-Uszynski, S. Stenger, O. Takeuchi et al., "Induction of direct antimicrobial activity through mammalian toll-like receptors," *Science*, vol. 291, no. 5508, pp. 1544–1547, 2001.

[28] S. E. Löfgren, L. C. Miletti, M. Steindel, E. Bachère, and M. A. Barracco, "Trypanocidal and leishmanicidal activities of different antimicrobial peptides (AMPs) isolated from aquatic animals," *Experimental Parasitology*, vol. 118, no. 2, pp. 197–202, 2008.

[29] M. M. Kulkarni, W. R. McMaster, E. Kamysz, W. Kamysz, D. M. Engman, and B. S. McGwire, "The major surface-metalloprotease of the parasitic protozoan, Leishmania, protects against antimicrobial peptide-induced apoptotic killing," *Molecular Microbiology*, vol. 62, no. 5, pp. 1484–1497, 2006.

[30] L. Helming, J. Böse, J. Ehrchen et al., "1α,25-dihydroxyvitamin D3 is a potent suppressor of interferon γ-mediated macrophage activation," *Blood*, vol. 106, no. 13, pp. 4351–4358, 2005.

[31] T. P. Staeva-Vieira and L. P. Freedman, "1,25-Dihydroxyvitamin D3 inhibits IFN-γ and IL-4 levels during in vitro polarization of primary murine CD4+ T cells," *Journal of Immunology*, vol. 168, no. 3, pp. 1181–1189, 2002.

[32] M. Cippitelli and A. Santoni, "Vitamin D3: a transcriptional modulator of the interferon-γ gene," *European Journal of Immunology*, vol. 28, no. 10, pp. 3017–3030, 1998.

[33] H. Reichel, H. P. Koeffler, A. Tobler, and A. W. Norman, "1α,25-Dihydroxyvitamin D3 inhibits γ-interferon synthesis by normal human peripheral blood lymphocytes," *Proceedings of the National Academy of Sciences of the United States of America*, vol. 84, no. 10, pp. 3385–3389, 1987.

[34] R. M. Martin, J. L. Brady, and A. M. Lew, "The need for IgG2c specific antiserum when isotyping antibodies from C57BL/6 and NOD mice," *Journal of Immunological Methods*, vol. 212, no. 2, pp. 187–192, 1998.

[35] C. M. Snapper and W. E. Paul, "Interferon-γ and B cell stimulatory factor-1 reciprocally regulate Ig isotype production," *Science*, vol. 236, no. 4804, pp. 944–947, 1987.

[36] J. M. Lemire, D. C. Archer, L. Beck, and H. L. Spiegelberg, "Immunosuppressive actions of 1,25-dihydroxyvitamin D3: preferential inhibition of Th1 functions," *Journal of Nutrition*, vol. 125, no. 6, 1995.

[37] T. A. Reinhardt, J. R. Stabel, and J. P. Goff, "1,25-dihydroxyvitamin D3 enhances milk antibody titers to Escherichia coli J5 vaccine," *Journal of Dairy Science*, vol. 82, no. 9, pp. 1904–1909, 1999.

[38] Y. Van Der Stede, T. Verfaillie, E. Cox, F. Verdonck, and B. M. Goddeeris, "1α,25-dihydroxyvitamin D3 increases IgA serum antibody responses and IgA antibody-secreting cell numbers in the Peyer's patches of pigs after intramuscular immunization," *Clinical and Experimental Immunology*, vol. 135, no. 3, pp. 380–390, 2004.

[39] D. Bruce, J. P. Whitcomb, A. August, M. A. McDowell, and M. T. Cantorna, "Elevated non-specific immunity and normal Listeria clearance in young and old vitamin D receptor knockout mice," *International Immunology*, vol. 21, no. 2, pp. 113–122, 2009.

[40] C. M. Elso, L. J. Roberts, G. K. Smyth et al., "Leishmaniasis host response loci (lmr1-3) modify disease severity through a Th1/Th2-independent pathway," *Genes and Immunity*, vol. 5, no. 2, pp. 93–100, 2004.

[41] C. Gysemans, E. Van Etten, L. Overbergh et al., "Unaltered diabetes presentation in NOD mice lacking the vitamin D receptor," *Diabetes*, vol. 57, no. 1, pp. 269–275, 2008.

[42] T. F. Meehan and H. F. DeLuca, "The vitamin D receptor is necessary for 1alpha,25-dihydroxyvitamin D(3) to suppress experimental autoimmune encephalomyelitis in mice," *Archives of Biochemistry and Biophysics*, vol. 408, no. 2, pp. 200–204, 2002.

[43] K. Skorija, M. Cox, J. M. Sisk et al., "Ligand-independent actions of the vitamin D receptor maintain hair follicle homeostasis," *Molecular Endocrinology*, vol. 19, no. 4, pp. 855–862, 2005.

3

Human Schistosome Infection and Allergic Sensitisation

Nadine Rujeni, David W. Taylor, and Francisca Mutapi

Institute of Immunology and Infection Research, Centre for Immunity, Infection, and Evolution, School of Biological Sciences, University of Edinburgh, Ashworth Laboratories, King's Buildings, West Mains Rd, Edinburgh EH9 3JT, UK

Correspondence should be addressed to Francisca Mutapi, f.mutapi@ed.ac.uk

Academic Editor: Maria Ilma Araujo

Several field studies have reported an inverse relationship between the prevalence of helminth infections and that of allergic sensitisation/atopy. Recent studies show that immune responses induced by helminth parasites are, to an extent, comparable to allergic sensitisation. However, helminth products induce regulatory responses capable of inhibiting not only antiparasite immune responses, but also allergic sensitisation. The relative effects of this immunomodulation on the development of protective schistosome-specific responses in humans has yet to be demonstrated at population level, and the clinical significance of immunomodulation of allergic disease is still controversial. Nonetheless, similarities in immune responses against helminths and allergens pose interesting mechanistic and evolutionary questions. This paper examines the epidemiology, biology and immunology of allergic sensitisation/atopy, and schistosome infection in human populations.

1. Introduction

The major human helminth parasites belong to two phyla, the nematodes (or roundworms) which include intestinal soil transmitted helminths (STH) and filarial worms (which cause lymphatic filariasis and onchocerciasis), and the platyhelminths (or flatworms) which include the flukes (or trematodes, including schistosomes) and the tapeworms (or cestodes). Although common in most parts of the world sixty years ago [1], these parasites are currently mainly prevalent in sub-Saharan Africa, Asia, and South America [2–4], where they are responsible for considerable disabilities including blindness and elephantiasis (filarial worms). Furthermore, helminth infections are responsible for morbidities that include anaemia, stunted growth, poor cognitive development, and malnutrition [5–7], hence exert a negative socioeconomic impact in some of the poorest communities in the world.

Immune-mediated diseases including auto-immune diseases (such as type 1 diabetes, inflammatory bowel diseases, and rheumatoid arthritis) and allergic diseases (such as asthma, allergic rhinitis, and atopic eczema) are reported to be more prevalent in developed countries and in urban areas of developing countries [8, 9]. But studies from Africa are demonstrating that allergic diseases are common, if not acknowledged, clinical problems in this region [10]. Immune disorders have been responsible for increased mortality and morbidity worldwide [11–13] and they negatively impact on economic growth due to their elevated cost of their treatment [14, 15]. There is also mounting evidence that allergic disorders, especially allergic rhinitis, are associated with attention deficit disorder and hyperactivity in children [16, 17].

Increasing rates of childhood allergies have long been a puzzle to epidemiologists [18, 19]. Thus, studying cohorts of children born in 1946, 1958, and 1970, concluded that a "new environmental agent," contained in breast milk and possibly infants' food was responsible for the increase in eczema. Emmanuel, reviewing medical literature published from 1820 to 1900, suggested that the hay fever "epidemic" was associated with the rapid industrial growth of the 19th century since this disorder was rarely described prior that period [19]. It was Strachan who in 1989, observing that the rate of hay fever and eczema was consistently negatively associated with family size and birth position in households, hypothesized that reduced exposure to childhood infections due to increased hygiene was responsible for the allergy epidemics. This hypothesis, currently referred to as

TABLE 1: Heterogeneity in studies investigating the effect of helminth infection on atopy.

Parasite spp, References	Atopy outcome	Association	Population age
Ascaris lumbrocoides			
[37][1]	Wheeze, SPT	Negative	1–4 years
[38][2]	IgE, PK	Negative	5–15 years
[39][1]	SPT, airway responsiveness	Positive	8–18 years
[40][2]	Allergen-induced Th2 cytokines	None	7–13 years
[41][2]	SPT, wheeze	None	9 years mean age
[42][2]	SPT	Negative	6–17 years
	Wheeze, eczema, EIB	None	
Trichuris trichiura			
[38][2]	IgE, PK	Negative	5–15 years
[37][1]	Wheeze, SPT	None	1–4 years
[43][#]	SPT	Negative	2–8 years
Hookworm			
[37][1]	Wheeze, SPT	None	1–4 years
[42][2]	SPT	Negative	6–17 years
	wheeze, eczema, EIB	None	
Schistosoma mansoni			
[44][1]	SPT, IgE	Negative	18 ± 9.7 years
[45][#]	SPT, asthma symptoms	Negative	15 years mean age
Schistosoma haematobium			
[46][1]	SPT	Negative	5–14 years

Cross-sectional [1] and treatment followup [2] studies are reported here.
[#]Longitudinal approach but treatment intervention was not the primary objective of the study. SPT: skin prick test; PK: Prausnitz-Kustner passive transfer test, EIB: exercise-induced bronchoconstriction.

the "hygiene hypothesis," was subsequently supported by some epidemiological studies [20, 21] but contradicted by others [22, 23] (see summary in Table 1). In a retrospective case control study on Italian military cadets, Matricardi and colleagues were able to show that cumulative exposures to foodborne and oral-faecal infections, but not infections transmitted via other routes, were associated with a reduced risk of being atopic [24]. They suggested that the mode of transmission of the pathogen was a determining factor in subsequent protection (or lack of protection) against atopy and asthma, hence explaining inconsistencies in previous studies.

2. Global Burden of Schistosomiasis and Atopy

2.1. Schistosomiasis. Schistosomiasis accounts for up to 70 million DALYs annually [6], with an estimated 15,000 deaths [4], and children carry the heaviest burden of infection [47]. With these figures, schistosomiasis is classified second only to malaria in terms of human morbidity and mortality due to parasitic diseases [48]. Schistosomiasis is caused by infection with blood-dwelling trematodes of the genus *Schistosoma*, of which *S. haematobium*, *S. mansoni*, and *S. japonicum* are the main human schistosomes [49]. It is typically prevalent in rural areas where natural streams, ponds, rivers, and lakes harbouring the infected intermediate host snails, are the main sources of water for domestic or occupational purposes

such as washing and fishing. School children usually become infected during swimming or collecting water, while younger children and infants become infected when accompanying adults (washing clothes or collecting water) or by being bathed in these water sources [50].

2.2. Atopy. Rising rates of atopic diseases have been reported in developed countries since the end of World War II [18] and currently constitute a major public health issue [51]. Demographic data in the US have shown an average increase in childhood asthma prevalence of 4.3% per year from 1980 to 1996, with associated deaths and hospitalisation increasing by 3.4% and 1.4%, respectively [52]. In the United Kingdom, according to the British Allergy Foundation, 1 in 3 people suffer from allergy at some time in their lives. This report indicates that 58% of allergic sensitisations are triggered by house dust mites (HDM), a known risk factor for developing asthma and allergic rhinitis [53, 54]. Increasing prevalence of asthma in adults over a period of 10 years and doubling in school children over 20 years have been reported in Australia [55]. A recent study involving 12 European countries and 19 centres reported incidences of asthma between 5 and 17% (average 8%), while allergic rhinitis varies between 23 and 44%, with an average of 30% [56].

In less affluent countries, comparable rates of atopic diseases are generally reported in urban and suburban areas. Thus, a prevalence of asthma of 9% was reported in urban areas of Rwanda [57] while the International Study of Asthma and Allergies in childhood (ISAAC) reported an overall prevalence of 10.9% across 22 centres in Africa [58]. Reported incidences of allergic rhinitis range from 14% to 54% in urban and suburban areas across African countries (reviewed by [59]). Importantly, according to the ISAAC phase three, although the prevalence was generally lower, there were more severe symptoms of rhinoconjnnctivitis reported in urban centres of developing countries compared to those reported in developed countries [8]. However, studies from Africa suggest that allergic conditions may be underdiagnosed in Africa due to "inappropriate" diagnostic tests and these studies call for component-resolved allergy testing in Africa [60]. Indeed a recent study in Zimbabwe showed that schistosome-infection resulted in impaired diagnosis of cat allergy [61].

3. Atopy and Schistosome Life-Cycle Stages

In schistosome infection, the human immune system is exposed to schistosome larvae (cercariae and schistosomula), adult worm, and egg antigens. Animal studies as well as *in vitro* studies have demonstrated immunological changes and regulatory mechanisms associated with these different life-cycle stages. The surface of cercariae (enriched in carbohydrates [62]) and the newly transformed schistosomula activate the complement cascade [63, 64] and eliciting proinflammatory responses [65, 66]. An excessive immunological reaction to skin stage cercarial antigens results in cercarial dermatitis or swimmer's itch [67], an allergic condition also occurring in contact with nonhuman schistosomes that is prevalent in developed countries [68–70]. This inflammatory reaction is rarely reported in populations in which schistosomiasis is endemic, possibly due to regulatory responses resulting from multiple exposures as has been demonstrated in mice [71]. Such regulatory responses may be induced by skin-stage schistosomula-derived molecules such as prostaglandin $E_2(PGE_2)$ which upregulates IL-10 production during skin penetration by the parasite [72]. The PGE_2 is also secreted by the lung-stage schistosomula during migration through the capillary beds of the lungs, and this is thought to diminish eosinophil infiltrates around the parasites (and thus inflammation [73]). In addition, these parasites are capable of inhibiting the expression of endothelial adhesion molecules such as E-selectin and VCAM-1, limiting leucocyte recruitment in the lungs [74]. These anti-inflammatory mechanisms in the lungs have been suggested as potential explanations for reduced severity of asthma symptoms in schistosome-infected asthmatic patients [45], although there are no mechanistic studies from human populations to support this.

Schistosome eggs are major Th2 triggers as demonstrated in murine studies [75, 76], and they induce formation of fibrotic lesions or granulomas [77–79]. Indeed, an *S. mansoni* egg-secreted glycoprotein, omega-1, has recently been identified that conditions dendritic cells for Th2 polarisation [25]. However, egg secretions are capable of inhibiting the specific binding of chemokines such as CXCL8 (IL-8) and CCL3 (MIP-1α), therefore blocking chemokine-elicited migration

of neutrophils and macrophages respectively during granuloma formation [80].

Schistosome adult worm antigens also induce Th2 responses and IgE in mice [81], baboons [82], and humans [83]. However, this parasite life stage elicits high levels of modulatory responses capable of inhibiting antiparasite [84] as well as allergic reactions [85]. The latter study demonstrated that worm infection induces IL-10, producing B cells that could protect mice against anaphylaxis. They later demonstrated that egg-laying worms exacerbate while single sex worms (precluding egg production) inhibit airway hyperresponsiveness [86].

Together, these studies show that the different parasite life-cycle stages are associated with different mechanisms of regulation and inflammation. Although concurrent exposure to all or most antigens is likely in endemic areas, and despite the fact that they may induce cross-reactive immune responses [87], the different parasite life-cycle stages may differentially affect atopic responses.

4. Epidemiology

4.1. Schistosomiasis. The epidemiological patterns of schistosomiasis differ from those of atopy, mainly because of their aetiology. Indeed, while schistosome infection is acquired as a result of exposure to parasites, atopy is a genetic predisposition (although the clinical manifestations are influenced by environmental factors). In schistosome endemic areas, infection levels follow a convex shape with host age, where infection intensity rises to peak in childhood-adolescence and decline in adulthood [88]. This peak was initially interpreted as arising from different water contact levels between age groups [89]. However, longitudinal studies showed that with the same exposure rate, "resistant" individuals were older than "susceptible" individuals [90–92], suggesting an age-dependent acquired resistance to reinfection. In addition, in communities of different parasite transmission, infection intensity peaks at a younger age in areas of high transmission compared to low-transmission areas [88, 93–95], a phenomenon referred to as a "peak shift." This phenomenon has been interpreted as reflecting different rates of development of acquired resistance to infection during schistosome infection as has been reported for *Plasmodium* infections (which cause malaria) [96, 97]. This interpretation is supported by age-related changes in immune responses, with the peak of antibodies and cytokine associated with protection coinciding with the decline in infection levels have been reported in *S. haematobium* endemic areas [83, 98, 99]. More recently, Black and colleagues observed that the rate of acquisition of antischistosome protective responses by adults occupationally exposed to *S. mansoni*, following treatment, is dependent on their history of exposure, being faster in those with a longer history [100]. This study, consistent with earlier studies [101, 102], demonstrated that resistance to schistosome infection/reinfection is acquired independent of age related physiological changes [103].

4.2. Atopic Diseases. Atopy is the genetic predisposition to become excessively sensitised and produce high levels of

IgE [104]. However, atopic diseases result from a genetic predisposition in combination with environmental stimuli such as allergens, smoke, diet, and/or infectious agents [105, 106]. The epidemiology of atopic diseases is complex as some diseases may become more prominent with age while others diminish or disappear [107]. The earliest phases of atopic diseases usually manifest during the first five years of life and the severity (and prevalence) of clinical symptoms seem to increase in late childhood/adolescence and plateau throughout adulthood [108–110] or decline for some conditions [111, 112].

It has been suggested that a natural history of allergy manifestations in atopic individuals involves progression from atopic eczema (below one year of age) to asthma or allergic rhinitis (late childhood/adolescence), a phenomenon referred to as the "atopic/allergic march" [112]. However, this is not always consistent as some children may develop atopic dermatitis long after the onset of asthma [113], while some atopic individuals may only develop one of these conditions throughout life. Longitudinal studies indicate that atopy in infancy predicts the occurrence and severity of asthma [114] and bronchial hyperresponsiveness [115] in later life. Total and allergen-specific IgE levels also seem to increase throughout childhood in allergic individuals [116]. However, a number of events occurring in the first few years of life and *in utero* are likely to influence the onset and persistence of disease. Thus, Klinnert and colleagues have shown that respiratory infections during the first year of life and parenting difficulties (e.g., postnatal maternal depression) were independent predictors of the onset of asthma during early (3 years) and late (6–8 years) childhood in children at risk [114, 117].

Microbial exposures and diet of pregnant mothers may also alter early gene expression in neonates, influencing the onset of allergy in childhood (see [106, 118]). Consistent with this hypothesis is the finding that maternal exposure to farm milk and farm animals during pregnancy was associated with demethylation within the *FOXP3* (Treg transcription factor) locus in cord blood and subsequent elevated levels of regulatory T cells (Tregs) (and their suppressive activity) in offspring [119]. Murine studies have also shown that, when exposed to a methyl rich diet during pregnancy (as may be the case for folate supplementation in humans), foetal DNA may undergo changes in methylation that results in decreased gene transcription activity, leading to subsequent enhanced development and severity of allergic diseases [120]. The study also showed that this diet-associated allergic phenotype was transgenerationally inheritable (persistence of high levels of IgE and eosinophilia into the F2 generation).

5. Effector Responses in Atopy and Schistosome Infection

5.1. Immunoglobulin E. Identified in the 1960s as a "carrier of reaginic activity" [121, 122], IgE is well known as a central player in atopic diseases and anaphylactic reactions. This antibody is part of a protein network involving its 3 receptors, namely, the FcεRI, the CD23 (or FcεRII) and galectin-3 [32], all of which can be found in soluble forms

[123, 124]. The FcεRI (also termed high-affinity receptor) is mainly expressed on mast cells and basophils but also on epidermal Langerhans cells [125] and eosinophils [126, 127]. Cross-linking of these high-affinity receptors by IgE induces activation of mast cells and basophils and their degranulation. The galectin-3 receptor is expressed on neutrophils and on trophoblast cells in placentas [128], where it is thought to facilitate IgE transport [129]. The CD23 receptor facilitates the transport of IgE-antigen complexes but is also involved in the regulation of IgE synthesis [32]. Highly conserved in mammalian lineages [130], IgE is thought to have evolved as a first line of defence against helminth parasites.

IgE antibodies are naturally strongly regulated and have the lowest concentrations of all antibodies in serum of healthy nonatopic individuals [32]. Mechanisms of regulation of IgE include its short half-life in serum (12 h for murine monoclonal antibodies [131]), the poor processing of mRNA for the membrane ε heavy chain [132], and the negative feedback regulation by the CD23 [133]. The latter has been a subject of investigations in terms of therapeutic application in atopic diseases but also in autoimmune diseases [134] and chronic lymphocytic leukaemia [135].

5.2. CD23. CD23 is the low-affinity receptor for IgE and differs from the high-affinity FcεRI receptor in structure and function. Thus, while cross-linking of the latter results in degranulation of mast cells and release of mediators, engagement of membrane-bound CD23 suppresses the production of IgE by B lymphocytes [33]. CD23 has long been proposed as a natural regulator for IgE synthesis [133] although elevated levels of CD23+ B cells have been reported in atopic patients [136]. As initially suggested by Aubry and colleagues [137], CD23 not only binds IgE but also CD21, a cell-surface protein expressed on T-cell, B-cell, and follicular dendritic cells, classically identified as a receptor for complement proteins [138] or Epstein-Barr virus [137]. The interaction between CD23, IgE, and CD21 may lead to either negative or positive regulation of IgE synthesis (reviewed in [34–36]). The binding of IgE stabilises membrane-bound CD23 and inhibits IgE synthesis from activated B cells, while in the absence of IgE binding, CD23 is cleaved by ADAM10 (a disintegrin and metalloprotease protein 10), and this destabilisation enhances IgE synthesis [32]. Soluble CD23 (sCD23) fragments resulting from the cleavage can bind to IgE with different affinities and outcomes for IgE synthesis depending on their oligomerization state. Trimers bind IgE with high affinity and enhance IgE synthesis by their ability to also bind the CD21 receptor while monomers bind with low affinity but do not bind CD21 and hence inhibit IgE synthesis [36, 139].

5.3. Immunoglobulin 4. Serum IgG4 antibodies, the least abundant among human IgG subclasses, have long been associated with IgE-mediated diseases [140–142]. However, rather than the cause of disease, these antibodies seem to be involved in the regulation of IgE-induced anaphylactic reactions [143]. IgG4 may interfere with antigen recognition by IgE due to their similar antigenic specificity [144], although different epitope-binding [145]. In a process that involves exchange of *fab* molecules, IgG4 are structurally hetero-bivalent (each heavy chain and light chain recognising a different epitope within a single IgG4 molecule) and often function as monovalent [141, 146], to bring about anti-inflammatory effects [147]. The interaction between IgG4 and a given antigen results in small and non-pathological immune complexes (since these antibodies cannot cross-link antigens) [146]. Furthermore, in contrast to other IgG subclasses, IgG4 cannot fix complement but inhibits complement activation by IgG1 [148]. IgG4 antibodies, in allergy or helminth infection, are secreted in response to high antigen loads [141, 149, 150] but levels of the antibodies are differentially regulated by the same cytokines [151] as those regulating IgE, suggesting an important homeostatic mechanism for controlling IgE-mediated responses.

6. Control of Effector Responses in Atopy and Schistosome Infection

In addition to the cross-regulation between Th1 and Th2 [152] (and potentially other T cell subsets), there is growing evidence that Th2 cells interact with a complex network of other T cell subsets as well as B cells and antibodies, naturally or during disease (atopic or infectious). Thus, it has emerged that Tregs play an important role in the tolerance of ubiquitous antigens and that alterations in Treg function [153, 154] and/or the fine balance between Tregs and Th2 cells [155, 156] determines the clinical manifestation of atopy. Indeed, in healthy (nonatopic) individuals T cell polarization occurs in contact with environmental allergens but higher levels of Tregs dampen the effect of Th2 cells, leading to peripheral tolerance [156]. Tregs modulate the activity of Th2 (and Th1) cells via several mechanisms including the secretion of anti-inflammatory cytokines such as IL-10 and TGF-β [155, 157]. As the description and role of other recently identified T-helper cells is clarified (e.g., Th17 cells shown to be important in nonatopic asthma) regulation of Th2 mediated responses will also become clearer [158]. The role of cells such as the T-helper cells recently shown to produce both IL-17 and Th2 cytokines (IL-4, IL-5, IL-9, and IL-13) [159] in pathogenesis is currently under intense investigation. Our own group has recently described a role for Th17 responses in human schistosome-acquired immunity (submitted).

IL-10 producing B cells (Bregs) are also involved in the recruitment of Tregs, hence contributing to the regulation of Th2 responses as demonstrated in murine models of helminth infection (see [31]). IL-10 can inhibit effector functions of mast cells and eosinophils, and regulate the growth of several cells including B cells, NK cells, mast cells, and dendritic cells. Furthermore, IL-10 modulates IgE : IgG4 ratios [154] possibly by indirectly inducing the antibody switch to IgG4 in the B-cell progeny while preventing IgE production [160].

IgG4 may control IgE-mediated histamine release as has been demonstrated in filarial infection [143]. Furthermore, it has recently been shown that the binding patterns of IgG4 antibodies correspond to natural recovery from childhood IgE-mediated milk allergy [161], suggesting their potential protective role in atopic diseases, although this is still controversial [141]. In addition, early observations that IgG4 antibodies were highly elevated in sera of patients receiving allergen immunotherapy [162] have prompted the use of IgE : IgG4 ratio as a marker for successful immunotherapy [142, 163, 164]. In helminth infections, high IgG4 : IgE ratio has been associated with reduced pathology while favouring a heavy worm load [150, 165, 166]. Interestingly, IgG4 may be one of the "regulatory antibodies" resulting from IgG syalilation involved in the control of immune disorders [167, 168].

7. Immune Responses in Atopy

The human immune system must distinguish between a dangerous pathogen and ubiquitous environmental allergens and has evolved to mount appropriate defensive responses to the first while tolerating (or ignoring) the latter. However, a certain proportion of individuals fail to tolerate environmental allergens and develop allergic diseases such as asthma, atopic dermatitis and allergic rhinitis. These result from excessive sensitisation to ordinary exposures to allergens [104]. IgE antibodies are critical effector molecules in the pathogenesis of these diseases [169]. Mast cells and basophils are coated with specific IgE antibodies and this results in immediate hypersensitivity (release of mediators) and/or late-phase inflammatory reaction (cytokine secretion and recruitment of leucocytes).

In atopic individuals, allergen products (e.g., cysteine proteases) activate epithelial cells, which produce thymic stromal lymphopoietin (TSLP), IL-25, and IL-33 which in turn initiate Th2 polarisation with increased production of IL-4, IL-5, IL-9, and IL-13 cytokines [170, 171]. Th2 cytokines are involved in the class-switching to IgE as well as the development and recruitment of basophils, mast cells, and eosinophils (see Figure 1(c)). IgE binds to the high-affinity FcεRI receptor on mast cells, basophils, and eosinophils which (upon exposure to allergens) results in their activation and degranulation (via cross-linking of allergens), with the release of preformed mediators such as histamine, cysteinyl leukotrienes, and prostaglandin D$_2$ [154, 172].

A complex interplay between innate and adaptive immune responses underlies the heterogeneous characteristics of atopic diseases. Thus, recruitment of eosinophils into the lungs of asthmatics may be promoted not just by Th2 (IL-5) alone but in conjunction with natural killer T cells (NKT) as well as CD8+ T cells (see [173]). In addition, IL-17-producing T cells (Th17 [174]) may be involved in the severity of asthma [175]. These promote the recruitment and activation of neutrophils and lead to corticosteroids—resistant asthma [175]. IL-9-producing T-cell subset (Th9), which probably derive from Th2 cells under the influence of TGFβ1 [176], may also be involved in the production of IgE and mast cell recruitment in the lungs [173]. In allergic rhinitis, mast cells accumulate in the epithelium of the nasal mucosa where they secrete inflammatory cytokines (IL-6, IL-8, and TNFα) in addition to Th2 cytokines [177].

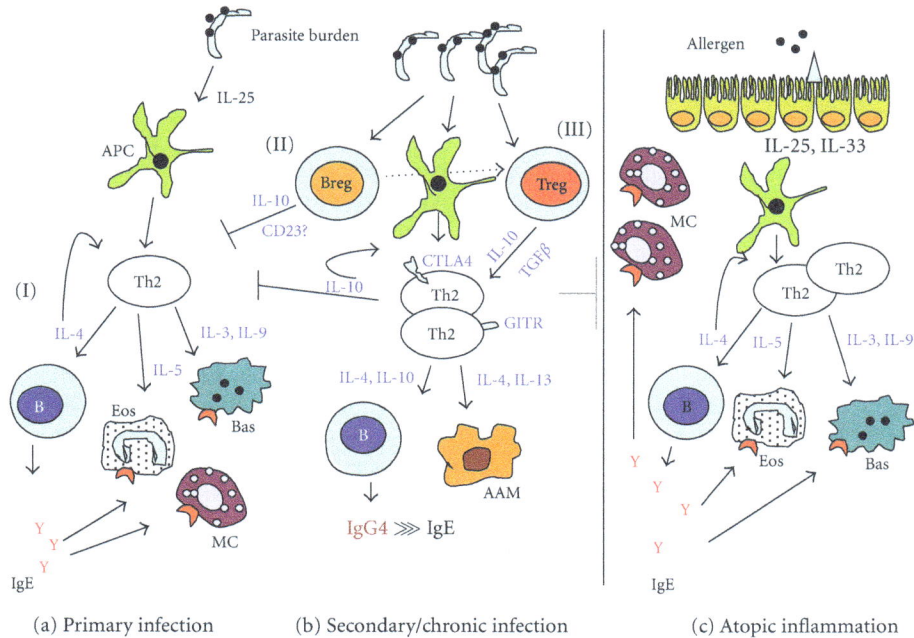

FIGURE 1: Possible regulatory mechanisms in helminth infections. Primary response (a) to parasite antigens involves Th2 polarization, IgE production, and eosinophil, mast cell, and basophil activation (I), mechanisms similar to those observed in allergic sensitisation (c). This Th2-response may be induced by parasite-secreted antigens such as the Omega-1 secreted by *S. mansoni* eggs [25]. However, with increasing parasite load or chronic infection (b), regulatory B cells are activated which suppress Th2 responses (II) via IL-10 secretion or CD23 expression [26], and/or contribute to the recruitment of Tregs [27]. Tregs (III), which may also be induced and expanded by parasite antigens [28, 29], either induce anergic Th2 cells (expressing GITR and CTLA4) unable to progress through to effector cells, or modify downstream effector functions such as B cell switch to IgG4 and/or alternative activation of macrophages, resulting in immunological tolerance (reviewed by [30]). This immunosuppression is induced in the context of helminth infection, but may also expand to allergen-induced inflammation (gray line), hence suppressing allergy. DC: dendritic cell; B: B cell, Eos: eosinophil; Bas: basophil; MC: mast cell; GITR: glucocorticoid-induced TNFα-related protein; CTLA4: cytotoxic T lymphocyte antigen 4; AAM: alternatively activated macrophage; Breg: regulatory B cell; Treg: regulatory T cell. The question mark (?) denotes lack of strong evidence. Figure adapted from [30, 31] and collated information from the cited references.

8. Immune Responses during Schistosome Infection

Acquired immunity to schistosome infection was first proposed by Fisher in the 1930s when analysing data from animal studies as well as those from hospital-diagnosed *S. haematobium* infected people [88]. Subsequently, the susceptibility of schistosome larvae to immune attack was demonstrated by in vitro studies showing that sera from *S. mansoni* infected individuals could damage schistosomula in the presence of normal human peripheral blood leucocytes [178]. This "antibody-dependent" killing was subsequently shown to be eosinophil mediated [178, 179], and studies on monoclonal antibodies led to the identification of IgE antibodies with the highest cytotoxicity for the schistosomula [180–182]. Field studies were conducted to identify antibody responses predictive of resistance to reinfection following chemotherapy. Hagan et al. [83] demonstrated in a multi-variate logistic regression that reinfection with *S. haematobium* was less likely in individuals producing high IgE levels against the worm antigens and more likely in those producing high levels of IgG4 against the worm or egg antigens. The role of IgE in resistance was also demonstrated by Rihet et al.

[183], who identified specific antigens (120–165 KDa and 85 KDa) to which IgE reacted (on immunoblots) and showed that these antibodies, in contrast to IgM and IgG, were significantly higher in the sera of the most resistant individuals. This study showed that some of the immunogenic antigens were readily accessible to IgE on living *S. mansoni* larvae as they were located on the outer membrane. However, Dunne and colleagues, working on *S. mansoni* as well, showed that IgE (produced following treatment) against adult worm antigens, particularly a 22 kDa tegumental antigen (Sm22), but not against any other life-cycle stages, were associated with resistance to reinfection following treatment [184]. Both antiadult worm and anti-schistosomula tegument IgE antibodies were associated with resistance to *S. mansoni* reinfection in another study in Brazil [185] while antiegg IgE antibodies also could confer protection against *S. japonicum* reinfection [186].

Collectively, these studies and several others [187–190] have led to the conclusion that resistance to schistosome infection/reinfection is dependent on IgE antibodies. However, data on other antibody isotypes have been reported which correlated with resistance to infection/reinfection. For example, IgG3 against the recombinant antigen Sh13 has been associated with resistance to *S. haematobium* infection [191], while antiworm and cercariae IgM were significantly higher in individuals more resistant to reinfection with *S. mansoni* [185]. Furthermore, a decline in IgA together with an increase in IgG1 were associated with resistance acquired with host age as well as following treatment in *S. haematobium* endemic area [192]. IgA against Sm28GST antigen has also been associated with reduced *S. mansoni* fecundity and increased host resistance to reinfection [193]. More recently, antiworm IgE antibodies, as well as eosinophilia and the low affinity receptor for IgE (the CD23) have been shown to correlate with resistance in individuals undergoing multiple rounds of treatment [100, 194], again suggesting that IgE may be directly involved in parasite killing via antibody-dependent cellular cytotoxicity (ADCC) *in vivo*. However, since schistosomula are more susceptible to ADCC, it is possible that adult-worm-specific antibody responses may rather target the incoming larvae, a process termed "concomitant immunity" (as predicted by Fisher [88]) and well demonstrated in rhesus monkeys [195].

As initially demonstrated by in vitro studies [196–198], ADCC is dependent on Th2 cytokines, and these have been involved in resistance to schistosome infection. Thus, higher ratios of IL-4/IFN-γ and IL-5/IFN-γ were produced by specific T-cell clones from *S. mansoni* resistant than susceptible individuals [199] and IL-5 correlated with lower levels of *S. haematobium* infection [99] and *S. mansoni* reinfection after treatment [200]. Furthermore, IL-4, IL-5, and IL-10 levels were associated with resistance posttreatment while IFN-γ was associated with susceptibility [201]. However, significantly higher levels of IFN-γ against adult worm and cercariae antigens by PBMCs from resistant individuals compared to those from susceptible individuals [202], suggesting that acquired resistance to human schistosomiasis cannot be exclusively classified into a single T helper cell subset.

Cellular immune responses, although involved in resistance, mediate most of schistosome-related pathology [203, 204], which can be divided into acute and chronic diseases based on disease progression. Acute schistosomiasis is a debilitating febrile disease which often occurs in individuals with no experience of infection. It is characterized by high percentage of eosinophilia, which may be reversed by chemotherapy [205], nausea, urticaria, dry cough, and fever [206, 207]. Anatomically, this stage is accompanied by a dissemination of large and destructive granulomas around the eggs [67, 208]. Chronic schistosomiasis is often referred to as a Th2 disease and accounts for most human immunopathologies in endemic areas [77, 204, 208–210]. As infection becomes chronic, schistosome eggs lodge in the liver, gut (*S. mansoni*), or bladder (*S. haematobium*), and the granulomatous response translates into extensive tissue damage and excessive extracellular matrix protein (ECMP) deposition, leading to fibrosis [210].

9. Immunological Interaction between Helminth Antigens and Allergens

9.1. Helminth Infection and the "Mast Cell Saturation" Hypothesis. The earliest protective mechanism of helminth infection suggested was "mast cell saturation," whereby helminths induce high levels of nonspecific IgE that saturate Fc receptors on mast cells, thus inhibiting hypersensitivity reactions [211, 212]. Further supportive evidence for the Fc saturation hypothesis came from a study showing that histamine release of human mast cells from lung fragments could be blocked by preexposure of these fragments to high total IgE [213]. However, more recent studies on basophils have shown that high levels of polyclonal IgE and polyclonal/specific IgE ratios from filarial- and hookworm-infected patients do not prevent antigen-induced histamine release [214, 215]. Nevertheless, Mitre and colleagues were able to show that extremely high ratios of polyclonal/specific IgE, enhanced with polyclonal myeloma IgE *in vitro*, could prevent histamine release [214]. Although basophils and mast cells may be differentially regulated [216], these experiments suggested that the FcεRI receptor saturation may not be the primary mechanism by which helminths "protect" against allergy.

9.2. Helminth Infection and Cross-Reactive IgE Responses. Another hypothesis suggested was that helminth parasites induce a "clinically irrelevant" allergen-specific IgE response, which would be cross-reactive between helminths and allergens [217]. Cross-reactive anti-tropomyosin IgE antibodies between helminths and allergens have recently been demonstrated, where monkeys infected with *Loa loa* (filarial parasites) mounted an IgE cross-reacting between filarial tropomyosin and Derp 1 allergen but not with timothy grass [218]. Furthermore, cross-reactivity between ascaris and mites has been reported [219]. However, field studies report mixed results on the effects of helminth infections on allergen-specific IgE in endemic areas [220–222]. Our recent study has demonstrated that the levels of anti-Derp1 IgE antibodies inversely correlate with *S. haematobium* infection intensity in a high schistosome infection area in Zimbabwe [223].

9.3. Helminth-Induced Immunomodulation. Technological and scientific advances such as genomic sequences and proteomic approaches have generated molecular and evolutionary information on the relationship between helminth parasites and allergic reactivity. Helminth infections are generally characterised by a Th2-polarised immune response [25, 84, 224], which is often associated with host resistance to infection/reinfection [30]. However, this Th2 response is also associated with pathology [204], consistent with the role for Th2 in allergic diseases [171]. Nevertheless, helminth parasites are capable of modulating this response to prolong their survival and minimize severe pathology in their host [76, 84, 225]. This immunomodulation is thought to affect unrelated antigens such as allergens, hence dampening the clinical manifestation of allergy. Indeed, experimental studies have demonstrated helminth-induced suppression of allergic responses via multiple pathways (Figure 1). However, these observations and mechanisms remain to be rigorously tested in humans. Furthermore, biological and evolutionary differences in the mouse experimental host and the natural human host must be taken into account when extrapolating

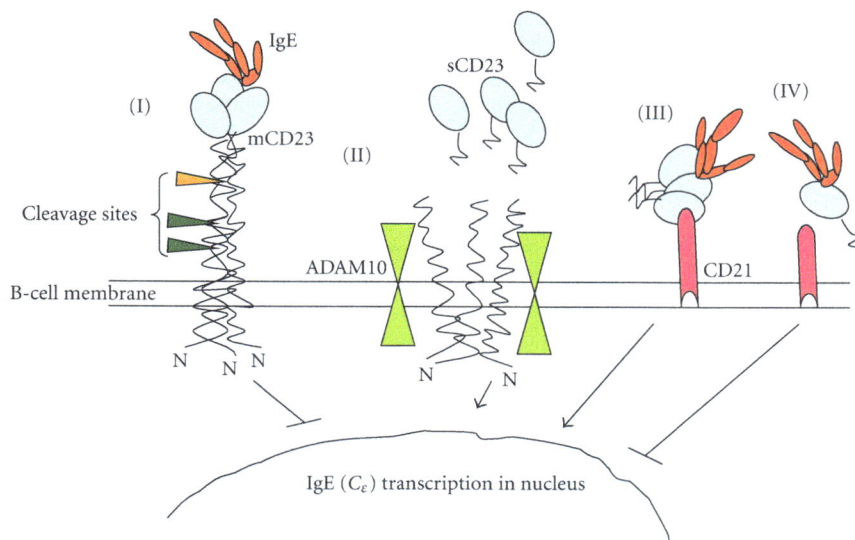

FIGURE 2: Interaction between CD23 and its ligands, IgE and CD21. Binding of IgE stabilises membrane-bound CD23 and inhibits IgE synthesis (I) from activated B cells while in the absence of IgE binding the CD23 is cleaved by ADAM10 (a disintegrin and metalloprotease protein 10) and this destabilisation enhances IgE synthesis (II). However, soluble CD23 (sCD23) fragments resulting from the cleavage have the ability to bind IgE with different affinities depending on their oligomerization state: trimers (III) bind IgE with high affinity while monomers (IV) bind with low affinity. Trimers enhance IgE synthesis by their ability to also bind CD21 receptor (III) while monomers fail to bind CD21 and inhibit IgE synthesis (IV) (adapted from [32–36]).

mechanistic and phenomenological results from the mouse to the human, for example, differences in the IgE receptors [226].

Human studies investigating the regulatory mechanisms underlying the protective effect of helminth infections on atopy have primarily focused on IL-10. Thus, parasite-induced IL-10 production and skin prick reactivity were negatively associated in *Ascaris lumbricoides* [42] and *Schistosoma haematobium* [46] infected populations. Furthermore, allergen-induced IL-10 was associated with reduced Th2 responses (IL-4 and IL-5) in asthmatic schistosome infected patients [227]. More recently, it has been shown that the frequency of PBMCs expressing cytotoxic-T-lymphocyte antigen 4 (CTLA-4) and monocytes expressing IL-10 from asthmatic patients infected with *S. mansoni* was significantly higher compared to their asthmatic uninfected counterparts [228]. However, a study on an Ecuadorian population showed no association between skin prick reactivity with either IL-10 or IL-10-producing T cells induced by *Ascaris lumbricoides* [40].

The TGFβ is another cytokine involved in the modulation of immune responses, and is secreted by antigen-presenting cells (APCs) or regulatory T cells [229]. However, there is a paucity of human studies on this cytokine in the context of atopy and helminth infections. Interestingly, we have observed a negative association between atopy and the levels of soluble CD23 in *S. haematobium* infected populations (Rujeni et al., manuscript in preparation). The CD23 is the low affinity receptor for IgE and is involved in the regulation of these antibodies [35]. As illustrated in Figure 2, the soluble CD23 can either upregulate or downregulate IgE synthesis depending on their size and oligomerization state.

Of note, expression of this receptor has been associated with resistance to schistosome [194] and *Ascaris* [230] infections in humans, and suppressed airway allergy in helminth-infected mice [231].

As illustrated in Figure 1, immunomodulation during chronic helminth infection is driven by regulatory T and B cells (Tregs and Bregs, resp.), which secrete the above mentioned anti-inflammatory cytokines. Treg cells are either recruited by Bregs or induced and expanded by helminth-derived products [28]. Both T and B regulatory cells can suppress Th2 cells thereby regulating atopy and helminth-induced pathology [31, 232]. Indeed, a study in our lab has shown that Treg proportions correlate with the levels of schistosome infection in young children actively acquiring infection [233].

Helminth molecules have been identified from excretory-secretory (ES) products that are associated with immunomodulation during helminth infection. Thus, the ES-62 is a phosphorylcholine-containing glycoprotein secreted by *Acanthocheilonema viteae*, a rodent filarial nematode [234]. This protein presents anti-inflammatory properties and has been successfully tested in mouse models of allergy and autoimmune diseases [235, 236], and it is currently being exploited as a potential therapeutic agent for inflammatory diseases in humans [237]. The anti-inflammatory properties of this molecule include modulation of B-cell proliferation and cytokine production as well as hyporesponsiveness and desensitization of mast-cell degranulation [237–240]. The interleukin-4-inducing principle from *S. mansoni* egg IPSE/alpha-1, identified as one of the most abundant proteins secreted by *S. mansoni* eggs [241], has also been associated with immunomodulation, possibly by inducing granulomatous responses [242]. Furthermore, IPSE/alpha-1 has been shown to induce antigen-independent IL-4 production by murine basophils in vivo [243]. The venom allergen-like (VAL) proteins are another group of helminth ES products involved in immunomodulation. Thus, Hewitson et al. have demonstrated that antibodies to these VAL antigens are dominant in susceptible mice in an *H. polygyrus* infection model [244]. Sj-VAL-1 is one of the VAL proteins identified in *S. japonicum* egg ES products inducing an antibody response during the first 6 weeks of infection in mice [245].

10. Convergence of Allergic and Antiparasite Responses

There is current interest in determining the common features in the induction of immune responses by allergens and by helminths as well as the evolutionary advantages of maintaining allergic responses. As illustrated above, several studies have suggested that similarities in antigens may underlie the commonality of Th2 responses elicited by allergens and by helminths. A recent review [246] of allergic responses indicated that there is relatively little structural similarity between different allergens (e.g., house dust mite, food allergens, and haematophagous fluids) and between allergens and helminth parasites. Instead, this paper suggests that the relationship between allergic and antiparasite Th2 responses arises from a common response to different classes of environmental challenges which include helminth parasites, venoms and haematophagous fluids, and environmental irritants such as carcinogens and noxious xenobiotics, so that this diverse group of stimuli activates responses collectively known as "allergic host defences" [246]. Within this paradigm, these environmental challenges are characterised only by the type of response they elicit with multiple pathways leading to the activation of Th2 responses with the result of protecting against environmental challenges by either reduced exposure to, or elimination of the "irritant." In this scenario, allergic reactivity is believed to have evolved as an important and essential mechanism against harm rather than a harmful overreaction of a misdirected immune system [247]. Studies in cancer patients also show a negative association between cancer and atopy which has led to the suggestion that allergy protects against some types of cancer [248, 249]. This suggests that the Th2 responses protecting against allergens, carcinogens, and helminths are complex. This presents a challenge for the development of therapeutics relying on helminth products to overcome allergic responses, since induction of allergic responses as well as the effector mechanisms maybe tightly regulated, and the effector responses they elicit may have been selected for redundancy.

11. Conclusions

We have shown similarities in the immunological responses to schistosome parasites and to allergens. Studies continue to determine the aetiology of the similar responses and the evolutionary pathways that may have led to the development

and maintenance of allergic responses which are paradoxically harmful to the host [246, 247], but may be essential to protect against harm from environmental challenges [246]. The clinical manifestation of atopy is complex with several studies from helminth endemic areas having shown that allergic sensitisation and clinical manifestation of allergy can be dissociated [222]. Furthermore, allergic disease and parasitic infections exist as comorbidities in many patients and are not mutually exclusive [250]. The role of impaired serological allergy diagnosis in parasitized allergy patients as well as under diagnosis in developing countries needs to be addressed to inform future studies. Detailed longitudinal and mechanistic studies relating atopy and clinical disease to schistosome infection and disease in human populations will be valuable to inform on not only the immunological process occurring, but more importantly on clinical management of allergy and schistosomiasis patients.

Acknowledgments

This paper contains some of our work which was funded by the World Health Organization (Grant no. RPC264), The Welcome Trust (Grant no. WT082028MA), the University of Edinburgh, and the Government of Rwanda. F. Mutapi is funded by the Thrasher Foundation. The authors are grateful to Laura Appleby for providing comments on the manuscript.

References

[1] N. R. Stoll, "This wormy world," *Journal of Parasitology*, vol. 33, pp. 1–18, 1947.

[2] GaHI, "Global Atlas of helminth infections," Consulted, Manta Ray Media Ltd, 2012, http://www.thiswormyworld.org/.

[3] P. J. Hotez, J. H. F. Remme, P. Buss, G. Alleyne, C. Morel, and J. G. Breman, "Combating tropical infectious diseases: report of the disease control priorities in developing countries project," *Clinical Infectious Diseases*, vol. 38, no. 6, pp. 871–878, 2004.

[4] P. J. Hotez, P. J. Brindley, J. M. Bethony, C. H. King, E. J. Pearce, and J. Jacobson, "Helminth infections: the great neglected tropical diseases," *Journal of Clinical Investigation*, vol. 118, no. 4, pp. 1311–1321, 2008.

[5] J. Bethony, S. Brooker, M. Albonico et al., "Soil-transmitted helminth infections: ascariasis, trichuriasis, and hookworm," *The Lancet*, vol. 367, no. 9521, pp. 1521–1532, 2006.

[6] C. H. King and M. Dangerfield-Cha, "The unacknowledged impact of chronic schistosomiasis," *Chronic Illness*, vol. 4, no. 1, pp. 65–79, 2008.

[7] C. H. King, K. Dickman, and D. J. Tisch, "Reassessment of the cost of chronic helmintic infection: a meta-analysis of disability-related outcomes in endemic schistosomiasis," *The Lancet*, vol. 365, no. 9470, pp. 1561–1569, 2005.

[8] N. Aït-Khaled, N. Pearce, H. R. Anderson et al., "Global map of the prevalence of symptoms of rhinoconjunctivitis in children: the International Study of Asthma and Allergies in Childhood (ISAAC) Phase Three," *Allergy*, vol. 64, no. 1, pp. 123–148, 2009.

[9] H. Williams, C. Robertson, A. Stewart et al., "Worldwide variations in the prevalence of symptoms of atopic eczema in the international study of asthma and allergies in childhood,"

Journal of Allergy and Clinical Immunology, vol. 103, no. 1 I, pp. 125–138, 1999.

[10] E. N. Sibanda, "Inhalant allergies in Zimbabwe: a common problem," *International Archives of Allergy and Immunology*, vol. 130, no. 1, pp. 2–9, 2003.

[11] R. Beasley, S. Nishima, N. Pearce, and J. Crane, "β-agonist therapy and asthma mortality in Japan," *The Lancet*, vol. 351, no. 9113, pp. 1406–1407, 1998.

[12] P. Davis, R. Jackson, and N. Pearce, "Asthma mortality," *New Zealand Medical Journal*, vol. 98, no. 783, p. 604, 1985.

[13] R. Jackson, "Undertreatment and asthma deaths," *The Lancet*, vol. 2, no. 8453, p. 500, 1985.

[14] R. Brown, F. Turk, P. Dale, and J. Bousquet, "Cost-effectiveness of omalizumab in patients with severe persistent allergic asthma," *Allergy*, vol. 62, no. 2, pp. 149–153, 2007.

[15] R. Gupta, A. Sheikh, D. P. Strachan, and H. R. Anderson, "Burden of allergic disease in the UK: secondary analyses of national databases," *Clinical and Experimental Allergy*, vol. 34, no. 4, pp. 520–526, 2004.

[16] P. Suwan, D. Akaramethathip, and P. Noipayak, "Association between allergic sensitization and attention deficit hyperactivity disorder (ADHD)," *Asian Pacific Journal of Allergy and Immunology*, vol. 29, no. 1, pp. 57–65, 2011.

[17] M. C. Tsai, H. K. Lin, C. H. Lin, and L. S. Fu, "Prevalence of attention deficit/hyperactivity disorder in pediatric allergic rhinitis: a nationwide population-based study," *Allergy and Asthma Proceedings*, vol. 32, no. 6, pp. 41–46, 2011.

[18] B. Taylor, J. Wadsworth, M. Wadsworth, and C. Peckham, "Changes in the reported prevalence of childhood eczema since the 1939–45 war," *The Lancet*, vol. 2, no. 8414, pp. 1255–1257, 1984.

[19] M. B. Emanuel, "Hay fever, a post industrial revolution epidemic: a history of its growth during the 19th century," *Clinical Allergy*, vol. 18, no. 3, pp. 295–304, 1988.

[20] F. D. Martinez, D. A. Stem, A. L. Wright, L. M. Taussig, and M. Halonen, "Association of non-wheezing lower respiratory tract illnesses in early life with persistently diminished serum IgE levels," *Thorax*, vol. 50, no. 10, pp. 1067–1072, 1995.

[21] S. O. Shaheen, P. Aaby, A. J. Hall et al., "Measles and atopy in Guinea-Bissau," *The Lancet*, vol. 347, no. 9018, pp. 1792–1796, 1996.

[22] J. S. Alm, G. Lilja, G. Pershagen, and A. Scheynius, "Early BCG vaccination and development of atopy," *The Lancet*, vol. 350, no. 9075, pp. 400–403, 1997.

[23] I. L. Strannegård, L. O. Larsson, G. Wennergren, and O. Strannegård, "Prevalence of allergy in children in relation to prior BCG vaccination and infection with atypical mycobacteria," *Allergy*, vol. 53, no. 3, pp. 249–254, 1998.

[24] P. M. Matricardi, F. Rosmini, S. Riondino et al., "Exposure to foodborne and orofecal microbes versus airborne viruses in relation to atopy and allergic asthma: epidemiological study," *British Medical Journal*, vol. 320, no. 7232, pp. 412–417, 2000.

[25] B. Everts, G. Perona-Wright, H. H. Smits et al., "Omega-1, a glycoprotein secreted by *Schistosoma mansoni* eggs, drives Th2 responses," *Journal of Experimental Medicine*, vol. 206, no. 8, pp. 1673–1680, 2009.

[26] M. S. Wilson, M. D. Taylor, M. T. O'Gorman et al., "Helminth-induced CD19+CD23hi B cells modulate experimental allergic and autoimmune inflammation," *European Journal of Immunology*, vol. 40, no. 6, pp. 1682–1696, 2010.

[27] S. Amu, S. P. Saunders, M. Kronenberg, N. E. Mangan, A. Atzberger, and P. G. Fallon, "Regulatory B cells prevent and reverse allergic airway inflammation via FoxP3-positive T regulatory cells in a murine model," *Journal of Allergy and*

Clinical Immunology, vol. 125, no. 5, pp. 1114.e8–1124.e8, 2010.

[28] H. J. McSorley, Y. M. Harcus, J. Murray, M. D. Taylor, and R. M. Maizels, "Expansion of Foxp3+ regulatory T cells in mice infected with the filarial parasite Brugia malayi," *Journal of Immunology*, vol. 181, no. 9, pp. 6456–6466, 2008.

[29] J. R. Grainger, K. A. Smith, J. P. Hewitson et al., "Helminth secretions induce de novo T cell Foxp3 expression and regulatory function through the TGF-β pathway," *Journal of Experimental Medicine*, vol. 207, no. 11, pp. 2331–2341, 2010.

[30] J. E. Allen and R. M. Maizels, "Diversity and dialogue in immunity to helminths," *Nature Reviews Immunology*, vol. 11, no. 6, pp. 375–388, 2011.

[31] L. Hussaarts, L. E. P. M. van der Vlugt, M. Yazdanbakhsh, and H. H. Smits, "Regulatory B-cell induction by helminths: implications for allergic disease," *Journal of Allergy and Clinical Immunology*, vol. 128, no. 4, pp. 733–739, 2011.

[32] H. J. Gould and B. J. Sutton, "IgE in allergy and asthma today," *Nature Reviews Immunology*, vol. 8, no. 3, pp. 205–217, 2008.

[33] M. Acharya, G. Borland, A. L. Edkins et al., "CD23/FcepsilonRII: molecular multi-tasking," *Clinical and Experimental Immunology*, vol. 162, no. 1, pp. 12–23, 2010.

[34] R. G. Hibbert, P. Teriete, G. J. Grundy et al., "The structure of human CD23 and its interactions with IgE and CD21," *Journal of Experimental Medicine*, vol. 202, no. 6, pp. 751–760, 2005.

[35] D. H. Conrad, J. W. Ford, J. L. Sturgill, and D. R. Gibb, "CD23: an overlooked regulator of allergic disease," *Current Allergy and Asthma Reports*, vol. 7, no. 5, pp. 331–337, 2007.

[36] S. L. Bowles, C. Jaeger, C. Ferrara et al., "Comparative binding of soluble fragments (derCD23, sCD23, and exCD23) of recombinant human CD23 to CD21 (SCR 1-2) and native IgE, and their effect on IgE regulation," *Cellular immunology*, vol. 271, no. 2, pp. 371–378, 2011.

[37] D. Dagoye, Z. Bekele, K. Woldemichael et al., "Wheezing, allergy, and parasite infection in children in urban and rural ethiopia," *American Journal of Respiratory and Critical Care Medicine*, vol. 167, no. 10, pp. 1369–1373, 2003.

[38] N. R. Lynch, I. Hagel, M. Perez, M. C. Di Prisco, R. Lopez, and N. Alvarez, "Effect of anthelmintic treatment on the allergic reactivity of children in a tropical slum," *Journal of Allergy and Clinical Immunology*, vol. 92, no. 3, pp. 404–411, 1993.

[39] L. J. Palmer, J. C. Celedón, S. T. Weiss, B. Wang, Z. Fang, and X. Xu, "Ascaris lumbricoides infection is associated with increased risk of childhood asthma and atopy in rural China," *American Journal of Respiratory and Critical Care Medicine*, vol. 165, no. 11, pp. 1489–1493, 2002.

[40] P. J. Cooper, E. Mitre, A. L. Moncayo, M. E. Chico, M. G. Vaca, and T. B. Nutman, "Ascaris lumbricoides-induced interleukin-10 is not associated with atopy in schoolchildren in a rural area of the tropics," *Journal of Infectious Diseases*, vol. 197, no. 9, pp. 1333–1340, 2008.

[41] P. J. Cooper, M. E. Chico, M. G. Vaca et al., "Effect of albendazole treatments on the prevalence of atopy in children living in communities endemic for geohelminth parasites: a cluster-randomised trial," *The Lancet*, vol. 367, no. 9522, pp. 1598–1603, 2006.

[42] C. Flohr, L. N. Tuyen, R. J. Quinnell et al., "Reduced helminth burden increases allergen skin sensitisation but not clinical allergy: a randomized, double-blind, placebo-controlled trial in Vietnam," *Clinical and Experimental Allergy*, vol. 40, no. 1, pp. 131–142, 2010.

[43] L. C. Rodrigues, P. J. Newcombe, S. S. Cunha et al., "Early infection with Trichuris trichiura and allergen skin test reactivity in later childhood," *Clinical and Experimental Allergy*, vol. 38, no. 11, pp. 1769–1777, 2008.

[44] M. I. Araujo, A. A. Lopes, M. Medeiros et al., "Inverse association between skin response to aeroallergens and *Schistosoma mansoni* infection," *International Archives of Allergy and Immunology*, vol. 123, no. 2, pp. 145–148, 2000.

[45] M. Medeiros Jr., J. P. Figueiredo, M. C. Almeida et al., "*Schistosoma mansoni* infection is associated with a reduced course of asthma," *Journal of Allergy and Clinical Immunology*, vol. 111, no. 5, pp. 947–951, 2003.

[46] A. H. J. van den Biggelaar, R. Van Ree, L. C. Rodrigues et al., "Decreased atopy in children infected with *Schistosoma haematobium*: a role for parasite-induced interleukin-10," *The Lancet*, vol. 356, no. 9243, pp. 1723–1727, 2000.

[47] P. J. Hotez and A. Fenwick, "Schistosomiasis in Africa: an emerging tragedy in our new global health decade," *PLoS Neglected Tropical Diseases*, vol. 3, no. 9, article e485, 2009.

[48] A. Fenwick, J. P. Webster, E. Bosque-Oliva et al., "The Schistosomiasis Control Initiative (SCI): rationale, development and implementation from 2002–2008," *Parasitology*, vol. 136, no. 13, pp. 1719–1730, 2009.

[49] B. Gryseels, K. Polman, J. Clerinx, and L. Kestens, "Human schistosomiasis," *The Lancet*, vol. 368, no. 9541, pp. 1106–1118, 2006.

[50] J. R. Stothard and A. F. Gabrielli, "Schistosomiasis in African infants and preschool children: to treat or not to treat?" *Trends in Parasitology*, vol. 23, no. 3, pp. 83–86, 2007.

[51] J. Ring, "Davos Declaration: allergy as a global problem," *Allergy*, vol. 67, no. 2, pp. 141–143, 2012.

[52] L. J. Akinbami and K. C. Schoendorf, "Trends in childhood asthma: prevalence, health care utilization, and mortality," *Pediatrics*, vol. 110, no. 2, part 1, pp. 315–322, 2002.

[53] J. M. Smith, M. E. Disney, J. D. Williams, and Z. A. Goels, "Clinical significance of skin reactions to mite extracts in children with asthma," *British Medical Journal*, vol. 1, no. 659, pp. 723–726, 1969.

[54] D. O. Miranda, A. O. Deise Silva, F. C. Jorge et al., "Serum and salivary IgE, IgA, and IgG4 antibodies to Dermatophagoides pteronyssinus and its major allergens, Der p1 and Der p2, in allergic and nonallergic children," *Clinical and Developmental Immunology*, vol. 2011, Article ID 302739, 11 pages, 2011.

[55] J. L. Hopper, M. A. Jenkins, J. B. Carlin, and G. G. Giles, "Increase in the self-reported prevalence of asthma and hay fever in adults over the last generation: a matched parent-offspring study," *Australian Journal of Public Health*, vol. 19, no. 2, pp. 120–124, 1995.

[56] D. Jarvis, R. Newson, J. Lotvall et al., "Asthma in adults and its association with chronic rhinosinusitis: the GA(2) LEN survey in Europe," *Allergy*, vol. 67, no. 1, pp. 91–98, 2011.

[57] S. Musafiri, J. van Meerbeeck, L. Musango et al., "Prevalence of atopy, asthma and COPD in an urban and a rural area of an African country," *Respiratory Medicine*, vol. 105, no. 11, pp. 1596–1605, 2011.

[58] C. K. W. Lai, R. Beasley, J. Crane et al., "Global variation in the prevalence and severity of asthma symptoms: phase Three of the International Study of Asthma and Allergies in Childhood (ISAAC)," *Thorax*, vol. 64, no. 6, pp. 476–483, 2009.

[59] C. H. Katelaris, B. W. Lee, P. C. Potter et al., "Prevalence and diversity of allergic rhinitis in regions of the world beyond Europe and North America," *Clinical and Experimental Allergy*, vol. 42, no. 2, pp. 186–207, 2011.

[60] K. Westritschnig, E. Sibanda, W. Thomas et al., "Analysis of the sensitization profile towards allergens in central Africa," *Clinical and Experimental Allergy*, vol. 33, no. 1, pp. 22–27, 2003.

[61] K. Arkestl, E. Sibanda, C. Thors et al., "Impaired allergy diagnostics among parasite-infected patients caused by IgE antibodies to the carbohydrate epitope galactose-α1,3-galactose," *Journal of Allergy and Clinical Immunology*, vol. 127, no. 4, pp. 1024–1028, 2011.

[62] J. C. Samuelson and J. P. Caulfield, "The cercarial glycocalyx of *Schistosoma mansoni*," *Journal of Cell Biology*, vol. 100, no. 5, pp. 1423–1434, 1985.

[63] J. T. Culbertson, "The cercaricidal action of normal serums," *The Journal of Parasitology*, vol. 22, no. 2, 1936.

[64] W. Dias Da Silva and M. D. Kazatchkine, "*Schistosoma mansoni*: activation of the alternative pathway of human complement by schistosomula," *Experimental Parasitology*, vol. 50, no. 2, pp. 278–286, 1980.

[65] S. J. Jenkins, J. P. Hewitson, S. Ferret-Bernard, and A. P. Mountford, "Schistosome larvae stimulate macrophage cytokine production through TLR4-dependent and -independent pathways," *International Immunology*, vol. 17, no. 11, pp. 1409–1418, 2005.

[66] R. A. Paveley, S. A. Aynsley, J. D. Turner et al., "The Mannose Receptor (CD206) is an important pattern recognition receptor (PRR) in the detection of the infective stage of the helminth *Schistosoma mansoni* and modulates IFNgamma production," *International Journal for Parasitology*, vol. 41, no. 13-14, pp. 1335–1345, 2011.

[67] J. R. Lambertucci, "Acute *Schistosomiasis mansoni*: revisited and reconsidered," *Memorias do Instituto Oswaldo Cruz*, vol. 105, no. 4, pp. 422–435, 2010.

[68] S. V. Brant and E. S. Loker, "Schistosomes in the southwest United States and their potential for causing cercarial dermatitis or swimmer's itch," *Journal of Helminthology*, vol. 83, no. 2, pp. 191–198, 2009.

[69] S. J. Fraser, S. J. R. Allan, M. Roworth et al., "Cercarial dermatitis in the UK," *Clinical and Experimental Dermatology*, vol. 34, no. 3, pp. 344–346, 2009.

[70] A. Soleng and R. Mehl, "Geographical distribution of cercarial dermatitis in Norway," *Journal of Helminthology*, vol. 85, no. 3, pp. 345–352, 2010.

[71] P. C. Cook, S. A. Aynsley, J. D. Turner et al., "Multiple helminth infection of the skin causes lymphocyte hypo-responsiveness mediated by Th2 conditioning of dermal myeloid cells," *PLoS Pathogens*, vol. 7, no. 3, Article ID e1001323, 2011.

[72] K. Ramaswamy, P. Kumar, and Y. X. He, "A role for parasite-induced PGE2 in IL-10-mediated host immunoregulation by skin stage schistosomula of *Schistosoma mansoni*," *Journal of Immunology*, vol. 165, no. 8, pp. 4567–4574, 2000.

[73] V. Angeli, C. Faveeuw, P. Delerive et al., "*Schistosoma mansoni* induces the synthesis of IL-6 in pulmonary microvascular endothelial cells: role of IL-6 in the control of lung eosinophilia during infection," *European Journal of Immunology*, vol. 31, no. 9, pp. 2751–2761, 2001.

[74] F. Trottein, S. Nutten, V. Angeli et al., "*Schistosoma mansoni* schistosomula reduce E-selectin and VCAM-1 expression in TNF-alpha-stimulated lung microvascular endothelial cells by interfering with the NF-kappaB pathway," *European Journal of Immunology*, vol. 29, no. 11, pp. 3691–3701, 1999.

[75] E. J. Pearce, C. M. Kane, J. Sun, J. J. Taylor, A. S. McKee, and L. Cervi, "Th2 response polarization during infection with

die helminth parasite *Schistosoma mansoni*," *Immunological Reviews*, vol. 201, pp. 117–126, 2004.

[76] E. J. Pearce and A. S. MacDonald, "The immunobiology of schistosomiasis," *Nature Reviews Immunology*, vol. 2, no. 7, pp. 499–511, 2002.

[77] H. M. Coutinho, L. P. Acosta, H. W. Wu et al., "Th2 cytokines are associated with persistent hepatic fibrosis in human *Schistosoma japonicum* infection," *Journal of Infectious Diseases*, vol. 195, no. 2, pp. 288–295, 2007.

[78] S. Henri, C. Chevillard, A. Mergani et al., "Cytokine regulation of periportal fibrosis in humans infected with *Schistosoma mansoni*: IFN-γ is associated with protection against fibrosis and TNF-α with aggravation of disease," *Journal of Immunology*, vol. 169, no. 2, pp. 929–936, 2002.

[79] A. Dessein, B. Kouriba, C. Eboumbou et al., "Interleukin-13 in the skin and interferon-γ in the liver are key players in immune protection in human schistosomiasis," *Immunological Reviews*, vol. 201, pp. 180–190, 2004.

[80] P. Smith, R. E. Fallon, N. E. Mangan et al., "*Schistosoma mansoni* secretes a chemokine binding protein with antiinflammatory activity," *Journal of Experimental Medicine*, vol. 202, no. 10, pp. 1319–1325, 2005.

[81] L. A. de Oliveira Fraga, E. W. Lamb, E. C. Moreno et al., "Rapid induction of IgE responses to a worm cysteine protease during murine pre-patent schistosome infection," *BMC Immunology*, vol. 11, article 56, 2010.

[82] M. Nyindo, T. M. Kariuki, P. W. Mola et al., "Role of adult worm antigen-specific immunoglobulin E in acquired immunity to *Schistosoma mansoni* infection in baboons," *Infection and Immunity*, vol. 67, no. 2, pp. 636–642, 1999.

[83] P. Hagan, U. J. Blumenthal, D. Dunn, A. J. G. Simpson, and H. A. Wilkins, "Human IgE, IgG4 and resistance to reinfection with *Schistosoma haematobium*," *Nature*, vol. 349, no. 6306, pp. 243–245, 1991.

[84] R. M. Maizels and M. Yazdanbakhsh, "Immune regulation by helminth parasites: cellular and molecular mechanisms," *Nature Reviews Immunology*, vol. 3, no. 9, pp. 733–744, 2003.

[85] N. E. Mangan, R. E. Fallon, P. Smith, N. Van Rooijen, A. N. McKenzie, and P. G. Fallon, "Helminth infection protects mice from anaphylaxis via IL-10-producing B cells," *Journal of Immunology*, vol. 173, no. 10, pp. 6346–6356, 2004.

[86] N. E. Mangan, N. Van Rooijen, A. N. J. McKenzie, and P. G. Fallon, "Helminth-modified pulmonary immune response protects mice from allergen-induced airway hyperresponsiveness," *Journal of Immunology*, vol. 176, no. 1, pp. 138–147, 2006.

[87] R. S. Curwen, P. D. Ashton, D. A. Johnston, and R. A. Wilson, "The *Schistosoma mansoni* soluble proteome: a comparison across four life-cycle stages," *Molecular and Biochemical Parasitology*, vol. 138, no. 1, pp. 57–66, 2004.

[88] A. C. Fisher, "A study of the schistosomiasis of the Stanleyville district of the Belgian congo," *Transactions of the Royal Society of Tropical Medicine and Hygiene*, vol. 28, no. 3, pp. 277–IN1, 1934.

[89] K. S. Warren, "Regulation of the prevalence and intensity of schistosomiasis in man: immunology or ecology?" *Journal of Infectious Diseases*, vol. 127, no. 5, pp. 595–609, 1973.

[90] A. E. Butterworth, M. Capron, and J. S. Cordingley, "Immunity after treatment of human *Schistosomiasis mansoni*. II. Identification of resistant individuals, and analysis of their immune responses," *Transactions of the Royal Society of Tropical Medicine and Hygiene*, vol. 79, no. 3, pp. 393–408, 1985.

[91] D. W. Dunne, A. J. Fulford, A. E. Butterworth, D. Koech, and J. H. Ouma, "Human antibody responses to *Schistosoma mansoni*: does antigen directed, isotype restriction result in the production of blocking antibodies?" *Memorias do Instituto Oswaldo Cruz*, vol. 82, supplement 4, pp. 101–104, 1987.

[92] A. E. Butterworth, A. J. Fulford, D. W. Dunne, J. H. Ouma, and R. F. Sturrock, "Longitudinal studies on human schistosomiasis," *Philosophical transactions of the Royal Society of London B*, vol. 321, no. 1207, pp. 495–511, 1988.

[93] R. M. Anderson and R. M. May, "Herd immunity to helminth infection and implications for parasite control," *Nature*, vol. 315, no. 6019, pp. 493–496, 1985.

[94] A. J. C. Fulford, A. E. Butterworth, R. F. Sturrock, and J. H. Ouma, "On the use of age-intensity data to detect immunity to parasitic infections, with special reference to *Schistosoma mansoni* in Kenya," *Parasitology*, vol. 105, no. 2, pp. 219–227, 1992.

[95] M. E. J. Woolhouse, P. Taylor, D. Matanhire, and S. K. Chandiwana, "Acquired immunity and epidemiology of *Schistosoma haematobium*," *Nature*, vol. 351, no. 6329, pp. 757–759, 1991.

[96] K. Marsh and R. W. Snow, "Host-parasite interaction and morbidity in malaria endemic areas," *Philosophical Transactions of the Royal Society B*, vol. 352, no. 1359, pp. 1385–1394, 1997.

[97] M. E. J. Woolhouse, "Patterns in parasite epidemiology: the peak shift," *Parasitology Today*, vol. 14, no. 10, pp. 428–434, 1998.

[98] F. Mutapi, P. D. Ndhlovu, P. Hagan, and M. E. J. Woolhouse, "A comparison of humoral responses to *Schistosoma haematobium* in areas with low and high levels of infection," *Parasite Immunology*, vol. 19, no. 6, pp. 255–263, 1997.

[99] F. Mutapi, G. Winborn, N. Midzi, M. Taylor, T. Mduluza, and R. M. Maizels, "Cytokine responses to *Schistosoma haematobium* in a Zimbabwean population: contrasting profiles for IFN-γ, IL-4, IL-5 and IL-10 with age," *BMC Infectious Diseases*, vol. 7, article 139, 2007.

[100] C. L. Black, P. N. Mwinzi, E. M. Muok et al., "Influence of exposure history on the immunology and development of resistance to human *Schistosomiasis mansoni*," *PLoS Neglected Tropical Diseases*, vol. 4, no. 3, article e637, 2010.

[101] D. M. S. Karanja, A. W. Hightower, D. G. Colley et al., "Resistance to reinfection with *Schistosoma mansoni* in occupationally exposed adults and effect of HIV-1 co-infection on susceptibility to schistosomiasis: a longitudinal study," *The Lancet*, vol. 360, no. 9333, pp. 592–596, 2002.

[102] M. Z. Satti, S. M. Sulaiman, M. M. A. Homeida, S. A. Younis, and H. W. Ghalib, "Clinical, parasitological and immunological features of canal cleaners hyper-exposed to *Schistosoma mansoni* in the Sudan," *Clinical and Experimental Immunology*, vol. 104, no. 3, pp. 426–431, 1996.

[103] A. J. C. Fulford, M. Webster, J. H. Ouma, G. Kimani, D. W. Dunne, and T. Fulford, "Puberty and age-related changes in susceptibility to schistosome infection," *Parasitology Today*, vol. 14, no. 1, pp. 23–26, 1998.

[104] S. G. O. Johansson, T. Bieber, R. Dahl et al., "Revised nomenclature for allergy for global use: report of the Nomenclature Review Committee of the World Allergy Organization, October 2003," *Journal of Allergy and Clinical Immunology*, vol. 113, no. 5, pp. 832–836, 2004.

[105] M. Herr, L. Nikasinovic, C. Foucault et al., "Can early household exposure influence the development of rhinitis symptoms in infancy? Findings from the PARIS birth cohort,"

Annals of Allergy, Asthma and Immunology, vol. 107, no. 4, pp. 303–309, 2011.

[106] S. L. Prescott, "The influence of early environmental exposures on immune development and subsequent risk of allergic disease," *Allergy*, vol. 66, no. 95, pp. 4–6, 2011.

[107] J. M. Spergel and A. S. Paller, "Atopic dermatitis and the atopic march," *Journal of Allergy and Clinical Immunology*, vol. 112, no. 6, supplement, pp. S118–S127, 2003.

[108] N. Bhattacharyya, J. Grebner, and N. G. Martinson, "Recurrent acute rhinosinusitis: epidemiology and health care cost burden," *Otolaryngology and Head and Neck Surgery*. In press.

[109] R. Sporik, S. T. Holgate, and J. J. Cogswell, "Natural history of asthma in childhood—a birth cohort study," *Archives of Disease in Childhood*, vol. 66, no. 9, pp. 1050–1053, 1991.

[110] M. L. Martinson, J. O. Teitler, and N. E. Reichman, "Health across the life span in the United States and England," *American Journal of Epidemiology*, vol. 173, no. 8, pp. 858–865, 2011.

[111] C. Y. Hwang, Y. J. Chen, M. W. Lin et al., "Prevalence of atopic dermatitis, allergic rhinitis and asthma in Taiwan: a national study 2000 to 2007," *Acta Dermato-Venereologica*, vol. 90, no. 6, pp. 589–594, 2010.

[112] E. G. Weinberg, "The atopic march," *Current Allergy & Clinical Immunology*, vol. 18, no. 1, pp. 4–5, 2005.

[113] G. Barberio, G. B. Pajno, D. Vita, L. Caminiti, G. W. Canonica, and G. Passalacqua, "Does a "reverse" atopic march exist?" *Allergy*, vol. 63, no. 12, pp. 1630–1632, 2008.

[114] M. D. Klinnert, H. S. Nelson, M. R. Price, A. D. Adinoff, D. Y. Leung, and D. A. Mrazek, "Onset and persistence of childhood asthma: predictors from infancy," *Pediatrics*, vol. 108, no. 4, p. E69, 2001.

[115] P. O. Van Asperen, A. S. Kemp, and A. Mukhi, "Atopy in infancy predicts the severity of bronchial hyperresponsiveness in later childhood," *Journal of Allergy and Clinical Immunology*, vol. 85, no. 4, pp. 790–795, 1990.

[116] P. M. Matricardi, A. Bockelbrink, C. Grüber et al., "Longitudinal trends of total and allergen-specific IgE throughout childhood," *Allergy*, vol. 64, no. 7, pp. 1093–1098, 2009.

[117] D. A. Mrazek, M. Klinnert, P. J. Mrazek et al., "Prediction of early-onset asthma in genetically at-risk children," *Pediatric Pulmonology*, vol. 27, no. 2, pp. 85–94, 1999.

[118] P. G. Holt and D. H. Strickland, "Soothing signals: transplacental transmission of resistance to asthma and allergy," *Journal of Experimental Medicine*, vol. 206, no. 13, pp. 2861–2864, 2009.

[119] B. Schaub, J. Liu, S. Höppler et al., "Maternal farm exposure modulates neonatal immune mechanisms through regulatory T cells," *Journal of Allergy and Clinical Immunology*, vol. 123, no. 4, pp. 774.e5–782.e5, 2009.

[120] J. W. Hollingsworth, S. Maruoka, K. Boon et al., "In utero supplementation with methyl donors enhances allergic airway disease in mice," *Journal of Clinical Investigation*, vol. 118, no. 10, pp. 3462–3469, 2008.

[121] K. Ishizaka and T. Ishizaka, "Physicochemical properties of reaginic antibody. I. Association of reaginic activity with an immunoglobulin other than γA- or γG-globulin," *Journal of Allergy*, vol. 37, no. 3, pp. 169–185, 1966.

[122] K. Ishizaka and T. Ishizaka, "Identification of gamma-E-antibodies as a carrier of reaginic activity," *Journal of Immunology*, vol. 99, no. 6, pp. 1187–1198, 1967.

[123] E. Dehlink, B. Platzer, A. H. Baker et al., "A soluble form of the high affinity IgE receptor, Fc-epsilon-RI, circulates in human serum," *PLoS ONE*, vol. 6, no. 4, Article ID e19098, 2011.

[124] B. Platzer, E. Dehlink, S. J. Turley, and E. Fiebiger, "How to connect an IgE-driven response with CTL activity?" *Cancer Immunol Immunother*. In press.

[125] T. Bieber, H. De la Salle, A. Wollenberg et al., "Human epidermal Langerhans cells express the high affinity receptor for immunoglobulin E (FcɛRI)," *Journal of Experimental Medicine*, vol. 175, no. 5, pp. 1285–1290, 1992.

[126] A. S. Gounni, B. Lamkhioued, E. Delaporte et al., "The high-affinity IgE receptor on eosinophils: from allergy to parasites or from parasites to allergy?" *Journal of Allergy and Clinical Immunology*, vol. 94, no. 6, part 2, pp. 1214–1216, 1994.

[127] A. S. Gounni, B. Lamkhioued, K. Ochiai et al., "High-affinity IgE receptor on eosinophils is involved in defence against parasites," *Nature*, vol. 367, no. 6459, pp. 183–186, 1994.

[128] U. Jeschke, D. Mayr, B. Schiessl et al., "Expression of galectin-1, -3 (gal-1, gal-3) and the thomsen-friedenreich (TF) antigen in normal, IUGR, preeclamptic and HELLP placentas," *Placenta*, vol. 28, no. 11-12, pp. 1165–1173, 2007.

[129] E. Rindsjö, M. Joerink, N. Papadogiannakis, and A. Scheynius, "IgE in the human placenta: why there?" *Allergy*, vol. 65, no. 5, pp. 554–560, 2010.

[130] M. Vernersson, M. Aveskogh, and L. Hellman, "Cloning of IgE from the echidna (*Tachyglossus aculeatus*) and a comparative analysis of ε chains from all three extant mammalian lineages," *Developmental and Comparative Immunology*, vol. 28, no. 1, pp. 61–75, 2004.

[131] P. Vieira and K. Rajewsky, "The half-lives of serum immunoglobulins in adult mice," *European Journal of Immunology*, vol. 18, no. 2, pp. 313–316, 1988.

[132] A. Karnowski, G. Achatz-Straussberger, C. Klockenbusch, G. Achatz, and M. C. Lamers, "Inefficient processing of mRNA for the membrane form of IgE is a genetic mechanism to limit recruitment of IgE-secreting cells," *European Journal of Immunology*, vol. 36, no. 7, pp. 1917–1925, 2006.

[133] P. Yu, M. Kosco-Vilbois, M. Richards, G. Kohler, and M. C. Lamers, "Negative feedback regulation of IgE synthesis by murine CD23," *Nature*, vol. 369, no. 6483, pp. 753–756, 1994.

[134] C. Plater-Zyberk and J. Y. Bonnefoy, "Marked amelioration of established collagen-induced arthritis by treatment with antibodies to CD23 in vivo," *Nature Medicine*, vol. 1, no. 8, pp. 781–785, 1995.

[135] J. C. Byrd, T. J. Kipps, I. W. Flinn et al., "Phase 1/2 study of lumiliximab combined with fludarabine, cyclophosphamide, and rituximab in patients with relapsed or refractory chronic lymphocytic leukemia," *Blood*, vol. 115, no. 3, pp. 489–495, 2010.

[136] N. Aberle, A. Gagro, S. Rabatić, Z. Reiner-Banovac, and D. Dekaris, "Expression of CD23 antigen and its ligands in children with intrinsic and extrinsic asthma," *Allergy*, vol. 52, no. 12, pp. 1238–1242, 1997.

[137] J. P. Aubry, S. Pochon, P. Graber, K. U. Jansen, and J. Y. Bonnefoy, "CD21 is a ligand for CD23 and regulates IgE production," *Nature*, vol. 358, no. 6386, pp. 505–507, 1992.

[138] J. J. Weis, T. F. Tedder, and D. T. Fearon, "Identification of a 145,000 M(r) membrane protein as the C3d receptor (CR2) of human B lymphocytes," *Proceedings of the National Academy of Sciences of the United States of America*, vol. 81, no. 3, pp. 881–885, 1984.

[139] N. McCloskey, J. Hunt, R. L. Beavil et al., "Soluble CD23 monomers inhibit and oligomers stimulate IGE synthesis in human B cells," *Journal of Biological Chemistry*, vol. 282, no. 33, pp. 24083–24091, 2007.

[140] F. X. Desvaux, G. Peltre, and B. David, "Characterization of grass pollen-specific IgE, IgA, IgM classes and IgG subclasses

in allergic patients," *International Archives of Allergy and Applied Immunology*, vol. 89, no. 2-3, pp. 281–287, 1989.

[141] R. C. Aalberse, S. O. Stapel, J. Schuurman, and T. Rispens, "Immunoglobulin G4: an odd antibody," *Clinical and Experimental Allergy*, vol. 39, no. 4, pp. 469–477, 2009.

[142] R. C. Aalberse, F. Van Milligen, K. Y. Tan, and S. O. Stapel, "Allergen-specific IgG4 in atopic disease," *Allergy*, vol. 48, no. 8, pp. 559–569, 1993.

[143] R. Hussain, R. W. Poindexter, and E. A. Ottesen, "Control of allergic reactivity in human filariasis: predominant localization of blocking antibody to the IgG4 subclass," *Journal of Immunology*, vol. 148, no. 9, pp. 2731–2737, 1992.

[144] P. Rihet, C. E. Demeure, A. J. Dessein, and A. Bourgois, "Strong serum inhibition of specific IgE correlated to competing IgG4, revealed by a new methodology in subjects from a S. mansoni endemic area," *European Journal of Immunology*, vol. 22, no. 8, pp. 2063–2070, 1992.

[145] F. Mutapi, C. Bourke, Y. Harcus et al., "Differential recognition patterns of *Schistosoma haematobium* adult worm antigens by the human antibodies IgA, IgE, IgG1 and IgG4," *Parasite Immunology*, vol. 33, no. 3, pp. 181–192, 2011.

[146] R. C. Aalberse and J. Schuurman, "IgG4 breaking the rules," *Immunology*, vol. 105, no. 1, pp. 9–19, 2002.

[147] M. V. D. N. Kolfschoten, J. Schuurman, M. Losen et al., "Anti-inflammatory activity of human IgG4 antibodies by dynamic Fab arm exchange," *Science*, vol. 317, no. 5844, pp. 1554–1557, 2007.

[148] J. S. van der Zee, P. Van Swieten, and R. C. Aalberse, "Inhibition of complement activation by IgG4 antibodies," *Clinical and Experimental Immunology*, vol. 64, no. 2, pp. 415–422, 1986.

[149] R. M. Maizels, E. Sartono, A. Kurniawan, F. Partono, M. E. Selkirk, and M. Yazdanbakhsh, "T-cell activation and the balance of antibody isotypes in human lymphatic filariasis," *Parasitology Today*, vol. 11, no. 2, pp. 50–56, 1995.

[150] T. Adjobimey and A. Hoerauf, "Induction of immunoglobulin G4 in human filariasis: an indicator of immunoregulation," *Annals of Tropical Medicine and Parasitology*, vol. 104, no. 6, pp. 455–464, 2010.

[151] J. Punnonen, G. Aversa, B. G. Cocks et al., "Interleukin 13 induces interleukin 4-independent IgG4 and IgE synthesis and CD23 expression by human B cells," *Proceedings of the National Academy of Sciences of the United States of America*, vol. 90, no. 8, pp. 3730–3734, 1993.

[152] T. R. Mosmann, H. Cherwinski, and M. W. Bond, "Two types of murine helper T cell clone. I. Definition according to profiles of lymphokine activities and secreted proteins," *Journal of Immunology*, vol. 136, no. 7, pp. 2348–2357, 1986.

[153] J. H. Lee, H. H. Yu, L. C. Wang, Y. H. Yang, Y. T. Lin, and B. L. Chiang, "The levels of CD4+CD25+ regulatory T cells in paediatric patients with allergic rhinitis and bronchial asthma," *Clinical and Experimental Immunology*, vol. 148, no. 1, pp. 53–63, 2007.

[154] S. Dimeloe, A. Nanzer, K. Ryanna, and C. Hawrylowicz, "Regulatory T cells, inflammation and the allergic response-the role of glucocorticoids and Vitamin D," *Journal of Steroid Biochemistry and Molecular Biology*, vol. 120, no. 2-3, pp. 86–95, 2010.

[155] M. Akdis, "Immune tolerance in allergy," *Current Opinion in Immunology*, vol. 21, no. 6, pp. 700–707, 2009.

[156] A. Taylor, J. Verhagen, C. A. Akdis, and M. Akdis, "T regulatory cells in allergy and health: a question of allergen specificity and balance," *International Archives of Allergy and Immunology*, vol. 135, no. 1, pp. 73–82, 2004.

[157] R. M. McLoughlin, A. Calatroni, C. M. Visness et al., "Longitudinal relationship of early life immunomodulatory T cell phenotype and function to development of allergic sensitization in an urban cohort," *Clinical and Experimental Allergy*, vol. 42, no. 3, pp. 392–404, 2011.

[158] L. Cosmi, F. Liotta, E. Maggi, S. Romagnani, and F. Annunziato, "Th17 cells: new players in asthma pathogenesis," *Allergy*, vol. 66, no. 8, pp. 989–998, 2011.

[159] L. Cosmi, L. Maggi, V. Santarlasci et al., "Identification of a novel subset of human circulating memory CD4+ T cells that produce both IL-17A and IL-4," *Journal of Allergy and Clinical Immunology*, vol. 125, no. 1, pp. 222.e4–230.e4, 2010.

[160] P. Jeannin, S. Lecoanet, Y. Delneste, J. F. Gauchat, and J. Y. Bonnefoy, "IgE versus IgG4 production can be differentially regulated by IL-10," *Journal of Immunology*, vol. 160, no. 7, pp. 3555–3561, 1998.

[161] E. M. Savilahti, V. Rantanen, J. S. Lin et al., "Early recovery from cow's milk allergy is associated with decreasing IgE and increasing IgG4 binding to cow's milk epitopes," *Journal of Allergy and Clinical Immunology*, vol. 125, no. 6, pp. 1315.e9–1321.e9, 2010.

[162] M. E. Devey, D. V. Wilson, and A. W. Wheeler, "The IgG subclasses of antibodies of grass pollen allergens produced in hay fever patients during hyposensitization," *Clinical Allergy*, vol. 6, no. 3, pp. 227–236, 1976.

[163] R. J. Bullock, D. Barnett, and M. E. H. Howden, "Immunologic and clinical responses to parenteral immunotherapy in peanut anaphylaxis—a study using IgE and IgG4 immunoblot monitoring," *Allergologia et Immunopathologia*, vol. 33, no. 5, pp. 250–256, 2005.

[164] M. T. Gallego, V. Iraola, M. Himly et al., "Depigmented and polymerised house dust mite allergoid: allergen content, induction of IgG4 and clinical response," *International Archives of Allergy and Immunology*, vol. 153, no. 1, pp. 61–69, 2010.

[165] A. Hoerauf, J. Satoguina, M. Saeftel, and S. Specht, "Immunomodulation by filarial nematodes," *Parasite Immunology*, vol. 27, no. 10-11, pp. 417–429, 2005.

[166] A. Kurniawan, M. Yazdanbakhsh, R. Van Ree et al., "Differential expression of IgE and IgG4 specific antibody responses in asymptomatic and chronic human filariasis," *Journal of Immunology*, vol. 150, no. 9, pp. 3941–3950, 1993.

[167] Y. Kaneko, F. Nimmerjahn, and J. V. Ravetch, "Anti-inflammatory activity of immunoglobulin G resulting from Fc sialylation," *Science*, vol. 313, no. 5787, pp. 670–673, 2006.

[168] D. N. Mekhaiel, C. T. Daniel-Ribeiro, P. J. Cooper, and R. J. Pleass, "Do regulatory antibodies offer an alternative mechanism to explain the hygiene hypothesis?" *Trends in Parasitology*, vol. 27, no. 12, pp. 523–529, 2011.

[169] T. H. T. Nguyen and T. B. Casale, "Immune modulation for treatment of allergic disease," *Immunological Reviews*, vol. 242, no. 1, pp. 258–271, 2011.

[170] A. B. Kay, "T lymphocytes and their products in atopic allergy and asthma," *International Archives of Allergy and Applied Immunology*, vol. 94, no. 1–4, pp. 189–193, 1991.

[171] M. Jutel and C. A. Akdis, "T-cell subset regulation in atopy," *Current Allergy and Asthma Reports*, vol. 11, no. 2, pp. 139–145, 2011.

[172] W. E. Paul and J. Zhu, "How are T(H)2-type immune responses initiated and amplified?" *Nature Reviews Immunology*, vol. 10, no. 4, pp. 225–235, 2010.

[173] C. M. Lloyd and E. M. Hessel, "Functions of T cells in asthma: more than just TH2 cells," *Nature Reviews Immunology*, vol. 10, no. 12, pp. 838–848, 2010.

[174] L. E. Harrington, R. D. Hatton, P. R. Mangan et al., "Interleukin 17-producing CD4$^+$ effector T cells develop via a lineage distinct from the T helper type 1 and 2 lineages," *Nature Immunology*, vol. 6, no. 11, pp. 1123–1132, 2005.

[175] W. Al-Ramli, D. Préfontaine, F. Chouiali et al., "TH17-associated cytokines (IL-17A and IL-17F) in severe asthma," *Journal of Allergy and Clinical Immunology*, vol. 123, no. 5, pp. 1185–1187, 2009.

[176] M. Veldhoen, C. Uyttenhove, J. van Snick et al., "Transforming growth factor-β "reprograms" the differentiation of T helper 2 cells and promotes an interleukin 9-producing subset," *Nature Immunology*, vol. 9, no. 12, pp. 1341–1346, 2008.

[177] R. Pawankar, S. Mori, C. Ozu, and S. Kimura, "Overview on the pathomechanisms of allergic rhinitis," *Asia Pacific Allergy*, vol. 1, no. 3, pp. 157–167, 2011.

[178] A. E. Butterworth, R. F. Sturrock, V. Houba, and P. H. Rees, "Antibody dependent cell mediated damage to schistosomula in vitro," *Nature*, vol. 252, no. 5483, pp. 503–505, 1974.

[179] P. Hagan, P. J. Moore, and A. B. Adjukiewicz, "In-vitro antibody-dependent killing of schistosomula of *Schistosoma haematobium* by human eosinophils," *Parasite Immunology*, vol. 7, no. 6, pp. 617–624, 1985.

[180] A. Capron, J. P. Dessaint, M. Capron, and H. Bazin, "Specific IgE antibodies in immune adherence of normal macrophages to *Schistosoma mansoni* schistosomules," *Nature*, vol. 253, no. 5491, pp. 474–475, 1975.

[181] A. Capron and J. P. Dessaint, "IgE: a molecule in search of a function," *Annales d'Immunologie*, vol. 132, no. 1, pp. 3–8, 1981.

[182] C. Verwaerde, J. M. Grzych, H. Bazin, M. Capron, and A. Capron, "Production of monoclonal anti-*Schistosoma mansoni* antibodies. Preliminary study of their biological activities," *Comptes Rendus des Seances de l''Academie des Sciences D*, vol. 289, no. 10, pp. 725–727, 1979.

[183] P. Rihet, C. E. Demeure, A. Bourgois, A. Prata, and A. J. Dessein, "Evidence for an association between human resistance to *Schistosoma mansoni* and high anti larval IgE levels," *European Journal of Immunology*, vol. 21, no. 11, pp. 2679–2686, 1991.

[184] D. W. Dunne, A. E. Butterworth, A. J. C. Fulford et al., "Immunity after treatment of human schistosomiasis: association between IgE antibodies to adult worm antigens and resistance to reinfection," *European Journal of Immunology*, vol. 22, no. 6, pp. 1483–1494, 1992.

[185] I. R. Caldas, R. Correa-Oliveira, E. Colosimo et al., "Susceptibility and resistance to *Schistosoma mansoni* reinfection: parallel cellular and isotypic immunologic assessment," *American Journal of Tropical Medicine and Hygiene*, vol. 62, no. 1, pp. 57–64, 2000.

[186] Z. Zhang, H. Wu, S. Chen et al., "Association between IgE antibody against soluble egg antigen and resistance to reinfection with *Schistosoma japonicum*," *Transactions of the Royal Society of Tropical Medicine and Hygiene*, vol. 91, no. 5, pp. 606–608, 1997.

[187] M. Jiz, J. F. Friedman, T. Leenstra et al., "Immunoglobulin E (IgE) responses to paramyosin predict resistance to reinfection with *Schistosoma japonicum* and are attenuated by IgG4," *Infection and Immunity*, vol. 77, no. 5, pp. 2051–2058, 2009.

[188] C. W. A. Naus, C. J. Van Dam, P. G. Kremsner, F. W. Krijger, and A. M. Decider, "Human IgE, IgG subclass, and IgM responses to worm and egg antigens in Schistosomiasis

haematobium: a 12-month study of reinfection in Cameroonian children," *Clinical Infectious Diseases*, vol. 26, no. 5, pp. 1142–1147, 1998.

[189] A. P. de Moira, A. J. C. Fulford, N. B. Kabatereine, J. H. Ouma, M. Booth, and D. W. Dunne, "Analysis of complex patterns of human exposure and immunity to *Schistosomiasis mansoni* the influence of age, sex, ethnicity and IgE," *PLoS Neglected Tropical Diseases*, vol. 4, no. 9, article e820, 2010.

[190] M. Z. Satti, P. Lind, B. J. Vennervald, S. M. Sulaiman, A. A. Daffalla, and H. W. Ghalib, "Specific immunoglobulin measurements related to exposure and resistance to *Schistosoma mansoni* infection in Sudanese canal cleaners," *Clinical and Experimental Immunology*, vol. 106, no. 1, pp. 45–54, 1996.

[191] F. Mutapi, T. Mduluza, N. Gomez-Escobar et al., "Immunoepidemiology of human *Schistosoma haematobium* infection: preferential IgG3 antibody responsiveness to a recombinant antigen dependent on age and parasite burden," *BMC Infectious Diseases*, vol. 6, article 96, 2006.

[192] F. Mutapi, P. D. Ndhlovu, P. Hagan et al., "Chemotherapy accelerates the development of acquired immune responses to *Schistosoma haematobium* infection," *Journal of Infectious Diseases*, vol. 178, no. 1, pp. 289–293, 1998.

[193] J. M. Grzych, D. Grezel, Chuan Bo Xu et al., "IgA antibodies to a protective antigen in human *Schistosomiasis mansoni*," *Journal of Immunology*, vol. 150, no. 2, pp. 527–535, 1993.

[194] P. N. M. Mwinzi, L. Ganley-Leal, C. L. Black, W. E. Secor, D. M. S. Karanja, and D. G. Colley, "Circulating CD23$^+$ B cell subset correlates with the development of resistance to *Schistosoma mansoni* reinfection in occupationally exposed adults who have undergone multiple treatments," *Journal of Infectious Diseases*, vol. 199, no. 2, pp. 272–279, 2009.

[195] S. R. Smithers and R. J. Terry, "Resistance to experimental infection with *Schistosoma mansoni* in rhesus monkeys induced by the transfer of adult worms," *Transactions of the Royal Society of Tropical Medicine and Hygiene*, vol. 61, no. 4, pp. 517–533, 1967.

[196] C. J. Sanderson, A. O'Garra, D. J. Warren, and G. G. B. Klaus, "Eosinophil differentiation factor also has B-cell growth factor activity: proposed name interleukin 4," *Proceedings of the National Academy of Sciences of the United States of America*, vol. 83, no. 2, pp. 437–440, 1986.

[197] C. J. Sanderson, D. J. Warren, and M. Strath, "Identification of a lymphokine that stimulates eosinophil differentiation in vitro. Its relationship to interleukin 3, and functional properties of eosinophils produced in cultures," *Journal of Experimental Medicine*, vol. 162, no. 1, pp. 60–74, 1985.

[198] M. Veith, D. W. Taylor, and K. Thorne, "Studies on the enhancement of human eosinophil function by mononuclear cell products in vitro," *Clinical and Experimental Immunology*, vol. 58, no. 3, pp. 603–610, 1984.

[199] P. Couissinier-Paris and A. J. Dessein, "Schistosoma-specific helper T cell clones from subjects resistant to infection by *Schistosoma mansoni* are Th0/2," *European Journal of Immunology*, vol. 25, no. 8, pp. 2295–2302, 1995.

[200] M. Roberts, A. E. Butterworth, G. Kimani et al., "Immunity after treatment of human schistosomiasis: association between cellular responses and resistance to reinfection," *Infection and Immunity*, vol. 61, no. 12, pp. 4984–4993, 1993.

[201] T. Mduluza, P. D. Ndhlovu, N. Midzi et al., "Contrasting cellular responses in *Schistosoma haematobium* infected and exposed individuals from areas of high and low transmission in Zimbabwe," *Immunology Letters*, vol. 88, no. 3, pp. 249–256, 2003.

[202] I. R. Viana, A. Sher, O. S. Carvalho et al., "Interferon-gamma production by peripheral blood mononuclear cells from residents of an area endemic for *Schistosoma mansoni*," *Transactions of the Royal Society of Tropical Medicine and Hygiene*, vol. 88, no. 4, pp. 466–470, 1994.

[203] I. R. Caldas, A. C. Campi-Azevedo, L. F. A. Oliveira, A. M. S. Silveira, R. C. Oliveira, and G. Gazzinelli, "Human *Schistosomiasis mansoni*: immune responses during acute and chronic phases of the infection," *Acta Tropica*, vol. 108, no. 2-3, pp. 109–117, 2008.

[204] C. N. L. De Morais, J. R. De Souza, W. G. Melo et al., "Cytokine profile associated with chronic and acute human *Schistosomiasis mansoni*," *Memorias do Instituto Oswaldo Cruz*, vol. 103, no. 6, pp. 561–568, 2008.

[205] G. Gazzinelli, J. R. Lambertucci, and N. Katz, "Immune responses during human *Schistosomiasis mansoni*. XI. Immunologic status of patients with acute infections and after treatment," *Journal of Immunology*, vol. 135, no. 3, pp. 2121–2127, 1985.

[206] A. D. Bastos and I. L. Brito, "Acute pulmonary schistosomiasis: HRCT findings and clinical presentation," *Jornal Brasileiro de Pneumologia*, vol. 37, no. 6, pp. 823–825, 2011.

[207] J. Clerinx, E. Bottieau, D. Wichmann, E. Tannich, and M. Van Esbroeck, "Acute schistosomiasis in a cluster of travelers from Rwanda: diagnostic contribution of schistosome DNA detection in serum compared to parasitology and serology," *Journal of Travel Medicine*, vol. 18, no. 6, pp. 367–372, 2011.

[208] A. W. Cheever, K. F. Hoffmann, and T. A. Wynn, "Immunopathology of *Schistosomiasis mansoni* in mice and men," *Immunology Today*, vol. 21, no. 9, pp. 465–466, 2000.

[209] S. Ikemoto, T. Kishimoto, S. Wada, S. Nishio, and M. Maekawa, "Clinical studies on cell-mediated immunity in patients with urinary bladder carcinoma: blastogenic response, interleukin-2 production and interferon-γ production of lymphocytes," *British Journal of Urology*, vol. 65, no. 4, pp. 333–338, 1990.

[210] M. S. Wilson, M. M. Mentink-Kane, J. T. Pesce, T. R. Ramalingam, R. Thompson, and T. A. Wynn, "Immunopathology of schistosomiasis," *Immunology and Cell Biology*, vol. 85, no. 2, pp. 148–154, 2007.

[211] M. Bazaral, H. A. Orgel, and R. N. Hamburger, "The influence of serum IgE levels of selected recipients, including patients with allergy, helminthiasis and tuberculosis, on the apparent P-K titre of a reaginic serum," *Clinical and Experimental Immunology*, vol. 14, no. 1, pp. 117–125, 1973.

[212] I. Hagel, N. R. Lynch, M. Perez, M. C. Di Prisco, R. Lopez, and E. Rojas, "Modulation of the allergic reactivity of slum children by helminthic infection," *Parasite Immunology*, vol. 15, no. 6, pp. 311–315, 1993.

[213] R. C. Godfrey and C. F. Gradidge, "Allergic sensitisation of human lung fragments prevented by saturation of IgE binding sites," *Nature*, vol. 259, no. 5543, pp. 484–486, 1976.

[214] E. Mitre, S. Norwood, and T. B. Nutman, "Saturation of immunoglobulin E (IgE) binding sites by polyclonal IgE does not explain the protective effect of helminth infections against atopy," *Infection and Immunity*, vol. 73, no. 7, pp. 4106–4111, 2005.

[215] D. I. Pritchard, D. S. W. Hooi, A. Brown, M. J. Bockarie, R. Caddick, and R. J. Quinnell, "Basophil competence during hookworm (Necator americanus) infection," *American Journal of Tropical Medicine and Hygiene*, vol. 77, no. 5, pp. 860–865, 2007.

[216] S. J. Galli, "Mast cells and basophils," *Current Opinion in Hematology*, vol. 7, no. 1, pp. 32–39, 2000.

[217] M. Yazdanbakhsh, P. G. Kremsner, and R. Van Ree, "Immunology: allergy, parasites, and the hygiene hypothesis," *Science*, vol. 296, no. 5567, pp. 490–494, 2002.

[218] H. C. Santiago, S. Bennuru, A. Boyd, M. Eberhard, and T. B. Nutman, "Structural and immunologic cross-reactivity among filarial and mite tropomyosin: implications for the hygiene hypothesis," *Journal of Allergy and Clinical Immunology*, vol. 127, no. 2, pp. 479–486, 2011.

[219] L. Caraballo and N. Acevedo, "Allergy in the tropics: the impact of cross-reactivity between mites and ascaris," *Frontiers in Bioscience*, vol. 3, pp. 51–64, 2011.

[220] N. R. Lynch, I. Hagel, M. Vargas et al., "Effect of age and helminthic infection on IgE levels in slum children," *Journal of Investigational Allergology & Clinical Immunology*, vol. 3, no. 2, pp. 96–99, 1993.

[221] N. R. Lynch, M. Palenque, I. Hagel, and M. C. Diprisco, "Clinical improvement of asthma after anthelminthic treatment in a tropical situation," *American Journal of Respiratory and Critical Care Medicine*, vol. 156, no. 1, pp. 50–54, 1997.

[222] A. M. J. van den Biggelaar, L. C. Rodrigues, R. Van Ree et al., "Long-term treatment of intestinal helminths increases mite skin-test reactivity in Gabonese schoolchildren," *Journal of Infectious Diseases*, vol. 189, no. 5, pp. 892–900, 2004.

[223] N. Rujeni, N. Nausch, C. D. Bourke et al., "Atopy is inversely related to Schistosome infection intensity: a comparative study in zimbabwean villages with distinct levels of *Schistosoma haematobium* infection," *International Archives of Allergy and Applied Immunology*, vol. 158, no. 3, pp. 288–298, 2012.

[224] C. J. Oliphant, J. L. Barlow, and A. N. McKenzie, "Insights into the initiation of type 2 immune responses," *Immunology*, vol. 134, no. 4, pp. 378–385, 2011.

[225] R. M. Maizels, "Parasite immunomodulation and polymorphisms of the immune system," *Journal of Biology*, vol. 8, no. 7, article 62, 2009.

[226] H. Kita and G. J. Gleich, "Eosinophils and IgE receptors: a continuing controversy," *Blood*, vol. 89, no. 10, pp. 3497–3501, 1997.

[227] M. I. A. S. Araujo, B. Hoppe, M. Medeiros et al., "Impaired T helper 2 response to aeroallergen in helminth-infected patiente with asthma," *Journal of Infectious Diseases*, vol. 190, no. 10, pp. 1797–1803, 2004.

[228] R. R. Oliveira, K. J. Gollob, J. P. Figueiredo et al., "*Schistosoma mansoni* infection alters co-stimulatory molecule expression and cell activation in asthma," *Microbes and Infection*, vol. 11, no. 2, pp. 223–229, 2009.

[229] R. M. Maizels and M. Yazdanbakhsh, "T-cell regulation in helminth parasite infections: implications for inflammatory diseases," *Chemical Immunology and Allergy*, vol. 94, pp. 112–123, 2008.

[230] I. Hagel, M. Cabrera, P. Sánchez, P. Rodríguez, and J. J. Lattouf, "Role of the low affinity IgE receptor (CD23) on the IgE response against Ascaris lumbricoides in Warao Amerindian children from Venezuela," *Investigacion Clinica*, vol. 47, no. 3, pp. 241–251, 2006.

[231] M. S. Wilson, M. D. Taylor, A. Balic, C. A. M. Finney, J. R. Lamb, and R. M. Maizels, "Suppression of allergic airway inflammation by helminth-induced regulatory T cells," *Journal of Experimental Medicine*, vol. 202, no. 9, pp. 1199–1212, 2005.

[232] Y. Osada and T. Kanazawa, "Parasitic helminths: new weapons against immunological disorders," *Journal of Biomedicine & Biotechnology*, vol. 2010, Article ID 743758, p. 9, 2010.

[233] N. Nausch, N. Midzi, T. Mduluza, R. M. Maizels, and F. Mutapi, "Regulatory and activated T cells in human *Schistosoma haematobium* infections," *PLoS ONE*, vol. 6, no. 2, Article ID e16860, 2011.

[234] W. Harnett and M. M. Harnett, "Helminth-derived immunomodulators: can understanding the worm produce the pill?" *Nature Reviews Immunology*, vol. 10, no. 4, pp. 278–284, 2010.

[235] A. J. Melendez, M. M. Harnett, P. N. Pushparaj et al., "Inhibition of Fc epsilon RI-mediated mast cell responses by ES-62, a product of parasitic filarial nematodes," *Nature Medicine*, vol. 13, pp. 1375–1381, 2007.

[236] I. B. McInnes, B. P. Leung, M. Harnett, J. A. Gracie, F. Y. Liew, and W. Harnett, "A novel therapeutic approach targeting articular inflammation using the filarial nematode-derived phosphorylcholine-containing glycoprotein ES-62," *The Journal of Immunology*, vol. 171, no. 4, pp. 2127–2133, 2003.

[237] M. M. Harnett, A. J. Melendez, and W. Harnett, "The therapeutic potential of the filarial nematode-derived immodulator, ES-62 in inflammatory disease," *Clinical & Experimental Immunology*, vol. 159, no. 3, pp. 256–267, 2010.

[238] W. Harnett, M. M. Harnett, B. P. Leung, J. A. Gracie, and I. B. McInnes, "The anti-inflammatory potential of the filarial nematode secreted product, ES-62," *Current topics in Medicinal Chemistry*, vol. 4, pp. 553–559, 2004.

[239] W. Harnett and M. M. Harnett, "Filarial nematode secreted product ES-62 is an anti-inflammatory agent: therapeutic potential of small molecule derivatives and ES-62 peptide mimetics," *Clinical and Experimental Pharmacology & Physiology*, vol. 33, no. 5-6, pp. 511–518, 2006.

[240] M. M. Harnett, D. E. Kean, A. Boitelle et al., "The phosphorycholine moiety of the filarial nematode immunomodulator ES-62 is responsible for its anti-inflammatory action in arthritis," *Annals of the Rheumatic Diseases*, vol. 67, no. 4, pp. 518–523, 2008.

[241] G. Schramm, F. H. Falcone, A. Gronow et al., "Molecular characterization of an interleukin-4-inducing factor from *Schistosoma mansoni* eggs," *The Journal of Biological Chemistry*, vol. 278, no. 20, pp. 18384–818392, 2003.

[242] M.-H. Abdulla, K.-C. Lim, J. H. McKerrow, and C. R. Caffrey, "Proteomic identification of IPSE/alpha-1 as a major hepatotoxin secreted by *Schistosoma mansoni* eggs," *PLoS Neglected Tropical Diseases*, vol. 5, no. 10, article e1368, 2011.

[243] G. Schramm, K. Mohrs, M. Wodrich et al., "Cutting edge: IPSE/alpha-1, a glycoprotein from Schistosoma mansoni eggs, induces IgE-dependent, antigen-independent IL-4 production by murine basophils in vivo," *The Journal of Immunology*, vol. 178, no. 10, pp. 6023–6027, 2007.

[244] J. P. Hewitson, K. J. Filbey, J. R. Grainger et al., "Heligmosomoides polygyrus elicits a dominant nonprotective antibody response directed against restricted glycan and peptide epitopes," *The Journal of Immunology*, vol. 187, no. 9, pp. 4764–4777, 2011.

[245] J. Chen, X. Hu, S. He et al., "Expression and immune response analysis of *Schistosoma japonicum* VAL-1, a homologue of vespid venom allergens," *Parasitology Research*, vol. 106, no. 6, pp. 1413–1418, 2010.

[246] N. W. Palm, R. K. Rosenstein, and R. Medzhitov, "Allergic host defences," *Nature*, vol. 484, no. 7395, pp. 465–472, 2012.

[247] D. Artis, R. M. Maizels, and F. D. Finkelman, "Forum: immunology: allergy challenged," *Nature*, vol. 484, no. 7395, pp. 458–459, 2012.

[248] P. W. Sherman, E. Holland, and J. S. Sherman, "Allergies: their role in cancer prevention," *Quarterly Review of Biology*, vol. 83, no. 4, pp. 339–362, 2008.

[249] B. E. Zacharia and P. Sherman, "Atopy, helminths, and cancer," *Medical Hypotheses*, vol. 60, no. 1, pp. 1–5, 2003.

[250] E. Sibanda, D. Gallerano, E. Wollmann, and R. Valenta, "EFIS-EJI African International Conference on Immunity (AICI)," *European Journal of Immunology*, vol. 42, no. 5, pp. 1070–1071, 2012.

4

Innate Immune Activation and Subversion of Mammalian Functions by *Leishmania* Lipophosphoglycan

Luis H. Franco,[1] **Stephen M. Beverley,**[2] **and Dario S. Zamboni**[1]

[1] *Department of Cell Biology, School of Medicine of Ribeirão Preto, University of São Paulo, FMRP/USP, 14049-900, Ribeirão Preto, SP, Brazil*

[2] *Department of Molecular Microbiology, Washington University School of Medicine, 660 S. Euclid Avenue, St. Louis, MO 63110, USA*

Correspondence should be addressed to Dario S. Zamboni, dszamboni@fmrp.usp.br

Academic Editor: Hugo D. Lujan

Leishmania promastigotes express several prominent glycoconjugates, either secreted or anchored to the parasite surface. Of these lipophosphoglycan (LPG) is the most abundant, and along with other phosphoglycan-bearing molecules, plays important roles in parasite infectivity and pathogenesis in both the sand fly and the mammalian host. Besides its contribution for parasite survival in the sand fly vector, LPG is important for modulation the host immune responses to favor the establishment of mammalian infection. This review will summarize the current knowledge regarding the role of LPG in *Leishmania* infectivity, focusing on the interaction of LPG and innate immune cells and in the subversion of mammalian functions by this molecule.

1. Introduction: *Leishmania* and Lipophosphoglycan

Leishmaniasis is caused by infection with protozoan parasites of the Trypanosomatid genus *Leishmania*. The disease is endemic in several regions, including west Asia, Africa, and South America. In humans, several disease manifestations have been observed, ranging from self-healing cutaneous lesions to progressive and fatal systemic infection [1]. Leishmaniasis is transmitted by the bite of phlebotomine sand flies and in most parts of the world is a zoonosis, although in some areas direct human-fly-human transmission has been reported [1].

The life cycle of *Leishmania* has two main morphological forms: flagellated promastigotes, which replicate and develop in the midgut of the sand fly vector, and rounded amastigotes, which live and multiply inside the macrophages of the vertebrate host. The establishment of the infection begins with the inoculation by the sand fly vector's bite of metacyclic promastigotes into the vertebrate host. From this wound site, the parasites encounter a variety of cell types including neutrophils, Langerhans and dendritic cells, keratinocytes, and tissue macrophages, all of which have been proposed to serve as the "first contact" host cell (reviewed in [2]). While *in vitro* and in some cases *in vivo* studies provide good support for these models, the complex nature of the sand fly bite makes it difficult to ascertain the quantitative importance of these to the final parasitic outcome. Ultimately, the metacyclic forms of the parasite are internalized and differentiate intracellularly to the amastigote form. In macrophages, amastigotes multiply inside the acidic vacuoles, and eventually are released after lysis, spreading the infection to uninfected cells [3]. Current knowledge about the steps leading to parasite escape is limited, for example, whether it is regulated by the parasite or occurs simply through overwhelming the capacity of the macrophage to harbor them.

Leishmania promastigotes are covered by a thick glyco-calyx comprised of abundant glycoconjugates important for parasite survival and pathogenesis. These molecules include Lipophosphoglycan (LPG), proteophosphoglycan (PPG), gp63 metalloproteinase, and glycophosphatidylinositol lipids (GIPLs). One notable feature distinguishing the *Leishmania* surface from that of the host is that most parasite molecules are linked to the parasite surface through glycosylphosphatidylinositol (GPI) lipid anchors [4–8]. *Leishmania* also secrete protein-linked phosphoglycans (PGs), such as the secreted proteophosphoglycan (sPPG) and secreted acid phosphatase (sAP) [9].

LPG is the most abundant glycoconjugate on the surface of *Leishmania* promastigotes. The GPI anchor which links LPG at surface of the parasite is constituted by a 1-*O*-alkyl-2-*lyso*-phosphatidyl(*myo*)inositol lipid anchor with a heptasaccharide glycan core, to which is joined a long PG polymer composed of 15–30 [6-Gal(β1,4)Man(α1)-PO$_4$−] repeating units, and terminated by a capping oligosaccharide (Figure 1). The PG repeating units are often modified by other sugars, which are typically species and stage specific. Procyclic and metacyclic promastigotes of all *Leishmania* species express high amounts of LPG on their surface, in contrast to amastigotes, whose LPG expression is highly downregulated [10]. In promastigotes, LPG plays an important role for parasite survival inside sand fly vector and for macrophage infection, as discussed below. In contrast, the survival of amastigotes inside host macrophages is improved by other PG-containing glycoconjugates, such as PPG, which are highly expressed on its surface. All of the LPG domains are shared with other parasite surface molecules, to varying extents and degrees of relatedness. The PG repeat, side chains, and caps can be found on PPG or sAP, and both the GPI glycan core and lipid anchor have similarities with those present in both GIPLs and GPI-anchored proteins [8, 11, 12]. As described below, the usual of mutants defective in specific steps of LPG biosynthesis have proven useful in resolving the role of LPG domains clearly from related ones borne by other molecules.

2. The Role(s) of LPG and PGs in the Sand Fly Vector

A number of obstacles present in the sand fly vector digestive tract are potentially able to impair the development of *Leishmania*, including digestive enzymes, the midgut peritrophic membrane barrier, avoidance of excretion along with the digested blood meal, and the anatomy and physiology of the anterior gut (Figure 2). These barriers have provided the evolutionary drive for expression of molecules by the parasite required for successful development in the sand fly vector. As in the mammalian stages emphasized in later sections, LPG and related PGs are key molecules important for survival inside the hostile environment of sand fly vector [9].

During the digestion of blood meal in the insect midgut, the intracellular amastigotes initiate their differentiation to the motile procyclic promastigotes. These forms of the parasite leave the macrophages and are exposed to the hostile environment of the midgut. The dense glycocalyx formed by LPG and PPG provides protection against the action of midgut hydrolytic enzymes and by inhibiting the release of midgut proteases [13]. Procyclic promastigotes are able to attach to midgut epithelial cells, which enable the parasite to be retained within the gut during excretion of the digested blood meal. Several findings have suggested that LPG plays an important role in attachment of promastigotes in midgut in some species or strains such as the *L. major* Friedlin

line [14–16], which binds to the sand fly midgut lectin PpGalec [17]. However, in other species, LPG appears to play less of a role in attachment, as LPG-deficient mutants retain the ability to bind [18, 19]. The molecules mediating this attachment are unknown although a role for parasite lectins has been suggested [20, 21]. For those strains/species dependent upon LPG for binding, the parasite must then find a way to release from the midgut in order to be free for subsequent transmission. To do this, metacyclic parasites synthesize an LPG unable to interact with host lectins. For *L. major* strain Friedlin, the procyclic Gal-β 1–3 PGs of LPG are"capped" with D-arabinopyranose, resulting in an LPG unable to bind PpGalec [17, 22]. In contrast, in *L. donovani* which synthesizes an LPG lacking PG modifications, binding through the terminal capping sugar is "masked" through elongation of the LPG chain [23].

The promastigote stage of many *Leishmania* species elaborates a thick mucoid "plug" during infections, comprised primarily of PPGs along with other shed parasite molecules. At the time of transmission by biting, the plug contents are inoculated along with parasites and saliva into the host. Seminal studies by Bates and collaborators have suggested that the PG repeats borne on PPGs within the plug play key roles in exacerbating the subsequent infections in *L. mexicana*, thereby implicating PGs synthesized and secreted by *Leishmania* in the fly as important immunomodulators of the host response [24, 25]. Notably sand fly saliva can exacerbate *Leishmania* infections as well. It is worth pointing out that most experimental studies of *Leishmania* transmission are compromised to some extent by the use of needle inoculated parasites, lacking these key biological mediators as well as differing in the amount of local tissue damage.

3. The Role(s) of LPG and Related PGs in Mammalian Infectivity

As seen with the sand fly stages, LPG and related PGs have been implicated in a variety of key steps required for infectivity of mammalian hosts (Figure 2). Here, we summarize the current information regarding the role of LPG for subversion of mammalian protective responses by the parasite, and the recognition of parasite LPG by the mammalian innate immune cells.

4. The Role of LPG for Avoidance of Lysis by Complement

Before the internalization by host cells, metacyclic promastigotes must evade lysis by the mammalian complement system. Several studies using purified LPG or LPG-deficient parasites have shown that this molecule defends against complement-mediated lysis [26, 27]. *L. major* metacyclic promastigotes, the infective forms for mammals, are resistant to complement-mediated lysis while the procyclic forms, which reside inside the sand fly vector, are highly susceptible [28]. This difference is conferred by changes in the length of the metacyclic LPG PG polymer domain, which bears about

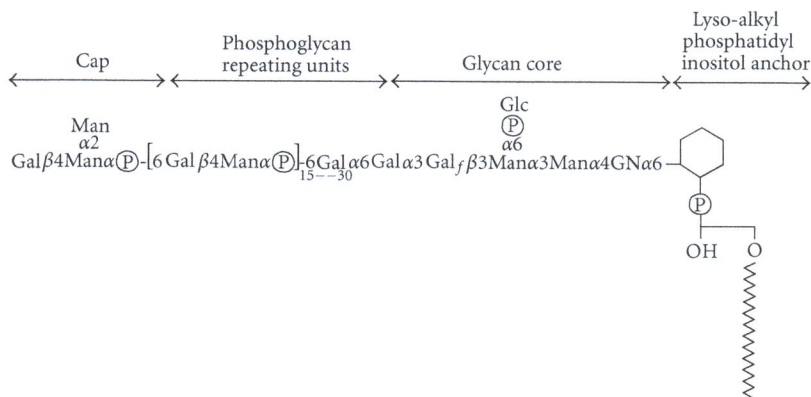

FIGURE 1: Structure of Lipophosphoglycan from *Leishmania donovani*. The four key domains (cap, phosphoglycan repeating units, glycan core and lipid anchor) are discussed further in the text. The number of phosphoglycan (PG) repeating units increases during metacyclogenesis, contributing to the role of LPG in complement resistance. In many *Leishmania* species, side chain modifications of the PG Gal residue are common, where they can play a role in sand fly transmission. The structure of the cap also differs amongst species. Gal, galactose; Man; Mannose; GN, glucosamine; Glc, glucose.

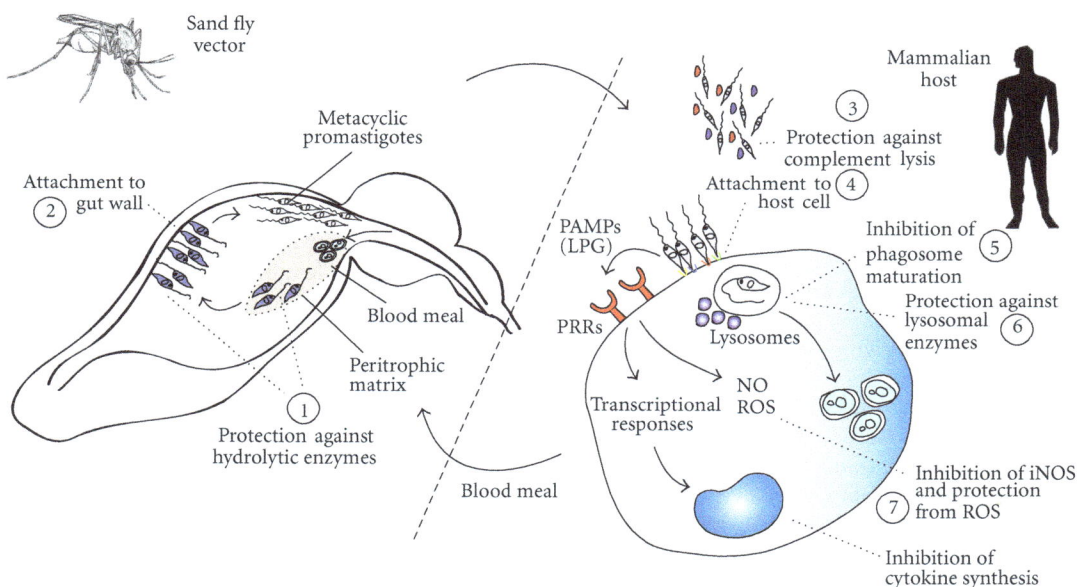

FIGURE 2: Role of LPG in *Leishmania* infectivity and virulence. Shown are putative and bona fide actions of *Leishmania* spp. LPG molecules in subversion of host and vector functions. These LPG functions include (1) physical protection to promastigotes against hydrolytic enzymes in the digestive tract of insect; (2) attachment of promastigotes to the gut wall; (3) In the mammalian host, promastigotes protection against lysis by complement proteins; (4) attachment of parasites to the macrophage membranes or alternative transiently infected cells, such as neutrophils, dendritic cells and perhaps others; (5) transient impairment of the phagosome maturation; (6) physical protection against degradation by lysosomal enzymes; (7) modulation of macrophages activation through impairing the synthesis of nitrogen species and cytokines related to the control of infection and protection from ROS.

twice as many repeating units as the procyclic promastigotes. This prevents the attachment of complement membrane attack complex (MAC) and pore formation on parasite surface [28]. However, earlier steps in the complement cascade may contribute in the entrance of *Leishmania* into macrophages through complement receptors. LPG, together with the protease gp63, is able to activate the complement system, leading to the generation of the C3b and C3bi opsonins. C3b and C3bi thus bind to *Leishmania* surface and mediate the parasite phagocytosis by complement receptor (CR) 1 and CR3 [29–35]. Phagocytosis of *Leishmania* via CR1 and CR3 receptors is considered as a means of "silent entry" into macrophages, because it does not prompt the oxidative burst and impairs the production of IL-12 [32, 36–38]. However, infections of CR3-deficient mice show very little attenuation of infection, suggesting that this step may be of lesser importance or redundant with other binding interactions in survival [39].

5. The Role of LPG in Parasite Invasion and Survival in Macrophages

After being inoculated into the mammalian host by the sand fly vector, the metacyclic promastigotes are internalized through interactions with a number of different receptors. While at various times interactions with one or another of these have been presumed or shown to be dominant in cellular or biochemical tests, genetic studies have typically led to conclusions that these interactions typically may instead be highly redundant in biological settings. In this scenario, the use of multiple receptors allows the promastigotes to be quickly internalized by macrophages (reviewed in [40]).

Importantly, the LPG plays an important role as a ligand during the attachment and invasion process of macrophages, either directly or indirectly through binding to other proteins.One example is the interaction of LPG with mannose-fucose receptor expressed by macrophages [41]. In addition, mannan-binding protein (MBP) is able to bind to mannose residues on LPG, enabling the formation of C3 convertase and generation of C3b, which helps promastigotes to attach to the macrophage as noted above [42]. C-reactive protein (CRP) binds to LPG of L. donovani metacyclic promastigotes triggering their phagocytosis by human macrophages via CRP receptor [43]. Commonly, the engagement of CRP receptor by its ligand leads to macrophage activation, resulting in proinflammatory cytokine production [44, 45]. However, phagocytosis of L. donovani by CRP receptor leads to an incomplete activation of macrophages, thus favoring parasite replication [46].

Following entry, promastigotes are contained in a phagosome known as parasitophorous vacuole (PV), which undergoes several fusion processes, giving rise to a phagolysosome-like organelle [47, 48]. During this process, LPG acts to delay PV fusion with lysosomes, promoting delay in PV acidification and acquisition of lysosomal enzymes [49]. Vacuoles harboring promastigotes of L. donovani and L. major genetically deficient for LPG fuse more extensively and rapidly with endosomes and lysosomes [27, 50]. While initially workers postulated that this delay protected promastigotes from acidic conditions and hydrolytic enzymes until they had differentiated to the more acidophilic amastigote stage, work with LPG-null L. major promastigotes provided little support for this model [27], as these parasites are able to survive under several conditions despite rapid fusion with host lysosomes. Instead, the delay in fusion reflects changes in membrane properties that result in delocalization of the host oxidative burst from its normal peri-PV location [51]. On top of this, LPG itself is able to interact and deflect oxidants directly [52]. Other roles of the delayed fusion may not only concern survival but immune recognition and antigen processing, which is dependent on host hydrolytic enzymes [50].

Whereas LPG seems to be important to protect Leishmania during differentiation from promastigote to amastigote forms, it does not play a significant role during the development of the amastigote form. Indeed, LPG expression on amastigotes of several species of Leishmania is highly downregulated (1000 fold or more) [10] suggesting that the protective role of LPG is transient and limited to the beginning of host cell infection. However, other PG-containing glycoconjugates and especially PPG are expressed at high levels in amastigotes, and act in PG-dependent manner to protect the amastigote [53, 54].

6. The Role of LPG for Inhibition of Macrophage Activation

Infected macrophages employ several microbicidal mechanisms to eliminate intracellular pathogens. When previously activated by interferon-gamma (IFN-γ) and tumor necrosis factor-α (TNF-α) or other microbial components, infected macrophages express high levels of the inducible nitric oxide synthase (NOS2), culminating with production of nitric oxide (NO), NO_2^-, and NO_3^- [55]. These nitrogen intermediates coordinate processes that lead to deprivation of important components, such as iron, which lead to restriction of intracellular parasites replication [56]. L. major is able to induce higher amounts of NOS2 in the cutaneous lesion and draining lymph nodes of the clinically resistant lineage C57BL/6 compared to the nonhealing BALB/c strain [57]. In addition, mice deficient for NOS2 are more susceptible to infection with Leishmania, compared to their littermate controls, as well as macrophages derived from these mice [58–60]. Thus, the production of NO is indispensable for the control of L. major infection and for maintaining life-long control of persisting Leishmania parasites [61–63].

In contrast, infection of unactivated macrophages typically leads to parasite survival and minimal levels of NOS2-dependent NO production, to the point that L. major was referred to as a "stealthy parasite" [64]. Thus, one of the challenges in experimental models is the need to distinguish infections where macrophages are naturally or experimentally activated from those situations where Leishmania exhibits successful parasitism and survival. Perusal of the literature suggests that many workers do not provide evidence about which fate meets Leishmania under their experimental infections, which may contribute occasionally to seemingly contradictory results.

Experimental studies have shown that similar to Leishmania, treatment of macrophages with Leishmania glycoconjugates can likewise regulate the activation of NOS2 and production of NO. LPG can synergize with IFN-γ for the induction of NO expression in murine macrophages in vitro. However, incubation of macrophages with LPG-derived PG before stimulation with LPG plus IFN-γ led to inhibition of NOS2 expression [65]. These studies provided evidence that the interaction between the macrophage and the parasite impairs the activation of the microbicidal mechanisms of macrophages after exposure to IFN-γ in a process that is replicated by PG treatments. Given these findings, it was surprising that despite the complete absence of LPG or all PGs in the lpg1$^-$ or lpg2$^-$ mutants (described further below), mutant parasites remained "stealthy" and able to down regulate host cell activation [27, 66]. A similar contradiction was seen in studies of the smaller GIPL, which are highly abundant in both parasites stages and had been shown to inhibit NO synthesis by macrophage in a dose- and time-dependent manner, impairing its leishmanicidal activity [67]. However, mutants defective in the synthesis of the ether lipid anchor and thus lacking both LPG and GIPLs resembled the *lpg1⁻* and *lpg2⁻* mutants in remaining "stealthy" and inhibiting host cell activation [68]. This apparent paradox has not been resolved and has led to proposals that avoidance of host cell activation may be highly redundant amongst many parasite surface molecules, perhaps through their ability to interact with secondary ligands/mediators such as complement or other serum proteins. One attractive model is that macrophage deactivation is independent of surface molecules, instead depending on other processes such as secretion of parasite molecules through an exosomes-like or other pathways (reviewed in [69]).

In addition to NO, activated macrophages employ other antimicrobial molecules, such as ROS or antimicrobial peptides, to kill intracellular parasites. ROS such as superoxide, hydrogen peroxide, and hydroxyl radicals, are produced after activation of NADPH oxidase and interact with pathogen phospholipid membranes, inducing damage and dead of pathogens [70]. Some evidence highlights the importance of ROS in control of *Leishmania* growth [71, 72]. Upon infection by *L. donovani* promastigotes, peritoneal macrophages elicit a strong respiratory burst with release of superoxide anion, thus favoring the elimination of the intracellular amastigotes. When infected in the presence of catalase, an enzyme that catalyze the decomposition of hydrogen peroxide to water and oxygen, macrophages lost their ability to kill *L. donovani* [73]. These results show that ROS are central compounds that act to eliminate intracellular *Leishmania in vitro*.

Respiratory burst activity and NO production are regulated by phosphorylation events mediated by protein kinase C (PKC) [74]. Infection with *Leishmania* is able to inhibit PKC activity in macrophages and several findings suggesting that LPG is related to this activity, thereby favoring intracellular survival of the parasite through inhibition of both oxidative burst and NO production [10, 75–79]. Besides PKC, production of cytokines such as IL-12 was inhibited in bone marrow-derived macrophages after infection with *L. major* [80]. Furthermore, purified LPG plays similar inhibitory effect over IL-12 production, probably thought the activation of the mitogen-activated protein kinase (MAPK) Erk 1/2, which suppresses IL-12 gene transcription [81]. Besides IL-12, purified LPG also suppressed IL-1β gene expression in THP-1 monocytes induced by endotoxin, TNF-α or *Staphylococcus* stimulation [82].

7. Recognition of LPG by Mammalian Innate Immune Receptors

The initiation of immune response against invading pathogens starts upon the interaction of microbial molecules with receptors of innate immune cells. Glycoconjugates expressed by protozoans interact with macrophage receptors and are recognized as foreign by immune system. Purified GPI-anchored surface proteins of *Plasmodium falciparum*, *Trypanosoma brucei* and *L. mexicana*, initiate the rapid

activation of macrophage protein tyrosine kinases (PTKs) [83–85]. GPI anchors expressed by protozoans, such as *Plasmodium* and *Trypanosoma*, can activate the secretion of cytokines, such as IL-12 and TNF-α, and NO synthesis by macrophages [83, 85–90].

The activation of Toll-like receptors (TLRs) by microbial ligands recruits the adaptor protein MyD88 (myeloid differentiation primary response gene 88) and triggers intracellular signaling events, culminating on the activation of the transcription factor NF-κB and its translocation to nucleus. NF-κB in turn induces innate immune mechanisms such as the production of reactive oxygen and nitrogen intermediates, chemokine/cytokine secretion, and cellular differentiation [91]. Several evidences have suggested that *Leishmania* expresses ligands able to stimulate the TLRs signaling pathways. RAW macrophages selectively upregulated the IL-1α mRNA expression in response to *L. major* infection, and this was not observed when macrophages were transfected with a dominant-negative of MyD88 or when peritoneal macrophages derived from MyD88-deficient mice were infected with *L. major* [92]. In addition, mice deficient for MyD88 infected with *L. major* showed an increase in lesion size compared to their littermate controls [93]. These results suggest that *L. major* may express ligands for TLR activation. Accordingly, *L. major* LPG activated NF-κB, the secretion of Th1-type cytokines, ROS, and NO by either human or murine macrophages in a mechanism dependent of TLR2 [93–95]. In addition, purified LPG upregulates TLR2 expression and stimulate IFN-γ and TNF-α secretion by human NK cells in a TLR2-dependent manner [96]. Thus, activation of TLR2 may contribute to host resistance against *Leishmania* and LPG is proposed to be a putative agonist for TLR activation.

In addition to macrophages and NK cells, LPG has been shown to exert stimulatory effects on dendritic cells (DCs). Purified *L. mexicana* LPG was able to induce the expression of CD86 and major histocompatibility complex class II (MHC-II) by DCs; furthermore, *L. major* LPG stimulated the expression of CD25, CD31, and vascular-endothelial cadherin by mouse Langerhans cells, albeit accompanied by inhibition of their migratory activity [97, 98]. Importantly, upregulation of stimulatory and costimulatory molecules in DCs occurs in response to activation of pattern recognition receptors; therefore, these studies corroborate the hypothesis that that *Leishmania* LPG triggers activation of these receptors.

Given the interaction of LPG with TLRs in the context of activated macrophages where this leads to a proinflammatory response and parasite control, an important but as yet unanswered question is how the LPG-TLR interaction fails to control parasite infection in unactivated macrophages. A variety of pathways are known which negatively regulate TLR signaling, and potentially one of these acts to mitigate TLR activation. A second question is whether LPG or related molecules are internalized into host cells, which would then place them in contact with variety internal sensors including the NOD-like receptors protein family in the cytosol. Early studies showed LPG trafficking into the interior of host cells [99] and recently several groups have provided evidence suggesting that *Leishmania* molecules may gain access to the host cytosol through some routes, potentially including an exosome-like pathway [100, 101]. Further work is needed to confirm these provocative hypotheses and explore the role of LPG and related glycoconjugates in this process.

8. The Assessment of LPG Functions by Using LPG-Defective Mutants

As mentioned above, many studies have used purified LPG, fragments thereof, or related molecules, to investigate their role in *Leishmania* pathogenesis and host response. However, LPG preparations can include contaminating molecules including proteins, unless proper precautions are taken, and use of exogenous LPG may not properly mimic the physiological location and concentrations of LPG delivered by infecting parasites. Moreover, as noted above, many LPG domains are shared by other parasite molecules, raising the possibility that functions attributed to LPG *in vitro* may actually be fulfilled by LPG-related molecules *in vivo*.

In the last few years, the generation of LPG mutants of *Leishmania* has provided powerful tools to identify the function of these molecules [102–105]. Of note, the recent identification of genes related to LPG biosynthetic pathways allowed the generation of "clean" LPG mutant strains by specific gene targeting. *Leishmania* are typically diploid although recent studies suggest that many chromosomes may be aneuploid, at least transiently [106]. Thus, two or more successive rounds of gene replacement are required to generate full homozygous null mutants, as while feasible in some cases sexual crossing remains challenging [107]. Importantly, the phenotypes of the mutants chosen for biological studies were rescued by complementation of the specific LPG gene into the parasite [108–110]. This rules out the well-known problem of loss of virulence during transfection or culture of *Leishmania*, which occurs sporadically in all species. Thus far nearly 20 genes affecting various steps of LPG biosynthesis have been described through complementation of LPG mutants or through various reverse genetic strategies.

For the study of virulence, this repertoire of LPG genes has enabled researchers to concentrate on key mutants that cleanly affect LPG or related molecules. The first genetic assessment of the role of LPG in parasite virulence and host immunity followed the identification of the *LPG1* gene, which was recovered following complementation of the LPG-deficient R2D2 mutant of *L. donovani* [110]. This gene encodes a putative galactofuranosyl transferase involved in biosynthesis of the LPG glycan core, but not other galactofuranosyl-containing glycoconjugates whose synthesis depends on other LPG1-related transferases [111]. *lpg1*⁻ mutants of *L. major* or *L. donovani* do not express LPG on their surface, while the expression of other glycoconjugates remains normal [112, 113], rendering these ideal for studies of the biological roles mediated exclusively by LPG. *lpg1*⁻ mutants are highly susceptible to lysis by complement; sensitive to oxidative stress, and they fail to even transiently inhibit phagolysosomal fusion immediately after invasion [27]. Moreover, *L. major lpg1*⁻ showed an impaired ability to survive inside macrophages [27, 113] and in mouse

infections were highly attenuated, as represented by an extreme delay in lesion progression [27, 113].

Interestingly, the generality of the role of LPG or even PGs in parasite survival in all *Leishmania* has been questioned based on similar genetic studies in *L. mexicana*, where a proper *lpg1*⁻ line shows no decrease in infectivity tests in macrophages or mice [112], although it is complement sensitive [114]. Despite these observations, the *lpg1*⁻ *L. mexicana* nonetheless showed some alterations in host response, with a poor ability to stimulate the expression of costimulatory molecules on mouse DCs, and it was found that *lpg1*⁻ *L. mexicana*-infected mice showed lower numbers of activated DCs in draining lymph nodes and were unable to control early parasite burden [97]. Thus, it appears that LPG plays a quantitatively or qualitatively different role in *L. mexicana* virulence, especially in directing the immune response. A similar contrast was found in studies of an *L. mexicana lpg2*⁻, discussed below [115]. Amongst many potential explanations, the architecture of the PV has been proposed to be a factor, as it exists as a "spacious, multiparasite" compartment in *L. mexicana* infections versus a "tight, uniparasitic" compartment in *L. major* and *L. donovani* [114]. Thus, the roles of LPG appear to differ both quantitatively and qualitative amongst species.

While the LPG-deficient *lpg1*⁻ parasites show severe attenuation in both *L. donovani* and *L. major,* studies in the latter species show that some parasites survive and go on to generate normal amastigotes, in keeping with the downregulation of LPG during development. Since other PG-containing glycoconjugates such as PPG are found throughout the life cycle, the role of the PG moieties generally was investigated by the use of a mutant globally affecting PGs. The *LPG2* gene was identified by complementation of the *L. donovani* C3PO mutant [109] and was shown in a series of seminal studies in Turco's laboratory to encode the Golgi GDP-mannose transporter [116–118], one of the founding members of what is now known to be a large family of nucleoside sugar transporters [119]. LPG2 was also the first multispecific nucleotide sugar transporters to be described, being able to carry both GDP-D-Arabinopyranose and GDP-Fucose in addition to GDP-Man [116]. As noted earlier, *L. major* utilizes D-Arabinopyranose as an LPG side chain "capping" sugar, but neither a role nor glycoconjugates bearing fucose has been described in *Leishmania*, although low levels of GDP-Fuc have been observed in promastigotes [120].

lpg2⁻ mutant parasites lack all PGs, including LPG and PPG, but synthesize normal levels of GIPLs and gp63 [66]. *L. major* and *L. donovanI lpg2*⁻ mutants failed to survive in the midgut of sand fly vector and were unable to establish infection in macrophages. In animal infections, *L. major* parasites showed "persistence without pathology," with parasites persisting at low levels for the life of infected animals—a situation reminiscent of the life-long infection following healing of *Leishmania* infections in experimental animals and humans [16, 66, 114]. This parallel was further extended by the demonstration that as in healed animals, *L. major lpg2*⁻ induced long-term immunity against challenge with a virulent strain of *L. major* [121]. Observations that lymphocytes isolated from *L. major lpg2⁻*-infected mice produced less IL-4 and IL-10 after stimulation *in vitro*, compared to cells isolated from *L. major* WT-infected mice, provided evidences about the anti-inflammatory properties of PGs over immune cells [122]. Importantly, similar effects on cytokine expression were seen in the *lpg*5A⁻/*lpg*5B⁻ double mutant, which also lacks all PGs, but through inactivation of Golgi UDP-Gal transporter activity [122, 123]. However, the *lpg*5A⁻/*lpg*5B⁻ mutant shows a virulence defect comparable to that of the *lpg1⁻* rather than *lpg2⁻* mutant [123].This suggests that the "persistence without pathology" phenotype of the *lpg2⁻* mutant may arise from effects on gylcoconjugates other than PGs [123]. Thus, comparison amongst the well-characterized collection of LPG/PG mutants provides a "genetic sieve", allowing assignment of the roles of LPG and PGs separately and in immune interaction from their roles in general parasite infectivity. These studies using *lpg2⁻* mutant parasites provided evidence that PGs, in addition to LPG, play important roles in *Leishmania* virulence [122, 123].

9. Concluding Remarks

LPG is a key molecule mediating many important steps essential for *Leishmania* virulence, in the hostile environment of the sand fly vector midgut, or in the mammalian host. The identification of genes related to LPG synthesis allowed the generation of *Leishmania* strains defective in LPG. The uses of these mutants have provided valuable clues about the role of this glycoconjugate in the biology of *Leishmania*. We envisage that further studies using these and new mutants may elucidate important issues related to innate immune recognition and host cell activation by protozoan parasites. This information will greatly increase our understanding of both *Leishmania* pathogenesis and the recognition of protozoan parasites by the mammalian innate immune system.

Acknowledgment

The authors are grateful for M. Michelle Favila for critical reading of the manuscript. This work was supported by grants of NIH R01 AI031078 (to S. M. Beverley); INCTV/CNPq and FAPESP Grant 10/50959-4 (to D. S. Zamvoni). L. H. Franco is the recipient of a postdoctoral fellowship from FAPESP (Grant 2009/50024-8) and D. S. Zamboni is a research fellow from CNPq.

References

[1] WHO, "Control of the leishmaniasis," WHO Tecnical Report Series, 2010.

[2] P. Kaye and P. Scott, "Leishmaniasis: complexity at the host-pathogen interface," *Nature Reviews Microbiology*, vol. 9, no. 8, pp. 604–615, 2011.

[3] J. C. Antoine, E. Prina, N. Courret, and T. Lang, "Leishmania spp.: on the interactions they establish with antigen-presenting cells of their mammalian hosts," *Advances in Parasitology*, vol. 58, pp. 1–68, 2004.

[4] L. L. Button and W. R. McMaster, "Molecular cloning of the major surface antigen of Leishmania," *Journal of Experimental Medicine*, vol. 167, no. 2, pp. 724–729, 1988.

[5] M. J. Elhay, M. J. McConville, and E. Handman, "Immunochemical characterization of a glyco-inositol-phospholipid membrane antigen of Leishmania major," *Journal of Immunology*, vol. 141, no. 4, pp. 1326–1331, 1988.

[6] R. Etges, J. Bouvier, and C. Bordier, "The major surface protein of Leishmania promastigotes is a protease," *Journal of Biological Chemistry*, vol. 261, no. 20, pp. 9098–9101, 1986.

[7] M. J. McConville and M. A. J. Ferguson, "The structure, biosynthesis and function of glycosylated phosphatidylinositols in the parasitic protozoa and higher eukaryotes," *Biochemical Journal*, vol. 294, no. 2, pp. 305–324, 1993.

[8] M. J. McConville, K. A. Mullin, S. C. Ilgoutz, and R. D. Teasdale, "Secretory pathway of trypanosomatid parasites," *Microbiology and Molecular Biology Reviews*, vol. 66, no. 1, pp. 122–154, 2002.

[9] D. Sacks and S. Kamhawi, "Molecular aspects of parasite-vector and vector-host interactions in Leishmaniasis," *Annual Review of Microbiology*, vol. 55, pp. 453–483, 2001.

[10] S. J. Turco and D. L. Sacks, "Expression of a stage-specific lipophosphoglycan in Leishmania major amastigotes," *Molecular and Biochemical Parasitology*, vol. 45, no. 1, pp. 91–99, 1991.

[11] T. Ilg, E. Handman, and Y. D. Stierhof, "Proteophosphoglycans from Leishmania promastigotes and amastigotes," *Biochemical Society Transactions*, vol. 27, no. 4, pp. 518–525, 1999.

[12] S. J. Turco, G. F. Späth, and S. M. Beverley, "Is lipophosphoglycan a virulence factor? A surprising diversity between Leishmania species," *Trends in Parasitology*, vol. 17, no. 5, pp. 223–226, 2001.

[13] Y. Schlein, L. F. Schnur, and R. L. Jacobson, "Released glycoconjugate of indigenous Leishmania major enhances survival of a foreign L. major in Phlebotomus papatasi," *Transactions of the Royal Society of Tropical Medicine and Hygiene*, vol. 84, no. 3, pp. 353–355, 1990.

[14] T. Ilg, "Lipophosphoglycan of the protozoan parasite Leishmania: stage- and species-specific importance for colonization of the sandfly vector, transmission and virulence to mammals," *Medical Microbiology and Immunology*, vol. 190, no. 1-2, pp. 13–17, 2001.

[15] P. F. P. Pimenta, S. J. Turco, M. J. McConville, P. G. Lawyer, P. V. Perkins, and D. L. Sacks, "Stage-specific adhesion of Leishmania promastigotes to the sandfly midgut," *Science*, vol. 256, no. 5065, pp. 1812–1815, 1992.

[16] D. L. Sacks, G. Modi, E. Rowton et al., "The role of phosphoglycans in Leishmania-sand fly interactions," *Proceedings of the National Academy of Sciences of the United States of America*, vol. 97, no. 1, pp. 406–411, 2000.

[17] S. Kamhawi, M. Ramalho-Ortigao, M. P. Van et al., "A role for insect galectins in parasite survival," *Cell*, vol. 119, no. 3, pp. 329–341, 2004.

[18] J. Myskova, M. Svobodova, S. M. Beverley, and P. Volf, "A lipophosphoglycan-independent development of Leishmania in permissive sand flies," *Microbes and Infection*, vol. 9, no. 3, pp. 317–324, 2007.

[19] A. Svárovská, T. H. Ant, V. Seblová, L. Jecná, S. M. Beverley, and P. Volf, "Leishmania major glycosylation mutants require phosphoglycans (lpg2-) but not lipophosphoglycan (lpg1-)

for survival in permissive sand fly vectors," *PLoS Neglected Tropical Diseases*, vol. 4, no. 1, article no. e580, 2010.

[20] M. Svobodová, P. A. Bates, and P. Volf, "Detection of lectin activity in Leishmania promastigotes and amastigotes," *Acta Tropica*, vol. 68, no. 1, pp. 23–35, 1997.

[21] M. Svobodova, P. Volf, and R. Killick-Kendrick, "Agglutination of Leishmania promastigotes by midgut lectins from various species of phlebotomine sandflies," *Annals of Tropical Medicine and Parasitology*, vol. 90, no. 3, pp. 329–336, 1996.

[22] S. M. Beverley and D. E. Dobson, "Flypaper for parasites," *Cell*, vol. 119, no. 3, pp. 311–312, 2004.

[23] P. F. P. Pimenta, E. M. B. Saraiva, E. Rowton et al., "Evidence that the vectorial competence of phlebotomine sand flies for different species of Leishmania is controlled by structural polymorphisms in the surface lipophosphoglycan," *Proceedings of the National Academy of Sciences of the United States of America*, vol. 91, no. 19, pp. 9155–9159, 1994.

[24] M. E. Rogers, M. L. Chance, and P. A. Bates, "The role of promastigote secretory gel in the origin and transmission of the infective stage of Leishmania mexicana by the sandfly Lutzomyia longipalpis," *Parasitology*, vol. 124, no. 5, pp. 495–507, 2002.

[25] Y. D. Stierhof, P. A. Bates, R. L. Jacobson et al., "Filamentous proteophosphoglycan secreted by Leishmania promastigotes forms gel like three-dimensional networks that obstruct the digestive tract of infected sandfly vectors," *European Journal of Cell Biology*, vol. 78, no. 10, pp. 675–689, 1999.

[26] S. M. Puentes, R. P. Da Silva, D. L. Sacks, C. H. Hammer, and K. A. Joiner, "Serum resistance of metacyclic stage Leishmania major promastigotes is due to release of C5b-9," *Journal of Immunology*, vol. 145, no. 12, pp. 4311–4316, 1990.

[27] G. F. Späth, L. A. Garraway, S. J. Turco, and S. M. Beverley, "The role(s) of lipophosphoglycan (LPG) in the establishment of Leishmania major infections in mammalian hosts," *Proceedings of the National Academy of Sciences of the United States of America*, vol. 100, no. 16, pp. 9536–9541, 2003.

[28] S. M. Puentes, D. M. Dwyer, P. A. Bates, and K. A. Joiner, "Binding and release of C3 from Leishmania donovani promastigotes during incubation in normal human serum," *Journal of Immunology*, vol. 143, no. 11, pp. 3743–3749, 1989.

[29] R. P. Da Silva, B. F. Hall, K. A. Joiner, and D. L. Sacks, "CR1, the C3b receptor, mediates binding of infective Leishmania major metacyclic promastigotes to human macrophages," *Journal of Immunology*, vol. 143, no. 2, pp. 617–622, 1989.

[30] D. M. Mosser and P. J. Edelson, "Activation of the alternative complement pathway by leishmania promastigotes: parasite lysis and attachment to macrophages," *Journal of Immunology*, vol. 132, no. 3, pp. 1501–1505, 1984.

[31] D. M. Mosser and P. J. Edelson, "The mouse macrophage receptor for C3bi (CR3) is a major mechanism in the phagocytosis of Leishmania promastigotes," *Journal of Immunology*, vol. 135, no. 4, pp. 2785–2789, 1985.

[32] D. M. Mosser and P. J. Edelson, "The third component of complement (C3) is responsible for the intracellular survival of Leishmania major," *Nature*, vol. 327, no. 6120, pp. 329–331, 1987.

[33] D. M. Mosser, T. A. Springer, and M. S. Diamond, "Leishmania promastigotes require opsonic complement to bind to the human leukocyte integrin Mac-1 (CD11b/CD18)," *Journal of Cell Biology*, vol. 116, no. 2, pp. 511–520, 1992.

[34] D. G. Russell and S. D. Wright, "Complement receptor type 3 (CR3) binds to an Arg-Gly-Asp-containing region of the major surface glycoprotein, gp63, of Leishmania promastigotes," *Journal of Experimental Medicine*, vol. 168, no. 1, pp. 279–292, 1988.

[35] M. E. Wilson and R. D. Pearson, "Roles of CR3 and mannose receptors in the attachment and ingestion of Leishmania donovani by human mononuclear phagocytes," *Infection and Immunity*, vol. 56, no. 2, pp. 363–369, 1988.

[36] T. Marth and B. L. Kelsall, "Regulation of interleukin-12 by complement receptor 3 signaling," *Journal of Experimental Medicine*, vol. 185, no. 11, pp. 1987–1995, 1997.

[37] F. S. Sutterwala, G. J. Noel, R. Clynes, and D. M. Mosser, "Selective suppression of interleukin-12 induction after macrophage receptor ligation," *Journal of Experimental Medicine*, vol. 185, no. 11, pp. 1977–1985, 1997.

[38] S. D. Wright and S. C. Silverstein, "Receptors for C3b and C3bi promote phagocytosis but not the release of toxic oxygen from human phagocytes," *Journal of Experimental Medicine*, vol. 158, no. 6, pp. 2016–2023, 1983.

[39] C. R. Carter, J. P. Whitcomb, J. A. Campbell, R. M. Mukbel, and M. A. McDowell, "Complement receptor 3 deficiency influences lesion progression during Leishmania major infection in BALB/c Mice," *Infection and Immunity*, vol. 77, no. 12, pp. 5668–5675, 2009.

[40] J. L. Stafford, N. F. Neumann, and M. Belosevic, "Macrophage-mediated innate host defense against protozoan parasites," *Critical Reviews in Microbiology*, vol. 28, no. 3, pp. 187–248, 2002.

[41] M. E. Wilson and R. D. Pearson, "Evidence that Leishmania donovani utilizes a mannose receptor on human mononuclear phagocytes to establish intracellular parasitism," *Journal of Immunology*, vol. 136, no. 12, pp. 4681–4688, 1986.

[42] P. J. Green, T. Feizi, M. S. Stoll, S. Thiel, A. Prescott, and M. J. McConville, "Recognition of the major cell surface glycoconjugates of Leishmania parasites by the human serum mannan-binding protein," *Molecular and Biochemical Parasitology*, vol. 66, no. 2, pp. 319–328, 1994.

[43] F. J. Culley, R. A. Harris, P. M. Kaye, K. P. W. J. McAdam, and J. G. Raynes, "C-reactive protein binds to a novel ligand on Leishmania donovani and increases uptake into human macrophages," *Journal of Immunology*, vol. 156, no. 12, pp. 4691–4696, 1996.

[44] S. P. Ballou and G. Lozanski, "Induction of inflammatory cytokine release from cultured human monocytes by C-reactive protein," *Cytokine*, vol. 4, no. 5, pp. 361–368, 1992.

[45] B. Galve-de Rochemonteix, K. Wiktorowicz, I. Kushner, and J. M. Dayer, "C-reactive protein increases production of IL-1α, IL-1β, and TNF-α, and expression of mRNA by human alveolar macrophages," *Journal of Leukocyte Biology*, vol. 53, no. 4, pp. 439–445, 1993.

[46] K. B. Bodman-Smith, M. Mbuchi, F. J. Culley, P. A. Bates, and J. G. Raynes, "C-reactive protein-mediated phagocytosis of Leishmania donovani promastigotes does not alter parasite survival or macrophage responses," *Parasite Immunology*, vol. 24, no. 9-10, pp. 447–454, 2002.

[47] M. Desjardins, "Biogenesis of phagolysosomes: the 'kiss and run' hypothesis," *Trends in Cell Biology*, vol. 5, no. 5, pp. 183–186, 1995.

[48] M. Desjardins, L. A. Huber, R. G. Parton, and G. Griffiths, "Biogenesis of phagolysosomes proceeds through a sequential series of interactions with the endocytic apparatus," *Journal of Cell Biology*, vol. 124, no. 5, pp. 677–688, 1994.

[49] A. Descoteaux and S. J. Turco, "Glycoconjugates in Leishmania infectivity," *Biochimica et Biophysica Acta*, vol. 1455, no. 2-3, pp. 341–352, 1999.

[50] M. Desjardins and A. Descoteaux, "Inhibition of phagolysosomal biogenesis by the Leishmania lipophosphoglycan," *Journal of Experimental Medicine*, vol. 185, no. 12, pp. 2061–2068, 1997.

[51] R. Lodge, T. O. Diallo, and A. Descoteaux, "Leishmania donovani lipophosphoglycan blocks NADPH oxidase assembly at the phagosome membrane," *Cellular Microbiology*, vol. 8, no. 12, pp. 1922–1931, 2006.

[52] J. Chan, T. Fujiwara, P. Brennan et al., "Microbial glycolipids: possible virulence factors that scavenge oxygen radicals," *Proceedings of the National Academy of Sciences of the United States of America*, vol. 86, no. 7, pp. 2453–2457, 1989.

[53] V. Bahr, Y. D. Stierhof, T. Ilg, M. Demar, M. Quinten, and P. Overath, "Expression of lipophosphoglycan, high-molecular weight phosphoglycan and glycoprotein 63 in promastigotes and amastigotes of Leishmania mexicana," *Molecular and Biochemical Parasitology*, vol. 58, no. 1, pp. 107–122, 1993.

[54] M. J. McConville and J. M. Blackwell, "Developmental changes in the glycosylated phosphatidylinositols of Leishmania donovani," *Journal of Biological Chemistry*, vol. 266, no. 23, pp. 15170–15179, 1991.

[55] J. MacMicking, Q. W. Xie, and C. Nathan, "Nitric oxide and macrophage function," *Annual Review of Immunology*, vol. 15, pp. 323–350, 1997.

[56] Z. Dong, X. Qi, K. Xie, and I. J. Fidler, "Protein tyrosine kinase inhibitors decrease induction of nitric oxide synthase activity in lipopolysaccharide-responsive and lipopolysaccharide-nonresponsive murine macrophages," *Journal of Immunology*, vol. 151, no. 5, pp. 2717–2724, 1993.

[57] S. Stenger, H. Thüring, M. Röllinghoff, and C. Bogdan, "Tissue expression of inducible nitric oxide synthase is closely associated with resistance to Leishmania major," *Journal of Experimental Medicine*, vol. 180, no. 3, pp. 783–793, 1994.

[58] R. M. Mukbel, C. Patten, K. Gibson, M. Ghosh, C. Petersen, and D. E. Jones, "Macrophage killing of Leishmania amazonensis amastigotes requires both nitric oxide and superoxide," *American Journal of Tropical Medicine and Hygiene*, vol. 76, no. 4, pp. 669–675, 2007.

[59] H. W. Murray and F. Nathan, "Macrophage microbicidal mechanisms in vivo: reactive nitrogen versus oxygen intermediates in the killing of intracellular visceral Leishmania donovani," *Journal of Experimental Medicine*, vol. 189, no. 4, pp. 741–746, 1999.

[60] X. Q. Wei, I. G. Charles, A. Smith et al., "Altered immune responses in mice lacking inducible nitric oxide synthase," *Nature*, vol. 375, no. 6530, pp. 408–411, 1995.

[61] A. Diefenbach, H. Schindler, N. Donhauser et al., "Type 1 interferon (IFNα/β) and type 2 nitric oxide synthase regulate the innate immune response to a protozoan parasite," *Immunity*, vol. 8, no. 1, pp. 77–87, 1998.

[62] F. Y. Liew, S. Millott, C. Parkinson, R. M. J. Palmer, and S. Moncada, "Macrophage killing of Leishmania parasite in vivo is mediated by nitric oxide from L-arginine," *Journal of Immunology*, vol. 144, no. 12, pp. 4794–4797, 1990.

[63] S. Stenger, N. Donhauser, H. Thüring, M. Röllinghoff, and C. Bogdan, "Reactivation of latent leishmaniasis by inhibition of inducible nitric oxide synthase," *Journal of Experimental Medicine*, vol. 183, no. 4, pp. 1501–1514, 1996.

[64] S. L. Reiner, S. Zheng, Z. E. Wang, L. Stowring, and R. M. Locksley, "Leishmania promastigotes evade interleukin 12 (IL-12) induction by macrophages and stimulate a broad range of cytokines from CD4[+] T cells during initiation of infection," *Journal of Experimental Medicine*, vol. 179, no. 2, pp. 447–456, 1994.

[65] L. Proudfoot, A. V. Nikolaev, G. J. Feng et al., "Regulation of the expression of nitric oxide synthase and leishmanicidal activity by glycoconjugates of Leishmania lipophosphoglycan in murine macrophages," *Proceedings of the National Academy of Sciences of the United States of America*, vol. 93, no. 20, pp. 10984–10989, 1996.

[66] G. F. Späth, L. F. Lye, H. Segawa, D. L. Sacks, S. J. Turco, and S. M. Beverley, "Persistence without pathology in phosphoglycan-deficient Leishmania major," *Science*, vol. 301, no. 5637, pp. 1241–1243, 2003.

[67] L. Proudfoot, C. A. O'Donnell, and F. Y. Liew, "Glycoinositolphospholipids of Leishmania major inhibit nitric oxide synthesis and reduce leishmanicidal activity in murine macrophages," *European Journal of Immunology*, vol. 25, no. 3, pp. 745–750, 1995.

[68] R. Zufferey, S. Allen, T. Barron et al., "Ether phospholipids and glycosylinositolphospholipids are not required for amastigote virulence or for inhibition of macrophage activation by Leishmania major," *Journal of Biological Chemistry*, vol. 278, no. 45, pp. 44708–44718, 2003.

[69] J. M. Silverman and N. E. Reiner, "Exosomes and other microvesicles in infection biology: organelles with unanticipated phenotypes," *Cellular Microbiology*, vol. 13, no. 1, pp. 1–9, 2011.

[70] B. M. Babior, "NADPH oxidase: an update," *Blood*, vol. 93, no. 5, pp. 1464–1476, 1999.

[71] M. Blos, U. Schleicher, F. J. Soares Rocha, U. Meißner, M. Röllinghoff, and C. Bogdan, "Organ-specific and stage-dependent control of Leishmania major infection by inducible nitric oxide synthase and phagocyte NADPH oxidase," *European Journal of Immunology*, vol. 33, no. 5, pp. 1224–1234, 2003.

[72] H. W. Murray, "Cell-mediated immune response in experimental visceral leishmaniasis. II. Oxygen-dependent killing of intracellular Leishmania donovani amastigotes," *Journal of Immunology*, vol. 129, no. 1, pp. 351–357, 1982.

[73] C. G. Haidaris and P. F. Bonventre, "A role for oxygen-dependent mechanisms in killing of Leishmania donovani tissue forms by activated macrophages," *Journal of Immunology*, vol. 129, no. 2, pp. 850–855, 1982.

[74] Y. Nishizuka, "The molecular heterogeneity of protein kinase C and its implications for cellular regulation," *Nature*, vol. 334, no. 6184, pp. 661–665, 1988.

[75] S. Bhattacharyya, S. Ghosh, P. L. Jhonson, S. K. Bhattacharya, and S. Majumdar, "Immunomodulatory role of interleukin-10 in visceral leishmaniasis: defective activation of protein kinase C-mediated signal transduction events," *Infection and Immunity*, vol. 69, no. 3, pp. 1499–1507, 2001.

[76] A. Descoteaux, G. Matlashewski, and S. J. Turco, "Inhibition of macrophage protein kinase C-mediated protein phosphorylation by Leishmania donovani lipophosphoglycan," *Journal of Immunology*, vol. 149, no. 9, pp. 3008–3015, 1992.

[77] A. Descoteaux and S. J. Turco, "The lipophosphoglycan of Leishmania and macrophage protein kinase C," *Parasitology Today*, vol. 9, no. 12, pp. 468–471, 1993.

[78] K. J. Moore, S. Labrecque, and G. Matlashewski, "Alteration of Leishmania donovani infection levels by selective impairment of macrophage signal transduction," *Journal of Immunology*, vol. 150, no. 10, pp. 4457–4465, 1993.

[79] M. Olivier, R. W. Brownsey, and N. E. Reiner, "Defective stimulus-response coupling in human monocytes infected with Leishmania donovani is associated with altered activation and translocation of protein kinase C," *Proceedings of the National Academy of Sciences of the United States of America*, vol. 89, no. 16, pp. 7481–7485, 1992.

[80] L. Carrera, R. T. Gazzinelli, R. Badolato et al., "Leishmania promastigotes selectively inhibit interleukin 12 induction in bone marrow-derived macrophages from susceptible and resistant mice," *Journal of Experimental Medicine*, vol. 183, no. 2, pp. 515–526, 1996.

[81] G. J. Feng, H. S. Goodridge, M. M. Harnett et al., "Extra- cellular signal-related kinase (ERK) and p38 mitogen-activated protein (MAP) kinases differentially regulate the lipopolysaccharide-mediated induction of inducible nitric oxide synthase and IL-12 in macrophages: Leishmania phosphoglycans subvert macrophage IL-12 production by targeting ERK MAP kinase," *Journal of Immunology*, vol. 163, no. 12, pp. 6403–6412, 1999.

[82] D. E. Hatzigeorgiou, J. Geng, B. Zhu et al., "Lipophosphoglycan from Leishmania suppresses agonist-induced interleukin 1β gene expression in human monocytes via a unique promoter sequence," *Proceedings of the National Academy of Sciences of the United States of America*, vol. 93, no. 25, pp. 14708–14713, 1996.

[83] C. Ropert and R. T. Gazzinelli, "Signaling of immune system cells by glycosylphosphatidylinositol (GPI) anchor and related structures derived from parasitic protozoa," *Current Opinion in Microbiology*, vol. 3, no. 4, pp. 395–403, 2000.

[84] S. D. Tachado, P. Gerold, R. Schwarz, S. Novakovic, M. Mcconville, and L. Schofield, "Signal transduction in macrophages by glycosylphosphatidylinositols of Plasmodium, Trypanosoma, and Leishmania: activation of protein tyrosine kinases and protein kinase C by inositolglycan and diacylglycerol moieties," *Proceedings of the National Academy of Sciences of the United States of America*, vol. 94, no. 8, pp. 4022–4027, 1997.

[85] S. D. Tachado and L. Schofield, "Glycosylphosphatidylinositol toxin of Trypanosoma brucei regulates IL-1α and TNF-α expression in macrophages by protein tyrosine kinase mediated signal transduction," *Biochemical and Biophysical Research Communications*, vol. 205, no. 2, pp. 984–991, 1994.

[86] I. C. Almeida, M. M. Camargo, D. O. Procópio et al., "Highly purified glycosylphosphatidylinositols from Trypanosoma cruzi are potent proinflammatory agents," *EMBO Journal*, vol. 19, no. 7, pp. 1476–1485, 2000.

[87] M. M. Camargo, I. C. Almeida, M. E. S. Pereira, M. A. J. Ferguson, L. R. Travassos, and R. T. Gazzinelli, "Glycosylphosphatidylinositol-anchored mucin-like glycoproteins isolated from Trypanosoma cruzi trypomastigotes initiate the synthesis of proinflammatory cytokines by macrophages," *Journal of Immunology*, vol. 158, no. 12, pp. 5890–5901, 1997.

[88] M. M. Camargo, A. C. Andrade, I. C. Almeida, L. R. Travassos, and R. T. Gazzinelli, "Glycoconjugates isolated from Trypanosoma cruzi but not from Leishmania species membranes trigger nitric oxide synthesis as well as microbicidal activity in IFN-γ-primed macrophages," *Journal of Immunology*, vol. 159, no. 12, pp. 6131–6139, 1997.

[89] S. Magez, B. Stijlemans, M. Radwanska, E. Pays, M. A. J. Ferguson, and P. De Baetselier, "The glycosyl-inositol-phosphate and dimyristoylglycerol moieties of the glycosylphosphatidylinositol anchor of the trypanosome variant-specific surface glycoprotein are distinct macrophage-activating factors," *Journal of Immunology*, vol. 160, no. 4, pp. 1949–1956, 1998.

[90] S. D. Tachado, P. Gerold, M. J. McConville et al., "Glycosylphosphatidylinositol toxin of Plasmodium induces nitric oxide synthase expression in macrophages and vascular endothelial cells by a protein tyrosine kinase-dependent and protein kinase C-dependent signaling pathway," *Journal of Immunology*, vol. 156, no. 5, pp. 1897–1907, 1996.

[91] S. Akira and K. Takeda, "Toll-like receptor signalling," *Nature Reviews Immunology*, vol. 4, no. 7, pp. 499–511, 2004.

[92] T. R. Hawn, A. Ozinsky, D. M. Underhill, F. S. Buckner, S. Akira, and A. Aderem, "Leishmania major activates IL-1α expression in macrophages through a MyD88-dependent pathway," *Microbes and Infection*, vol. 4, no. 8, pp. 763–771, 2002.

[93] M. J. de Veer, J. M. Curtis, T. M. Baldwin et al., "MyD88 is essential for clearance of Leishmania major: possible role for lipophosphoglycan and Toll-like receptor 2 signaling," *European Journal of Immunology*, vol. 33, no. 10, pp. 2822–2831, 2003.

[94] G. Kavoosi, S. K. Ardestani, and A. Kariminia, "The involvement of TLR2 in cytokine and reactive oxygen species (ROS) production by PBMCs in response to Leishmania major phosphoglycans (PGs)," *Parasitology*, vol. 136, no. 10, pp. 1193–1199, 2009.

[95] G. Kavoosi, S. K. Ardestani, A. Kariminia, and M. H. Alimohammadian, "Leishmania major lipophosphoglycan: discrepancy in toll-like receptor signaling," *Experimental Parasitology*, vol. 124, no. 2, pp. 214–218, 2010.

[96] I. Becker, N. Salaiza, M. Aguirre et al., "Leishmania lipophosphoglycan (LPG) activates NK cells through toll-like receptor-2," *Molecular and Biochemical Parasitology*, vol. 130, no. 2, pp. 65–74, 2003.

[97] T. Aebischer, C. L. Bennett, M. Pelizzola et al., "A critical role for lipophosphoglycan in proinflammatory responses of dendritic cells to Leishmania mexicana," *European Journal of Immunology*, vol. 35, no. 2, pp. 476–486, 2005.

[98] A. Ponte-Sucre, D. Heise, and H. Moll, "Leishmania major lipophosphoglycan modulates the phenotype and inhibits migration of murine Langerhans cells," *Immunology*, vol. 104, no. 4, pp. 462–467, 2001.

[99] D. L. Tolson, S. J. Turco, and T. W. Pearson, "Expression of a repeating phosphorylated disaccharide lipophosphoglycan epitope on the surface of macrophages infected with Leishmania donovani," *Infection and Immunity*, vol. 58, no. 11, pp. 3500–3507, 1990.

[100] J. M. Silverman, J. Clos, C. C. De'Oliveira et al., "An exosome-based secretion pathway is responsible for protein export from Leishmania and communication with macrophages," *Journal of Cell Science*, vol. 123, no. 6, pp. 842–852, 2010.

[101] J. M. Silverman, J. Clos, E. Horakova et al., "Leishmania exosomes modulate innate and adaptive immune responses through effects on monocytes and dendritic cells," *Journal of Immunology*, vol. 185, no. 9, pp. 5011–5022, 2010.

[102] B. A. Butcher, S. J. Turco, B. A. Hilty, P. F. Pimentai, M. Panunzio, and D. L. Sacks, "Deficiency in β1,3-galactosyltransferase of a Leishmania major lipophosphoglycan mutant adversely influences the Leishmania-sand fly

interaction," *Journal of Biological Chemistry*, vol. 271, no. 34, pp. 20573–20579, 1996.

[103] A. Descoteaux, B. J. Mengeling, S. M. Beverley, and S. J. Turco, "Leishmania donovani has distinct mannosylphosphoryltransferases for the initiation and elongation phases of lipophosphoglycan repeating unit biosynthesis," *Molecular and Biochemical Parasitology*, vol. 94, no. 1, pp. 27–40, 1998.

[104] D. L. King and S. J. Turco, "A ricin agglutinin-resistant clone of Leishmania donovani deficient in lipophosphoglycan," *Molecular and Biochemical Parasitology*, vol. 28, no. 3, pp. 285–293, 1988.

[105] T. B. McNeeley, D. L. Tolson, T. W. Pearson, and S. J. Turco, "Characterization of Leishmania donovani variant clones using anti-lipophosphoglycan monoclonal antibodies," *Glycobiology*, vol. 1, no. 1, pp. 63–69, 1990.

[106] Y. Sterkers, L. Lachaud, L. Crobu, P. Bastien, and M. Pagès, "FISH analysis reveals aneuploidy and continual generation of chromosomal mosaicism in Leishmania major," *Cellular Microbiology*, vol. 13, no. 2, pp. 274–283, 2011.

[107] N. S. Akopyants, N. Kimblin, N. Secundino et al., "Demonstration of genetic exchange during cyclical development of Leishmania in the sand fly vector," *Science*, vol. 324, no. 5924, pp. 265–268, 2009.

[108] S. M. Beverley and S. J. Turco, "Lipophosphoglycan (LPG) and the identification of virulence genes in the protozoan parasite Leishmania," *Trends in Microbiology*, vol. 6, no. 1, pp. 35–40, 1998.

[109] A. Descoteaux, Y. Luo, S. J. Turco, and S. M. Beverley, "A specialized pathway affecting virulence glycoconjugates of Leishmania," *Science*, vol. 269, no. 5232, pp. 1869–1872, 1995.

[110] K. A. Ryan, L. A. Garraway, A. Descoteaux, S. J. Turco, and S. M. Beverley, "Isolation of virulence genes directing surface glycosyl-phosphatidylinositol synthesis by functional complementation of Leishmania," *Proceedings of the National Academy of Sciences of the United States of America*, vol. 90, no. 18, pp. 8609–8613, 1993.

[111] K. Zhang, T. Barron, S. J. Turco, and S. M. Beverley, "The LPG1 gene family of Leishmania major," *Molecular and Biochemical Parasitology*, vol. 136, no. 1, pp. 11–23, 2004.

[112] T. Ilg, "Lipophosphoglycan is not required for infection of macrophages or mice by Leishmania mexicana," *EMBO Journal*, vol. 19, no. 9, pp. 1953–1962, 2000.

[113] G. F. Späth, L. Epstein, B. Leader et al., "Lipophosphoglycan is a virulence factor distinct from related glycoconjugates in the protozoan parasite Leishmania major," *Proceedings of the National Academy of Sciences of the United States of America*, vol. 97, no. 16, pp. 9258–9263, 2000.

[114] U. Gaur, M. Showalter, S. Hickerson et al., "Leishmania donovani lacking the Golgi GDP-Man transporter LPG2 exhibit attenuated virulence in mammalian hosts," *Experimental Parasitology*, vol. 122, no. 3, pp. 182–191, 2009.

[115] T. Ilg, M. Demar, and D. Harbecke, "Phosphoglycan repeat-deficient Leishmania mexicana parasites remain infectious to macrophages and mice," *Journal of Biological Chemistry*, vol. 276, no. 7, pp. 4988–4997, 2001.

[116] K. Hong, D. Ma, S. M. Beverley, and S. J. Turco, "The Leishmania GDP-mannose transporter is an autonomous, multispecific, hexameric complex of LPG2 subunits," *Biochemistry*, vol. 39, no. 8, pp. 2013–2022, 2000.

[117] D. Ma, D. G. Russell, S. M. Beverley, and S. J. Turco, "Golgi GDP-mannose uptake requires leishmania LPG2: a member

of a eukaryotic family of putative nucleotide-sugar transporters," *Journal of Biological Chemistry*, vol. 272, no. 6, pp. 3799–3805, 1997.

[118] H. Segawa, R. P. Soares, M. Kawakita, S. M. Beverley, and S. J. Turco, "Reconstitution of GDP-mannose transport activity with purified Leishmania LPG2 protein in liposomes," *Journal of Biological Chemistry*, vol. 280, no. 3, pp. 2028–2035, 2005.

[119] C. E. Caffaro and C. B. Hirschberg, "Nucleotide sugar transporters of the Golgi apparatus: from basic science to diseases," *Accounts of Chemical Research*, vol. 39, no. 11, pp. 805–812, 2006.

[120] D. C. Turnock and M. A. J. Ferguson, "Sugar nucleotide pools of Trypanosoma brucei, Trypanosoma cruzi, and Leishmania major," *Eukaryotic Cell*, vol. 6, no. 8, pp. 1450–1463, 2007.

[121] J. E. Uzonna, G. F. Späth, S. M. Beverley, and P. Scott, "Vaccination with phosphoglycan-deficient Leishmania major protects highly susceptible mice from virulent challenge without inducing a strong Th1 response," *Journal of Immunology*, vol. 172, no. 6, pp. 3793–3797, 2004.

[122] D. Liu, C. Kebaier, N. Pakpour et al., "Leishmania major phosphoglycans influence the host early immune response by modulating dendritic cell functions," *Infection and Immunity*, vol. 77, no. 8, pp. 3272–3283, 2009.

[123] A. A. Capul, S. Hickerson, T. Barron, S. J. Turco, and S. M. Beverley, "Comparisons of mutants lacking the golgi UDP-galactose or GDP-mannose transporters establish that phosphoglycans are important for promastigote but not amastigote virulence in Leishmania major," *Infection and Immunity*, vol. 75, no. 9, pp. 4629–4637, 2007.

5

Reactive Oxygen Species and Nitric Oxide in Cutaneous Leishmaniasis

Maria Fátima Horta,[1] **Bárbara Pinheiro Mendes,**[1] **Eric Henrique Roma,**[1]
Fátima Soares Motta Noronha,[2] **Juan Pereira Macêdo,**[1] **Luciana Souza Oliveira,**[1]
Myrian Morato Duarte,[2] **and Leda Quercia Vieira**[1,3]

[1] *Departamento de Bioquímica e Imunologia, Instituto de Ciências Biológicas, Universidade Federal de Minas Gerais,*
31270-901 Belo Horizonte, MG, Brazil
[2] *Departamento de Microbiologia, Instituto de Ciências Biológicas, Universidade Federal de Minas Gerais,*
31270-901 Belo Horizonte, MG, Brazil
[3] *Núcleo de Pesquisas em Ciências Biológicas (NUPEB), Instituto de Ciências Biológicas e Exatas, Universidade Federal de*
Ouro Preto, Morro do Cruzeiro, 35400-000 Ouro Preto, MG, Brazil

Correspondence should be addressed to Leda Quercia Vieira, lqvieira@icb.ufmg.br

Academic Editor: Dario Zamboni

Cutaneous leishmaniasis affects millions of people around the world. Several species of *Leishmania* infect mouse strains, and murine models closely reproduce the cutaneous lesions caused by the parasite in humans. Mouse models have enabled studies on the pathogenesis and effector mechanisms of host resistance to infection. Here, we review the role of nitric oxide (NO), reactive oxygen species (ROS), and peroxynitrite ($ONOO^-$) in the control of parasites by macrophages, which are both the host cells and the effector cells. We also discuss the role of neutrophil-derived oxygen and nitrogen reactive species during infection with *Leishmania*. We emphasize the role of these cells in the outcome of leishmaniasis early after infection, before the adaptive T_h-cell immune response.

1. Introduction

More than 20 *Leishmania* species cause leishmaniasis in people with different genetic backgrounds and general states of health. Further, the diversity of clinical manifestations, epidemiology, and immunopathology makes leishmaniasis a complex disease to study. Clinical manifestations include ulcerative skin lesions, destructive mucosal inflammation, and disseminated visceral infection (kala azar). Morbidity includes disfigurement and disability. However, some features are shared by all forms of infection by these protozoan parasites: parasitism is persistent, tissue macrophages are the main parasitized cell, and the host immune response defines the outcome of the disease [1].

Cutaneous leishmaniasis is caused by several species of the genus *Leishmania*, including *L. major, L. tropica, L. aethiopica, L. mexicana, L. braziliensis, L. guyanensis, L. panamensis, L. peruviana,* and *L. amazonensis.* The *Leishmania* genus is divided in two subgenera, *Leishmania* and *Viannia.* In the subgenus *Leishmania, L. amazonensis, L. mexicana* (complex *L. mexicana*), and *L. major* (complex *L. major*) are by far the most studied species that cause cutaneous leishmaniasis. The subgenus *Viannia* comprises two important species that cause cutaneous leishmaniasis, *L. guyanensis* (complex *L. guyanensis*) and *L. braziliensis* (complex *L. braziliensis*) [2, 3].

The promastigote stage of the parasite lives in the gut of sandflies (*Phlebotomus* in the Old World and *Lutzomyia* in the New World) [4]. In the insect gut, *Leishmania* promastigotes develop into metacyclic (infective) forms and enter the vertebrate host when female sandflies take a blood meal. In the vertebrate host, phagocytic cells ingest the metacyclic promastigotes that, inside the phagolysosome, differentiate into the amastigote form and replicate. The amastigotes rupture the macrophage and proceed to infect other macrophages in the tissue, and, if unchecked by the immune system, they will replicate indefinitely. The parasites rely on macrophages for successful replication, although they can also be taken up by neutrophils [5, 6] and dendritic cells [7]. *Leishmania* do not enter cells actively; thus, they are macrophage obligatory parasites, and the mechanism of entrance is accepted to be phagocytosis [7]. The exit of parasites from the macrophage is less clear. It is becoming apparent that the release of intracellular pathogens is not simply a consequence of a physical or metabolic burden imposed on the host cell, but rather of particular exit strategies governed by the microorganisms (reviewed in [8]). In *Leishmania*, parasite-derived pore-forming cytolysins, which we call leishporin, may be involved [8–13]. The life cycle of *Leishmania* is complete when sandflies feed on infected hosts, ingesting infected cells.

Although the immune response induced by infection with *Leishmania* has been the subject of many investigations, the mechanisms that underlie host resistance and pathogenesis in leishmaniasis are not entirely understood. During the late 80s and early 90s, the discovery of two distinct subpopulations of CD4+ T helper cells based on their cytokine production, Th1 and Th2 [14], finally explained resistance and susceptibility to *L. major* in the murine model. The resistance of C57BL/6 and the susceptibility of BALB/c mice were shown to be the result of the development of a Th1 or Th2 response, respectively. IFN-γ produced by Th1 cells induces the expression of inducible nitric oxide synthase (iNOS or NOS2) by macrophages. This enzyme catalyzes the oxidation of the guanidino nitrogen of L-arginine to produce nitric oxide (NO), which kills the parasite. In contrast, the Th2 response not only activates macrophages to produce arginase (by the action of IL-4, IL-13, and IL-10), which competes with iNOS for the same substrate, but also inhibits the ability to produce NO [15–19]. For some time, Th1 cells and NO were thought to be the sole protagonists of mouse resistance to leishmaniasis, until other reports (referred below) showed that the polarization of the response to Th1 or to Th2 does not explain host resistance or susceptibility to all species of *Leishmania* and does not occur in all host/parasite combinations. Hence, infection with *L. amazonensis* is an example of the still controversial nature of protective immunity in mice. The disease caused in C57BL/6 mice by *L. amazonensis*, for instance, appears to depend on Th1 cells [20], and lesions in C3HeB/FeJ mice do not heal after induction of a Th1 response during chronic infection [21]. However, Th1 cells help mice control *L. amazonensis* infection established by promastigotes, but not by amastigotes [22], and a Th1 response elicited by *L. major* confers resistance in C3HeB/FeJ and C57BL/6 mice to *L. amazonensis* challenge [23, 24]. Likewise, the lack of resistance of C57BL/10 to *L. amazonensis* infection [25] and of BALB/c to *L. mexicana* [26] does not correlate with the presence of a typical Th2 response, suggesting that susceptibility to these species of *Leishmania* is due to a failure to mount a Th1 response, rather than the presence of a Th2 response. Conversely, the resistance of BALB/c to *L. braziliensis* appears to be due to the absence of a Th2

response rather than to the presence of a Th1 response [27]. The inconsistency of the pattern protection/Th1 and pathogenesis/Th2 to all species of *Leishmania* was recently reviewed [28].

Indeed, except for a few references [29–31], innate immunity has largely been overlooked with respect to the mechanism of host resistance to *Leishmania* infection. Dendritic cells, macrophages, and neutrophils, along with their early-produced cytokines and reactive nitrogen and oxygen species, have not been spotlighted as effector cells during the initial stages of infection. Even the leishmanicidal competence of macrophages has mostly been described as a T-cell-dependent event, even though inducers of NO are available very early after infection, namely, type 1 interferons (IFN-α and IFN-β) and type 2 (IFN-γ) interferons. While IFNs-α and -β have been shown to be secreted by macrophages [32], IFN-γ is produced by NK cells [16, 30, 33, 34] and possibly by γ/δ T cells [35], NKT cells [35], or even macrophages [36, 37], although the latter is still controversial [38]. More recently, however, innate immunity effector cells have been suggested to be coparticipants in the maintenance or elimination of the parasites, acting in the early stages of infection in the absence of a T_h-cell response.

In this paper, we highlight the participation of both NO and reactive oxygen species (ROS) in the resistance and pathogenesis of cutaneous leishmaniasis. We first address the fate of promastigotes in the initial phase of the infection, discussing the role of these leishmanicidal molecules in eliminating part of the parasite burden while the adaptive response is still absent (innate immunity). We also discuss the role of these molecules at later phases of the disease, when T_h cells are available (adaptive immunity). In both circumstances, we emphasize the differences among the various *Leishmania* species and mouse strains. The mechanisms that *Leishmania* utilize to evade killing by NO and ROS have been the subject of a recent review and will not be discussed here [39].

2. ROS and NO

Neutrophils and macrophages produce ROS in response to phagocytosis and ligands of pattern recognition receptors (PRRs). The patterns recognized by PRRs can be either of pathogenic origin (pathogen-associated molecular patterns (PAMPs)) or induced by danger patterns (damage-associated molecular patterns (DAMPs)) that signal tissue damage, which are generally hidden from PRRs, such as ATP [40–42]. Moreover, endothelial activation can also induce ROS production by neutrophils [43]. In response to these signals, nicotinamide adenine dinucleotide phosphate- (NADPH-) dependent phagocyte oxidase (Nox2, also known as phox or gp91^{phox}) is assembled, and superoxide is produced from molecular oxygen [44, 45]. Superoxide may be dismutated into hydrogen peroxide, which can, in turn, generate hydroxyl radicals and other ROS. Macrophages produce ROS in higher quantities than neutrophils [43, 46, 47].

NO is also produced by neutrophils and macrophages in response to IFN-γ and a second signal provided by a

PAMP ligand or TNF-α. iNOS expression is induced by these signals. iNOS promotes the oxidation of the guanidino nitrogen of L-arginine, resulting in the production of NO and citrulline [47].

In activated macrophages, superoxide and NO are produced in nearly equimolar quantities and generate peroxynitrite ($ONOO^-$), a free radical that is also highly toxic to pathogens [48].

3. First Encounters—The Neutrophils

As early as 30 seconds after exposure of C57BL/6 mice to L. major through the bite of infected sandflies or needle inoculation of promastigotes, the injected area is infiltrated by neutrophils, which has been elegantly visualized by two-photon intravital microscopy [49]. Recruited neutrophils readily phagocytose promastigotes, which remain viable, although it is not known to what extent parasites are taken up or survive. In fact, it has been reported that during the first 24 h, most parasites are localized extracellularly and can be taken up later by macrophages [49]. The above report showed that parasites taken up by the early neutrophil migration are kept alive inside these cells and do not suffer from oxidative stress. However, another study showed that at later time points, neutrophils might play a role in parasite attrition [50], and, within 2 days, parasites inside neutrophils show a wide variation in their morphology from healthy to completely destroyed forms [50]. Killing of intracellular parasites has been identified by severe signs of damage, such as aggregated cytoplasm and extended vacuolization or complete lysis [50], indicating that neutrophils can act as parasite killers within the first few days of infection. Neutrophils act through an array of microbicidal mechanisms, of which the ability to produce NO [51] and ROS [52] are the most studied in leishmaniasis. Indeed, L. major has been shown to induce NO production by mouse neutrophils in vitro [53] and to stimulate the respiratory burst in mouse [54], rabbit [55], and human [56] neutrophils. Another study, however, showed that L. major failed to induce a respiratory burst in human neutrophils, and L. major-containing phagosomes did not colocalize with granules involved in superoxide production [57]. However, work by Peters et al. [49] has very eloquently shown that there is no oxidative stress within the first hours of infection.

Inflammatory neutrophils harvested from BALB/c mice four hours after i.p. infection with L. major harbor more parasites than C57BL/6 cells, which, in turn, produce considerably higher amounts of NO than BALB/c in response to L. major and IFN-γ [53]. In agreement with these data, we have shown that neutrophils from uninfected C57BL/6 mice express much more iNOS and produce more NO than cells from BALB/c mice when stimulated with IFN-γ in vitro, indicating that the ability of these cells to be activated to produce NO is inherent to each strain. These data suggest that NO produced by neutrophils may help to control infection with L. major in very early disease stages. In vitro, however, iNOS expression and NO production can be inhibited in neutrophils from both mouse strains by live,

but not dead, promastigotes of L. major (our unpublished results).

In BALB/c mice, an iron-induced oxidative burst appears to prevent the growth of L. major, protecting the animals from developing the typical large lesions. This oxidative burst has mainly been attributed to neutrophils [58, 59]. However, C57BL/6 resistance and BALB/c susceptibility inversely correlate with the ability of their neutrophils to generate ROS since BALB/c neutrophils produce more ROS than C57BL/6 neutrophils when stimulated with phorbol myristate acetate (PMA). L. major has also been shown to inhibit a PMA-induced respiratory burst in neutrophils from both strains of mice (our unpublished results).

Interestingly, the rapid recruitment of neutrophils to L. major-induced lesions was previously reported to follow different kinetics in susceptible BALB/c and resistant C57BL/6 mice, which might account for these opposite outcomes. In susceptible mice, almost 100% of the initial cellular infiltrate is composed of neutrophils, half of which is replaced by mononuclear phagocytes in 2-3 days. Neutrophils comprise the other half of the cellular infiltrate for at least 12 days after infection. In contrast, in resistant mice, only about 60% of the initial cellular infiltrate is composed of neutrophils, and the number of these cells drastically decreases to only 1-2% at later time points. In resistant mice, mononuclear phagocytes predominate at later time points, comprising more than 70–80% of the cells [49]. Notably, infection with L. major also results in the differentiation of distinct neutrophil populations in BALB/c and C57BL/6 mice. The parasite induces CD49d expression in BALB/c, but not in C57BL/6, neutrophils. The levels of Toll-like receptor (TLR) 2, TLR7, and TLR9 mRNA are significantly higher in C57BL/6 cells than in BALB/c cells. Moreover, C57BL/6, but not BALB/c, neutrophils secrete biologically active IL-12p70 and IL-10. BALB/c neutrophils instead transcribe and secrete high levels of IL-12p40, which forms homodimers with inhibitory activity. In C57BL/6 mice, neutrophils may constitute one of the earliest sources of IL-12, while in BALB/c mice, secretion of IL-12p40 may contribute to impaired early IL-12 signaling [53]. Furthermore, C57BL/6 neutrophils were found to release 2-3-fold more elastase than BALB/c cells, which contributes to parasite killing through activation of TLR4 [60]. These distinct neutrophil phenotypes may thus influence both the early resistance or susceptibility and the development of an L. major-specific immune response. The role of these different populations of neutrophils on resistance to parasites through reactive nitrogen and oxygen species production deserves further investigation.

Recently, the interaction of neutrophils and macrophages has been investigated in vitro (reviewed in [5]). Dead neutrophils from C57BL/6 mice can activate infected macrophages to kill L. major. In this system, activation is mediated by the induction of TNF-α by neutrophil elastase, but NO is not involved in parasite killing. Rather, superoxide is partially responsible for parasite killing, as evidenced by the partial inhibition of this effect when catalase was added to this in vitro system [60, 61]. The same results were obtained with dead human neutrophils and L. amazonensis-infected human macrophages [62]. In another study, live murine neutrophils induced killing of *L. braziliensis*, but not *L. major*, by infected macrophages. Superoxide production was detected in this system, and killing of parasites was inhibited by *N*-acetylcysteine, a superoxide scavenger. Killing of *L. braziliensis* by macrophages cocultured with live neutrophils was also independent of NO [63]. Neutrophil-induced killing of *L. amazonensis* by macrophages from resistant and susceptible mouse strains was also described and is mediated by neutrophil elastase, TNF-α, and platelet-activating factor (PAF), but not by NO or reactive oxygen species [64].

In response to pathogens, neutrophils may release the so-called neutrophil extracellular traps (NETs), which are fibrous nets composed of decondensed chromatin, histones, and granule antimicrobial proteins that trap and kill microbes extracellularly [65, 66]. NETs extruded by human neutrophils cultured *in vitro* were shown to kill *L. amazonensis, L. major*, and *L. chagasi*. These NETs were found in lesions from patients. Killing of parasites was found to be mediated mainly by histones [67]. Importantly, NET formation is defective in patients suffering from chronic granulomatous disease, who lack Nox2 activity [68]. In fact, reactive oxygen species are required to initiate NETs. Oxidative stress ruptures neutrophil elastase and mieloperoxidase-containing granules, and neutrophil elastase binds to chromatin and cleaves histones, a reaction that is further enhanced by mieloperoxidase, independent of its enzymatic activity. This enzyme promotes chromatin decondensation, which culminates in NET release due to cellular rupture [69]. The molecular mechanism linking ROS production to chromatin decondensation and binding to antimicrobial proteins is still unknown.

Although several *in vivo* studies have addressed the role of neutrophils during infection with *L. major*, their function in resistance to the parasite is not totally understood and is still a subject of debate. Due to the heterogeneous models used to study the role of neutrophils in experimental leishmaniasis, it is still unknown whether these cells have a protective or pathogenic role. Like other immune responses in murine models, the neutrophil function appears to depend on the species and even the strain of *Leishmania* and the genetic background of mice used as host (thoroughly reviewed in [70]). Hence, even less clear is the *in vivo* role of reactive oxygen and nitrogen species from neutrophils in *Leishmania* resistance or pathology caused by the parasites. However, *in vitro* evidence suggests that ROS from neutrophils are involved in killing of the parasite, suggesting that ROS may be important for resistance to parasites early in infection.

4. Latecomers—The Macrophages

Like neutrophils, macrophages are microbicidal cells that are able to produce NO and ROS [47]. Paradoxically, these cells are also the long-term host cell for *Leishmania*. In experimental leishmaniasis, macrophages are as crucial for parasite survival as for its elimination [71]. The role played by these cells depends on the type of activation and the vulnerability of the parasite to the effector mechanisms.

The mechanism by which macrophages are responsible for resistance to *Leishmania* was first characterized by *in vitro* experiments using murine macrophages infected with *L. major*. In this model, killing of parasites is dependent on the activation of macrophages by IFN-γ and a second signal that triggers TNF-α. This signal is given by amastigotes, promastigotes, or parasite-derived glycoinositolphospholipids (GIPLs) and lipophosphoglycan (LPG), but not by killed cells or cellular lysates. Once these two signals are present, iNOS is induced and NO is produced [72–74]. The clear role of NO in killing *L. major* was established by pharmacological inhibition of the production of NO *in vitro* and by the observation of a higher susceptibility of iNOS knockout mice to infections with *L. major* [16, 74–76]. It was further confirmed by the inability of macrophages from iNOS knockout mice to be activated and kill *L. major* by IFN-γ [77]. Hence, NO clearly has a crucial role in killing of *L. major* by IFN-γ-activated macrophages.

During *L. amazonensis* infection, IFN-γ and TNF-α are not produced at high levels as in *L. major* infection [25, 78]. Therefore, infection of *L. major*-resistant mice with *L. amazonensis* leads to chronic lesions and inefficient control of parasites at the site of infection. IFN-γ-activated macrophages from CBA/J mice infected with either *L. major* or *L. amazonensis* are able to kill the former, but not the latter. When very high concentrations of NO were generated *in vitro*, axenic *L. amazonensis* amastigotes succumbed. In addition, macrophages infected with *L. amazonensis* produce less TNF-α when compared to those infected with *L. major* [79]. However, macrophages infected with either *L. major* or *L. amazonensis* produce similar levels of NO (measured as nitrite in culture supernatants) and express similar levels of iNOS message when activated with IFN-γ [79]. Corroborating these data, we found lower levels of TNF (α and β were measured collectively) from *L. amazonensis*-infected macrophages from C57BL/10 mice than from *L. major*-infected macrophages (Figure 1(a)). In addition, two days after infection in the hind footpad, popliteal lymph node cells from C3H/HeN, C57BL/10 (mouse strains resistant to *L. major*), and BALB/c mice produced more TNF *ex vivo* when infected with *L. major* than with *L. amazonensis* (Figure 1(b)). Interestingly, *L. amazonensis*-infected CBA/J macrophages also produce less reactive oxygen species than *L. major*-infected cells [79], which could be, in part, responsible for the different abilities of macrophages to kill these two species of *Leishmania*. The mechanism by which *L. amazonensis* resists killing remains unknown.

Even more intriguing is the observation that low doses of IFN-γ actually promote amastigote growth within macrophages [22]. In accordance with this observation, at later stages of infection, increased amounts of NO were found in the more susceptible BALB/c mice than in C57BL/6 mice infected with *L. amazonensis* as lesions progressed and parasites expanded because C57BL/6 mice partially control lesions and parasite growth [80].

IFN-γ-activated macrophages represent the host-parasite interaction in which T cells are already producing a large amount of this cytokine. During the first 2 days after infection with *L. major*, nearly all macrophages recruited to

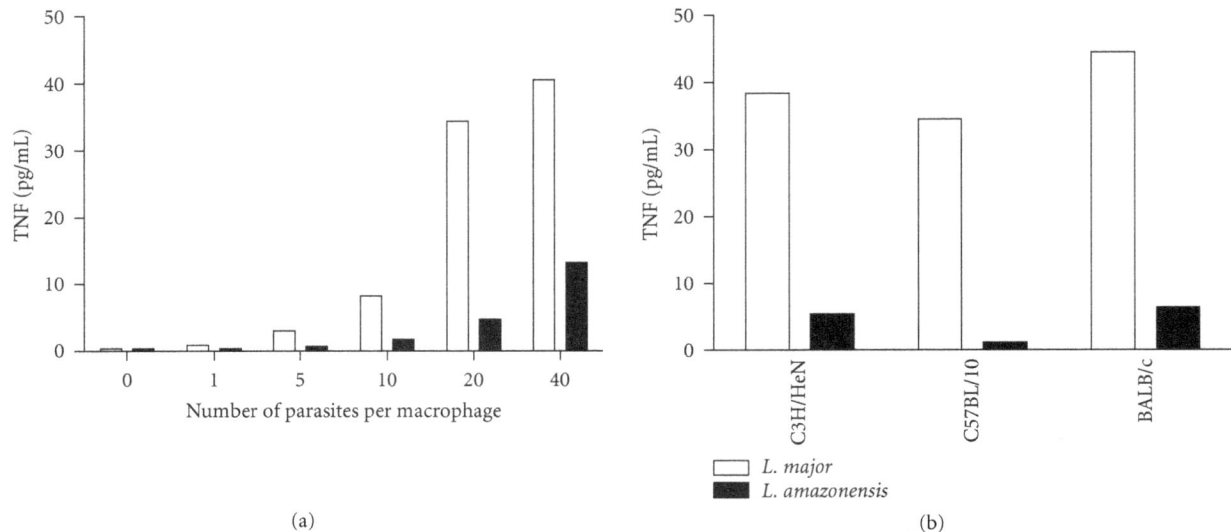

FIGURE 1: Infection with *L. major* induces more TNF than infection with *L. amazonensis*. (a) TNF production by inflammatory macrophages from C57BL/10, mice infected *in vitro* with *L. major* or *L. amazonensis*. (b) Production of TNF *ex vivo* by lymph node cells from C3H/HeN, C57BL/10 and BALB/c mice infected with *L. major* or *L. amazonensis*, 2 days after infection. A biological assay that does not distinguish between TNF-α or TNF-β was used in these experiments. These are representative experiments of more than five performed experiments (L. Q. Vieira and P. Scott, unpublished).

the site of infection contain phagocytosed parasites, both in C57BL/6 and in BALB/c mice. However, the percentage of cells (mostly neutrophils and mononuclear phagocytes) containing intact parasites in BALB/c mice is higher than that in C57BL/6 cells (mostly mononuclear cells), and the elimination of parasites from the site of infection is higher in resistant mice [50]. This suggests that parasites may also be killed by tissue mononuclear cells well before the onset of a T-cell response. Whether this killing is mediated by reactive oxygen and nitrogen species remains unknown.

Isolated macrophages from C57BL/6 mice produce more NO than macrophages from susceptible strains when stimulated with IFN-γ [81–84], TNF-α [81, 85], or LPS [83, 85–89]. This is an interesting but poorly explored aspect of the murine models of resistance/susceptibility to microbial infections, which is clearly independent of the development of an adaptive Th1 or Th2 response. Mills et al. [90] systematically tested this observation and generalized it to other strains of mice. They showed that macrophages from strains that are typical Th1 responders (termed M-1) or typical Th2 responders (termed M-2) differ qualitatively in their ability to be activated, as measured by their arginine metabolic programs. M-2 macrophages from BALB/c mice (prototypes of Th2 responders) stimulated with a particular concentration of LPS not only produce little or no NO, but increase arginine metabolism to ornithine. In contrast, M-1 cells from C57BL/6 mice (prototypes of Th1 responders) generate a strong NO and citrulline response and appear to decrease their production of ornithine.

We investigated the molecular basis of the differential production of NO by macrophages from mice with resistant or susceptible phenotypes to *L. major* by *in vitro* stimulation with IFN-γ and LPS. We have shown that M-1 macrophages

show a remarkably strong expression of the enzyme iNOS upon stimulation when compared to M-2 cells [84]. The accumulation of iNOS mRNA is also higher in M-1 cells. Interestingly, however, we found that the accumulation of the iNOS protein is more dramatic than the accumulation of iNOS mRNA. The accumulation of both iNOS mRNA and protein is not a consequence of a higher stability of the molecule. The data showed that iNOS gene expression is differentially regulated in M-1 and M-2 macrophages and suggested that it is transcribed and translated at different rates in these two types of cells [84]. Recent results from our group indicate that the higher iNOS expression in M-1 macrophages may be multifactorial and may be regulated by higher levels of TNF-α, IL-12, and IFN-β (unpublished data).

The intrinsic differential sensitivity to IFN-γ and LPS of M-1 or M-2 cells has led to two important observations regarding the *in vivo* infection.

(1) Small amounts of IFN-γ (from NK, NKT, or γ/δ T cells) or other pathogen-derived inducers may induce M-1, but not M-2 cells, to kill the pathogen through NO, before T cells differentiate into the IFN-γ-Th1 subpopulation. In fact, larger numbers of *L. major* are found in iNOS-deficient macrophages than in wild-type macrophages 72 hours after infection, indicating that some NO is produced by macrophages that have not been activated with IFN-γ and that NO, even if not detectable, exerts some control of parasite growth [75, 77]. Further evidence of a NO-dependent T_h-cell-independent mechanism was obtained when resting human macrophages were infected with NO-susceptible and NO-resistant *L. amazonensis* and *L. braziliensis* isolates and selected *in vitro* with increasing concentrations of $NaNO_2$: NO-resistant parasites grew better in resting macrophages than the NO-susceptible isolates [91].

(2) Activated M-1 and M-2 cells can distinctly affect subsequent production of Th1-dominant or Th2-dominant cytokines (IFN-γ or TGF-β1, resp.), positioning macrophages as key performers in directing the Th1 or Th2 outcome. M-1 and M-2 macrophages differentially influence the Th lymphocyte response, and how macrophages are stimulated determines the route that Th responses will take [90]. These observations indicate that macrophages may contribute to the outcome of an immune response through mechanisms other than by acting as established NO-producing cells and that their role in determining the resistant/susceptible phenotype in mice may be significant. M-1 macrophages not only can mount an early (innate) resistance, but also can consolidate the status of resistance by favoring a Th1 adaptive response.

In addition to NO, ROS are considered to be a major macrophage effector mechanism induced by IFN-γ to control infections. Upon bacteria or other pathogen engulfment by a phagocytic cell, ROS are rapidly produced by NADPH oxidase, an enzymatic complex comprised of membrane bound (p22^{phox} and gp91^{phox}) and cytosolic (p40^{phox}, p47^{phox}, p67^{phox}, and Rac-1/2) proteins [45, 92], which may be assembled after TLR stimulation by bacterial products via MyD88-dependent p38 MAPK activation [93].

Macrophages [54, 76] and neutrophils [54] produce ROS in response to *Leishmania in vitro*. Killing of *L. major* by IFN-γ-activated macrophages is dependent on NO production, but not on the production of superoxide or peroxynitrite [76]. Lesions in Nox2 knockout mice [94] (Nox2 mice are genetically deficient in the NADPH-dependent phagocyte oxidase. These mice were originally described as a model for chronic granulomatous disease and are more susceptible to bacterial infection, and neither neutrophils nor macrophages present respiratory burst oxidase activity [94].) infected with *L. major* are similar to those in wild-type C57BL6 mice. Nox2 knockout mice control *L. major* at the site of infection at early time points, but display an unexpected reactivation of *L. major* infection after long periods of observation (more than 200 days of infection). Further, they show deficient control of parasite replication in draining lymph nodes and spleens, suggesting that Nox2 is important for the control of *L. major in vivo* at later times of infection by preventing visceralization [54]. The participation of ROS in killing of *L. amazonensis* by mouse [95, 96] or human [97] macrophages has been reported. Our preliminary data suggest that macrophages from Nox2 knockout mice behave similarly to macrophages from wild-type mice when infected with *L. amazonensis*. Moreover, similar to infection with *L. major*, Nox2 knockout mice control parasites at the site of infection as well as wild-type mice (Figure 2). Surprisingly, at earlier times of infection, lesions are larger in Nox2 knockout mice, and, at later times of infection, they become smaller than in wild-type mice (Figure 2(a)). This indicates that the differences in Ros activity on macrophage behavior at different stages of infection may be due to differences in the inflammatory infiltrate. The contradictions between the *in vitro* evidence for a role for ROS in resistance to *L. amazonensis* and *in vivo* data remain to be explained.

Although BALB/c mice are the prototype model of susceptibility to most species of *Leishmania* (such as *L. major* and *L. amazonensis*), *L. braziliensis* [27, 98] and *L. guyanensis* [99] do not cause large skin lesions in this mouse strain. Our studies using *L. guyanensis* have shown that BALB/c mice develop minor or no lesions, do not enable parasite replication, and do not die of the infection. In addition, *L. guyanensis* [99] and *L. braziliensis* [100], unlike *L. amazonensis*, fail to survive within nonactivated peritoneal macrophages *in vitro*. *In vitro* infection of BALB/c macrophages with *L. guyanensis* does not activate the production of NO; instead, it activates a respiratory burst that is exceptionally higher than that activated by infection with *L. amazonensis*. We have further shown that the production of ROS is responsible for the elimination of *L. guyanensis* by macrophages. We have also shown that *L. guyanensis* amastigotes die inside BALB/c macrophages through an apoptosis-like process mediated by parasite-induced ROS [99]. These findings demonstrate an important killing mechanism of *L. guyanensis* amastigotes. ROS are probably involved in resistance to infection with this species because mice that are unable to activate the respiratory burst by the regular administration of apocynin, an inhibitor of NADPH oxidase, do not control the infection as in untreated animals (our preliminary results). Together, our results suggest that the elimination of *L. guyanensis in vivo* may occur in early infection due to ROS production, before the development of an adaptive T$_h$ response.

There is evidence that peroxynitrite (ONOO$^-$) is not involved in the killing of *L. major* [54, 76], but the role of this important oxidant has not been thoroughly explored. In contrast, the production of nitric oxide and ONOO$^-$ has been shown during infection with *L. amazonensis* in BALB/c (more susceptible to infection) and C57BL/6 mice (more resistant to infection). The production of nitric oxide *in vivo* was detected as the nitrosyl hemoglobin complex by electron paramagnetic resonance analysis of nitrosyl hemoglobin in blood drawn from mice and in infected footpads at several time points, and ONOO$^-$ formation was inferred from immunodetection of nitrotyrosine [101, 102]. C57BL/6 mice presented higher levels of nitrosyl complexes than BALB/c mice at 6 weeks of infection, at which point lesions became chronic in this partially resistant mouse strain. Nitrosyl complexes increased in BALB/c mice, which was dependent on lesion size. iNOS and nitrotyrosine-containing complexes colocalize in lesion macrophages from both mouse strains, and the most probable agent of protein nitration is ONOO$^-$ [102]. Peroxynitrite killed *L. amazonensis* axenic amastigotes *in vitro* more efficiently than nitric oxide [102]. The authors proposed that in the susceptible mouse strain, ONOO$^-$ is involved in tissue damage. It is possible that the delayed production of ONOO$^-$ impairs the capacity of BALB/c mice to control *L. amazonensis*. Treatment of C57BL/6 mice with Tempol, a stable cyclic nitroxide radical that protects cells from damage due to oxidative stress, promoted larger lesions, parasite growth, and lower levels of nitric oxide products and nitrotyrosine [103]. Albeit transient, this effect of Tempol provides further evidence that ONOO$^-$ is involved in the control of *L. amazonensis in vivo*.

(a)

(b)

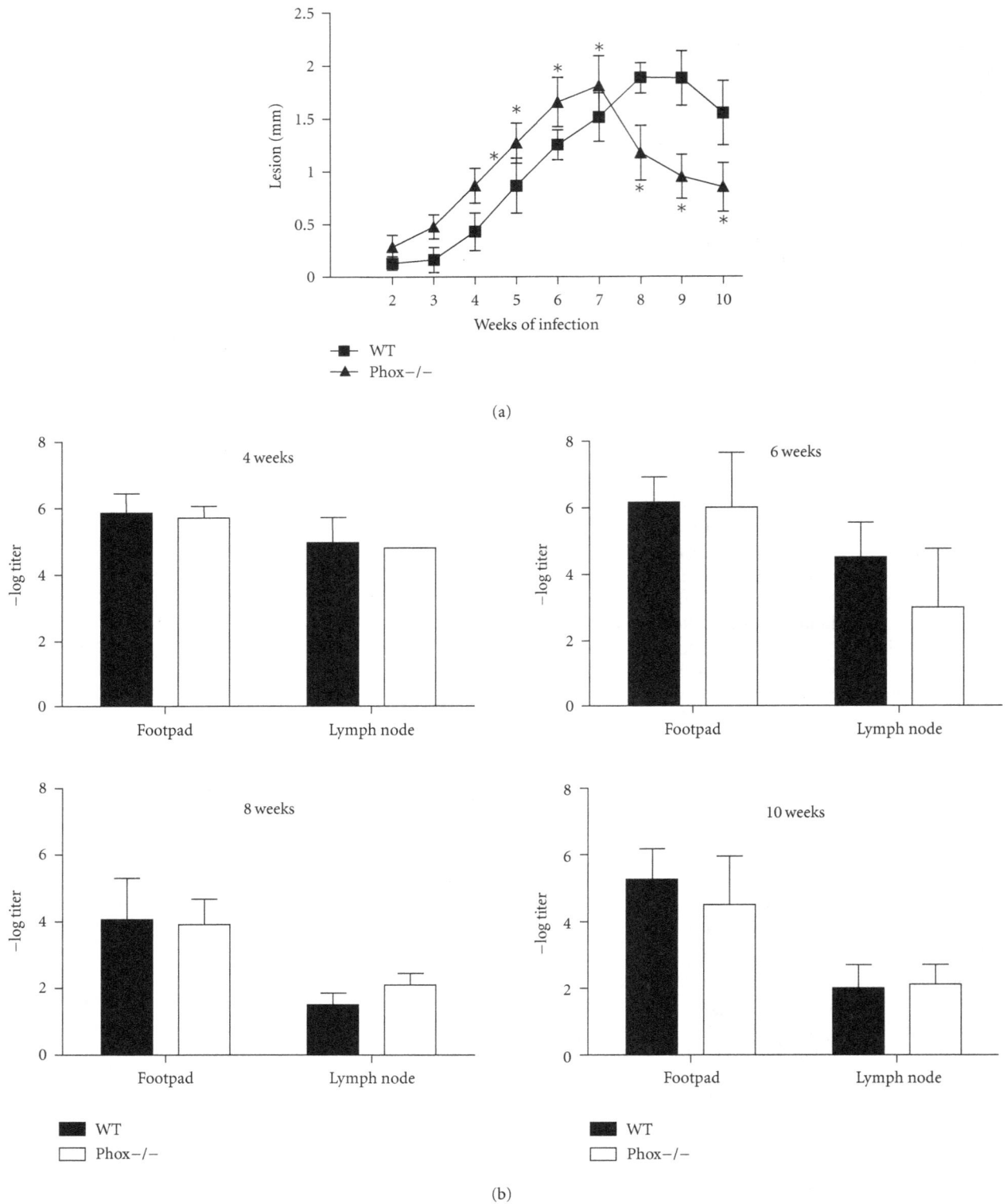

FIGURE 2: Course of infection with *L. amazonensis* in wild-type C57BL/6 and Nox2 knockout mice (a) and parasite quantitation using a limiting dilution analysis (b). *indicates statistical difference by Student's *t* test, *P* < 0.05 (E. H. Roma and J. P. Macedo, unpublished).

5. Concluding Remarks

The role of reactive oxygen and nitrogen species in killing of *Leishmania* has been the subject of many studies, but there is still much that is not understood. The following questions remain: why do some species of parasites resist oxidative stress? Why do cells that can kill parasites with reactive species harbor live parasites? Is there some attrition when parasites enter neutrophils and macrophages? What is the role of peroxinitrite? What is the reason for the differences in the oxidative responses among different species of parasites? What is the role of reactive oxygen and nitrogen species in the inflammatory response? Collective efforts to fully comprehend the mechanisms that produce disease upon infection with *Leishmania* and the strategies hosts employ to avoid them have been made. However, leishmaniasis persists without safe treatments or effective vaccines. Perhaps the recent attention paid to components of the innate immune system might help to unravel this complex parasite-host relationship.

Acknowledgments

L. Q. Vieira is a member of INCT de Processos Redox em Biomedicina-Redoxoma (CNPq/FAPESP/ CAPES 573530/2008-4). The authors' results were supported by grants from Conselho Nacional de Desenvolvimento Científico e Tecnológico (CNPq), Coordenação de Aperfeiçoamento de Pessoal de Nível Superior (CAPES), and Fundação de Amparo à Pesquisa do Estado de Minas Gerais (FAPEMIG). The authors are grateful to FUNDEP for expert administration of grants. M. F. Horta and L. Q. Vieira are CNPq research fellows.

References

[1] H. W. Murray, J. D. Berman, C. R. Davies, and N. G. Saravia, "Advances in leishmaniasis," *Lancet*, vol. 366, no. 9496, pp. 1561–1577, 2005.

[2] L. Kedzierski, Y. Zhu, and E. Handman, "*Leishmania* vaccines: progress and problems," *Parasitology*, vol. 133, no. 2, pp. S87–S112, 2006.

[3] A. L. Bañuls, M. Hide, and F. Prugnolle, "*Leishmania* and the Leishmaniases: a parasite genetic update and advances in taxonomy, epidemiology and pathogenicity in humans," *Advances in Parasitology*, vol. 64, pp. 1–109, 2007.

[4] R. Killick-Kendrick, "The life-cycle of *Leishmania* in the sandfly with special reference to the form infective to the vertebrate host," *Annales de Parasitologie Humaine et Comparee*, vol. 65, no. 1, supplement 1, pp. 37–42, 1990.

[5] A. A. Filardy, D. R. Pires, and G. A. Dosreis, "Macrophages and neutrophils cooperate in immune responses to *Leishmania* infection," *Cellular and Molecular Life Sciences*, vol. 68, no. 11, pp. 1863–1870, 2011.

[6] N. C. Peters and D. L. Sacks, "The impact of vector-mediated neutrophil recruitment on cutaneous leishmaniasis," *Cellular Microbiology*, vol. 11, no. 9, pp. 1290–1296, 2009.

[7] D. Sacks and N. Noben-Trauth, "The immunology of susceptibility and resistance to *Leishmania major* in mice," *Nature Reviews Immunology*, vol. 2, no. 11, pp. 845–858, 2002.

[8] M. F. Horta, "Pore-forming proteins in pathogenic protozoan parasites," *Trends in Microbiology*, vol. 5, no. 9, pp. 363–366, 1997.

[9] F. S. M. Noronha, F. J. Ramalho-Pinto, and M. F. Horta, "Cytolytic activity in the genus *Leishmania*: involvement of a putative pore-forming protein," *Infection and Immunity*, vol. 64, no. 10, pp. 3975–3982, 1996.

[10] F. S. M. Noronha, J. S. Cruz, P. S. L. Beirão, and M. F. Horta, "Macrophage damage by *Leishmania amazonensis* cytolysin: evidence of pore formation on cell membrane," *Infection and Immunity*, vol. 68, no. 8, pp. 4578–4584, 2000.

[11] F. R. Almeida-Campos and M. F. Horta, "Proteolytic activation of leishporin: evidence that *Leishmania amazonensis* and *Leishmania guyanensis* have distinct inactive forms," *Molecular and Biochemical Parasitology*, vol. 111, no. 2, pp. 363–375, 2000.

[12] T. Castro-Gomes, F. R. Almeida-Campos, C. E. Calzavara-Silva, R. A. da Silva, F. Frézard, and M. F. Horta, "Membrane binding requirements for the cytolytic activity of *Leishmania amazonensis* leishporin," *FEBS Letters*, vol. 583, no. 19, pp. 3209–3214, 2009.

[13] F. S. Noronha, F. J. Ramalho-Pinto, and M. F. Horta, "Identification of a putative pore-forming hemolysin active at acid pH in *Leishmania amazonensis*," *Brazilian Journal of Medical and Biological Research*, vol. 27, no. 2, pp. 477–482, 1994.

[14] T. R. Mosmann, H. Cherwinski, and M. W. Bond, "Two types of murine helper T cell clone. I. Definition according to profiles of lymphokine activities and secreted proteins," *Journal of Immunology*, vol. 136, no. 7, pp. 2348–2357, 1986.

[15] I. M. Corraliza, G. Soler, K. Eichmann, and M. Modolell, "Arginase induction by suppressors of nitric oxide synthesis (IL-4, IL-10 and PGE2) in murine bone-marrow-derived macrophages," *Biochemical and Biophysical Research Communications*, vol. 206, no. 2, pp. 667–673, 1995.

[16] F. Y. Liew, S. Millott, C. Parkinson, R. M. J. Palmer, and S. Moncada, "Macrophage killing of *Leishmania* parasite in vivo is mediated by nitric oxide from L-arginine," *Journal of Immunology*, vol. 144, no. 12, pp. 4794–4797, 1990.

[17] R. M. Locksley, F. P. Heinzel, M. D. Sadick, B. J. Holaday, and K. D. Gardner, "Murine cutaneous leishmaniasis: susceptibility correlates with differential expansion of helper T-cell subsets," *Annales de l'Institut Pasteur. Immunology*, vol. 138, no. 5, pp. 744–749, 1987.

[18] P. Scott, P. Natovitz, R. L. Coffman, E. Pearce, and A. Sher, "Immunoregulation of cutaneous leishmaniasis. T cell lines that transfer protective immunity or exacerbation belong to different T helper subsets and respond to distinct parasite antigens," *Journal of Experimental Medicine*, vol. 168, no. 5, pp. 1675–1684, 1988.

[19] S. J. Green, M. S. Meltzer, J. B. Hibbs, and C. A. Nacy, "Activated macrophages destroy intracellular *Leishmania major* amastigotes by an L-arginine-dependent killing mechanism," *Journal of Immunology*, vol. 144, no. 1, pp. 278–283, 1990.

[20] L. Soong, C. H. Chang, J. Sun et al., "Role of CD4+ T cells in pathogenesis associated with *Leishmania amazonensis* infection," *Journal of Immunology*, vol. 158, no. 11, pp. 5374–5383, 1997.

[21] Y. F. Vanloubbeeck, A. E. Ramer, F. Jie, and D. E. Jones, "CD4+ Th1 cells induced by dendritic cell-based immunotherapy in mice chronically infected with *Leishmania amazonensis* do not promote healing," *Infection and Immunity*, vol. 72, no. 8, pp. 4455–4463, 2004.

[22] H. Qi, J. Ji, N. Wanasen, and L. Soong, "Enhanced replication of *Leishmania amazonensis* amastigotes in gamma interferon-stimulated murine macrophages: implications for the pathogenesis of cutaneous leishmaniasis," *Infection and Immunity*, vol. 72, no. 2, pp. 988–995, 2004.

[23] Y. Vanloubbeeck and D. E. Jones, "Protection of C3HeB/FeJ mice against *Leishmania amazonensis* challenge after previous *Leishmania major* infection," *American Journal of Tropical Medicine and Hygiene*, vol. 71, no. 4, pp. 407–411, 2004.

[24] C. Z. González-Lombana, H. C. Santiago, J. P. Macedo et al., "Early infection with *Leishmania major* restrains pathogenic response to *Leishmania amazonensis* and parasite growth," *Acta Tropica*, vol. 106, no. 1, pp. 27–38, 2008.

[25] L. C. C. Afonso and P. Scott, "Immune responses associated with susceptibility of C57BL/10 mice to *Leishmania amazonensis*," *Infection and Immunity*, vol. 61, no. 7, pp. 2952–2959, 1993.

[26] O. Guevara-Mendoza, C. Une, P. Franceschi Carreira, and A. Örn, "Experimental infection of Balb/c mice with *Leishmania panamensis* and *Leishmania mexicana* induction of early IFN-γ but not IL-4 is associated with the development of cutaneous lesions," *Scandinavian Journal of Immunology*, vol. 46, no. 1, pp. 35–40, 1997.

[27] G. K. DeKrey, H. C. Lima, and R. G. Titus, "Analysis of the immune responses of mice to infection with *Leishmania braziliensis*," *Infection and Immunity*, vol. 66, no. 2, pp. 827–829, 1998.

[28] D. Mahon-Pratt and J. Alexander, "Does the *Leishmania major* paradigm of pathogenesis and protection hold for New World cutaneous leishmaniases or the visceral disease?" *Immunological Reviews*, vol. 201, pp. 206–224, 2004.

[29] J. L. Stafford, N. F. Neumann, and M. Belosevic, "Macrophage-mediated innate host defense against protozoan parasites," *Critical Reviews in Microbiology*, vol. 28, no. 3, pp. 187–248, 2002.

[30] T. M. Scharton and P. Scott, "Natural killer cells are a source of interferon γ that drives differentiation of CD4+ T cell subsets and induces early resistance to *Leishmania major* in mice," *Journal of Experimental Medicine*, vol. 178, no. 2, pp. 567–577, 1993.

[31] R. Birnbaum and N. Craft, "Innate immunity and *Leishmania* vaccination strategies," *Dermatologic Clinics*, vol. 29, no. 1, pp. 89–102, 2011.

[32] A. Diefenbach, H. Schindler, N. Donhauser et al., "Type 1 interferon (IFNα/β) and type 2 nitric oxide synthase regulate the innate immune response to a protozoan parasite," *Immunity*, vol. 8, no. 1, pp. 77–87, 1998.

[33] G. J. Bancroft, R. D. Schreiber, and G. C. Bosma, "A T cell-independent mechanism of macrophage activation by interferon-γ," *Journal of Immunology*, vol. 139, no. 4, pp. 1104–1107, 1987.

[34] K. N. Dileepan, K. M. Simpson, and D. J. Stechschulte, "Modulation of macrophage superoxide-induced cytochrome c reduction by mast cells," *Journal of Laboratory and Clinical Medicine*, vol. 113, no. 5, pp. 577–585, 1989.

[35] J. Hao, X. Wu, S. Xia et al., "Current progress in γδ T-cell biology," *Cellular and Molecular Immunology*, vol. 7, no. 6, pp. 409–413, 2010.

[36] M. Munder, M. Mallo, K. Eichmann, and M. Modolell, "Murine macrophages secrete interferon γ upon combined stimulation with interleukin (IL)-12 and IL-18: a novel pathway of autocrine macrophage activation," *Journal of Experimental Medicine*, vol. 187, no. 12, pp. 2103–2108, 1998.

[37] P. Puddu, M. Carollo, I. Pietraforte et al., "IL-2 induces expression and secretion of IFN-γ in murine peritoneal macrophages," *Journal of Leukocyte Biology*, vol. 78, no. 3, pp. 686–695, 2005.

[38] S. Fujii, S. Motohashi, K. Shimizu, T. Nakayama, Y. Yoshiga, and M. Taniguchi, "Adjuvant activity mediated by iNKT cells," *Seminars in Immunology*, vol. 22, no. 2, pp. 97–102, 2010.

[39] T. van Assche, M. Deschacht, R. A.I. Da Luz, L. Maes, and P. Cos, "*Leishmania*-macrophage interactions: insights into the redox biology," *Free Radical Biology and Medicine*, vol. 51, no. 2, pp. 337–351, 2011.

[40] E. Ogier-Denis, S. B. Ogier-Denis, and A. Vandewalle, "NOX enzymes and Toll-like receptor signaling," *Seminars in Immunopathology*, vol. 30, no. 3, pp. 291–300, 2008.

[41] G. Y. Chen and G. Nuñez, "Sterile inflammation: sensing and reacting to damage," *Nature Reviews Immunology*, vol. 10, no. 12, pp. 826–837, 2010.

[42] S. Carta, P. Castellani, L. Delfino, S. Tassi, R. Venè, and A. Rubartelli, "DAMPs and inflammatory processes: the role of redox in the different outcomes," *Journal of Leukocyte Biology*, vol. 86, no. 3, pp. 549–555, 2009.

[43] W. M. Nauseef, "How human neutrophils kill and degrade microbes: an integrated view," *Immunological Reviews*, vol. 219, no. 1, pp. 88–102, 2007.

[44] B. M. Babior, "NADPH oxidase: an update," *Blood*, vol. 93, no. 5, pp. 1464–1476, 1999.

[45] A. Mizrahi, Y. Berdichevsky, Y. Ugolev et al., "Assembly of the phagocyte NADPH oxidase complex: chimeric constructs derived from the cytosolic components as tools for exploring structure-function relationships," *Journal of Leukocyte Biology*, vol. 79, no. 5, pp. 881–895, 2006.

[46] M. B. Hampton, A. J. Kettle, and C. C. Winterbourn, "Inside the neutrophil phagosome: oxidants, myeloperoxidase, and bacterial killing," *Blood*, vol. 92, no. 9, pp. 3007–3017, 1998.

[47] C. Nathan and M. U. Shiloh, "Reactive oxygen and nitrogen intermediates in the relationship between mammalian hosts and microbial pathogens," *Proceedings of the National Academy of Sciences of the United States of America*, vol. 97, no. 16, pp. 8841–8848, 2000.

[48] R. Radi, G. Peluffo, M. N. Alvarez, M. Naviliat, and A. Cayota, "Unraveling peroxynitrite formation in biological systems," *Free Radical Biology and Medicine*, vol. 30, no. 5, pp. 463–488, 2001.

[49] N. C. Peters, J. G. Egen, N. Secundino et al., "In vivo imaging reveals an essential role for neutrophils in leishmaniasis transmitted by sand flies," *Science*, vol. 321, no. 5891, pp. 970–974, 2008.

[50] W. J. Beil, G. Meinardus-Hager, D. C. Neugebauer, and C. Sorg, "Differences in the onset of the inflammatory response to cutaneous leishmaniasis in resistant and susceptible mice," *Journal of Leukocyte Biology*, vol. 52, no. 2, pp. 135–142, 1992.

[51] A. K. Nussler and T. R. Billiar, "Inflammation, immunoregulation, and inducible nitric oxide synthase," *Journal of Leukocyte Biology*, vol. 54, no. 2, pp. 171–178, 1993.

[52] A. W. Segal, "How neutrophils kill microbes," *Annual Review of Immunology*, vol. 23, pp. 197–223, 2005.

[53] M. Charmoy, R. Megnekou, C. Allenbach et al., "*Leishmania major* induces distinct neutrophil phenotypes in mice that are resistant or susceptible to infection," *Journal of Leukocyte Biology*, vol. 82, no. 2, pp. 288–299, 2007.

[54] M. Blos, U. Schleicher, F. J. S. Rocha, U. Meißner, M. Röllinghoff, and C. Bogdan, "Organ-specific and stage-dependent control of *Leishmania major* infection by

inducible nitric oxide synthase and phagocyte NADPH oxidase," *European Journal of Immunology*, vol. 33, no. 5, pp. 1224–1234, 2003.

[55] D. J. Mallinson, J. M. Lackie, and G. H. Coombs, "The oxidative response of rabbit peritoneal neutrophils to leishmanias and other trypanosomatids," *International Journal for Parasitology*, vol. 19, no. 6, pp. 639–645, 1989.

[56] H. Laufs, K. Müller, J. Fleischer et al., "Intracellular survival of *Leishmania major* in neutrophil granulocytes after uptake in the absence of heat-labile serum factors," *Infection and Immunity*, vol. 70, no. 2, pp. 826–835, 2002.

[57] F. Mollinedo, H. Janssen, J. De La Iglesia-Vicente, J. A. Villa-Pulgarin, and J. Calafat, "Selective fusion of azurophilic granules with *Leishmania*-containing phagosomes in human neutrophils," *Journal of Biological Chemistry*, vol. 285, no. 45, pp. 34528–34536, 2010.

[58] S. Bisti, G. Konidou, J. Boelaert, M. Lebastard, and K. Soteriadou, "The prevention of the growth of *Leishmania major* progeny in BALB/c iron-loaded mice: a process coupled to increased oxidative burst, the amplitude and duration of which depend on initial parasite developmental stage and dose," *Microbes and Infection*, vol. 8, no. 6, pp. 1464–1472, 2006.

[59] S. Bisti and K. Soteriadou, "Is the reactive oxygen species-dependent-NF-κB activation observed in iron-loaded BALB/c mice a key process preventing growth of *Leishmania major* progeny and tissue-damage?" *Microbes and Infection*, vol. 8, no. 6, pp. 1473–1482, 2006.

[60] F. L. Ribeiro-Gomes, M. C. A. Moniz-de-Souza, M. S. Alexandre-Moreira et al., "Neutrophils activate macrophages for intracellular killing of *Leishmania major* through recruitment of TLR4 by neutrophil elastase," *Journal of Immunology*, vol. 179, no. 6, pp. 3988–3994, 2007.

[61] F. L. Ribeiro-Gomes, A. C. Otero, N. A. Gomes et al., "Macrophage Interactions with Neutrophils Regulate *Leishmania major* Infection," *Journal of Immunology*, vol. 172, no. 7, pp. 4454–4462, 2004.

[62] L. Afonso, V. M. Borges, H. Cruz et al., "Interactions with apoptotic but not with necrotic neutrophils increase parasite burden in human macrophages infected with *Leishmania amazonensis*," *Journal of Leukocyte Biology*, vol. 84, no. 2, pp. 389–396, 2008.

[63] F. O. Novais, R. C. Santiago, A. Bafica et al., "Neutrophils and macrophages cooperate in host resistance against *Leishmania braziliensis* infection," *Journal of Immunology*, vol. 183, no. 12, pp. 8088–8098, 2009.

[64] E. V. de Souza Carmo, S. Katz, and C. L. Barbiéri, "Neutrophils reduce the parasite burden in *Leishmania (Leishmania) amazonensis* macrophages," *PLoS ONE*, vol. 5, no. 11, Article ID e13815, 2010.

[65] V. Brinkmann, U. Reichard, C. Goosmann et al., "Neutrophil extracellular traps kill bacteria," *Science*, vol. 303, no. 5663, pp. 1532–1535, 2004.

[66] V. Papayannopoulos and A. Zychlinsky, "NETs: a new strategy for using old weapons," *Trends in Immunology*, vol. 30, no. 11, pp. 513–521, 2009.

[67] A. B. Guimarães-Costa, M. T. C. Nascimento, G. S. Froment et al., "*Leishmania amazonensis* promastigotes induce and are killed by neutrophil extracellular traps," *Proceedings of the National Academy of Sciences of the United States of America*, vol. 106, no. 16, pp. 6748–6753, 2009.

[68] T. A. Fuchs, U. Abed, C. Goosmann et al., "Novel cell death program leads to neutrophil extracellular traps," *Journal of Cell Biology*, vol. 176, no. 2, pp. 231–241, 2007.

[69] V. Papayannopoulos, K. D. Metzler, A. Hakkim, and A. Zychlinsky, "Neutrophil elastase and myeloperoxidase regulate the formation of neutrophil extracellular traps," *Journal of Cell Biology*, vol. 191, no. 3, pp. 677–691, 2010.

[70] U. Ritter, F. Frischknecht, and G. van Zandbergen, "Are neutrophils important host cells for *Leishmania* parasites?" *Trends in Parasitology*, vol. 25, no. 11, pp. 505–510, 2009.

[71] C. Bogdan and M. Röllinghoff, "The immune response to *Leishmania*: mechanisms of parasite control and evasion," *International Journal for Parasitology*, vol. 28, no. 1, pp. 121–134, 1998.

[72] A. Piani, T. Ilg, A. G. Elefanty, J. Curtis, and E. Handman, "*Leishmania major* proteophosphoglycan is expressed by amastigotes and has an immunomodulatory effect on macrophage function," *Microbes and Infection*, vol. 1, no. 8, pp. 589–599, 1999.

[73] G. Kavoosi, S. K. Ardestani, A. Kariminia, and Z. Tavakoli, "Production of nitric oxide by murine macrophages induced by lipophosphoglycan of *Leishmania major*," *The Korean Journal of Parasitology*, vol. 44, no. 1, pp. 35–41, 2006.

[74] S. J. Green, R. M. Crawford, J. T. Hockmeyer, M. S. Meltzer, and C. A. Nacy, "*Leishmania major* amastigotes initiate the L-arginine-dependent killing mechanism in IFN-γ-stimulated macrophages by induction of tumor necrosis factor-α1," *Journal of Immunology*, vol. 145, no. 12, pp. 4290–4297, 1990.

[75] X. Q. Wei, I. G. Charles, A. Smith et al., "Altered immune responses in mice lacking inducible nitric oxide synthase," *Nature*, vol. 375, no. 6530, pp. 408–411, 1995.

[76] J. Assreuy, F. Q. Cunha, M. Epperlein et al., "Production of nitric oxide and superoxide by activated macrophages and killing of *Leishmania major*," *European Journal of Immunology*, vol. 24, no. 3, pp. 672–676, 1994.

[77] X. Q. Wei, B. P. Leung, W. Niedbala et al., "Altered immune responses and susceptibility to *Leishmania major* and Staphylococcus aureus infection in IL-18-deficient mice," *Journal of Immunology*, vol. 163, no. 5, pp. 2821–2828, 1999.

[78] J. Ji, J. Sun, and L. Soong, "Impaired expression of inflammatory cytokines and chemokines at early stages of infection with *Leishmania amazonensis*," *Infection and Immunity*, vol. 71, no. 8, pp. 4278–4288, 2003.

[79] I. N. Gomes, A. F. Calabrich, R. S. Tavares, J. Wietzerbin, L. A. Rodrigues De Freitas, and P. S. Tavares Veras, "Differential properties of CBA/J mononuclear phagocytes recovered from an inflammatory site and probed with two different species of *Leishmania*," *Microbes and Infection*, vol. 5, no. 4, pp. 251–260, 2003.

[80] S. Giorgio, E. Linares, H. Ischiropoulos, F. J. Von Zuben, A. Yamada, and O. Augusto, "In vivo formation of electron paramagnetic resonance-detectable nitric oxide and of nitrotyrosine is not impaired during murine leishmaniasis," *Infection and Immunity*, vol. 66, no. 2, pp. 807–814, 1998.

[81] F. Y. Liew, Y. Li, D. Moss, C. Parkinson, M. V. Rogers, and S. Moncada, "Resistance to *Leishmania major* infection correlates with the induction of nitric oxide synthase in murine macrophages," *European Journal of Immunology*, vol. 21, no. 12, pp. 3009–3014, 1991.

[82] S. Stenger, H. Thüring, M. Röllinghoff, and C. Bogdan, "Tissue expression of inducible nitric oxide synthase is closely associated with resistance to *Leishmania major*," *Journal of Experimental Medicine*, vol. 180, no. 3, pp. 783–793, 1994.

[83] Z. Zidek, D. Frankova, and M. Boubelik, "Genetic variation in in-vitro cytokine-induced production of nitric oxide by

murine peritoneal macrophages," *Pharmacogenetics*, vol. 10, no. 6, pp. 493–501, 2000.

[84] J. L. Santos, A. A. Andrade, A. A. M. Dias et al., "Differential sensitivity of C57BL/6 (M-1) and BALB/c (M-2) macrophages to the stimuli of IFN-γ/LPS for the production of NO: correlation with iNOS mRNA and protein expression," *Journal of Interferon and Cytokine Research*, vol. 26, no. 9, pp. 682–688, 2006.

[85] J. Lima-Santos, I. S. Jardim, S. M. Teixeira, and M. F. Horta, "Diffferential induction of nitric oxide synthase in C57BL/6 and BALB/c macrophages by IFN-gamma and TNF-alpha or LPS," *Abstracts of the Brazilian Society of Immunolgy Meeting*, vol. 24, p. 122, 1999.

[86] K. N. Dileepan, J. C. Page, Y. Li, and D. J. Stechschulte, "Direct activation of murine peritoneal macrophages for nitric oxide production and tumor cell killing by interferon-γ," *Journal of Interferon and Cytokine Research*, vol. 15, no. 5, pp. 387–394, 1995.

[87] I. P. Oswald, S. Afroun, D. Bray, J. F. Petit, and G. Lemaire, "Low response of BALB/c macrophages to priming and activating signals," *Journal of Leukocyte Biology*, vol. 52, no. 3, pp. 315–322, 1992.

[88] I. S. Jardim, J. L. Santos, and M. F. Horta, "Differential production of nitric oxide by murine macrophages from *Leishmania* resistant and susceptible mice strains," *Memórias do Instituto Oswaldo Cruz*, vol. 94, supplement, p. 197, 1999.

[89] I. S. Jardim, J. L. Santos, M. F. Horta, and F. J. Ramalho-Pinto, "Inhibition of the production of nitric oxide impairs cytotoxicity of macrophages to *Leishmania amazonensis*," *Memórias do Instituto Oswaldo Cruz*, vol. 92, supplement, p. 217, 2011.

[90] C. D. Mills, K. Kincaid, J. M. Alt, M. J. Heilman, and A. M. Hill, "M-1/M-2 macrophages and the Th1/Th2 paradigm," *Journal of Immunology*, vol. 164, no. 12, pp. 6166–6173, 2000.

[91] A. Giudice, I. Camada, P. T. G. Leopoldo et al., "Resistance of *Leishmania (Leishmania) amazonensis* and *Leishmania (Viannia) braziliensis* to nitric oxide correlates with disease severity in Tegumentary Leishmaniasis," *BMC Infectious Diseases*, vol. 7, article no. 7, 2007.

[92] B. M. Babior, "NADPH oxidase," *Current Opinion in Immunology*, vol. 16, no. 1, pp. 42–47, 2004.

[93] S. Laroux, X. Romero, L. Wetzler, P. Engel, and C. Terhorst, "Cutting edge: MyD88 controls phagocyte NADPH oxidase function and killing of gram-negative bacteria," *Journal of Immunology*, vol. 175, no. 9, pp. 5596–5600, 2005.

[94] J. D. Pollock, D. A. Williams, M. A. C. Gifford et al., "Mouse model of X-linked chronic granulomatous disease, an inherited defect in phagocyte superoxide production," *Nature Genetics*, vol. 9, no. 2, pp. 202–209, 1995.

[95] A. Degrossoli, W. W. Arrais-Silva, M. C. Colhone, F. R. Gadelha, P. P. Joazeiro, and S. Giorgio, "The influence of low oxygen on macrophage response to *Leishmania* infection," *Scandinavian Journal of Immunology*, vol. 74, no. 2, pp. 165–175, 2011.

[96] R. M. Mukbel, C. Patten Jr., K. Gibson, M. Ghosh, C. Petersen, and D. E. Jones, "Macrophage killing of *Leishmania amazonensis* amastigotes requires both nitric oxide and superoxide," *American Journal of Tropical Medicine and Hygiene*, vol. 76, no. 4, pp. 669–675, 2007.

[97] R. Khouri, A. Bafica, M. D. P. P. Silva et al., "IFN-β impairs superoxide-dependent parasite killing in human macrophages: evidence for a deleterious role of SOD1 in cutaneous leishmaniasis," *Journal of Immunology*, vol. 182, no. 4, pp. 2525–2531, 2009.

[98] R. A. Neal and C. Hale, "A comparative study of susceptibility of inbred and outbred mouse strains compared with hamsters to infection with New World cutaneous leishmaniases," *Parasitology*, vol. 87, no. 1, pp. 7–13, 1983.

[99] J. Sousa-Franco, E. Araujo-Mendes, I. Silva-Jardim et al., "Infection-induced respiratory burst in BALB/c macrophages kills *Leishmania guyanensis* amastigotes through apoptosis: possible involvement in resistance to cutaneous leishmaniasis," *Microbes and Infection*, vol. 8, no. 2, pp. 390–400, 2006.

[100] P. Scott and A. Sher, "A spectrum in the susceptibility of Leishmanial strains to intracellular killing by murine macrophages," *Journal of Immunology*, vol. 136, no. 4, pp. 1461–1466, 1986.

[101] S. Giorgio, E. Linares, M. D. L. Capurro, A. G. De Bianchi, and O. Augusto, "Formation of nitrosyl hemoglobin and nitrotyrosine during murine leishmaniasis," *Photochemistry and Photobiology*, vol. 63, no. 6, pp. 750–754, 1996.

[102] E. Linares, S. Giorgio, R. A. Mortara, C. X. C. Santos, A. T. Yamada, and O. Augusto, "Role of peroxynitrite in macrophage microbicidal mechanisms in vivo revealed by protein nitration and hydroxylation," *Free Radical Biology and Medicine*, vol. 30, no. 11, pp. 1234–1242, 2001.

[103] E. Linares, S. Giorgio, and O. Augusto, "Inhibition of in vivo leishmanicidal mechanisms by tempol: nitric oxide downregulation and oxidant scavenging," *Free Radical Biology and Medicine*, vol. 44, no. 8, pp. 1668–1676, 2008.

6

Host-Parasite Interactions in Some Fish Species

R. A. Khan

Department of Biology, Memorial University of Newfoundland, St. John's, NL, Canada A1B 3X9

Correspondence should be addressed to R. A. Khan, rakhan@mun.ca

Academic Editor: Renato A. Mortara

Host-parasite interactions are complex, compounded by factors that are capable of shifting the balance in either direction. The host's age, behaviour, immunological status, and environmental change can affect the association that is beneficial to the host whereas evasion of the host's immune response favours the parasite. In fish, some infections that induce mortality are age and temperature dependent. Environmental change, especially habitat degradation by anthropogenic pollutants and oceanographic alterations induced by climatic, can influence parasitic-host interaction. The outcome of these associations will hinge on susceptibility and resistance.

1. Introduction

Interaction between hosts and parasites is a complex relationship that can favour one or the other depending on a number of factors. Initially, the parasite attempts to establish itself in the host while the latter resists the infection via its defense mechanisms. Consequently, host susceptibility and resistance will determine whether or not the infection becomes established. In some fish species, the host's age, behaviour, physiological and immunological condition, proximity to shore, location in the water column, and feeding habits could affect the relationship while the parasite's mode of entry, ability to evade its host's defense, nutritional requirements, and living in a site where the immune response is reduced and mimicking its host's protein composition are factors that influence susceptibility and infectivity. There are also environmental variables such as water temperature, crowding, and habitat changes that could affect the interaction. Moreover, this interrelationship between hosts and parasites has evolved some associations resulting in host specificity, latitudinal gradients, and diversity in communities and siblings within one species [1–4]. Other associations can exist in a harmonious compromise whereas host avoidance of the parasite to prevent an infection or evasion by the parasite of the host's immune system can occur. For example, certain song birds in Hawaii avoid *Plasmodium* sp. infections from ground-feeding mosquitoes

by remaining high up in the forest canopy [5]. In contrast, an African trypanosome, *Trypanosoma brucei* complex, sheds its surface coats in response to its host's antibodies [6]. Some parasites have evolved methods, such as circadian rhythms, to maximize transmission to uninfected hosts. Synchrony of peak abundance of microfilaria of *Wucheraria bancrofti* and mature gamonts of *Plasmodium* species in the peripheral blood of humans, when female mosquitoes are actively seeking a blood meal at night, increases infectivity and development in the vector [7]. McCarthy [8] also noted that a hemoflagellate, *T. murmanensis*, was more abundant in blood from the gills of Atlantic cod, *Gadus morhua*, at night when the vector, a marine leech, *Johanssonia arctica*, fed on its piscine hosts than during daylight. The leech, a deep-water (>90 meters) species inhabiting the benthic zone where light is probably negligible, was significantly more abundant than at shallow depths (30 meters) where some light penetrates [9]. Some parasites can also avoid their host's defence mechanisms. Cysticerci of *Taenia solium* which infect the human brain, a site impervious to the immune response, are protected from destruction by the host. Similarly, plerocercoids of the fish tapeworm, *Proteocephalus ambloplitis*, avoid shedding from the intestine of its host by migrating into parenteral sites as the water temperature declines in winter, a period when feeding is reduced and worms voided because of a lack of nutrients, [10]. However, the parasites reenter the intestine as the water temperature increases in spring to mature as adult tapeworms. Some tissue-invading parasites can become encapsulated by fibrous tissue in their hosts as a result of the latter's defence mechanism and appear as cysts. They might remain viable for lengthy periods until acquired by new and/or appropriate definitive hosts. Xenomas of microsporans such as *Loma branchialis* in salmonids, species of *Sarcocystis* in birds and mammals, and cysticercoids of tapeworms as *Taenia saginata* in cattle and *Trichinella spiralis* in pigs use this strategy to infect new hosts following ingestion. Encapsulation can also be viewed as a defence mechanism by the host to curtail tissue damage by some migrating parasites. Mathematical models have been proposed to explain some host-parasite interactions [11]. However, Sures [12] discussed host-parasite interactions based on observations that some intestinal fish parasites accumulate heavy metals and can be useful as diagnostic tools to determine bioavailability. The approach, in this presentation will focus on some factors that influence fish-parasite interactions including disease aspects, climatic change, and environmental pollution.

The immune response to foreign proteins in fish is lower in magnitude compared to mammals. Some of the defense barriers include mucous in the skin and gills, bile, digestive enzymes, and immunological barriers, primarily cellular and antibody responses. Some of these hinge on age of the host and ambient water temperature. A parasite's specificity will also determine if an infection becomes established. For example, a hemoflagellate, *Cryptobia salmositica*, is infective to some salmonid species but *Salvelinus namaycush* exhibits innate resistance [13]. Lytic antibodies were responsible for resistance but these apparently were absent in *S. fontinalis* that was susceptible to the infection. Cellular response is observed when some tissue-invading parasites become encapsulated. The immune response is temperature dependent as larval anisakine nematodes ingested in late autumn-winter, when water temperatures range from 0 to 4 EC, remain free in the tissues of Atlantic cod, *Gadus morhua*, whereas in summer, most become encapsulated (Khan, unpubl. data).

Some environmental factors have a profound effect on several fish-parasite interactions. Generally, ectoparasites differ from endoparasites as the defence mechanisms tend to be reduced externally in fish. The interaction between an ectoparasitic ciliate, *Trichodina murmanica*, and its host, the Atlantic cod, is age and temperature-dependent. Prevalence and abundance of the parasite on the skin of 1-year juvenile fish in nature are extremely low and rare in older fish. However, outbreaks have occurred in fingerling and 1-year fish cultured cod held in over-stocked conditions during winter when water temperature was 0-1°C [14]. The infection and mortality declined with increasing water temperature and were rarely seen during summer at 8–14°C. Infection of Atlantic cod with *Trypanosoma murmanensis* was also age and temperature dependant as mortality was greater in younger than older fish [15]. Moreover, the infection persisted for longer periods (6–8 weeks) at lower (0–2°C) than higher (10–12°C) temperature (Khan, unpubl. data). It is probable that the host's immunity is temperature-dependent or the parasite is adapted to low temperatures as

noted in some subarctic marine leeches [16]. In contrast, a microsporan, *Loma branchialis*, appearing as macroscopic cysts on the gills of fingerling and juvenile cod, caused die-offs only in summer-autumn when water temperatures were high [17]. Die-offs have also occurred in commercial-size cod held in sea pens for market in summer (Barker, unpubl. data). Xenomas resembling tumours, occurred in all the internal organs and moribund fish, in an emaciated condition, succumbed in a matter of weeks. Similar temperature-related die-offs have occurred in cultured juvenile Arctic charr, *Salvelinus alpinus*, only in summer following infection with a myxozoan, *Tetracapsuloides bryosalmonae*, held in earthen ponds [18]. Outbreaks of disease caused by plerocer-coids of a cestode, *Diphyllobothrium dendriticum*, occurred in cultured rainbow trout, *S. namaycush*, after transfer from a hatchery to an embayment only in summer [18]. Host response to the parasites was minimal except for *Loma branchialis* as encapsulation of the xenomas did occur but some ruptured releasing spores that presumably infected other organs and tissues. Influence of water temperature on host-parasite interactions has also been reported previously [19].

Some parasitic crustaceans interacting with their fish hosts can have a profound impact on their health [20]. The pennellid copepod, *Lernaeocera branchialis*, anchors its holdfast into one of the branchial blood vessels causing anemia and mortality depending on the age of the fish and the number of infecting parasites [21]. Mortality was high in juvenile cod about 3 years old but declined with increasing age [21]. Additionally, fish with multiple numbers of parasites were also likely to succumb. However, some fish that shed their parasites previously exhibited complete recovery. Prevalence and abundance of the infection varied with location in coastal Newfoundland [22]. Commercial-size Atlantic cod that were held in sea cages during the summer also succumbed when water temperatures were high but declined during winter [23]. Smith et al. [24] reported that the parasite induced extensive hyperplasia in the gills, intravascular thrombus formation and moderate cellular response in the cardiac and branchial tissues. In contrast, hundreds of larval stages that attach to the gills of the intermediate host, the lumpfish, *Cyclopterus lumpus*, cause no effect [22].

Another gadid fish, the rock cod, *Gadus ogac*, that inhabits inshore embayments off St. Lewis, Labrador (52°22′N, 55°41′W), appeared to tolerate multiple numbers of parasites without exhibiting debility in contrast to Atlantic cod. (Khan, unpubl. data). Young fish were infected with a fewer mean number of parasites than older cod, some with as many as 7 parasites per older host (Table 1). Prevalence of the infection was 84% in 1976 and mean abundance increased with the length of the fish. Although fewer samples were caught in 1986 (as a result of a population decline triggered by climatic changes), there was a slight increase in prevalence but no significant change in mean abundance. Unlike Atlantic cod, the infected rock cod appeared robust and the gills pink to red without any indication of an anaemic condition observed in other gadids [21]. Prevalence of the infection was considerably lower in Atlantic cod captured

TABLE 1: Abundance ($\bar{x} \pm$ s.e.) and prevalence (%) of the copepod, *Lernaeocera branchialis*, on the gills of the rock cod, *Gadus ogac*, sampled in Labrador in 1976 and 1986.

Year	Fish length (cm class)	No. infected/ no. examined	$\bar{x} \pm$ s.e.	%
1976	21–30	9/17	1.8 ± 0.2	53
	31–40	6/12	2.1 ± 0.2	50
	41–50	21/21	3.3 ± 0.4	100
	51–60	24/24	3.1 ± 0.3	100
1986	31–40	2/3	3.5 ± 0.5	67
	41–50	4/4	3.1 ± 0.4	100
	51–60	3/3	3.0 ± 0.4	100

inshore by cod trap in both 1976 (4% of 48 fish) and in 1986 (5% of 36 fish). The rock cod is cold-water-adapted fish that lives inshore beneath the ice in winter and apparently was not affected by oceanographic changes that occurred from the mid-1980s off eastern Canada [25].

Marine hematophagous leeches (Hirudinea: Piscicolidae) exhibit an interesting interrelationship with their fish hosts. Some species of leeches feed on a variety of teleosts while others tend to be host specific. *Johanssonia arctica*, a deep-sea species adapted to subarctic conditions, fed on several species of teleosts while two species of *Malmiana*, *M. scorpii* and *M. brunnea*, were found only on shorthorn (*Myoxocephalus scorpii*) and longhorn sculpins (*M. octodecemspinosus*), respectively, in the NW Atlantic Ocean [26, 27]. Several species of marine leeches attach to the skin of their hosts to feed on blood but others, such as *Oxytonstoma microstoma* and *O. sexoculata*, adhere to the gills and on the angles of the oral cavity, respectively [26, 27]. These previously mentioned leeches remain permanently attached to their hosts, feeding intermittently until maturity, copulation, and cocoon deposition [26, 27]. Others, such as J. *arctica*, *Myzobdella lugubris*, and *Notostomum cyclostomum*, after feeding on fish, detach and reattach to crabs (Crustacea: Decapoda) for transport and deposition of cocoons [9, 16, 28, 29]. Locating fish hosts for a blood meal is increased by the foraging activities of the crabs. Other leeches deposit their cocoons on the egg masses of their piscine hosts or on rocks frequented by fish [16]. Synchronous hatching of larval fish and young leeches ensures that the latter can locate a host for their subsequent blood meals. The quantity of blood extracted by leeches varies considerably depending on size and species [9]. Mace and Davis [30] reported slow growth rate and energy loss in shorthorn sculpins infected with *M. scorpii*. Hematophagous leeches can induce anemia, subcutaneous hemorrhage, and inflammation especially when a heavy infestation occurs [31]. Moreover, leeches can also transmit blood parasites during hematophagy. *J. arctica* is capable of transmitting a trypanosome, *T. murmanensis*, a piroplasm, *Haemohormidium beckeri*, and probably a hemogregarine, *Haemogregarina uncinata* [32–34]. Another genus of hemoflagellates, species of *Trypanoplasma* (*Cryptobia*) is transmitted by leeches [35]. *Trypanoplasma bullocki* has caused mortality in summer flounder, *Paralichthys*

dentatus, populations in the Middle Atlantic Bight [36]. Consequently, interaction between hematophagous leeches and fish can result in stress as a result of blood loss and also in the transmission of pathogenic parasites [32].

Some parasites are known to predispose their hosts for predation by alteration of their behavior as in some carnivore-herbivore interactions. In the three-spine stickleback, *Gasterosteus aculeatus*, a larval cestode, the plerocercoid of *Schistocephalus solidus*, infects the body cavity and can impair swimming. It has been reported that infected fish are more likely to be predated than uninfected sticklebacks [37, 38]. Some fish with distended abdomens have been observed swimming near the surface in ponds in Newfoundland (Khan, unpubl. data). Examination of the fish with swollen abdomens revealed at least two large larvae per host. It is likely that sea gulls, *Larus* spp., that frequented these areas, were feeding on the fish. The Arctic tern (*Sterna paradisaea*) is also a definitive host of *S. solidus*. Nestlings on a small island near Cow Head (49°55′N, 57°53′W), Newfoundland, were fed sticklebacks by the parental birds, and during one wet and cool summer, several nestlings were observed in an emaciated condition. Predation of nestlings by sea gulls was apparent in the area. Examination of nine freshly dead birds revealed cestode larvae in the coelomic cavity of six carcasses ($\bar{x}$ 3.9 ± 1.2/nestling), all exhibiting evidence of hemorrhage in the body cavity and an absence of food in the digestive tract (Khan, unpubl. data). It is likely that the parasite, lacking nutrients in the digestive tract, migrated from this site through the coelomic wall, causing the lesions observed and also predisposed them to predation by foraging gulls.

An unusual difference in the abundance and prevalence of parasites was observed in two populations of landlocked Arctic charr inhabiting different habitats in a pristine deep-water lake in Gander (48°58′N, 54°57′W), Newfoundland (Khan, unpubl. data). One of these, a pale-colored morph, pelagic and living in shallow water, fed primarily on mayfly nymphs, *Heptagenia* spp. (Ephemeroptera: Heptageniidae) and other insect larvae were more parasitized than the dark morph inhabiting a mid-water-benthic zone feeding on macroinvertebrates and fish such as sticklebacks. DNA evidence has revealed that the two populations were distinct and might have been separated a long time previously probably during the postglacial period [39]. Meristic results and colouration have revealed that mixing was rare, with each group occupying different niches. Species diversity, abundance, and prevalence of the parasitic helminth taxa, trematodes, cestodes, and nematodes were significantly greater in the pelagic than in the mid-water-benthic group (Table 2). These results are reminiscent of a hybrid salmonid, the splake (*Salvelinus fontinalis* × *S. namaycush*), a cross between a brook trout (*S. fontinalis*) and lake trout (*S. namaycush*), that was bred to avoid lamprey (*Petromyzon marinus*) predation by inhabiting mid-water rather that the benthic area where lake trout frequented [40]. Fewer parasitic species (23 spp.) were observed in the splake than in the lake trout (75 spp.) [41]. The Arctic charr is a host to several species of metazoan parasites [41]. Factors responsible for the separation of the two Arctic charr populations in Gander

TABLE 2: Comparison of the abundance ($\bar{x} \pm$ s.e.) and prevalence (%) of some metazoan parasites found in the digestive tract of two populations of Arctic charr ($n = 25$) living near the surface and in the midwater benthic zone of Gander Lake, Newfoundland.

Parasite taxa	Parasite species	Fish groups			
		Surface		Midwater/benthic	
		$\bar{x} \pm$ s.e.	%	$\bar{x} \pm$ s.e.	%
Trematoda	*Bunodera lucioperca*	<0.1	12	—	—
	Crepidostomum sp.	<0.1	8	—	—
Cestoidea	*Eubothrium salvalini*	3.2 ± 0.4	100	<0.1	12
	Proteocephalus sp.	<0.1	32	—	—
Nematoda	*Pseudocapillaria salvelini*	2.1 ± 0.2	100	<0.1	8
Acanthocephala	*Echinorhynchus lateralis*	<0.1	20	—	—

Lake remain enigmatic but they suggest an example of parasite paucity resulting from habitat selection.

Oceanographic changes caused by a series of adverse climatic events have also had an impact on host-parasite interactions especially on the abundance and prevalence of metazoan parasites in the digestive tract of Atlantic cod occurring off the coast of Labrador, Canada [42]. Changes in the climate caused water temperatures to decrease resulting in a decline of ocean fish and sea birds [25]. Prior to this time, the abundance of an acanthocephalan parasite, *E. gadi*, was high but, following a chain of cascading events during the mid-1980s, it decreased to extremely low levels [43]. Outmigration of the main food source, the capelin, *Mallotus villosus*, and also the paratenic host of the infection in older Atlantic cod, was the underlying cause [43]. Decline of the abundance of *E. gadi* should favour its fish host as the spines on the proboscis of some acanthocephalans are known to cause lesions in the intestinal wall and ultimately affect growth [44]. However, the abundance of cod in the area continues to be low as a result of a sparcity of capelin [45].

Fish parasites can also be useful as bioindicators of habitat degradation caused by anthropogenic contaminants especially when sensitive species are sampled as sentinels. These bioindicators include abundance, prevalence, and species diversity. These variables might increase or decrease following long-term exposure. Effluent, discharged by two pulp and paper mills in Newfoundland, caused both external and internal lesions, disrupted gonadal development, and altered length-class distribution in all age groups of winter flounder, a sediment-inhabiting flatfish species [46, 47]. In both inlets, the fish were infected with large numbers of metacercaria of a digenetic trematode, *Cryptocotyle lingua*, on the body, head, and fins compared to reference samples. Low levels of lymphocytes in the heavily parasitised fish were most likely indicative of a compromised immune system [48]. Abundance of *C. lingua* was also high in nursery areas of the flounder where untreated municipal effluent, containing sewage- and crank-case petroleum waste, was discharged (Khan, unpubl. data). It appears that an abundance of food for fish and sea gulls (*Larus* spp.), definitive hosts of the parasite, attracted them to these areas. Additionally, macroscopic xenomas of another parasite, a microsporan, *Glugea stephani*, occurred in the internal organs including the heart, liver, spleen, kidneys, intestine, and gonads in samples taken near the mill whereas they were restricted to the wall of the digestive tract of reference samples [49]. Lower than normal lymphocyte levels associated with host resistance probably provided an opportunity for the parasite to metastasise following release of spores from ruptured xenomas to infect other sites. Another study reported reduced numbers of digeneans and myxozoans but increased numbers of acanthocephalans in roach (*Rutilus rutilus*) and perch (*Perca fluviatilis*) in a lake receiving effluent from a pulp mill when compared to samples from two less polluted oligotrophic lakes [50]. Changes in the density of the intermediate hosts, toxic effect on the ectoparasites, and impairment of the immune response were suggested as the underlying causes.

Both field and laboratory studies have revealed that some parasites of winter flounder respond differently at various concentrations to discharges from a pulp and paper mill [47, 51, 52]. Gradient sampling of winter flounder inhabiting a fjord where pulp and paper mill effluent had been discharged for several decades revealed that two selected helminths, a digenean, *S. furciger*, and an acanthocephalan, *E. gadi*, increased in abundance down current from the outfall [47]. External and internal lesions, low body condition, and organosomatic indices, but elevated levels of detoxifying enzymes in the liver, were also noted in the affected fish [53]. Winter flounder captured from a pristine site were exposed to sediment collected at the four sites down current from the discharge. The results provided evidence to support the field study that enteric parasites were more abundant in flounder taken from the farthest location than others originating from the proximity of the paper mill [52]. Histopathological changes confirmed that the fish were exposed to toxic chemicals. These results suggest that host-parasite interactions can be affected after chronic exposure to anthropogenic discharges.

The balance between host and parasite interactions was also affected in sculpins and winter flounder living in coastal habitats where untreated domestic sewage and polychlorinated biphenyls (PCBs) were disposed. Untreated domestic sewage was responsible for an increase in the abundance of *Trichodina* spp. and *Gyrodactylus pleuronecti* on the secondary gill lamellae of shorthorn sculpins sampled in an embayment located in eastern Newfoundland. The trichodinids were more abundant ($\bar{x}$, 3.6±0.4; $n = 16$) where the sewage was discharged than at one ($\bar{x}$, 4.1 ± 1.3; $n = 14$) or 5 km ($\bar{x}$, 2 ± 0.3, $n = 23$) offshore (Khan unpubl.

data). Khan and Hooper [54] noted that abundance and prevalence of ectoparasitic ciliates and enteric helminths increased with distance from the point of discharge of thermal effluent in winter flounder. However, myxozoans in the gall bladder were more prevalent at the discharge location than down current. Sculpins sampled at a site where PCBs had been discharged exhibited also a greater abundance of trichodinids on the gills and myxozoans in the gall bladder of the sculpin, *M. scorpius*, than at the reference site, while enteric helminthes and ectoparasitic leeches were fewer than in latter fish [55]. Moreover, 14 species of parasites occurred in the PCB-affected sculpins in contrast to 11 in the reference samples [55]. External and toxicopathic lesions in several tissues, significantly lower body condition and organ somatic indices as well as lower hemoglobin and lymphocyte levels, were noted in the PCB-contaminated fish when compared to reference fish. Lack of parasite diversity was likely associated with the brackish water conditions that affected the transmission of some sensitive parasite species in the reference samples.

Studies on winter flounder exposed to petroleum aromatic hydrocarbons (PAHs), both in the field and in the laboratory in a dose-response trial, revealed a similar result [56]. Ectoparasites such as trichodinid ciliates and a monogenean increased to a peak but declined as the concentration increased while enteric helminthes declined progressively. Similar results were observed when species of sculpins (*Myoxocephalus* spp.) were exposed to PAHs. Trichodinid ciliates and monogeneans infecting the gills of sculpins and Atlantic cod, respectively, also increased in abundance or prevalence following chronic exposure to PAHs ([57, 58] and references therein). Additionally, exposure to PAHs and a concurrent infection with the hemoflagellate, *T. murmanensis*, caused greater mortality and a greater abundance of parasites in both Atlantic cod and winter flounder than in fish infected only with the parasite or exposed only to PAHs [59]. In contrast, endoparasitic helminthes declined after exposure, with the PAHs probably simulating antihelminthic drug action and/or changes in host physiology [56, 60]. However, ectoparasitic abundance hinged on the concentration of the pollutant as it declined after a peak with increasing levels of the PAHs [56]. Fewer species and their abundance of both groups of parasites were also observed in sculpins exposed to PCBs at an impacted military dockyard than at a reference site as noted previously [55]. Increase of ectoparasites was probably associated with a decline of the host's immune response, hyperplasia in the secondary gill lamellae, excessive epidermal mucus secretion, and exfoliation that attracted opportunistic bacteria which served as additional food for them. Changes in host physiology, toxicity of the contaminants to the larval and adult stages of the helminths as well as to their intermediate hosts, especially in the field, might have been responsible for their decline. Sanchez-Ramirez et al. [61] also reported that a monogenean *Cichlidogyrus sclerosus* increased in abundance after exposure of Nile tilapia (*Oreochromis niloticus*) to sediment contaminated with PAHs, PCBs, and heavy metals in a static system for 15 days. The sediment induced immunosuppression and caused histological anomalies in the gills and spleen. Future

studies on fish parasites as bioindicators should include additional information on the fish's body condition, organ somatic indices, histopathological effects, and also hepatic detoxifying enzymes [51].

Acid precipitation has also affected host-parasite interactions in fish. Parasite richness in eels (*Anguilla rostrata*), including monogeneans and digeneans, was greater in less acidified locations than in more acidic sites in Nova Scotia [62]. Other parasites including acanthocephalans and copepods did not appear to be affected.

In summary, observations on host-parasite interactions are complex, at times difficult to interpret on account of a number of variables that can shift the balance one way or the other. Factors such as host's age, behaviour, immunological competence, and environmental change can play a role in the association. Alternatively, establishment and evasion by the parasite of the host's responses appear to be significant factors. Consequently, the outcome in this interaction will hinge on host susceptibility and resistance and the parasite's ability to infect its host. It is suggested that future studies, investigating host-parasite interactions in habitats degraded by anthropogenic contaminants, should consider sampling multiple sites, especially along a gradient, and include more than one bioindicator and sensitive fish species.

References

[1] K. Rohde, "Latitudinal gradients in species diversity and their causes. I. A review of the hypotheses explaining the gradients," *Biologisches Zentralblatt*, vol. 97, no. 4, pp. 393–403, 1978.

[2] D. J. Marcogliese and D. K. Cone, "Comparison of richness and diversity of macroparasite communities among eels from Nova Scotia, the United Kingdom and Australia," *Parasitology*, vol. 116, no. 1, pp. 73–83, 1998.

[3] M. George-Nascimento, "Geographical variations in the jack mackerel *Trachurus symmetricus* murphyi populations in the southeastern Pacific ocean as evidenced from the associated parasite communities," *Journal of Parasitology*, vol. 86, no. 5, pp. 929–932, 2000.

[4] S. Mattiucci and G. Nascetti, "Molecular systematics, phylogeny and ecology of anisakid nematodes of the genus Anisakis Dujardin, 1845: an update," *Parasite*, vol. 13, no. 2, pp. 99–113, 2006.

[5] R. E. Warner, "The role of introduced diseases in the extinction of the endemic Hawaiian avifauna," *The Condor*, vol. 70, pp. 101–120, 1968.

[6] M. A. J. Ferguson, M. G. Low, and G. A. M. Cross, "Glycosyl-sn-1,2-dimyristylphosphatidylinositol is covalently linked to *Trypanosoma brucei* variant surface glycoprotein," *Journal of Biological Chemistry*, vol. 260, no. 27, pp. 14547–14555, 1985.

[7] F. Hawking, M. J. Worms, and K. Gammage, "24- and 48-hour cycles of malaria parasites in the blood; their purpose, production and control," *Transactions of the Royal Society of Tropical Medicine and Hygiene*, vol. 62, no. 6, pp. 731–738, 1968.

[8] R. P. McCarthy, *Periodicity in Trypanosomes of Cod [Hons Thesis]*, Memorial University of Newfoundland, Newfoundland, Canada, 1978.

[9] R. A. Khan, "Biology of the marine piscicolid leech *Johanssonia arctica* (Johansson) from Newfoundland," *Proceedings of The*

Helminthological Society of Washington, vol. 49, pp. 266–278, 1982.

[10] H. Fischer and R. S. Freeman, "The role of plerocercoids in the biology of *Proteocephalus ambloplitis* (Cestoda) maturing in smallmouth bass," *Canadian Journal of Zoology*, vol. 51, no. 2, pp. 133–141, 1973.

[11] M. J. Hatcher, J. T. A. Dick, and A. M. Dunn, "How parasites affect interactions between competitors and predators," *Ecology Letters*, vol. 9, no. 11, pp. 1253–1271, 2006.

[12] B. Sures, "Host-parasite interactions from an ecotoxicological perspective," *Parassitologia*, vol. 49, no. 3, pp. 173–176, 2007.

[13] B. F. Ardelli and P. T. K. Woo, "Protective antibodies and anamnestic response in *Salvelinus fontinalis* to *Cryptobia salmositica* and innate resistance of *Salvelinus namaycush* to the hemoflagellate," *Journal of Parasitology*, vol. 83, no. 5, pp. 943–946, 1997.

[14] R. A. Khan, "Disease outbreaks and mass mortality in cultured Atlantic cod, *Gadus morhua* L., associated with *Trichodina murmanica* (Ciliophora)," *Journal of Fish Diseases*, vol. 27, no. 3, pp. 181–184, 2004.

[15] R. A. Khan, "The life cycle of *Trypanosoma murmanensis* Nikitin," *Canadian Journal of Zoology*, vol. 54, no. 11, pp. 1840–1845, 1976.

[16] R. A. Khan and A. J. Paul, "Life cycle studies on arcto-boreal leeches (Hirudinea)," *Journal of the Helminthological Society of Washington*, vol. 62, no. 2, pp. 105–110, 1995.

[17] R. A. Khan, "Prevalence and influence of *Loma branchialis* (*Microspora*) on growth and mortality in Atlantic cod (*Gadus morhua*) in coastal Newfoundland," *Journal of Parasitology*, vol. 91, no. 5, pp. 1230–1232, 2005.

[18] R. A. Khan, "Parasites causing disease in wild and cultured fish in Newfoundland," *Icelandic Agricultural Sciences*, vol. 22, pp. 29–35, 2009.

[19] R. M. Overstreet, "Abiotic factors affecting marine parasitism," in *Parasites Their World and Ours*, D. F. Mettrick and S. S. Desser, Eds., vol. 2 of *Fifth International Congress of Parasitology Proceedings and Abstracts*, pp. 36–39, 1982.

[20] Z. Kabata, "Crustacea as enemies of fishes," in *Diseases of Fishes. Book 1*, S. F. Snieszko and H. R. Axelrod, Eds., TFH, Jersey City, NJ, USA, 1970.

[21] R. A. Khan, "Experimental transmission, development, and effects of a parasitic copepod, *Lernaeocera branchialis*, on Atlantic cod, *Gadus morhua*," *Journal of Parasitology*, vol. 74, no. 4, pp. 586–599, 1988.

[22] W. Templeman, V. M. Hodder, and A. M. Fleming, "Infection of lumpfish (*Cyclopterus lumpus*) with larvae and of Atlantic cod (*Gadus morhua*) with adults of the copepod, *Lernaeocera branchialis*, in and adjacent to the Newfoundland area, and inferences there from on inshore-offshore migration of cod," *Journal of the Fisheries Research Board of Canada*, vol. 33, pp. 711–731, 1976.

[23] R. A. Khan, E. M. Lee, and D. Barker, "*Lernaeocera branchialis* a potential pathogen to cod ranching," *Journal of Parasitology*, vol. 76, no. 6, pp. 911–917, 1990.

[24] J. L. Smith, R. Wootten, and C. Sommerville, "The pathology of the early stages of the crustacean parasite, *Lernaeocera branchialis* (L.), on Atlantic cod, *Gadus morhua* L," *Journal of Fish Diseases*, vol. 30, no. 1, pp. 1–11, 2007.

[25] J. E. Carscadden, K. T. Frank, and W. C. Leggett, "Ecosystem changes and the effects on capelin (*Mallotus villosus*), a major forage species," *Canadian Journal of Fisheries and Aquatic Sciences*, vol. 58, no. 1, pp. 73–85, 2001.

[26] R. A. Khan and M. C. Meyer, "Taxonomy and biology of some Newfoundland marine leeches (Rhynchobdellae: Piscicolidae)," *Journal of the Fisheries Research Board of Canada*, vol. 33, pp. 1699–1714, 1976.

[27] M. C. Meyer and R. A. Khan, "Taxonomy, biology, and occurrence of some marine leeches in Newfoundland waters," *Proceedings of the Helminthological Society of Washington*, vol. 46, no. 2, pp. 254–264, 1979.

[28] M. C. Meyer and A. A. Barden, "Leeches symbiotic on Arthropoda, especially decapods crustacean," *Wasmann Journal of Biology*, vol. 13, pp. 297–311, 1955.

[29] B. A. Daniels and R. T. Sawyer, "The biology of the leech *Myzobdella lugubris* infesting blue crabs and catfish," *Biological Bulletin*, vol. 148, no. 2, pp. 193–198, 1975.

[30] T. F. Mace and C. C. Davis, "Energetics of a host-parasite relationship as illustrated by the leech, *Malmia nuda*, and the shorthorn sculpin, *Myoxocephalus scorpius*," *Oikos*, vol. 23, pp. 336–343, 1972.

[31] I. Paperna and D. E. Zwerner, "Massive leech infestation on a white catfish (*Ictalurus catus*): a histopathological consideration," *Proceedings of the Helminthological Society of Washington*, vol. 41, pp. 64–67, 1974.

[32] R. A. Khan, "Susceptibility of marine fish to trypanosomes," *Canadian Journal of Zoology*, vol. 55, no. 8, pp. 1235–1241, 1977.

[33] R. A. Khan, "A new hemogregarine from marine fishes," *Journal of Parasitology*, vol. 64, no. 1, pp. 35–44, 1978.

[34] R. A. Khan, "The leech as a vector of a fish piroplasm," *Canadian Journal of Zoology*, vol. 58, pp. 1631–1637, 1980.

[35] C. D. Becker, "Haematozoa of fishes, with emphasis on North American records," in *A Symposium on Diseases of Fishes and Shellfishes*, S. F. Snieszko, Ed., Special Publication No. 5, pp. 82–100, American Fisheries Society, Washington, DC, USA, 1970.

[36] E. M. Burreson, "The life cycle of *Trypanoplasma bullocki* (Zoomastigophorea: Kinetoplastida)," *Journal of Eukaryotic Microbiology*, vol. 29, pp. 72–77, 1982.

[37] J. F. Tierney, "Effects of *Schistocephalus solidus* (Cestoda) on the food intake and diet of the three-spinked stickleback, *Gasterosteus aculeatus*," *Journal of Fish Biology*, vol. 44, no. 4, pp. 731–735, 1994.

[38] D. C. Heins and J. A. Baker, "Do heavy burdens of *Schistocephalus solidus* in juvenile threespine stickleback result in disaster for the parasite?" *Journal of Parasitology*, vol. 97, pp. 775–778, 2011.

[39] D. Gomez-Uchida, K. P. Dunphy, M. F. O'Connell, and D. E. Ruzzante, "Genetic divergence between sympatric Arctic charr *Salvelinus alpinus* morphs in Gander Lake, Newfoundland: roles of migration, mutation and unequal effective population sizes," *Journal of Fish Biology*, vol. 73, no. 8, pp. 2040–2057, 2008.

[40] J. M. Thompson, "Prey strategies of fishes in evolution and ecology-or how to stay alive long enough to fertilize some eggs," *Environmental Biology of Fishes*, vol. 1, no. 1, pp. 93–100, 1976.

[41] T. M. McDonald and L. Margolis, "Synopsis of the parasites of fishes of Canada: supplement (1978–1993)," *Canadian Special Publication of Fisheries and Aquatic Sciences*, vol. 122, article 265, 1995.

[42] R. A. Khan and C. V. Chandra, "Influence of climatic changes on the parasites of Atlantic cod *Gadus morhua* off coastal Labrador, Canada," *Journal of Helminthology*, vol. 80, no. 2, pp. 193–197, 2006.

[43] R. A. Khan, "Influence of environmental changes in the north-western Atlantic ocean on a parasite, *Echinorhynchus gadi* (*Acanthocephala*) of Atlantic cod (*Gadus morhua*) occurring off coastal Labrador, Canada," *Journal of Helminthology*, vol. 82, no. 3, pp. 203–209, 2008.

[44] K. Buchmann, "Ecological implications of *Echinorhynchus gadi* parasitism of Baltic cod (*Gadus morhua*)," *Journal of Fish Biology*, vol. 46, no. 3, pp. 539–540, 1995.

[45] R. A. Khan, "Effect of environmental change on parasites of Atlantic cod (*Gadus morhua*) as bioindicators of populations in the north-western Atlantic ocean," *Journal of Helminthology*, vol. 81, no. 2, pp. 129–135, 2007.

[46] R. A. Khan, "Assessment of stress-related bioindicators in winter flounder (*Pleuronectes americanus*) exposed to discharges from a pulp and paper mill in Newfoundland: a 5-year field study," *Archives of Environmental Contamination and Toxicology*, vol. 51, no. 1, pp. 103–110, 2006.

[47] R. A. Khan and S. M. Billiard, "Parasites of winter flounder (*Pleuronectes americanus*) as an additional bioindicator of stress-related exposure to untreated pulp and paper mill effluent: a 5-year field study," *Archives of Environmental Contamination and Toxicology*, vol. 52, no. 2, pp. 243–250, 2007.

[48] D. E. Barker, R. A. Khan, and R. Hooper, "Bioindicators of stress in winter flounder, *Pleuronectes americanus*, captured adjacent to a pulp and paper mill in St. George's bay, Newfoundland," *Canadian Journal of Fisheries and Aquatic Sciences*, vol. 51, no. 10, pp. 2203–2209, 1994.

[49] R. A. Khan, "Effect, distribution, and prevalence of *Glugea stephani* (*Microspora*) in winter flounder (*Pleuronectes americanus*) living near two pulp and paper mills in Newfoundland," *Journal of Parasitology*, vol. 90, no. 2, pp. 229–233, 2004.

[50] E. T. Valtonen, J. C. Holmes, and M. Koskivaara, "Eutrophication, pollution, and fragmentation: effects on parasite communities in roach (*Rutilus rutilus*) and perch (*Perca fluviatilis*) in four lakes in central Finland," *Canadian Journal of Fisheries and Aquatic Sciences*, vol. 54, no. 3, pp. 572–585, 1997.

[51] R. A. Khan and J. F. Payne, "A multidisciplinary approach using several biomarkers, including a parasite, as indicators of pollution: a case history from a paper mill in Newfoundland," *Parassitologia*, vol. 39, no. 3, pp. 183–188, 1997.

[52] R. A. Khan, "Influence of sediment contaminated with untreated pulp and paper mill effluent on winter flounder, *Pleuronectes americanus*," *Archives of Environmental Contamination and Toxicology*, vol. 58, no. 1, pp. 158–164, 2010.

[53] R. A. Khan and J. F. Payne, "Some factors influencing EROD activity in winter flounder (*Pleuronectes americanus*) exposed to effluent from a pulp and paper mill," *Chemosphere*, vol. 46, no. 2, pp. 235–239, 2002.

[54] R. A. Khan and R. G. Hooper, "Influence of a thermal discharge on parasites of a cold-water flatfish, *Pleuronectes americanus*, as a bioindicator of subtle environmental change," *Journal of Parasitology*, vol. 93, no. 5, pp. 1227–1230, 2007.

[55] R. A. Khan, "Chronic exposure and decontamination of a marine sculpin (*Myoxocephalus scorpius*) to polychlorinated biphenyls using selected body indices, blood values, histopathology, and parasites as bioindicators," *Archives of Environmental Contamination and Toxicology*, vol. 60, no. 3, pp. 479–485, 2011.

[56] R. A. Khan and J. F. Payne, "Comparative study of oil well drill cuttings and polycyclic aromatic hydrocarbons on parasitism in winter flounder: a dose-response study," *Bulletin of Environmental Contamination and Toxicology*, vol. 73, no. 4, pp. 652–658, 2004.

[57] R. A. Khan and J. Thulin, "Influence of pollution on parasites of aquatic animals," *Advances in Parasitology*, vol. 30, pp. 201–238, 1991.

[58] R. A. Khan, D. E. Barker, K. Williams-Ryan, and R. G. Hooper, "Influence of crude oil and pulp and paper mill effluent on mixed infections of *Trichodina cottidarium* and *T.saintjohnsi* (*Ciliophora*) parasitizing *Myoxocephalus octodecemspinosus* and *M.scorpius*," *Canadian Journal of Zoology*, vol. 72, no. 2, pp. 247–251, 1994.

[59] R. A. Khan, "Effets of chronic exposure to petroleum hydrocarbons on two species of marine fish infected with a hemoprotozoan, *Trypanosoma murmanensis*," *Canadian Journal of Zoology*, vol. 65, no. 11, pp. 2703–2709, 1987.

[60] R. A. Khan, "Influence of petroleum at a refinery terminal on feral winter flounder, *Pleuronectes americanus*," *Bulletin of Environmental Contamination and Toxicology*, vol. 61, no. 6, pp. 770–777, 1998.

[61] C. Sanchez-Ramirez, V. M. Vidal-Martinez, M. L. Aguirre-Macedo, R. P. Rodriguez-Canul, G. Gold-Bouchot, and B. Sures, "*Cichlidogyrus sclerosus* (*Monogenea: Ancyrocephalinae*) and its host, the Nile tilapia (*Oreochromis niloticus*), as bioindicators of chemical pollution," *Journal of Parasitology*, vol. 93, no. 5, pp. 1097–1106, 2007.

[62] D. J. Marcogliese and D. K. Cone, "On the distribution and abundance of eel parasites in Nova Scotia: influence of pH," *Journal of Parasitology*, vol. 82, no. 3, pp. 389–399, 1996.

7

Prevalence and Level of Antibodies Anti-*Plasmodium* spp. in Travellers with Clinical History of Imported Malaria

Rita Medina Costa, Karina Pires de Sousa, Jorge Atouguia, Luis Távora Tavira, and Marcelo Sousa Silva

Unidade de Ensino e Investigação de Clínica Tropical, Centre for Malaria and Tropical Diseases, Instituto de Higiene e Medicina Tropical, Universidade Nova de Lisboa, Rua da Junqueira 100, 1349-008 Lisbon, Portugal

Correspondence should be addressed to Marcelo Sousa Silva; mssilva@ihmt.unl.pt

Academic Editor: Dave Chadee

In this study, we show that 40.29% of travellers with a possible history of malaria exposure were positive for anti-*Plasmodium* spp. antibodies, while these individuals were negative by microscopy. The antibody test described here is useful to elucidate malaria exposure in microscopy-negative travellers from endemic countries.

1. Introduction

Malaria is an infectious disease caused by a protozoan parasite of the genus *Plasmodium*, which is transmitted between humans by the bite of infected female *Anopheles* mosquitoes. These parasites have a complex life cycle, both in the invertebrate vector and vertebrate hosts. In the human body, parasites multiply in hepatocytes, and then invade red blood cells (RBCs), initiating blood stage infection, which corresponds to the symptomatic period of the disease [1].

Malaria remains one of the most serious public health problems not only in endemic countries, where 2 billion people (approximately 40% of the world's population) are at risk of contracting the disease, but also in nonendemic areas, where the increasing number of imported malaria cases is worrying [2]. In developed countries, imported malaria predominates in tourists and immigrants who travel to their home countries to visit friends and relatives. Every year, approximately 125 million international travellers visit malaria endemic areas, and 30,000 of them contract the disease [3, 4]. In Portugal, the occurrence of 50 such cases per year [5] is estimated according to the National Public Health System.

Following infection with any of the five species of *Plasmodium* that are capable of infecting humans, *P. falciparum*, *P. ovale*, *P. vivax*, *P. malariae*, and *P. knowlesi*, specific antibodies are produced one or two weeks after the initial infection and persist for three to six months after parasite clearance [6]. These antibodies may endure for months or years in semi-immune patients in endemic countries where reinfection is frequent. However, in a naïve patient, antibody levels fall more rapidly. Reinfection or relapse leads to a secondary response with a high and rapid rise in antibody titres [6, 7].

Thus, in the present study, we aim to evaluate the prevalence and the level of anti-*Plasmodium* spp. antibodies in serum samples from travellers with possible clinical signals and symptoms of malaria. Using an ELISA-based commercial immunoassay kit to measure antimalarial antibodies, we determined the raw serological profile of these individuals. Additionally, we compare the latter serological profile with the gold-standard laboratory diagnosis, based on direct microscopy.

2. Materials and Methods

2.1. Study Population. The population for this study consisted of 335 individuals with possible clinical history of malaria and 23 healthy individuals (healthy Portuguese individuals who have never been in malaria-endemic countries). All of the 435 subjects who have had potential exposure to *Plasmodium* spp. travelled back to Portugal from malaria-endemic regions of

Africa, Brazil, Ecuador, India, Indonesia, Thailand, and Haiti, either as residents or tourists, and most of them are adults. Subjects for this study were actively recruited after being seen for symptoms of malaria at the Clinical Unit for Tropical Diseases (IHMT, Portugal).

Following microscopic examination of Giemsa-stained blood films, subjects who were potentially exposed to the parasite and had concomitant positive microscopy were categorized into group 1 (n = 45); subjects potentially exposed to the parasite but displayed negative microscopy were categorized into group 2 (n = 290); and finally, healthy naïve subjects were categorized into group 3 (n = 23).

2.2. Microscopic Diagnosis of Malaria. From each patient was obtained blood by venipuncture (5 mL of blood in anticoagulant), and two blood smears were prepared (thick and thin blood films). The haematological data was obtained from an automatic Coulter Sysmex K-1000 analyzer (Emílio de Azevedo Campos). Both blood films were stained by Giemsa's staining method and were observed on an optical microscope. The thick blood film was used to attain a qualitative diagnosis for malarial infection, and the thin blood film was used to identify the *Plasmodium* species, when infection was present. Moreover, when infection was established, the thin blood film was also used to count the number of parasites in 200 leucocytes, and this number was then converted to number of parasites in one microliter of blood [8]. Samples with no visible parasites after scoring 100 fields were considered to be negative for this test. These procedures were used as the diagnostic test for malaria. This clinical study protocol was approved by the Institutional Ethics Committee of the Instituto de Higiene e Medicina Tropical, Universidade Nova de Lisboa, Portugal (clinical study registration 4, 2012, PN, February 2012).

2.3. Serological Measurement of Antimalarial Antibodies. Total anti-*Plasmodium* spp. (antimalarial) antibodies were analysed from serum samples collected from all individuals (n = 358). The Newmarket Laboratories Malaria EIA kit (Bio-Rad, USA) was used in this study for evaluating the prevalence of total antimalarial antibodies in the depicted groups of subjects. This system is based on the binding of anti-*Plasmodium* spp. antibodies (IgG, IgM, and IgA) by use of four recombinant antigens that detect antigens from *P. falciparum*, *P. ovale*, *P. vivax*, and *P. malariae*. The test was performed as recommended by the manufacturer, as follows. 50 μL of individual undiluted sera samples were added in each single well. For each assay plate, 50 μL of positive and negative controls were also dispensed. The negative control was tested in triplicate and the positive control in duplicates. After mixing on a plate shaker for 30 seconds, the plate was covered and incubated for 30 minutes at 37°C before being washed 5 times with wash buffer. 50 μL of diluted horseradish peroxidase-conjugated antibody were added to each well, and the plates were incubated for 30 minutes at 37°C. The wells were washed again 5 times, and 50 μL of substrate solution were added to each well. The plate was then covered and incubated in the dark for 30 minutes. Finally, 50 μL of 0.5 M

sulphuric acid were added to each well to stop the reaction, and the absorbance was read at 450 nm, with a reference wavelength of 620 nm. The antibody index was obtained by dividing the OD value of each sample (at 490 nm) by the cut-off value, which was calculated as the mean of the negative control value plus 0.100, according to the manufacturer.

2.4. Data Analysis. After establishing the study groups, a commercial immunoassay Malaria EIA kit (Bio-Rad, USA) was used to determine total anti-*Plasmodium* spp. antibodies. The Malaria EIA kit is based on presence of antibodies (IgM, IgG, and IgA) reactive to four recombinant antigens to detect *P. falciparum*, *P. ovale*, *P. vivax*, and *P. malariae*. The cut-off value was calculated as the mean of the negative control value plus 0.100. To validate the assay, the optical density (OD) of each negative control should be lower or equal to 0.080, and the OD of each positive control should be greater than or equal to 1.000. The antibody index was obtained by dividing the OD value of each sample (at 490 nm) by the cut-off value. The samples with indexes lesser than or equal to 1 were considered negative. Moreover, the samples with an antibody index greater than 1 were considered positive for the presence of antimalarial antibodies. This immunoassay does not distinguish between IgG, IgM, and IgA antibodies, or between antibodies to *P. falciparum*, *P. vivax*, *P. ovale*, and *P. malariae*.

3. Results and Discussion

In this study, the prevalence and level of total antimalarial antibodies (anti-*Plasmodium* spp.) were determined in patients with possible clinical history of malaria. These patients were actively recruited in the Clinical Unit for Tropical Diseases (IHMT, Portugal). The malaria antibody EIA (Newmarket, UK; Bio-Rad) is based on binding of anti-*Plasmodium* antibodies present in a serum sample to antigens immobilized on a solid phase. The antigens are four recombinant types specific for *P. falciparum* with cross-reactivity for *P. ovale* and *P. malariae* and one specific antigen for *P. vivax*. The test detects total immunoglobulin (Ig) antibodies against *P. falciparum* and *P. vivax* and shows 80% cross-reactivity with *P. ovale* and 67% with *P. malariae*. The malaria antibody EIA was performed as recommended by the manufacturer.

Table 1 shows the distribution and characterization of two groups of individuals potentially exposed to *Plasmodium* spp. based on microscopic diagnosis for malaria (blood films), compared with a third group consisting of nonexposed individuals.

Figure 1 and Table 1 show the distribution and level of total antimalarial antibodies in three groups: group no. 1 (travellers potentially exposed to *Plasmodium* spp. and microscopically positive for malaria), group no. 2 (travellers potentially exposed to *Plasmodium* spp. and microscopically negative for malaria), and group no. 3 (nonexposed individuals, healthy individuals who reside in Portugal). From all individuals potentially exposed to *Plasmodium* spp. (n = 335), 13.43% (n = 45) had positive blood films. On the

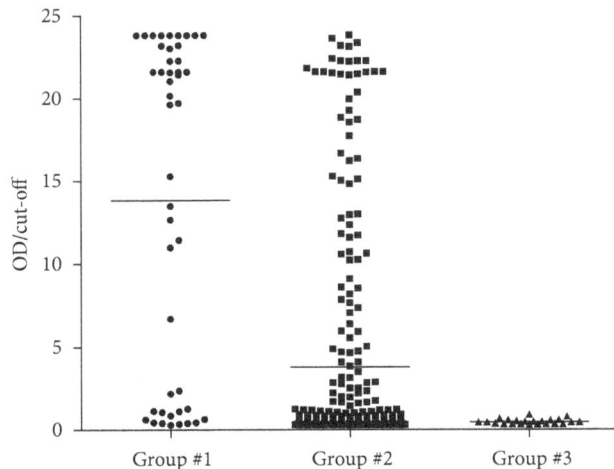

FIGURE 1: Distribution of antimalarial antibodies in subjects with possible clinical history of imported malaria. Antibody index represents the ratio OD/cut-off for each sample. Group no. 1: travellers potentially exposed to *Plasmodium* spp. and microscopically positive for malaria ($n = 45$); group no. 2: travellers potentially exposed to *Plasmodium* spp. and microscopically negative for malaria ($n = 290$); and group no. 3: control, healthy subjects ($n = 23$).

TABLE 1: Characterization of the individual groups used for the evaluation of antimalarial antibodies.

	Number of individuals (n)	Microscopic diagnosis	Serological diagnosis
Group no. 1	45	Positive	Positive ($n = 36$) Negative ($n = 9$)
Group no. 2	290	Negative	Positive ($n = 99$) Negative ($n = 191$)
Group no. 3	23	Negative	Positive ($n = 0$) Negative ($n = 23$)

other hand, 40.29% ($n = 135$) had positive ELISA serological reactions to *Plasmodium* spp. (Table 1). As expected, naïve individuals (group no. 3) had serological and microscopically negative results to malaria.

Results regarding the prevalence and level of antimalarial antibodies displayed by subjects in group no. 1 (positive blood film) and group no. 2 (negative blood film) are detailed in Figures 2(a) and 2(b), respectively. Both results are compared with the ones from group no. 3 (nonexposed individuals). Of all microscopically positive individuals ($n = 45$), 80% ($n = 36$) showed some level of antimalarial antibodies (Figure 2(a)). Furthermore, of all individuals microscopically negative for malaria ($n = 290$), 34.1% ($n = 127$) had antimalarial antibodies and 65.9% ($n = 191$) had no serological reaction to *Plasmodium* spp. (Figure 2(b)).

The use of immunofluorescence to determine the prevalence of antimalarial antibodies has been used to estimate malaria endemicity [9, 10], but its use was limited by dependence on cultured parasites, expensive fluorescence microscopes, and the subjective nature of slide reading. However, determination of antimalarial antibodies by ELISA has been shown to be a potentially useful epidemiological tool [11, 12]. For example, antibodies to the circumsporozoite antigen have been associated with transmission intensity [10] and with cumulative (age related) exposure to infection in both Brazil [13] and Sri Lanka [14]. In addition, serological data are relatively simple to collect: blood processed or blood spots collected. Processed blood or blood spots onto filter paper are a suitable source of serum if appropriately collected and stored [15, 16], and ELISA-based antibody assays are robust, relatively low tech with high throughput and inexpensive.

However, antibody detection is not a substitute for blood film examination in the diagnosis of an acute attack of

malaria, and it is mainly used in the screening of prospective blood donors to avoid transfusion-transmitted malaria [17, 18]. A blood film should be taken from anyone whose symptoms and history of travel suggest malaria. In clinical practice, serology has no place in diagnosing acute malaria, but in certain circumstances, it can be useful for better understanding a possible history of malaria when no parasites have been found through microscopy, which is the gold standard method for the diagnosis of malaria. That may happen when a blood film from the subject was never analysed or because they had taken antimalarial drugs at a dosage sufficient to depress parasitemia to undetectable levels before the contact with the malaria parasite. In such cases, serological investigations may confirm or exclude malaria, according to previous reports suggesting that an antibody test is effective in detecting malaria infection in travellers returning from overseas [19] and in nonimmune visitors to endemic areas after their departure [19, 20].

Thirty-six out of 45 (80%) subjects from group 1 displayed levels of anti-*Plasmodium* spp. antibodies when titrated by a commercial immunoassay Malaria EIA kit, while the remaining 9 (20%) subjects had no significant levels of anti-*Plasmodium* spp. antibodies, when compared to the control group (group no. 3). These 9 individuals might be at the initial phase of infection or at an acute phase of the disease, which would explain why no antibodies have been detected in their serum, while they had microscopically positive slides or a false positive microscopically. In any of these cases, antibody levels are too low to be detected by this method or have not been yet produced. On the other hand, 200 out of 335 (66%) subjects did not show significant levels of anti-*Plasmodium* spp. antibodies with the commercial kit used in this study. Moreover, 135 (40.29%) subjects had significant levels of anti-*Plasmodium* spp. antibodies, when compared to the control group (group no. 3). In our study, the presence of antibodies in 34.1% of the subjects with negative microscopy confirms the microscopy limitations, and that an additional antibody test is an asset to know about contact with malaria parasite in travellers from endemic countries. Furthermore, it is often difficult to distinguish between species, especially if the patient has already been treated or has done chemoprophylaxis.

This work has been complemented with new studies that evaluated the serological reactivity in these patients and were identified major antigenic proteins that are responsible for the production of the antibodies detected in this study [21].

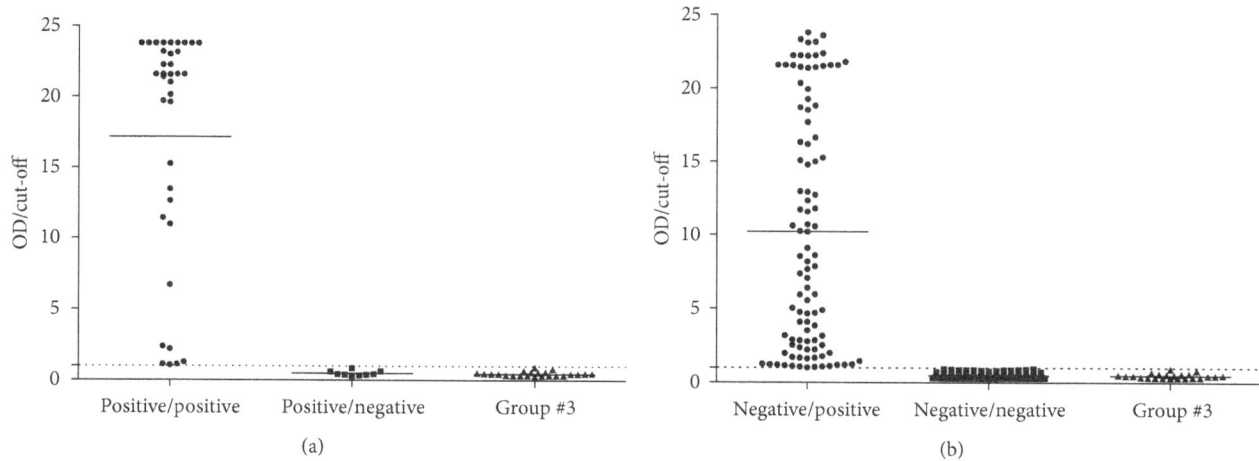

FIGURE 2: Distribution of antimalarial antibodies in subjects with possible clinical history of imported malaria. (a) (Group no. 1): individuals who were potentially exposed to the parasites and have as a positive blood film diagnostic for malaria ($n = 45$). (b) (Group no. 2): subjects potentially exposed to the parasite but have negative blood film diagnostic for malaria ($n = 290$). Group no. 3: control, healthy subjects ($n = 23$).

4. Conclusions

This study showed that 40.29% of travellers with a possible history of malaria exposure were positive for anti-*Plasmodium* spp. antibodies, while 40.29% of these individuals were negative by microscopy. The antibody test described here is useful to elucidate malaria exposure in microscopy-negative travellers from endemic countries.

Conflict of Interests

The authors declare that they have no conflict of interests.

Acknowledgments

The authors would like to thank all the subjects who have agreed to adhere to the study and the laboratorial support from Paula Maduro and Laura Cravo.

References

[1] R. Tuteja, "Malaria—an overview," *FEBS Journal*, vol. 274, no. 18, pp. 4670–4679, 2007.

[2] WHO, "World Health Organization/Global Malaria Programme," World Malaria Report, Geneva, Switzerland, 2010.

[3] K. C. Kain and J. S. Keystone, "Malaria in travelers. Epidemiology, disease, and prevention," *Infectious Disease Clinics of North America*, vol. 12, no. 2, pp. 267–284, 1998.

[4] WHO, "International travel and health," Tech. Rep., World Health Organization, Geneva, Switzerland, 2011.

[5] Estatísticas de Saúde, "Doenças de Declaração Obrigatória, 2004–2008," http://www.dgs.pt/.

[6] O. Garraud, S. Mahanty, and R. Perraut, "Malaria-specific antibody subclasses in immune individuals: a key source of information for vaccine design," *Trends in Immunology*, vol. 24, no. 1, pp. 30–35, 2003.

[7] C. R. Seed, A. Kitchen, and T. M. E. Davis, "The current status and potential role of laboratory testing to prevent transfusion-transmitted malaria," *Transfusion Medicine Reviews*, vol. 19, no. 3, pp. 229–240, 2005.

[8] K. De Sousa, M. S. Silva, and L. T. Tavira, "Variation of nitric oxide levels in imported *Plasmodium falciparum* malaria episodes," *African Journal of Biotechnology*, vol. 7, no. 6, pp. 796–799, 2008.

[9] A. Voller and P. O'Neill, "Immunofluorescence method suitable for large-scale application to malaria," *Bulletin of the World Health Organization*, vol. 45, pp. 524–529, 1971.

[10] P. Druilhe, O. Pradier, and J. P. Marc, "Levels of antibodies to *Plasmodium falciparum* sporozoite surface antigens reflect malaria transmission rates and are persistent in the absence of reinfection," *Infection and Immunity*, vol. 53, no. 2, pp. 393–397, 1986.

[11] F. Esposito, S. Lombardi, D. Modiano et al., "Prevalence and levels of antibodies to the circumsporozoite protein of *Plasmodium falciparum* in an endemic area and their relationship to resistance against malaria infection," *Transactions of the Royal Society of Tropical Medicine and Hygiene*, vol. 82, no. 6, pp. 827–832, 1988.

[12] R. Ramasamy, K. Nagendran, and M. S. Ramasamy, "Antibodies to epitopes on merozoite and sporozoite surface antigens as serologic markers of malaria transmission: studies at a site in the dry zone of Sri Lanka," *American Journal of Tropical Medicine and Hygiene*, vol. 50, no. 5, pp. 537–547, 1994.

[13] M. E. De Carvalho, M. U. Ferreira, M. R. De Souza et al., "Malaria seroepidemiology: comparison between indirect fluorescent antibody test and enzyme immunoassay using bloodspot eluates," *Memorias do Instituto Oswaldo Cruz*, vol. 87, no. 2, pp. 205–208, 1992.

[14] C. Mendis, G. Del Giudice, A. C. Gamage-Mendis et al., "Anti-circumsporozoite protein antibodies measure age related exposure to malaria in Kataragama, Sri Lanka," *Parasite Immunology*, vol. 14, no. 1, pp. 75–86, 1992.

[15] J. V. Mei, J. R. Alexander, B. W. Adam, and W. H. Hannon, "Use of filter paper for the collection and analysis of human whole

blood specimens," *Journal of Nutrition*, vol. 131, pp. 1631S–1636S, 2001.

[16] R. F. Helfand, H. L. Keyserling, I. Williams et al., "Comparative detection of measles and rubella IgM and IgG derived from filter paper blood and serum samples," *Journal of Medical Virology*, vol. 65, no. 4, pp. 751–757, 2001.

[17] A. Kitchen, A. Mijovic, and P. Hewitt, "Transfusion-transmitted malaria: current donor selection guidelines are not sufficient," *Vox Sanguinis*, vol. 88, no. 3, pp. 200–201, 2005.

[18] H. W. Reesink, "European strategies against the parasite transfusion risk," *Transfusion Clinique et Biologique*, vol. 12, no. 1, pp. 1–4, 2005.

[19] M. Knappik, G. Peyerl-Hoffmann, and T. Jelinek, "*Plasmodium falciparum*: use of a NANP19 antibody-test for the detection of infection in non-immune travellers," *Tropical Medicine and International Health*, vol. 7, no. 8, pp. 652–656, 2002.

[20] T. Jelinek, A. Blüml, T. Löscher, and H. D. Nothdurft, "Assessing the incidence of infection with *Plasmodium falciparum* among international travelers," *The American Journal of Tropical Medicine and Hygiene*, vol. 59, pp. 35–37, 1998.

[21] R. Medina Costa, F. Nogueira, K. P. De Sousa, R. Vitorino, and M. S. Silva, "Immunoproteomic analysis of *Plasmodium falciparum* antigens using sera from patients with clinical history of imported malaria," *Malaria Journal*, vol. 12, article 100, 2013.

8

Molecular and Clinical Characterization of *Giardia duodenalis* Infection in Preschool Children from Lisbon, Portugal

Filipa Santana Ferreira,[1] Rita Alexandre dos Santos Soares de Bellegarde Machado Sá da Bandeira,[2] Cláudia Alexandra Cecílio de Sampaio Ferreira Constantino,[2,3] Ana Maria Teixeira Duarte Cancela da Fonseca,[1,4] Joana da Graça Matias Gomes,[1] Rúben Miguel Lopes Rodrigues,[1] Jorge Luís Marques da Silva Atouguia,[1] and Sónia Chavarria Alves Ferreira Centeno-Lima[1]

[1] *Unidade de Clínica Tropical e Centro de Malária e Doenças Tropicais-LA, Instituto de Higiene e Medicina Tropical, Universidade Nova de Lisboa, Rua da Junqueira 100, 1349-008 Lisboa, Portugal*

[2] *Departamento de Pediatria Médica, Hospital Dona Estefânia, Rua Jacinta Marto, 1169-045 Lisboa, Portugal*

[3] *Serviço de Pediatria Instituto Português de Oncologia de Lisboa Francisco Gentil, Rua Professor Lima Bastos, 1099-023 Lisboa, Portugal*

[4] *Graduated Program in Areas of Basic and Applied Biology, Instituto de Ciências Biomédicas Abel Salazar, University of Porto, Rua Dr. Roberto Frias, s/n, 4200-465 Porto, Portugal*

Correspondence should be addressed to Filipa Santana Ferreira; fsferreira@ihmt.unl.pt

Academic Editor: C. Genchi

Giardia duodenalis is the most prevalent intestinal protozoan infection especially in children. In Portugal scarce data are available relative to this infection in preschoolers. The present study was conducted from April to July 2009 in public preschools in Lisbon enrolling 316 children. Stool examination was performed through microscopy. Molecular analysis was conducted in all positive samples for *G. duodenalis* in order to determine the assemblage and subassemblage of this parasite. Eight of the preschoolers studied children (2.5%, 8/316) were infected with *G. duodenalis*. Additionally the brother of one of the infected children was also infected. Genotyping analysis targeting *ssu-rRNA* and *β-giardin* loci revealed six infections with assemblage A and 3 with assemblage B. Subassemblage determination was possible in four of the samples, with three A2 and one A3. The limited number of cases precluded an association of a determined symptom with an assemblage. The data presented here show the relevance of considering *G. duodenalis* analysis in children with intestinal complaints even in developed countries.

1. Introduction

Giardiasis is a widespread intestinal disease caused by *Giardia duodenalis*. This protozoan parasite has a global distribution, infecting humans and a wide range of mammalian hosts [1]. The prevalence of giardiasis in humans in developed countries is 2–7% [2], while it may vary between 20 and 30% in developing countries [3].

The spectrum of clinical manifestations in human giardiasis is relatively variable, ranging from the absence of symptoms to acute or chronic diarrhea, dehydration, abdominal pain, nausea, vomiting, and weight loss [4]. Children are especially affected, with more severe consequences than adults. However, the impact of *Giardia* infection in children development is not clear. Some studies showed detrimental effects on nutritional status and poorer cognitive function on children with giardiasis [5–8], while others showed that giardiasis did not affect childhood growth [9]. Host factors, such as immune status, nutritional status, and age, are recognized as important determinants for the severity of infection

[10]. However, studies on the possible association between *G. duodenalis* assemblages and the severity of the disease have proved thus far to be inconsistent [11].

Molecular tools have demonstrated that *G. duodenalis* is a species complex comprising at least eight assemblages (A–H), among which only A and B were found infecting humans [12].

Previous studies focused on the prevalence of giardiasis in preschool children (3-4%) [13, 14], on *G. duodenalis* assemblages determination in humans [13, 15], animals, and water [15, 16], and, more recently, on prevalence and risk factors for *G. duodenalis* infection among children [17].

The aim of this study was to determine whether there was any *G. duodenalis* in preschool children enrolled, the assemblages and their relation to clinical data from the city of Lisbon.

2. Material and Methods

2.1. Study Design, Population, and Sample. A cross-sectional study was conducted from April to July 2009. The population in study were all preschoolers (3306 children) attending that year the public preschools under the supervision of the Lisbon City Hall. The selection of the kindergartens where the study was conducted was decided by the Lisbon City aiming at reflecting a wider socioeconomic of the families and geographic dispersion of the children. The number of children involved was of 685.

A total of 316 (46.1%, 316/685) preschool children, aged 3–6, attending the selected kindergartens of the network of public schools, were enrolled in this study. All children in the schools were invited to participate. The enrolled children were those whose parents collected the stool samples and signed the informed consent.

2.2. Sample Collection. A meeting was held with the director of each school as well as with the parents to explain the study objectives, prior to sample collection.

The stool containers were delivered to the parents or guardians on a Friday. The parents/guardians were told to collect three stool samples in consecutive days without any pharmacologic induction and stored at $4°C$ till Monday morning, when the team went to the schools to collect all the samples.

2.3. Microscopy. The fresh stool samples were screened for *G. duodenalis* and other intestinal parasites through microscopic analysis in saline and also in iodine. Furthermore, the formolether concentration method was also performed to increase the sensitivity of the detection. All samples were screened by three different microscopists, for cross-check results. Positive samples for *G. duodenalis* were kept in filter paper (Generation Card Kit, Qiagen) and preserved at $-20°C$ for further analysis.

2.4. DNA Extraction and PCR Amplification. DNA was extracted from samples preserved in filter paper. A DNA extraction protocol for dried blood spots (Generation Capture Card Kit, Qiagen) was adapted for stool samples. Changes to the original protocol included tripling the solution volumes

used, except for the final elution step ($100 \mu L$) [18]. For some samples, DNA was reextracted using elution volumes of $25 \mu L$. All samples were amplified using primers targeting the small subunit ribosomal RNA (*ssu-rRNA*) [19, 20] and *β-giardin* (*bg*) loci [21, 22].

Amplification reactions were performed using $2 \mu L$ of DNA template in a final volume of $25 \mu L$, using illustra PuReTaq Ready-To-Go PCR beads (GE Healthcare, UK). Both positive (DNA isolated from the Portland-1 strain (ATCC 30888DLGC Promochem) and negative controls (no template added) were included in each series of PCR reactions. PCR products were visualized on 2% agarose gel stained with ethidium bromide.

2.5. Sequence Analysis. For sequence analysis, PCR products were purified using JETQUICK Gel Extraction Spin Kit/50 (Genomed, Germany) according to the manufacturer's instructions. DNA sequencing reactions were carried out in both directions using primers GiarF/GiarR for *ssu-rRNA* gene fragment (175 bp) [20] and those described previously [22] for *β-giardin* gene fragment (511 bp). Sequences obtained in this study were aligned with previously published sequences of *G. duodenalis* isolates available in the GenBank database, using ClustalW.

2.6. Clinical Data and Treatment. A questionnaire for clinical data relative to the children was filled by each participant's parent or guardian.

All children infected with *G. duodenalis* were assisted by a paediatric doctor from the Tropical Diseases Clinic of the Institute of Hygiene and Tropical Medicine and treated for the infection with 15 mg/kg/day of metronidazole divided in three daily doses, for seven days. Three weeks after the treatment new stool samples were obtained in order to confirm the treatment efficacy. If a child was infected with any intestinal parasite, the remaining members of the household were invited to collect their stools that were also screened for intestinal parasites.

2.7. Ethical Considerations. The present study was submitted and approved by the Ethical Committee of the Institute of Hygiene and Tropical Medicine, Lisbon. Written informed consent was obtained from all parents or the legal guardians of the children participating in the study.

3. Results

3.1. Parasitological Results. From the 316 children participating in this study, 166 (52.5%) were males and 150 (47.5%) were females. The mean age was 5.03, ranging between three and six years old. *G. duodenalis* was the only pathogenic parasite found in the faeces. *G. duodenalis* cysts were found in 8 out of the 316 samples examined by microscopy (2.5%), corresponding to four of the schools included in this work. Furthermore one family member (brother), from one of the infected children, was also infected with *G. duodenalis*.

3.2. Genotyping Characterization. The nine positive samples for *G. duodenalis* were successfully amplified for *ssu-rRNA*

TABLE 1: Clinical data and *G. duodenalis* assemblages of the infected children.

Case	Age (years)	School	Present symptoms	Medical examination	Assemblages	
					ssu	*bg*
1	5.8	Musgueira (JI77)	Lack of appetite	Abdominal distension	B	NA
2	4.8	Horta Nova	No symptoms observed	Abdominal distension	A	NA
3	6.0	Horta Nova	Flatulence	Normal	A	A2
4*	9.5	Horta Nova	Abdominal pain, lack of appetite, and flatulence	Pain on deep palpation and abdominal distention	A	A2
5	6.3	Alto da Faia	Abdominal pain, lack of appetite	Abdominal distension	A	A2
6	6.3	Alto da Faia	No symptoms observed	Normal	A	A3
7	3.9	Ameixoeira	No symptoms observed	Normal	B	B**
8	5.8	Ameixoeira	No symptoms observed	Normal	B	B**
9	4.3	Ameixoeira	No symptoms observed	Abdominal distension	A	NI

*This children was a family member (brother) of participant 3.
**It was not possible to subtype assemblage B due to high level of polymorphism observed.
NA: not amplified.
NI: not identified. Although successfully amplified for *bg* gene, this sample did not present enough quality for sequentiation.

fragment gene, and for the *bg* gene only seven samples were amplified (77.8%, 7/9) (Figures 1 and 2).

Ssu-rRNA sequences obtained in this study were compared with homologous sequences found in GenBank using BLAST. Six samples belong to assemblage A and three to B (Table 1). Sequences obtained for *bg* gene were also compared with public sequences from GenBank using BLAST. Additionally, *bg* sequences were analysed for subassemblage discrimination according to the genetic polymorphisms described elsewhere [23]. Isolates 3, 4 and 5 belong to subassemblage A2, while the other isolate (6) belongs to subassemblage A3 (Table 1).

For the remaining two isolates, 7 and 8, belonging to assemblage B, it was not possible to determine the respective subassemblage due to the high nucleotide variability observed in the chromatogram.

New stool samples were collected from all infected children after the complete treatment and analyzed. No parasite was detected by microscopy.

3.3. Clinical Data.
At the time of the stool sample collection only four children reported symptoms (1, 3, 4, 5), including lack of appetite, abdominal pain, and flatulence (Table 1).

Case number 9, the one with intermittent diarrhea, was the only case of moderate malnutrition (BMI 12; $-2 <$ z score < -3).

The medical examination was normal in four children (3, 6, 7, and 8). Abdominal distension and pain on deep palpation were observed in the remaining five (Table 1).

3.4. Genetic Assemblage and Clinical Presentation.
From the three children infected with *G. duodenalis* assemblage B, two

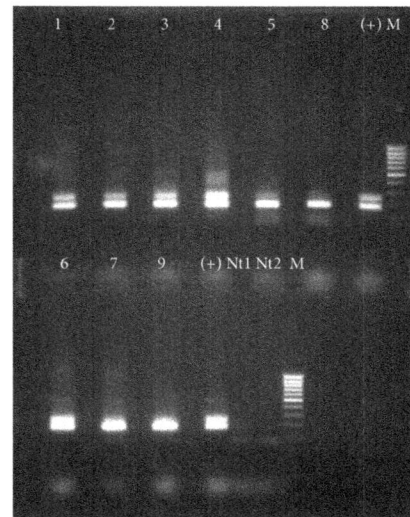

FIGURE 1: Electrophoretic separation of *ssu-rRNA* PCR products (175 bp). Lanes 1–9, *G. duodenalis* positive samples through microscopy; lanes (+), positive control (*G. duodenalis* DNA, strain Portland-1, ATCC 30888DTM LGC Promochem); lanes M, 100 bp ladder; lanes Nt1 and Nt2, negatives controls (no DNA) from the first and the second PCR reaction, respectively.

presented a normal medical examination with no symptoms (7 and 8), while the other one (1) presented abdominal distension and referred lack of appetite as a symptom. For the six children infected with assemblage A, two had a normal medical examination (3 and 6) and three referred no symptoms (2, 6, and 9). All these results are described in Table 1.

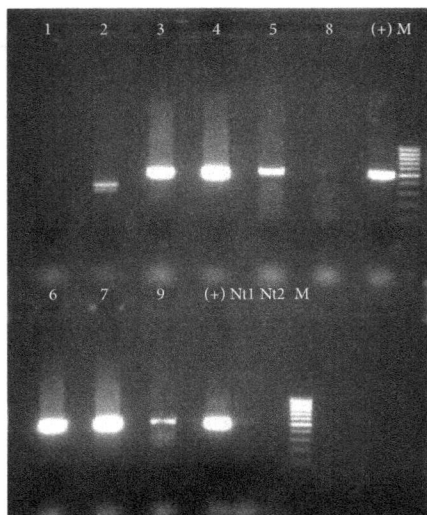

FIGURE 2: Electrophoretic separation of β-giardin PCR products (511 bp). Lanes 1–9, *G. duodenalis* positive samples through microscopy; lanes (+), positive control (*G. duodenalis* DNA, strain Portland-1, ATCC 30888DTM LGC Promochem); lanes M, 100 bp ladder; lanes Nt1 and Nt2, negatives controls (no DNA) from the first and the second PCR reaction, respectively.

4. Discussion

The number of children found infected with *G. duodenalis* in this study (8/316; 2.5%) was similar to other studies conducted in Northern and Centre of Portugal where 3%, 3.7%, and 1.9%, respectively [13, 14, 17] were infected. These results are in agreement with the reported prevalence for this parasite in developed countries [12].

The use of microscopy as the only diagnostic procedure for detecting *G. duodenalis* may be considered as a limitation of this study. However, the use of three stool samples allow the detection of over 90% of infection [24], which is very similar to the sensitivities recently reported for rapid diagnostic tests [25], while one stool sample will allow the detection of 60 to 80%, and the analysis of two stool samples will allow the detection of 80 to 90% [24]. In this study at least two stool samples were obtained from all the children and three samples for more than 80% of the enrolled children. Furthermore microscopy has the additional advantage of allowing the detection of other intestinal parasites [26].

In our study assemblage A was more frequent, with six isolates, while assemblage B was detected in three children, which is the opposite pattern reported worldwide with assemblage B appearing more common [11, 12]. Subassemblage determination was only possible for assemblage A positive samples, as B samples were impossible to subtype due to the presence of double peaks at specific position in the chromatogram. The difficulty of subtyping assemblage B has been reported [18, 27]. The importance of being able to subtype is especially relevant when a source of infection must be traced or when a distinction between reinfection/new infection is mandatory for therapeutic control. For instance, in this study,

two brothers were infected with the same subassemblage (A2) of *G. duodenalis*, suggesting a common source of infection.

Another interesting data obtained from this work was the finding of a child (case 9) with malnutrition (moderate), which is a common situation in developing countries [6]. This child also corresponds to the only one complaining of intermittent diarrhea. In this case chronic infection could have contributed to the child nutritional status.

While there has been a growing interest in the molecular characterization of *G. duodenalis*, there is still a lack of clear association between the assemblage and the clinical outcome, with contradictory results. A study conducted in the Netherlands found a strong correlation between assemblage A and mild, intermittent diarrhea and assemblage B with severe and persistent diarrhea [10]. Other studies conducted in Ethiopia and Saudi Arabia also suggest a correlation between the presence of symptoms and infection with the assemblage [28, 29]. On the other hand, a study performed in Australia revealed that children infected with *G. duodenalis* isolates from assemblage A were 26 times more likely to have diarrhea than children with assemblage B [20]. Supporting these results other works in Bangladesh and Spain also showed a statistical association between assemblage A and symptomatic infections and between assemblage B and asymptomatic infections [30, 31]. In what concerns to Portugal, the data obtained by Sousa et al. [15] supports that *G. duodenalis* belonging to assemblage A is strongly associated with symptomatic cases. In agreement with this, other work revealed a higher prevalence of assemblage B in asymptomatic children [13]. None of the children studied presented diarrhea at the moment of the stool sample collection. However, of the four symptomatic children at enrolment, three were infected with A2 subassemblage with more evident symptoms when compared with the other symptomatic children carrying B assemblage.

5. Conclusions

This study contributed with new and relevant data for the epidemiology of giardiasis in Portugal and challenged the pattern of assemblage B being more frequent than assemblage A, eventually suggesting that a different epidemiological profile is detected in developed countries. The fact that a brother from an infected child was also infected reinforced the importance of when testing a child or a family member; if positive, then the entire household should be screened. The study did not show an association between the clinical pattern observed and the assemblages. The low number of children infected with *G. duodenalis* does not justify the need for consistent request of stool parasitological analysis. However, special attention should be given to children reporting abdominal complains, considering that 4/9 of the infected children were symptomatic.

Conflict of Interests

The authors declare that they have no conflict of interests.

Acknowledgments

The authors thank Rosalia Vargas, deputy mayor for education of the Lisbon City Hall, for allowing and supporting this work; all head directors of the participating kindergartens for their support; and Laura Cravo for technical assistance.

References

[1] R. C. A. Thompson and P. T. Monis, "Variation in *Giardia*: implications for taxonomy and epidemiology," *Advances in Parasitology*, vol. 58, pp. 69–137, 2004.

[2] B. W. Furness, M. J. Beach, and J. M. Roberts, "Giardiasis surveillance—United States, 1992–1997," *Morbidity and Mortality Weekly Report*, vol. 49, no. 7, pp. 1–13, 2000.

[3] Y. R. Ortega and R. D. Adam, "*Giardia*: overview and update," *Clinical Infectious Diseases*, vol. 25, no. 3, pp. 545–550, 1997.

[4] L. Eckmann, "Mucosal defences against *Giardia*," *Parasite Immunology*, vol. 25, no. 5, pp. 259–270, 2003.

[5] D. S. Berkman, A. G. Lescano, R. H. Gilman, S. L. Lopez, and M. M. Black, "Effects of stunting, diarrheal disease, and parasitic infection during infancy on cognition in late childhood: a follow-up study," *Lancet*, vol. 359, no. 9306, pp. 564–571, 2002.

[6] J. H. Botero-Garcés, G. M. García-Montoya, D. Grisales-Patiño, D. C. Aguirre-Acevedo, and M. C. Álvarez-Uribe, "*Giardia intestinalis* and nutritional status in children participating in the complementary nutrition program, Antioquia, Colombia, May to October 2006," *Revista do Instituto de Medicina Tropical de Sao Paulo*, vol. 51, no. 3, pp. 155–162, 2009.

[7] R. D. Newman, S. R. Moore, A. A. M. Lima, J. P. Nataro, R. L. Guerrant, and C. L. Sears, "A longitudinal study of *Giardia lamblia* infection in north-east Brazilian children," *Tropical Medicine and International Health*, vol. 6, no. 8, pp. 624–634, 2001.

[8] WHO, *Intestinal Parasites Control: Burden and Trends*, WHO Division of Control of Tropical Diseases, World Health Organization, Geneva, Switzerland, 1998.

[9] M. G. Hollm-Delgado, R. H. Gilman, C. Bern et al., "Lack of an adverse effect of *Giardia intestinalis* infection on the health of Peruvian children," *American Journal of Epidemiology*, vol. 168, no. 6, pp. 647–655, 2008.

[10] W. L. Homan and T. G. Mank, "Human giardiasis: genotype linked differences in clinical symptomatology," *International Journal for Parasitology*, vol. 31, no. 8, pp. 822–826, 2001.

[11] S. M. Cacciò and U. Ryan, "Molecular epidemiology of giardiasis," *Molecular and Biochemical Parasitology*, vol. 160, no. 2, pp. 75–80, 2008.

[12] F. Yaoyu and L. Xiao, "Zoonotic potential and molecular epidemiology of *Giardia* species and giardiasis," *Clinical Microbiology Reviews*, vol. 24, no. 1, pp. 110–140, 2011.

[13] A. A. Almeida, M. L. Delgado, S. C. Soares et al., "Genotype analysis of *Giardia* isolated from asymptomatic children in northern Portugal," *Journal of Eukaryotic Microbiology*, vol. 53, no. 1, pp. S177–S178, 2006.

[14] A. Sarmento, J. M. Costa, C. A. P. Valente, and M. E. Teixeira, "Infecção por parasitas intestinais na população pediátrica," *Acta Pediátrica Portuguesa*, vol. 35, pp. 307–311, 2004.

[15] M. C. Sousa, J. B. Morais, J. E. Machado, and J. Poiares-Da-Silva, "Genotyping of *Giardia lamblia* human isolates from Portugal by PCR-RFLP and sequencing," *Journal of Eukaryotic Microbiology*, vol. 53, no. 1, pp. S174–S176, 2006.

[16] F. S. Ferreira, P. Pereira-Baltasar, R. Parreira et al., "Intestinal parasites in dogs and cats from the district of Évora, Portugal," *Veterinary Parasitology*, vol. 179, no. 1–3, pp. 242–245, 2011.

[17] C. Júlio, A. Vilares, M. Oleastro et al., "Prevalence and risk factors for *Giardia duodenalis* infection among children: a case study in Portugal," *Parasites and Vectors*, vol. 5, no. 1, article 22, 2012.

[18] F. S. Ferreira, S. Centeno-Lima, J. Gomes et al., "Molecular characterization of *Giardia duodenalis* in children from the Cufada Lagoon Natural Park, Guinea-Bissau," *Parasitology Research*, vol. 111, no. 5, pp. 2173–2177, 2012.

[19] R. M. Hopkins, B. P. Meloni, D. M. Groth, J. D. Wetherall, J. A. Reynoldson, and R. C. A. Thompson, "Ribosomal RNA sequencing reveals differences between the genotypes of *Giardia* isolates recovered from humans and dogs living in the same locality," *Journal of Parasitology*, vol. 83, no. 1, pp. 44–51, 1997.

[20] C. Read, J. Walters, I. D. Robertson, and R. C. A. Thompson, "Correlation between genotype of *Giardia duodenalis* and diarrhoea," *International Journal for Parasitology*, vol. 32, no. 2, pp. 229–231, 2002.

[21] S. M. Cacciò, M. De Giacomo, and E. Pozio, "Sequence analysis of the β-giardin gene and development of a polymerase chain reaction-restriction fragment length polymorphism assay to genotype *Giardia duodenalis* cysts from human faecal samples," *International Journal for Parasitology*, vol. 32, no. 8, pp. 1023–1030, 2002.

[22] M. Lalle, E. Jimenez-Cardosa, S. M. Cacciò, and E. Pozio, "Genotyping of Giardia duodenalis from humans and dogs from Mexico using a β-giardin nested polymerase chain reaction assay," *Journal of Parasitology*, vol. 91, no. 1, pp. 203–205, 2005.

[23] S. M. Cacciò, R. Beck, M. Lalle, A. Marinculic, and E. Pozio, "Multilocus genotyping of *Giardia duodenalis* reveals striking differences between assemblages A and B," *International Journal for Parasitology*, vol. 38, no. 13, pp. 1523–1531, 2008.

[24] T. B. Gardner and D. R. Hill, "Treatment of giardiasis," *Clinical Microbiology Reviews*, vol. 14, no. 1, pp. 114–128, 2001.

[25] B. Swierczewski, E. Odundo, J. Ndonye, R. Kirera, C. Odhiambo, and E. Oaks, "Comparison of the triage micro parasite panel and microscopy for the detection of *Entamoeba histolytica/Entamoeba dispar*, *Giardia lamblia*, and *Cryptosporidium parvum* in stool samples collected in Kenya," *Journal of Tropical Medicine*, vol. 2012, Article ID 564721, 5 pages, 2012.

[26] T. Schuurman, P. Lankamp, A. van Belkum, M. Kooistra-Smid, and A. van Zwet, "Comparison of microscopy, real-time PCR and a rapid immunoassay for the detection of *Giardia lamblia* in human stool specimens," *Clinical Microbiology and Infection*, vol. 13, no. 12, pp. 1187–1191, 2007.

[27] J. Bonhomme, L. Le Goff, V. Lemée, G. Gargala, J. J. Ballet, and L. Favennec, "Limitations of tpi and bg genes sub-genotyping for characterization of human *Giardia duodenalis* isolates," *Parasitology International*, vol. 60, no. 3, pp. 327–330, 2011.

[28] T. Gelanew, M. Lalle, A. Hailu, E. Pozio, and S. M. Cacciò, "Molecular characterization of human isolates of *Giardia duodenalis* from Ethiopia," *Acta Tropica*, vol. 102, no. 2, pp. 92–99, 2007.

[29] H. I. Al-Mohammed, "Genotypes of *Giardia intestinalis* clinical isolates of gastrointestinal symptomatic and asymptomatic Saudi children," *Parasitology Research*, vol. 108, no. 6, pp. 1375–1381, 2011.

[30] R. Haque, S. Roy, M. Kabir, S. E. Stroup, D. Mondal, and E. R. Houpt, "*Giardia* assemblage A infection and diarrhea in

Bangladesh," *Journal of Infectious Diseases*, vol. 192, no. 12, pp. 2171–2173, 2005.

[31] J. Sahagún, A. Clavel, P. Goñi et al., "Correlation between the presence of symptoms and the *Giardia duodenalis* genotype," *European Journal of Clinical Microbiology and Infectious Diseases*, vol. 27, no. 1, pp. 81–83, 2008.

9

Interactions between *Leishmania braziliensis* and Macrophages Are Dependent on the Cytoskeleton and Myosin Va

Elisama Azevedo,[1,2] **Leandro Teixeira Oliveira,**[3] **Ana Karina Castro Lima,**[1,2] **Rodrigo Terra,**[1,4] **Patrícia Maria Lourenço Dutra,**[1] **and Verônica P. Salerno**[3]

[1] *Laboratório de Imunologia e Bioquímica de Protozoários, Departamento de Microbiologia, Imunologia e Parasitologia, FCM, UERJ, Avenida Professor Manuel de Abreu 444 5° andar. Vila Isabel, 20550-170 Rio de Janeiro, RJ, Brazil*
[2] *Programa de Pós-Graduação em Microbiologia Médica, Faculdade de Ciências Médicas, UERJ, 20550-170 Rio de Janerio, RJ, Brazil*
[3] *Departamento Biociências, Escola de Educação Física e Desportos, Universidade Federal do Rio de Janeiro, 21941-599 Rio de Janerio, RJ, Brazil*
[4] *Programa de Pós-Graduação em Biodinâmica do Movimento, EEFD, UFRJ, 21941-599 Rio de Janerio, RJ, Brazil*

Correspondence should be addressed to Patrícia Maria Lourenço Dutra, pmldutra@gmail.com

Academic Editor: Barbara Papadopoulou

Leishmaniasis is a neglected tropical disease with no effective vaccines. Actin, microtubules and the actin-based molecular motor myosin Va were investigated for their involvement in *Leishmania braziliensis* macrophage interactions. Results showed a decrease in the association index when macrophages were without F-actin or microtubules regardless of the activation state of the macrophage. In the absence of F-actin, the production of NO in non-activated cells increased, while in activated cells, the production of NO was reduced independent of parasites. The opposite effect of an increased NO production was observed in the absence of microtubules. In activated cells, the loss of cytoskeletal components inhibited the release of IL-10 during parasite interactions. The production of IL-10 also decreased in the absence of actin or microtubules in non-activated macrophages. Only the disruption of actin altered the production of TNF-α in activated macrophages. The expression of myosin Va tail resulted in an acute decrease in the association index between transfected macrophages and *L. braziliensis* promastigotes. These data reveal the importance of F-actin, microtubules, and myosin-Va suggesting that modulation of the cytoskeleton may be a mechanism used by *L. braziliensis* to overcome the natural responses of macrophages to establish infections.

1. Introduction

Leishmaniasis is caused by several different species of protozoan parasites from the genus *Leishmania*. *Leishmania* parasites maintain a life cycle consisting of a phase in a dipteran insect (sandflies) and a phase in a mammalian host. Transmission occurs when an infected sandfly bites a human. This can lead to infection of macrophages, in which the parasite thrives inside the hostile environment of the phagolysosomes [1]. Leishmaniasis is one of the most important of the neglected tropical diseases, with 350 million people in 88 countries worldwide living at risk of developing one of the many forms of the disease [2].

The numerous forms of leishmaniasis are dependent on factors that are not well understood, including the species of the parasite and the health of the host upon initial infection. The parasitosis can vary from self-healing dermal lesions to generalised organ infection, which can lead to death. *Leishmania braziliensis* is the causative agent of mucocutaneous disease in the Americas. Despite its great importance, it has been less studied than other strains because of difficulties in *in vitro* cultivation [3, 4].

The *Leishmania* parasites display multiple forms that are distinct in morphology, biochemistry, intracellular organisation, and behaviour. In the sandfly, the replicating form of *Leishmania* spp., the promastigote, is flagellated and motile.

A subset of promastigotes progress through differentiation to become the nondividing, infectious metacyclic promastigotes. Following a bite by the sandfly, these metacyclic promastigotes are transmitted to the mammalian host. The process of infection begins when the parasites undergo conventional phagocytosis by macrophages that are recruited to the site of the bite. After phagocytosis, the parasites are located within classic phagolysosomes and undergo differentiation into the amastigote form, which is resistant to the acidic pH and lysosomal enzymes present in these cellular structures [5]. Amastigotes do not have an exterior flagellum and live as intracellular parasites in a variety of mammalian cells, most notably within professional phagocytes such as macrophages [6].

Phagocytosis occurs by the extension of the plasma membrane around an extracellular particle, followed by internalisation of the particle into a membrane-bounded intracellular vesicle, the phagosome. In macrophages, different cell surface receptors stimulate various types of phagocytic responses [7]. Macrophage Fc receptors mediate the phagocytosis of IgG-coated particles. Ligation of Fc receptors initiates an intracellular signalling cascade that ultimately impinges on the actin cytoskeleton [8]. With regards to *Leishmania*, the phagocytic response is coupled to other cellular events that prevent the activation of deadly antimicrobial agents such as nitric oxide (NO) and many of the cytokine-inducible macrophages, which are necessary for the development of an effective immune response. This enables the parasite to evade the innate immune response and to divide within the phagolysosome of the infected macrophage, from whence it can spread and propagate the disease within the host [9].

It is well known that lipopolysaccharide (LPS) and IFN-γ promote classical macrophage activation (i.e., the activation of M1 macrophages). The phenotype of M1 macrophages includes high production of IL-12 and IL-23 and low production of IL-10, an anti-inflammatory cytokine. These cells are able to produce effectors molecules, such as reactive species of oxygen and nitric oxide (NO), and inflammatory cytokines, such as IL-1β, TNF-α and IL-6. M1 macrophages function as cellular immune response inductors, promoting T helper 1 lymphocyte (Th1) differentiation, and mediate resistance to intracellular pathogens and tumour cells [10].

The resistance or susceptibility to all forms of leishmaniasis has been associated to a balance between cellular and humoral immunity responses [11]. Several studies on *L. major* using models of infection in BALB/c and C57/BL6 mice have shown a good prognosis associated with immune responses that are predominantly Th1 as determined by the production levels of IL-12 and TNF-α cytokines. Susceptibility to more serious manifestations of leishmaniasis was associated with the activation of T helper 2 lymphocytes (Th2) based on the production of the anti-inflammatory cytokines IL-4 and IL-10 [12–16]. In cases of leishmaniasis caused by *L. braziliensis*, a major difference was observed. An exacerbated Th1 response (hiperergy) can be observed that promotes an increase in tissue damage near regions displaying high levels of parasitic antigens. This observation is characteristic of a classical mucosal leishmaniasis that is associated with high levels of IFN-γ and TNF-α and low

levels of IL-10. In addition, cells collected from individuals presenting this type of leishmaniasis display a poor response to cytokines that inhibit IFN-γ secretion [17].

There are no effective vaccines available for leishmaniasis, and treatments rely on parenteral drugs that present high toxicity, low efficacy, and, in some cases, widespread resistance [18, 19]. The most common drugs are nephro-, hepato-, and cardiotoxic [20]. Investigative studies of the relationship between *Leishmania* and its host cells are imperative for identifying potential targets that can interfere with the process of invasion. Numerous studies have shown that interference with the parasite-host interaction can be effective. For example, a decrease in the association index between *Leishmania* and macrophages was observed when 4,5,6,7-tetrabromobenzotriazole (TBB), a specific casein kinase 2 (CK2) inhibitor, was used [21]. TBB was also able to reverse the positive platelet-activating factor (PAF) effect on this type of cellular interaction [22]. A similar profile of inhibition was reported using 1,10-phenanthroline (phen) and 1,10-phenanthroline-5,6-dione (phendio), which are ion chelators that inhibit metal-dependent peptidases [23].

TBB also induced changes in cell shape and the cytoskeleton, which are important to the process of phagocytosis. Particle ingestion by phagocytosis results from sequential rearrangements of the actin cytoskeleton and the overlying membrane [24]. Inhibitors and/or enzymes capable of interfering with the dynamics of cytoskeletal components, such as latrunculin A, which acts specifically to disrupt the actin cytoskeleton [25], and nocodazole, which depolymerises microtubules [26], are good candidates for the study of parasite-host cell interactions. In addition, the molecular motor myosin Va has been observed to be associated with phagosomes [27].

Latrunculin A (2-thiazolidinone macrolide) is a toxin purified from the red sea sponge *Latrunculia magnifica*. This substance sequesters G-actin and prevents F-actin assembly. It binds monomeric actin with 1 : 1 stoichiometry and can be used to block the polymerisation of purified actin (Kd = 0.2 μM). The effects of this toxin are observed in cell cultures when used at a range of 0.1 and 10 μM [28, 29]. These effects appear to occur rapidly, with the toxin promoting the depolymerisation of tumour cell cytoskeletons within 10 minutes [29].

Nocodazol is an antimitotic agent that disrupts microtubules by binding to β-tubulin [30], thereby promoting the inhibition of microtubule dynamics. These effects can be observed in cells after 5 minutes of treatment [31]. Nocodazol promotes disruption of the mitotic spindle [31, 32] and fragmentation of the Golgi complex [33].

In the present study, we investigated the involvement of actin and the actin-based molecular motor myosin Va in the interaction between *Leishmania braziliensis* and macrophages.

2. Material and Methods

2.1. Chemicals. 5-(and-6)-Carboxyfluorescein diacetate, succinimidyl ester (green CFSE), chloromethyl tetramethylrhodamine (orange CMTMR), and latrunculin A were purchased from Invitrogen (Eugene, Oregon, USA). Foetal calf serum (FCS) was purchased from Cultilab Co (Campinas, São Paulo, Brazil). Nocodazol and all other chemicals used in this work were purchased from Sigma (St. Louis, MO, USA).

2.2. Parasites.
The strain L. (V.) braziliensis (MHOM/BR/2002/EMM-IOC-L2535) was maintained in the promastigote form by culturing at $26°C$ in Schneider's medium supplemented with 2 mM glutamine, 100 units/mL penicillin, 100 mg/mL streptomycin, and 10% FCS. For the interaction assays, the promastigotes were first labelled by incubation with $5 \mu M$ green CFSE for 10 min at $37°C$ or with $5 \mu M$ orange CMTMR for 20 min, followed by two washes with PBS by centrifugation, and the final pellet was suspended in Dulbecco's Modified Eagle Medium (DMEM) supplemented with 10% FCS.

2.3. Cell Culture.
The RAW 264.7 macrophage cell line (kindly supplied by Dr. Marcia Paes—Laboratório de Bioquímica, Universidade do Estado do Rio de Janeiro, UERJ, Rio de Janeiro, Brazil) was maintained at $37°C$ in DMEM medium supplemented with 10% foetal calf serum, penicillin (100 units/mL), and streptomycin (100 mg/mL) in a humidified atmosphere of 4% CO_2.

2.4. Leishmania braziliensis-Macrophages Interactions.
Macrophages were seeded onto 24-well plates containing glass coverslips for 2 h at $37°C$ in a 4% CO_2 atmosphere, washed once with DMEM and incubated for 17 h in the presence or absence of lipopolysaccharide (LPS) (100 ng/mL) and INF-γ (1 μg/mL) at $37°C$ in a 4% CO_2 atmosphere, as previously described [34, 35]. At the moment of the experiment, the culture medium was replaced with DMEM containing either latrunculin A (5 μM), nocodazole (5 μM), or no drug at $37°C$ in a humidified atmosphere of 4% CO_2 for 10 minutes. Then, promastigotes that had been previously marked with CSFE were added and allowed to interact with the macrophages at a 5 : 1 ratio for 30 min at $37°C$ in a 4% CO_2 atmosphere. After incubation, the coverslips were washed 3 times with PBS, fixed and stained with Giemsa, and the percentage of infected macrophages was determined by counting 600 cells in triplicate coverslips, as described. The association index was determined by multiplying the percentage of infected macrophages (cells with at least one associated or intracellular parasite) by the mean number of parasites per cell (internalised or simply attached to the cell) [36]. The time of latrunculin and nocodazole incubation used in this work were chosen in accordance with the cellular viability test. This time did not affect the macrophages viability.

For myosin Va detection and actin staining, the parasites were treated with 5 μM green CFSE or with 5 μM orange CMTMR (only in the case of myosin Va) and allowed to interact with the macrophages under the conditions described above. After interaction, the cells were fixed with 4% paraformaldehyde, permeabilised with 0.1% Triton X-100, and stained with 1 : 250 Phalloidin Alexa 546 for 30 min at room temperature (for actin staining) or blocked using 1% bovine serum albumin (BSA) and 20% sheep serum (for

myosin Va detection). These cells were incubated for 1 h with rabbit antimyosin Va primary antibodies (1 : 1000—Santa Cruz Biotech) and washed with PBS. The secondary antibody, anti-rabbit IgG conjugated to Alexa-546 (1 : 1000—Molecular Probes), was added. The cells used for myosin Va detection were placed on glass slides using *ProLong Antifade* mounting media (Molecular Probes). Coverslips containing the cells used for actin staining were treated with a saturated solution of n-propyl-gallate in PBS, followed by mounting on glass slides for microscopy. The images were collected by a Confocal Zeiss microscope (LSM 510 Meta, Zeiss, Germany) (software Zeiss Vision Release 4.6).

2.5. Nitrite Determination.
The nitrite that had accumulated in culture medium was measured as an indicator of NO production based on the Griess reaction. The macrophages were allowed to adhere to 24-well culture plates for 2 h at $37°C$ in a 4% CO_2 atmosphere. Adherent cells were washed with DMEM and incubated for 17 h in fresh medium in the presence or absence of LPS (100 ng/mL) and IFN-γ (1 μg/mL), as previously described [34, 35]. Next, the culture medium was removed, and the macrophages were incubated with latrunculin (5 μM), nocodazole (5 μM), or DMEM for 40 min at $37°C$ in a 4% CO_2 atmosphere. The time of latrunculin and nocodazole incubation used in this work were chosen in accordance to the cellular viability test. This time did not affect the macrophages viability. For the *Leishmania*-macrophage interaction experiments, 10 min after the addition of latrunculin A and nocodazole, the promastigotes were washed with PBS and allowed to interact with the macrophages at a 5 : 1 ratio for 30 min at $37°C$ in a 4% CO_2 atmosphere. In all cases, a sample of the supernatant was removed to measure the nitrite content by the Griess method [37]. Briefly, 50 μL of cell culture medium was mixed with 50 μL of Griess reagent and incubated at room temperature for 10 min, and then the absorbance was measured at 540 nm in a microplate reader (TP reader Thermo Plate). Fresh culture medium was used as the blank in all experiments. The amount of nitrite in the samples was calculated with a standard curve obtained from a serial dilution of sodium nitrite.

2.6. Cytokine Analysis.
Macrophages were seeded onto 24-well culture plates and allowed to adhere for 2 h at $37°C$ in a 4% CO_2 atmosphere, washed in DMEM and incubated for 17 h in this medium in the presence of LPS (100 ng/mL) and IFN-γ (1 μg/mL) at $37°C$ in a 4% CO_2 atmosphere, as previously described [34, 35]. Next, latrunculin (5 μM) or nocodazole (5 μM) was added directly to the incubation medium, and the cells were incubated for 1 h at $37°C$ in a 4% CO_2 atmosphere before the addition of parasites. The time of latrunculin and nocodazole incubation used in this work were chosen in accordance with the cellular viability test. This time did not affect the macrophages viability. The supernatant of macrophages was recovered and used to resuspend the promastigotes, and this suspension was added to the macrophages at a ratio of 5 : 1 (parasites : macrophages). These cells were allowed to interact for 30 min at $37°C$ in

a 4% CO_2 atmosphere. The supernatants were recovered for IL-10 or TNF-α measurements. Uninfected macrophages that were treated in the same manner were used as controls. All supernatants used to measure the cytokines were the result of 18 h and 30 min incubation. The level of cytokines was determined in the supernatants by ELISA (R&D Systems, USA) using recombinant murine cytokines and antibodies according to the manufacturer's instructions. The absorbance was measured at 450 nm in a microplate reader (TP reader Thermo Plate). The cytokine concentrations were evaluated using IL-10 and TNF-α standard curves.

2.7. Cellular Viability Test. Macrophages treated with latrunculin A and nocodazole for 1 hour and 30 minutes (the maximal incubation time used in this work) had their cellular viability assessed by an MTT [3-(4,5-dimethylthiazol-2-yl)-2,5-diphenyl tetrazolium bromide] assay as previously described [38].

2.8. Transfection of Macrophages with eGFP-DB (Myosin Va Brain-Isoform Tail) for Leishmania braziliensis-Macrophage Interaction Assays. The macrophages were seeded onto poly-L-lysine-coated coverslips in 24-well plates in DMEM without antibiotics. Lipofectamine 2000 was diluted into 25 μL of Opti-MEM for 5 minutes at room temperature and then mixed with plasmid DNA that had been diluted in Opti-MEM for 20 min. After this period of incubation, the DMEM was removed from the macrophage cultures, and the transfection mixture was layered on top of the cells. The cells were incubated for 1 h at 37°C in a humidified atmosphere of 5% CO_2, and then 1 mL DMEM supplemented with 10% foetal calf serum was added. The cells were incubated for an additional 18 h at 37°C in a humidified atmosphere of 5% CO_2 to permit expression from the plasmids. Subsequently, an interaction assay was carried out as previously described. The cells were observed on a Zeiss confocal microscope (LSM 510 Meta, Zeiss, Germany) and imaged using the Zeiss Vision Release 4.6 software.

2.9. Statistical Analysis. All results are presented as the mean and standard error of the mean (SEM). Normalised data were analysed by a one-way analysis of variance (ANOVA), and differences between groups were assessed with the Tukey posttest. We used the software GraphPad Prism 4.0, and values of $P \leq 0.05$ were considered significant.

3. Results

3.1. Effect of Latrunculin and Nocodazole on the Interaction of Leishmania braziliensis with RAW 264.7 Macrophages Activated by LPS and IFN-γ. Studies with other *Leishmania* strains have suggested that the cytoskeleton is involved in the interactions between the parasite and the host macrophages. However, the different strains also exhibit variations in the dynamics of cell invasion. To understand the role of the actin and microtubule cytoskeleton of macrophages during interactions with *L. braziliensis*, the cytoskeleton-disrupting drugs latrunculin A [25] and nocodazole [26] were used to

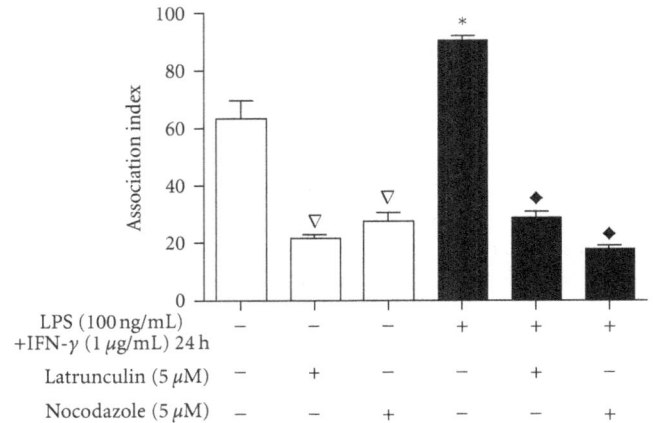

FIGURE 1: Effect of latrunculin and nocodazole on the interaction of *Leishmania braziliensis* with RAW 264.7 macrophages. Adherent macrophages, with or without 17 h previous activation by LPS and IFN-γ, were treated with latrunculin (5 μM) or nocodazole (5 μM) for 10 min and then incubated with promastigotes for 30 min. The association indices were determined by multiplying the percentage of infected macrophages by the mean number of parasites per cell. The bars represent the mean ± standard errors of mean (SEM) of at least three independent experiments performed in triplicate. *$P \leq$ 0.05 in relation to control inactivated macrophages, $^\Delta P \leq 0.05$ in relation to control inactivated macrophages, $^\blacklozenge P \leq 0.05$ in relation to control activated macrophages.

depolymerise filamentous actin and microtubules, respectively. The cellular viability of macrophages was not altered by drug treatment (data not shown). The drugs were added 10 min before the addition of parasites to prevent artefacts arising from the kinetics of depolymerisation. In the absence of F-actin, the association index of parasites to macrophages decreased 66.45% when compared to the nontreated controls. The absence of microtubules decreased the association index by 57.04% (Figure 1).

Because of the high potential for parasites to interact with activated macrophages *in vivo*, similar experiments were performed after 24 h of stimulating macrophages with LPS (100 ng/mL) and IFN-γ (1.0 μg/mL). Activated macrophages (M1) that were not treated with the cytoskeleton-disrupting agents showed a significant increase in the association index compared to the nonactivated macrophages. The loss of both F-actin and microtubules decreased the association index significantly. In the presence of latrunculin A, the association index between *L. braziliensis* and M1 macrophages was inhibited by 68.71%, while in the absence of microtubules as a result of nocodazole treatment, the inhibition was 80.17% (Figure 1).

3.2. Effect of Latrunculin and Nocodazole on NO Production by RAW 264.7 Macrophages. NO production is the main leishmanicidal process performed by macrophages, and the *Leishmania* genus has developed evasion mechanisms that interfere with NO production [39]. To determine the influence of the cytoskeleton on the production of NO

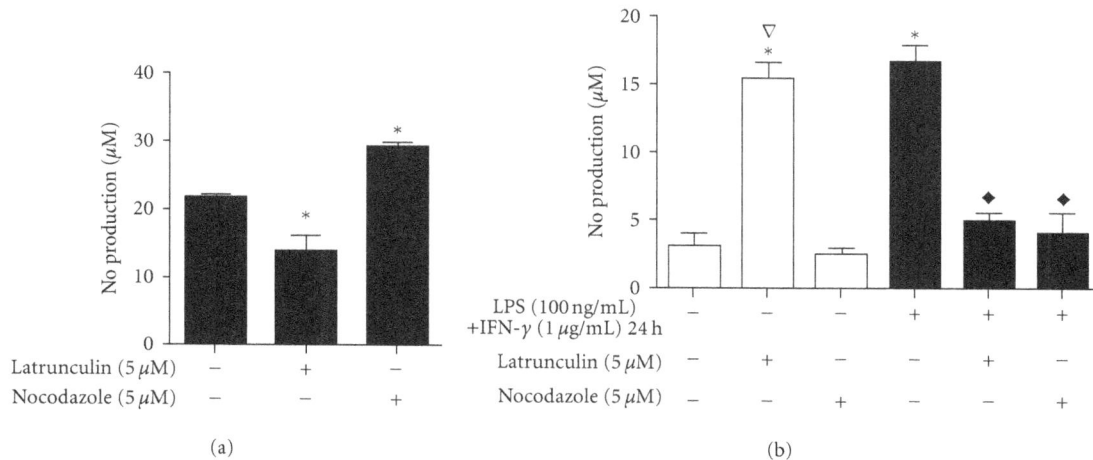

(a)   (b)

FIGURE 2: Effect of latrunculin and nocodazole on the production of NO by RAW 264.7 macrophages after interaction with *L. brasiliensis*. NO levels were inferred from the level of nitrite measured in the media after various treatments. Adherent macrophages were activated (black bars) for 17 h with LPS (100 ng/mL) + IFN-γ (1 μg/mL) or remained untreated (open bars). For the measurement of NO after interactions with parasites in (a), macrophages were treated with latrunculin (5 μM) or nocodazole (5 μM) for 10 min and then exposed to promastigotes for 30 min. In the absence of parasites (b), activated and nonactivated adherent macrophages were treated with latrunculin (5 μM) or nocodazole (5 μM) for 10 min. The bars represent the mean $\pm$ standard errors of mean (SEM) of at least three independent experiments performed in triplicate. *$P \leq 0.05$ compared to control activated macrophages in (a) and compared to control inactivated macrophages in (b); $^{\nabla}P \leq 0.05$ compared to control nonactivated macrophages; $^{\blacklozenge}P \leq 0.05$ compared to control activated macrophages.

by macrophages, nitrite was measured following disruption of F-actin and microtubules before and after interactions with the parasites (Figure 2). In control M1 macrophages, the presence of *L. braziliensis* promoted a small increase (approximately 20%) in the level of NO (Figure 2(a)) when compared to uninfected M1 macrophages (Figure 2(b)). The NO production in response to the parasites decreased by 36.86% after latrunculin A treatment. Treatment with nocodazole had the opposite effect, as the level of NO production increased by 36.17% (Figure 2(a)).

The effect of cytoskeleton disruption was dependent on the activation state of macrophages in addition to the presence of parasites. The NO production in nonactivated macrophages was significantly increased by latrunculin treatment (390%), reaching the same levels as that produced by the M1 macrophages. Conversely, latrunculin A treatment decreased NO production by 70% in activated cells. Nocodazole had no effect in non-activated cells, while in M1 macrophages, this substance inhibited NO production by 75% (Figure 2(b)).

3.3. Effect of Latrunculin and Nocodazole on the Production of IL-10 and TNF-α Cytokines by RAW 264.7 Macrophages.

Cytokine modulation is another macrophage response to activation or the internalisation of pathogens. In this work, the effect of the release of IL-10 and TNF-α on the cytoskeleton and interactions with parasites was evaluated. We observed an inhibition of IL-10 release by M1 macrophages, and interactions with *L. braziliensis* promoted a further decrease in IL-10 release (Figures 3(a) and 3(b)). Treatment with latrunculin during parasite interactions with M1 macrophages reduced the release of IL-10 by 79.17%.

Nocodazole treatment had a greater effect and abrogated IL-10 release. Similar results were observed in non-activated macrophages in the absence of parasite interactions (Figure 3(b)). Latrunculin and nocodazole were able to decrease the release of IL-10 by non-activated macrophages by 84.5% and 83%, respectively. No significant changes were detected in uninfected M1 macrophages.

Nocodazole and latrunculin A treatment had no effect on the release of TNF-α by infected M1 macrophages (Figure 3(c)) or by uninfected and non-activated cells (Figure 3(d)). However, treatment with latrunculin promoted the inhibition of TNF-α release by 82.3% in uninfected M1 macrophages (Figure 3(d)).

3.4. Effect of Latrunculin and Nocodazole on RAW 264.7 Macrophage Actin Organisation during Parasite-Host Interactions.

The observed changes in the association indexes, NO production, and cytokine production prompted us to assess morphological changes in the macrophages. The actin cytoskeleton was stained in macrophages after interaction with *L. braziliensis* (Figure 4). In control macrophages without drug treatment, highly organised F-actin was observed, with prominent staining in the periphery of the cell and numerous instances of filopodial structures. In the representative image (Figure 4(a)), internalised parasites can be observed in orange, the result of the superposition of red and green staining, showing a colocalisation between actin cytoskeleton and the parasite, confirming the involvement of actin on this process as showed before. In the absence of actin filaments resulting from latrunculin A treatment (Figure 4(b)), the macrophages were observed to be much more round in shape and lacked filopodial structures.

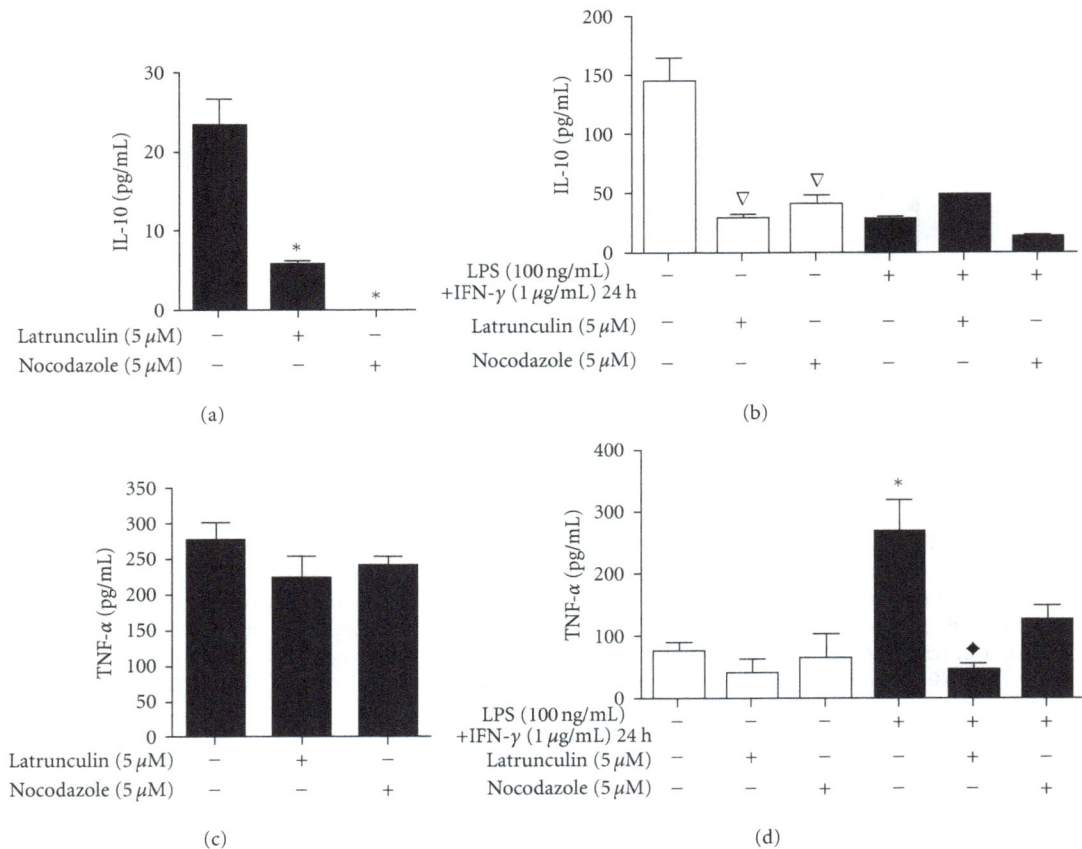

FIGURE 3: Effect of latrunculin and nocodazole on cytokine production in RAW 264.7 macrophages. Levels of IL-10 (a and b) and TNF-α (c and d) were measured in the media after various treatments. Adherent macrophages were treated with LPS (100 ng/mL) + IFN-γ (1 μg/mL) for 17 h before treatment with latrunculin (5 μM) or nocodazole (5 μM) for 1 h and then exposed to promastigotes for 30 min (a and c). Adherent macrophages not exposed to parasites were incubated for 17 h in the absence or in the presence of LPS (100 ng/mL) plus IFN-γ (1 μg/mL) and then treated with latrunculin (5 μM) or nocodazole (5 μM) for 1 h (b and d). The bars represent the mean $\pm$ standard errors of mean (SEM) of at least three independent experiments performed in triplicate. $^*P \leq 0.05$ compared to control infected macrophages (a) and compared to control non-activated and uninfected macrophages; $^\nabla P \leq 0.05$ compared to control non-activated and uninfected macrophages; $^\blacklozenge P \leq 0.05$ compared to control activated uninfected macrophages.

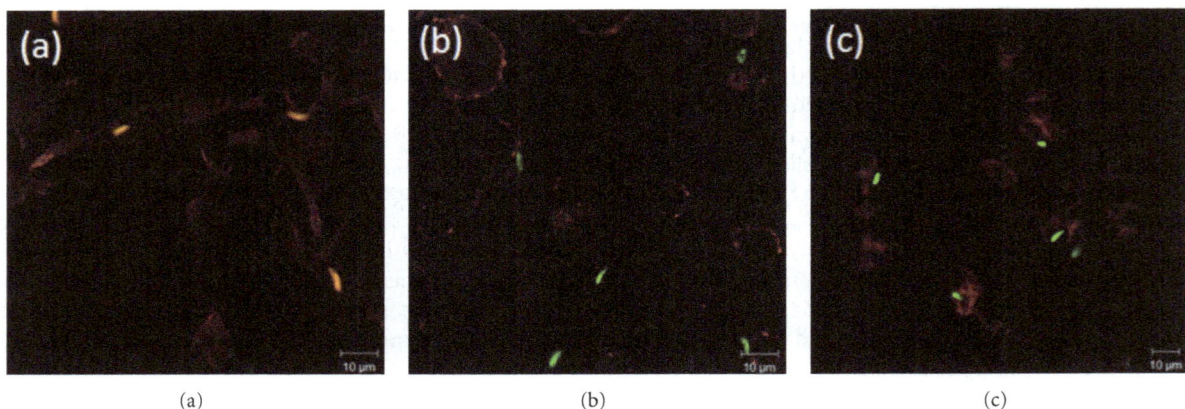

FIGURE 4: Integrity of the cytoskeleton influences the interaction between *L. braziliensis* and macrophages. Adherent Raw 264.7 macrophages were incubated for 10 min with (a) fresh DMEM, (b) 5 μM latrunculin A to disrupt actin filaments, or (c) 5 μM nocodazole to disrupt microtubules before exposure for 30 min to parasites that had been previously marked with CFSE. The panels are representative fields of colour-combined fluorescent images displaying parasites in green and F-actin (phalloidin staining) in red. Scale bars indicate imaging for all panels at 63x magnification.

There were no cells with internalised parasites; instead, parasites can be observed, in green, attached to the macrophages or free in the extracellular milieu. Treatment with nocodazole to disrupt microtubules had a profound effect on the cell organisation of the macrophages. The organisation of stained F-actin was random, without prominent structures in the periphery of the cell and without the presence of filopodia. Similar to macrophages treated with latrunculin, there were no cells with internalised parasites, which can be observed in green attached to the macrophages. Some parasites appeared to be semi-internalised (Figure 4(c)).

3.5. Transfection of Macrophages with eGFP-DB (Myosin Va Tail). Because of the high dependence on an intact cytoskeleton and the dynamic nature of the parasite/macrophage interactions, molecular motors were considered to be potentially important host cell factors. Myosin Va was the first candidate considered because of its wide range of expression and diverse functions, including involvement in phagosome movement. Using an antibody against MVa, we observed the presence of this protein in macrophages and the colocalisation of myosin with the parasite. Internalised parasites can be observed in orange, the result of the superposition of red and green staining (Figure 5(a)). To evaluate the role of myosin Va, a construct consisting of the myosin Va tail (brain isoform) fused to eGFP was transfected into macrophages before parasites were added. There was an abrupt decrease in the association index between transfected macrophages and *L. braziliensis* promastigotes (Figure 5(b)). The mock-transfected macrophages did not interfere with parasite interaction, and several promastigotes can be observed attached to or internalised by these cells (Figure 5(c)). The association index of the wild type macrophage was 53.25 ± 4.49 and that of the mock macrophages was 42.42 ± 3.91, while in the transfected macrophages, this index was 2.96 ± 0.94 (data not shown).

4. Discussion

The establishment of an *L. braziliensis* infection in humans, and the beginning of leishmaniasis, starts with the uptake of the parasite by macrophages recruited to the site of a sandfly bite [40]. As a professional phagocytic cell, the macrophage plays an important role in the immune response to foreign material, including pathogens. Normally, phagocytosis by macrophages initiates a destructive process that destroys the internalised microorganism, principally through NO production [41]. However, *Leishmania* has evolved mechanisms to neutralise the normal processes of macrophages, allowing it to survive within the cell and effectively avoid the immune system, thereby propagating the infection [39]. Understanding the pathways within the host cell that are modulated by the parasite has major implications for combating the consequences of an infection.

Considering the importance of phagocytosis, this study focused on the underlying cytoskeletal requirements, specifically F-actin and microtubules. Actin microfilaments, intermediate filaments, and microtubules are the major cytoskeleton components in eukaryotic cells. The cytoskeleton is a highly dynamic and responsive structure that is regulated by several proteins that are involved in a wide range of events such as signal transduction [42], cellular proliferation, cellular migration [29], phagocytosis and ROI, NO and cytokine production [43, 44]. Our initial experiments focused on the role of the cytoskeleton and the molecular motor myosin Va in the interaction between parasites and macrophages. Studies have shown the importance of the initial polymerisation of actin to the invasion of the host cells through phagocytosis [45–47] and the importance of interactions between the cytoskeleton and motor proteins such as myosin V [48].

Latrunculin A was chosen to depolymerise actin filaments because of its well-established ability to rapidly and reversibly modulate the organisation of actin in adherent cells [25, 49, 50]. In non-activated macrophages, the association index was significantly decreased after the reduction of F-actin (Figure 1). The index was also reduced in activated M1 macrophages. Although latrunculin A disrupts the actin cytoskeleton, at the concentration used in this work, some actin filaments still persist [51], which could explain the residual internalisation of *Leishmania*. The absolute index was greater in the activated macrophages than in non-activated macrophages, but the percent reduction from the nontreated controls was nearly equal. The observed higher index for the activated macrophages can be explained by the increase in phagocytosis that was promoted by LPS and IFN-γ treatment [52]. Despite the capacity of activated macrophages to kill parasites, the time frame of the experiments (30 min) was insufficient to observe any long-term effects.

The organisation of microtubules was disrupted through treatment with nocodazole [53, 54]. Nocodazole promoted high inhibition of parasite interactions with both M1 and non-activated macrophages, as measured by the association index. The effect was more pronounced in the M1 macrophages. This is in agreement with a previous observation that treatment of IFN-γ-activated RAW macrophages with nocodazole promoted the inhibition (40%) of phagocytosis indices [55]. Nocodazole was previously observed to inhibit by 40% the infection of intestinal epithelial cells (INT407) by *Enterobacter sakazakii* [56]. This low range of inhibition could explain the residual parasite-macrophage association observed in the presence of nocodazole. In addition, the nocodazole results suggest that the first steps in the internalisation of *Leishmania* by a macrophage are associated more with actin filaments than with microtubules.

It has long been observed that the treatment of macrophages with microfilament-inhibiting drugs such as cytochalasin D promotes a drastic decrease in *Leishmania* parasite binding. Both the number of parasites attached to macrophages and the proportion of infected macrophages diminish when macrophages were treated with 10 μg/mL of this substance [57]. Based on studies using cytochalasin, the dissociation of parasite attachment from the subsequent entry into host cells has been reported since the 1970s in several parasitic protozoa, including *Leishmania* [58], *Trypanosoma* [59], and *Toxoplasma* [60].

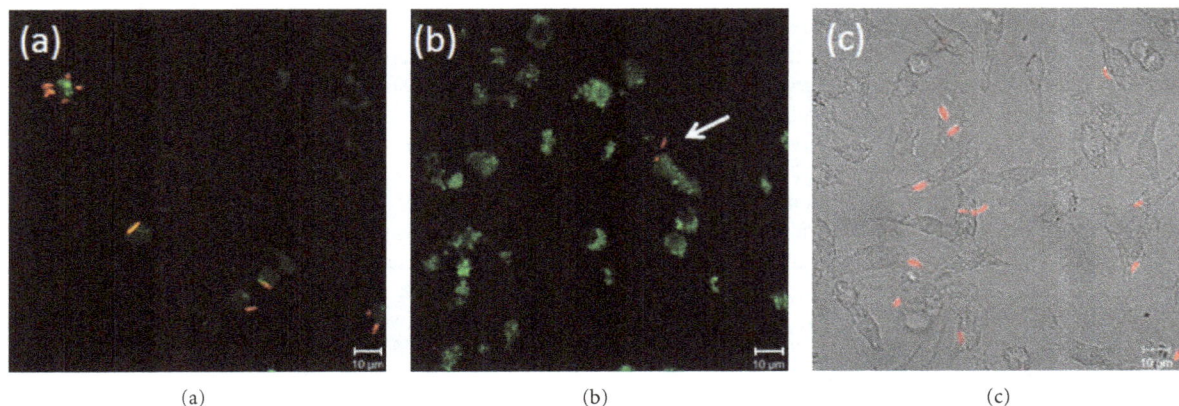

(a)   (b)   (c)

FIGURE 5: Internalisation of *L. braziliensis* requires myosin Va. The role of myosin Va was examined by observing the distribution of endogenous myosin Va in relation to parasites and the effect of the expression of a dominant negative tail domain fused to eGFP. Adherent Raw 264.7 macrophages were incubated with fluorescently marked parasites for 30 min before fixation, preparation and imaging by confocal microscopy. (a) shows endogenous myosin Va stained by an antibody (green) and parasites (red). (b) shows the myosin Va tail (green) and parasites (red). (c) shows mock-transfected macrophages by phase microscopy overlaid with a fluorescent image of parasites (red). Scale bars indicate imaging for all panels at 63x magnification.

In addition to phagocytosis interference, the possibility of these compounds to influence other microbiocidal macrophage events was tested. In this work, latrunculin has been shown to increase the NO production of non-activated and uninfected macrophages to the same extent as LPS and IFN-γ activation [44, 61, 62]. The monomeric form of actin (G-actin) promotes the stimulation of eNOS (endothelium NO synthase) [63]. In endothelial cells, NO production has been found to be directly related to changes in cell elasticity as a result of actin cytoskeleton reorganisation [64]. The increase in cell elasticity that occurs when actin is polymerised is accompanied by a decrease in NO production, while depolymerisation and, consequently, an increase in G-actin promote an increase in NO production [65].

Although the NO production of macrophages has primarily been associated with iNOS (induced NO synthase) activity, which is induced by the products of microorganisms such as LPS and a variety of inflammatory cytokines [66], RAW 264.7 mouse macrophages have been shown to constitutively express eNOS, and LPS-stimulation increased the activity of eNOS via changes in intracellular Ca^{2+} levels [67]. Because latrunculin associates with actin monomers, which prevent G-actin from polymerising [68], the accumulation of G-actin could be the key mechanism that causes the observed changes in the non-activated, uninfected macrophages. However, in the activated, infected macrophages, latrunculin promoted the inhibition of NO production. Likewise, activated macrophages treated with latrunculin or cytochalasin (another substance that disrupts the actin cytoskeleton) exhibited a decrease in NO production [69]. In contrast, nocodazole promoted an increase in the production of NO by infected M1 macrophages but inhibited NO production in uninfected macrophages (regardless of the activation status). Pulmonary artery cells (PAECs) treated with nocodazole have exhibited a decrease in NO production, but the exact mechanism remains unknown [53].

As already discussed, classical macrophage activation results in cells (M1) that are capable of producing high levels of tumour necrosis factor (TNF), reactive oxygen intermediates (ROIs), and iNOS expression. This activation is promoted by IFN-γ, and the pro-inflammatory response that is triggered results in the development of M1 macrophages. During infection by obligatory intracellular protozoan parasites such as *Leishmania*, *Trypanosoma*, *Toxoplasma*, and *Plasmodium*, these cells are necessary to control parasitemia, especially during the acute phase of the parasitosis [70–72]. However, depending on parameters such as host genotype, parasite virulence, and stage of infection, the hosts can also produce anti-inflammatory cytokines (Th2). These cytokines (IL-4 and IL-13) antagonise M1 macrophages, suppressing NO production by iNOS from L-arginine, and induce an alternative metabolic pathway of L-arginine catalysed by the enzyme arginase-1. This is the alternative pathway for macrophage activation, which involves type 2 responses and partly overlapping phenotypes, including IL-10 secretion. Therefore, these alternatively activated macrophages have been generically called M2 or AAM [10, 73, 74].

Latrunculin and nocodazole were observed to inhibit IL-10 secretion in infected M1 macrophages, as well as in uninfected and non-activated macrophages. There were no significant changes measured in the uninfected M1 macrophages. TNF-α secretion by infected M1 macrophages was not affected by either treatment, and only latrunculin was able to inhibit TNF-α secretion in uninfected M1 macrophages. Because both actin and microtubules are involved in the establishment of cell polarity and the directed secretion of cytokines and cytolytic granules [74], the data suggest that interference with the cytoskeleton is primarily responsible for the inhibition of cytokine secretion.

Another possibility is the inhibition of cytokine production. Colchicine, a drug that is also capable of promoting the depolymerisation of microtubules, was observed to cause a strong reduction in the accumulation of LPS-induced TNF-α mRNA. This event suggests that a pretranslational effect may represent the primary mechanism by which colchicine reduces TNF-α production [75]. Similarly, latrunculin and nocodazole could affect the mRNA of these cytokines, especially considering the role of the cytoskeleton in mRNA localisation [76, 77].

In addition, the disruption of microtubules can be associated with cytokine signal transduction. Toll-like receptors (TLRs) are involved in proinflammatory cytokine production following the recognition of several pathogen-derived components. These receptors activate the conserved MyD88 pathway, triggering the transcription factors NFκB and AP-1, which are essential factors in the production of inflammatory cytokines [78]. In dendritic cells, TLR2 and TLR4 are not present on the cell surface. They are associated with tubovesicular structures close to the Golgi complex and colocalise with microtubules. This suggests that TLR-decorated vesicles move along these structures. Supporting this, the depolymerisation of the microtubule network has been shown to disrupt the intracellular distribution of TLR2 and TLR4, which inhibits the production of IL-12 and TNF-α in response to *Neisseria meningitidis*. However, phagocytosis was not affected [79].

Myosin V motor proteins are responsible for cargo transportation and interact not only with actin filaments but also with several other components of the cytoskeleton, including microtubules, kinesins, and intermediate filaments [48]. Studies using RNA interference, gene deletion, and the expression of dominant-negative myosin tail constructs have been used to evaluate the role of this protein in the transport of motor organelles. These studies have shown a decrease in the speed or a total blockage of the cargo transport that is dependent on myosin Va [80–82]. Myosin Va appears to be involved in phagosome transportation, while other myosin classes appear to be involved in phagosome formation [24, 83, 84]. Araki [83] showed that myosin Va binds to the phagosome and F-actin, which restricts the movement of phagosomes.

Here, the presence of myosin Va in macrophages was confirmed, and colocalisation was observed between myosin Va and *L. braziliensis* (Figure 5(a)). To characterise the involvement of myosin Va with *L. braziliensis*, macrophages were transfected with a myosin Va tail construct to generate a dominant-negative effect by disengaging myosin Va cargo from actin filaments and thereby interfering with transport. As shown in Figure 5(b), the presence of the myosin Va tail reduced the association between macrophages and *L. braziliensis*. This result suggests that myosin Va plays a role in the association of *L. braziliensis* with macrophages.

Two important aspects of macrophage function that can be impacted by disrupting the cytoskeleton and myosin Va function are phagosome activity and receptor recycling. Recently, actin and microtubules were shown to have an important role in recycling from the phagosome [85]. In other cell types, myosin Va has been shown to contribute to membrane recycling and exocytosis [86, 87]. Considering the importance of the recycling of the mannose/fucose receptor

(MFR) to the plasma membrane and its role in the ingestion of *L. donovani* [88], the decreased *L. braziliensis* association observed in dominant-negative myosin Va macrophages could be caused by the absence or reduction of important receptors such as MFR on the macrophage membrane. Further investigations are necessary to characterise the role of myosin Va in the association of *L. braziliensis* with macrophages.

Overall, the data clearly show the importance of F-actin, microtubules, and myosin Va during the interactions between macrophages and *L. braziliensis*. Furthermore, the observations of the changes in the production of NO, IL-10, and TNF-α suggest that modulation of the cytoskeleton could be a mechanism for *L. braziliensis* to overcome the natural responses of macrophages and establish an infection.

Acknowledgments

The authors would like to thank Dr. Léa Cysne and Dr. Alda Maria da Cruz for providing the virulent strain stock MHOM/BR/2002/EMM-IOC-L2535 used in this study. This work was supported by grants from the Brazilian Agencies CNPq, FINEP, FAPERJ, and CAPES. Dr. P. Dutra is supported by a grant from the Third World Academy of Science (TWAS), Trieste, Italy (RGA no. 01-110 RG/BIO/LA), PADCT (CNPq/FAPERJ—no. 170.391/02) and CNPq (Edital Universal—no. 477562/2003-5, no. 480681/2004-0 and no. 478835/2008-6). Dr. P. Dutra and Dr. V. Salerno are supported by a grant from FAPERJ (Programa de Apoio aos Grupos Emergentes-E-26/111.544/2008), Brazil.

References

[1] P. A. Bates, "Housekeeping by *Leishmania*," *Trends in Parasitology*, vol. 22, no. 10, pp. 447–448, 2006.

[2] L. Kedzierski, "Leishmaniasis vaccine: Where are we today?" *Journal of Global Infectious Diseases*, vol. 2, no. 2, pp. 177–185, 2010.

[3] A. Q. Sousa and R. Pearson, "Drought, smallpox, and emergence of *Leishmania braziliensis* in northeastern Brazil," *Emerging Infectious Diseases*, vol. 15, no. 6, pp. 916–921, 2009.

[4] M. C. Brelaz, A. P. de Oliveira, A. F. de Almeida, M. de Assis Souza, M. E. de Brito, and V. R. Pereira, "Antigenic fractions of *Leishmania (Viannia) braziliensis*: the immune response characterization of patients at the initial phase of disease," *Parasite Immunology*, vol. 34, no. 4, pp. 183–239, 2012.

[5] J. Maüel, "Intracellular survival of protozoan parasites with special reference to *Leishmania* spp., *Toxoplasma gondii* and *Trypanosoma cruzi*," *Advances in Parasitology*, vol. 38, pp. 1–51, 1996.

[6] P. Kaye and P. Scott, "Leishmaniasis: complexity at the host-pathogen interface," *Nature Reviews Microbiology*, vol. 9, no. 8, pp. 604–615, 2011.

[7] A. Aderem and D. M. Underhill, "Mechanisms of phagocytosis in macrophages," *Annual Review of Immunology*, vol. 17, pp. 593–623, 1999.

[8] R. C. May and L. M. Machesky, "Phagocytosis and the actin cytoskeleton," *Journal of Cell Science*, vol. 114, supplement 6, pp. 1061–1077, 2001.

[9] K. S. Ravichandran and U. Lorenz, "Engulfment of apoptotic cells: signals for a good meal," *Nature Reviews Immunology*, vol. 7, no. 12, pp. 964–974, 2007.

[10] G. Raes, A. Beschin, G. H. Ghassabeh, and P. De Baetselier, "Alternatively activated macrophages in protozoan infections," *Current Opinion in Immunology*, vol. 19, no. 4, pp. 454–459, 2007.

[11] M. Belosevic, D. S. Finbloom, P. H. Van der Meide, M. V. Slayter, and C. A. Nacy, "Administration of monoclonal anti-IFN-γ antibodies in vivo abrogates natural resistance of C3H/HeN mice to infection with *Leishmania major*," *Journal of Immunology*, vol. 143, no. 1, pp. 266–274, 1989.

[12] C. Bogdan, A. Gessner, S. Werner, and R. Martin, "Invasion, control and persistence of *Leishmania* parasites," *Current Opinion in Immunology*, vol. 8, no. 4, pp. 517–525, 1996.

[13] S. Romagnani and A. K. Abbas, "IV International Conference on Cytokines. Ares-Serono Foundation (under the auspices of the European Cytokine Society)," *European Cytokine Network*, vol. 7, no. 4, pp. 801–827, 1996.

[14] P. Launois, F. Tacchini-Cottier, C. Parra-Lopez, and J. A. Louis, "Cytokines in parasitic diseases: the example of cutaneous Leishmaniasis," *International Reviews of Immunology*, vol. 17, no. 1–4, pp. 157–180, 1998.

[15] D. Sacks and N. Noben-Trauth, "The immunology of susceptibility and resistance to *Leishmania major* in mice," *Nature Reviews Immunology*, vol. 2, no. 11, pp. 845–858, 2002.

[16] P. Mansueto, G. Vitale, G. Di Lorenzo, G. B. Rini, S. Mansueto, and E. Cillari, "Immunopathology of Leishmaniasis: an update," *International Journal of Immunopathology and Pharmacology*, vol. 20, no. 3, pp. 435–445, 2007.

[17] M. D. Saúde and S. D. V. E. Saúde, "Manual de Vigilância da Leishmaniose Tegumentar Americana," 2007.

[18] S. L. Croft and G. H. Coombs, "Leishmaniasis—current chemotherapy and recent advances in the search for novel drugs," *Trends in Parasitology*, vol. 19, no. 11, pp. 502–508, 2003.

[19] L. Kedzierski, A. Sakthianandeswaren, J. M. Curtis, P. C. Andrews, P. C. Junk, and K. Kedzierska, "Leishmaniasis: current treatment and prospects for new drugs and vaccines," *Current Medicinal Chemistry*, vol. 16, no. 5, pp. 599–614, 2009.

[20] D. C. Arruda, F. L. D'Alexandri, A. M. Katzin, and S. R. B. Uliana, "AntiLeishmanial activity of the terpene nerolidol," *Antimicrobial Agents and Chemotherapy*, vol. 49, no. 5, pp. 1679–1687, 2005.

[21] A. A. Kramerov, A. G. Golub, V. G. Bdzhola et al., "Treatment of cultured human astrocytes and vascular endothelial cells with protein kinase CK2 inhibitors induces early changes in cell shape and cytoskeleton," *Molecular and Cellular Biochemistry*, vol. 349, no. 1-2, pp. 125–137, 2011.

[22] P. M. L. Dutra, D. P. Vieira, J. R. Meyer-Fernandes, M. A. C. Silva-Neto, and A. H. Lopes, "Stimulation of *Leishmania tropica* protein kinase CK2 activities by platelet-activating factor (PAF)," *Acta Tropica*, vol. 111, no. 3, pp. 247–254, 2009.

[23] A. K. C. Lima, C. G. R. Elias, J. E. O. Souza, A. L. S. Santos, and P. M. L. Dutra, "Dissimilar peptidase production by avirulent and virulent promastigotes of *Leishmania braziliensis*: inference on the parasite proliferation and interaction with macrophages," *Parasitology*, vol. 136, no. 10, pp. 1179–1191, 2009.

[24] M. Diakonova, G. Bokoch, and J. A. Swanson, "Dynamics of cytoskeletal proteins during Fcγ receptor-mediated phagocytosis in macrophages," *Molecular Biology of the Cell*, vol. 13, no. 2, pp. 402–411, 2002.

[25] T. Bretschneider, K. Anderson, M. Ecke et al., "The three-dimensional dynamics of actin waves, a model of cytoskeletal self-organization," *Biophysical Journal*, vol. 96, no. 7, pp. 2888–2900, 2009.

[26] N. Sharma, Z. A. Kosan, J. E. Stallworth, N. F. Berbari, and B. K. Yoder, "Soluble levels of cytosolic tubulin regulate ciliary length control," *Molecular Biology of the Cell*, vol. 22, no. 6, pp. 806–816, 2011.

[27] E. A. Roberts, J. Chua, G. B. Kyei, and V. Deretic, "Higher order Rab programming in phagolysosome biogenesis," *Journal of Cell Biology*, vol. 174, no. 7, pp. 923–929, 2006.

[28] M. Coue, S. L. Brenner, I. Spector, and E. D. Korn, "Inhibition of actin polymerization by latrunculin A," *FEBS Letters*, vol. 213, no. 2, pp. 316–318, 1987.

[29] C. Hayot, O. Debeir, P. Van Ham, M. Van Damme, R. Kiss, and C. Decaestecker, "Characterization of the activities of actin-affecting drugs on tumor cell migration," *Toxicology and Applied Pharmacology*, vol. 211, no. 1, pp. 30–40, 2006.

[30] R. F. Luduena and M. C. Roach, "Tubulin sulfhydryl groups as probes and targets for antimitotic and antimicrotubule agents," *Pharmacology and Therapeutics*, vol. 49, no. 1-2, pp. 133–152, 1991.

[31] R. J. Vasquez, B. Howell, A. M. C. Yvon, P. Wadsworth, and L. Cassimeris, "Nanomolar concentrations of nocodazole alter microtubule dynamic instability in vivo and in vitro," *Molecular Biology of the Cell*, vol. 8, no. 6, pp. 973–985, 1997.

[32] M. A. Jordan, D. Thrower, and L. Wilson, "Effects of vinblastine, podophyllotoxin and nocodazole on mitotic spindles. Implications for the role of microtubule dynamics in mitosis," *Journal of Cell Science*, vol. 102, part 3, pp. 401–416, 1992.

[33] B. Storrie and W. Yang, "Dynamics of the interphase mammalian Golgi complex as revealed through drugs producing reversible Golgi disassembly," *Biochimica et Biophysica Acta*, vol. 1404, no. 1-2, pp. 127–137, 1998.

[34] E. Gebru, E. H. Kang, D. Damte et al., "The role of Janus kinase 2 (JAK2) activation in pneumococcal EstA protein-induced inflammatory response in RAW 264.7 macrophages," *Microbial Pathogenesis*, vol. 51, no. 4, pp. 297–303, 2011.

[35] N. J. Laparra III and B. L. Kelly, "Supression of LPS-induced inflammatory responses in macrophages infected with *Leishmania*," *Jornal of Inflamtion*, vol. 7, article 8, 2010.

[36] M. D. S. S. Rosa, R. B. Vieira, A. F. Pereira, P. M. L. Dutra, and A. H. C. S. Lopes, "Platelet-activating factor (PAF) modulates peritoneal mouse macrophage infection by *Leishmania amazonensis*," *Current Microbiology*, vol. 43, no. 1, pp. 33–37, 2001.

[37] L. Viinikka, "Nitric oxide as a challenge for the clinicalchemistry laboratory," *Scandinavian Journal of Clinical and Laboratory Investigation*, vol. 56, no. 7, pp. 577–581, 1996.

[38] H. Tominaga, M. Ishiyama, F. Ohseto et al., "A water-soluble tetrazolium salt useful for colorimetric cell viability assay," *Analytical Communications*, vol. 36, no. 2, pp. 47–50, 1999.

[39] P. M. L. Dutra, S. A. G. Silva, A. M. da Cruz, and F. L. Dutra, "Leishmaniose Americana," in *Bases Moleculares em Clinica Médica*, Editora Atheneu, 1st edition, 2010.

[40] A. H. Lopes, T. Souto-Padón, F. A. Dias et al., "Trypanosomatids: odd organisms, devastating diseases," *The Open Parasitology Journal*, vol. 4, pp. 30–59, 2010.

[41] T. Bosschaerts, M. Guilliams, W. Noel et al., "Alternatively activated myeloid cells limit pathogenicity associated with african trypanosomiasis through the IL-10 inducible gene selenoprotein P," *Journal of Immunology*, vol. 180, no. 9, pp. 6168–6175, 2008.

[42] M. Kumeta, S. H. Yoshimura, J. Hejna, and K. Takeyasu, "Nucleocytoplasmic shuttling of cytoskeletal proteins: molecular mechanism and biological significance," *International Journal of Cell Biology*, vol. 2012, Article ID 494902, 12 pages, 2012.

[43] M. T. Shio, K. Hassani, A. Isnard et al., "Host cell signalling and *Leishmania* mechanisms of evasion," *Journal of Tropical Medicine*, vol. 2012, Article ID 819512, 14 pages, 2012.

[44] H. E. Cummings, R. Tuladhar, and A. R. Satoskar, "Cytokines and their STATs in cutaneous and visceral Leishmaniasis," *Journal of biomedicine & biotechnology*, vol. 2010, Article ID 294389, 6 pages, 2010.

[45] O. Poupel and I. Tardieux, "*Toxoplasma gondii* motility and host cell invasiveness are drastically impaired by jasplakinolide, a cyclic peptide stabilizing F-actin," *Microbes and Infection*, vol. 1, no. 9, pp. 653–662, 1999.

[46] J. M. Dobrowolski and L. D. Sibley, "Toxoplasma invasion of mammalian cells is powered by the actin cytoskeleton of the parasite," *Cell*, vol. 84, no. 6, pp. 933–939, 1996.

[47] E. W. Eng, A. Bettio, J. Ibrahim, and R. E. Harrison, "MTOC reorientation occurs during FcγR-mediated phagocytosis in macrophages," *Molecular Biology of the Cell*, vol. 18, no. 7, pp. 2389–2399, 2007.

[48] J. L. Maravillas-Montero and L. Santos-Argumedo, "The myosin family: unconventional roles of actin-dependent molecular motors in immune cells," *Journal of Leukocyte Biology*, vol. 91, pp. 35–46, 2012.

[49] Z. H. Németh, E. A. Deitch, M. T. Davidson, C. Szabó, E. S. Vizi, and G. Haskó, "Disruption of the actin cytoskeleton results in nuclear factor-κB activation and inflammatory mediator production in cultured human intestinal epithelial cells," *Journal of Cellular Physiology*, vol. 200, no. 1, pp. 71–81, 2004.

[50] Z. Feng, B. Chen, S. C. Tang, K. Liao, W. N. Chen, and V. Chan, "Effect of cytoskeleton inhibitors on deadhesion kinetics of HepG2 cells on biomimetic surface," *Colloids and Surfaces B*, vol. 75, no. 1, pp. 67–74, 2010.

[51] T. Wakatsuki, B. Schwab, N. C. Thompson, and E. L. Elson, "Effects of cytochalasin D and latrunculin B on mechanical properties of cells," *Journal of Cell Science*, vol. 114, no. 5, pp. 1025–1036, 2001.

[52] Y. Goto, K. Ogawa, A. Hattori, and M. Tsujimoto, "Secretion of endoplasmic reticulum aminopeptidase 1 is involved in the activation of macrophages induced by lipopolysaccharide and interferon-γ," *The Journal of Biological Chemistry*, vol. 286, no. 24, pp. 21906–21914, 2011.

[53] S. Yunchao, S. I. Zharikov, and E. R. Block, "Microtubule-active agents modify nitric oxide production in pulmonary artery endothelial cells," *American Journal of Physiology*, vol. 282, no. 6, pp. L1183–L1189, 2002.

[54] T. H. Sulahian, A. Imrich, G. DeLoid, A. R. Winkler, and L. Kobzik, "Signaling pathways required for macrophage scavenger receptor-mediated phagocytosis: analysis by scanning cytometry," *Respiratory Research*, vol. 9, article 59, 2008.

[55] A. Khandani, E. Eng, J. Jongstra-Bilen et al., "Microtubules regulate PI-3K activity and recruitment to the phagocytic cup during Fcγ receptor-mediated phagocytosis in nonelicited macrophages," *Journal of Leukocyte Biology*, vol. 82, no. 2, pp. 417–428, 2007.

[56] N. M. K. Mohan and K. Venkitanarayanan, "Role of bacterial OmpA and host cytoskeleton in the invasion of human intestinal epithelial cells by *Enterobacter sakazakii*," *Pediatric Research*, vol. 62, no. 6, pp. 664–669, 2007.

[57] A. Zenian, P. Rowles, and D. Gingell, "Scanning electron-microscopic study of the uptake of *Leishmania* parasites by macrophages," *Journal of Cell Science*, vol. 39, pp. 187–199, 1979.

[58] J. Alexander, "Effect of the antiphagocytic agent cytochalasin B on macrophage invasion by *Leishmania mexicana* promastigotes and *Trypanosoma cruzi* epimastigotes," *Journal of Protozoology*, vol. 22, no. 2, pp. 237–240, 1975.

[59] N. Nogueira and Z. Cohn, "*Trypanosoma cruzi*: mechanism of entry and intracellular fate in mammalian cells," *Journal of Experimental Medicine*, vol. 143, no. 6, pp. 1402–1420, 1976.

[60] F. W. Ryning and J. S. Remington, "Effect of cytochalasin D on *Toxoplasma gondii* cell entry," *Infection and Immunity*, vol. 20, no. 3, pp. 739–743, 1978.

[61] S. J. Green, R. M. Crawford, J. T. Hockmeyer, M. S. Meltzer, and C. A. Nacy, "*Leishmania major* amastigotes initiate the L-arginine-dependent killing mechanism in IFN-γ-stimulated macrophages by induction of tumor necrosis factor-α1," *Journal of Immunology*, vol. 145, no. 12, pp. 4290–4297, 1990.

[62] P. Tripathi, P. Tripathi, L. Kashyap, and V. Singh, "The role of nitric oxide in inflammatory reactions," *Immunology and Medical Microbiology*, vol. 51, no. 3, pp. 443–452, 2007.

[63] Y. Su, S. Edwards-Bennett, M. R. Bubb, and E. R. Block, "Regulation of endothelial nitric oxide synthase by the actin cytoskeleton," *American Journal of Physiology*, vol. 284, no. 6, pp. C1542–C1549, 2003.

[64] J. Fels, H. Oberleithner, and K. Kusche-Vihrog, "Ménage à trois: aldosterone, sodium and nitric oxide in vascular endothelium," *Biochimica et Biophysica Acta*, vol. 1802, no. 12, pp. 1193–1202, 2010.

[65] A. M. Szczygiel, G. Brzezinka, M. Targosz-Korecka, S. Chlopicki, and M. Szymonski, "Elasticity changes anti-correlate with NO production for human endothelial cells stimulated with TNF-α," *European Journal of Physiology*, vol. 463, no. 3, pp. 487–496, 2011.

[66] J. MacMicking, Q. W. Xie, and C. Nathan, "Nitric oxide and macrophage function," *Annual Review of Immunology*, vol. 15, pp. 323–350, 1997.

[67] H. H. H. W. Schmidt, T. D. Warner, M. Nakane, U. Forstermann, and F. Murad, "Regulation and subcellular location of nitrogen oxide synthases in RAW264.7 macrophages," *Molecular Pharmacology*, vol. 41, no. 4, pp. 615–624, 1992.

[68] I. J. Shin, Y. T. Ahn, Y. Kim, J. M. Kim, and W. G. An, "Actin disruption agents induce phosphorylation of histone H2AX in human breast adenocarcinoma MCF-7 cells," *Oncology Reports*, vol. 25, no. 5, pp. 1313–1319, 2011.

[69] S. M. Eswarappa, V. Pareek, and D. Chakravortty, "Role of actin cytoskeleton in LPS-induced NF-κB activation and nitric oxide production in murine macrophages," *Innate Immunity*, vol. 14, no. 5, pp. 309–318, 2008.

[70] P. Kropf, M. A. Freudenberg, M. Modolell et al., "Toll-like receptor 4 contributes to efficient control of infection with the protozoan parasite *Leishmania major*," *Infection and Immunity*, vol. 72, no. 4, pp. 1920–1928, 2004.

[71] M. Plebanski and A. V. S. Hill, "The immunology of malaria infection," *Current Opinion in Immunology*, vol. 12, no. 4, pp. 437–441, 2000.

[72] M. Walther, J. Woodruff, F. Edele et al., "Innate immune responses to human malaria: heterogeneous cytokine responses to blood-stage *Plasmodium falciparum* correlate with parasitological and clinical outcomes," *Journal of Immunology*, vol. 177, no. 8, pp. 5736–5745, 2006.

[73] A. Mantovani, A. Sica, S. Sozzani, P. Allavena, A. Vecchi, and M. Locati, "The chemokine system in diverse forms of

macrophage activation and polarization," *Trends in Immunology*, vol. 25, no. 12, pp. 677–686, 2004.

[74] T. S. Gomez and D. D. Billadeau, "T cell activation and the cytoskeleton: you can't have one without the other," *Advances in Immunology*, vol. 97, pp. 1–64, 2008.

[75] Z. Li, G. S. Davis, C. Mohr, M. Nain, and D. Gemsa, "Inhibition of LPS-induced tumor necrosis factor-α production by colchicine and other microtubule disrupting drugs," *Immunobiology*, vol. 195, no. 4-5, pp. 624–639, 1996.

[76] E. Louvet and P. Percipalle, "Transcriptional control of gene expression by actin and myosin," *International Review of Cell and Molecular Biology*, vol. 272, pp. 107–147, 2008.

[77] R. H. Singer, "Highways for mRNA transport," *Cell*, vol. 134, no. 5, pp. 722–723, 2008.

[78] T. Kawai and S. Akira, "TLR signaling," *Cell Death and Differentiation*, vol. 13, no. 5, pp. 816–825, 2006.

[79] H. Uronen-Hansson, J. Allen, M. Osman, G. Squires, N. Klein, and R. E. Callard, "Toll-like receptor 2 (TLR2) and TLR4 are present inside human dendritic cells, associated with microtubules and the Golgi apparatus but are not detectable on the cell surface: Integrity of microtubules is required for interleukin-12 production in response to internalized bacteria," *Immunology*, vol. 111, no. 2, pp. 173–178, 2004.

[80] W. Wagner, E. Fodor, A. Ginsburg, and J. A. Hammer 3rd, "The binding of DYNLL2 to myosin Va requires alternatively spliced exon B and stabilizes a portion of the myosin's coiled-coil domain," *Biochemistry*, vol. 45, no. 38, pp. 11564–11577, 2006.

[81] Z. Hódi, A. L. Németh, L. Radnai et al., "Alternatively spliced exon B of myosin Va is essential for binding the tail-associated light chain shared by dynein," *Biochemistry*, vol. 45, no. 41, pp. 12582–12595, 2006.

[82] J. A. Hammer and J. R. Sellers, "Walking to work: roles for class V myosins as cargo transporters," *Nature Reviews*, vol. 13, pp. 13–26, 2012.

[83] N. Araki, "Role of microtubules and myosins in Fc gamma receptor-mediated phagocytosis," *Frontiers in Bioscience*, vol. 11, no. 2, pp. 1479–1490, 2006.

[84] J. A. Swanson, M. T. Johnson, K. Beningo, P. Post, M. Mooseker, and N. Araki, "A contractile activity that closes phagosomes in macrophages," *Journal of Cell Science*, vol. 112, part 3, pp. 307–316, 1999.

[85] M. T. Damiani and M. I. Colombo, "Microfilaments and microtubules regulate recycling from phagosomes," *Experimental Cell Research*, vol. 289, no. 1, pp. 152–161, 2003.

[86] I. V. Röder, Y. Petersen, K. R. Choi, V. Witzemann, J. A. Hammer 3rd, and R. Rudolf, "Role of myosin Va in the plasticity of the vertebrate neuromuscular junction in vivo," *PLoS ONE*, vol. 3, no. 12, Article ID e3871, 2008.

[87] R. Rudolf, C. M. Bittins, and H. H. Gerdes, "The role of myosin v in exocytosis and synaptic plasticity," *Journal of Neurochemistry*, vol. 116, no. 2, pp. 177–191, 2011.

[88] M. E. Wilson and R. D. Pearson, "Evidence that *Leishmania donovani* utilizes a mannose receptor on human mononuclear phagocytes to establish intracellular parasitism," *Journal of Immunology*, vol. 136, no. 12, pp. 4681–4688, 1986.

10

Enterocytozoon bieneusi Identification Using Real-Time Polymerase Chain Reaction and Restriction Fragment Length Polymorphism in HIV-Infected Humans from Kinshasa Province of the Democratic Republic of Congo

Roger Wumba,[1] Menotti Jean,[2] Longo-Mbenza Benjamin,[3] Mandina Madone,[4] Kintoki Fabien,[4] Zanga Josué,[1] Sala Jean,[1] Kendjo Eric,[5] Guillo-Olczyk AC,[6] and Thellier Marc[5,6]

[1] *Département de Médecine Tropicale, Maladies Infectieuses et Parasitaires, Cliniques Universitaires de Kinshasa, Université de Kinshasa, 747 Kinshasa XI, Democratic Republic of Congo*
[2] *Service de Parasitologie-Mycologie, Hôpital Saint-Louis, Assistance Publique-Hôpitaux de Paris et Faculté de Médecine Lariboisière-Saint-Louis, Université Paris VII, 75010 Paris, France*
[3] *Faculty of Health Sciences, Walter Sisulu University, Private Bag XI, Mthatha, Eastern Cape 5117, South Africa*
[4] *Department of Internal Medicine, University of Kinshasa, 783 Kinshasa XI, Democratic Republic of Congo*
[5] *National Center for Malaria Research, AP-HP, CHU Pitié Salpetrière, 75013 Paris, France*
[6] *AP-HP, Groupe hospitalier Pitié-Salpêtrière, Service de Parasitologie-Mycologie, Université Pierre et Marie Curie, 75013 Paris, France*

Correspondence should be addressed to Longo-Mbenza Benjamin, longombenza@gmail.com

Academic Editor: Francisca Mutapi

Objective. To determine the prevalence and the genotypes of *Enterocytozoon bieneusi* in stool specimens from HIV patients. *Methods*. This cross-sectional study was carried out in Kinshasa hospitals between 2009 and 2012. Detection of microsporidia including *E. bieneusi* and *E. intestinalis* was performed in 242 HIV-infected patients. Typing was based on DNA polymorphism of the ribosomal DNA ITS region of *E. bieneusi*. PCRRFLP generated with two restriction enzymes (Nla III and Fnu 4HI) in PCR-amplified ITS products for classifying strains into different lineages. The diagnosis performance of the indirect immune-fluorescence-monoclonal antibody (IFI-AcM) was defined in comparison with real-time PCR as the gold standard. *Results*. Out of 242 HIV-infected patients, using the real-time PCR, the prevalence of *E. bieneusi* was 7.9% ($n = 19$) among the 19 *E. bieneusi*, one was coinfected with *E. intestinalis*. In 19 *E. bieneusi* persons using PCR-RFLP method, 5 type I strains of *E. bieneusi* (26.3%) and 5 type IV strains of *E. bieneusi* (26.3%) were identified. The sensitivity of IFI-AcM was poor as estimated 42.1%. *Conclusion*. Despite different PCR methods, there is possible association between HIVinfection, geographic location (France, Cameroun, Democratic Republic of Congo), and the concurrence of type I and type IV strains.

1. Introduction

It is established that *Enterocytozoon bieneusi* (*E. bieneusi*) is the most commonly characterized microsporidia species among human beings. Microsporidia, obligate intracellular parasites, lack eukaryotic ribosomal features and peroxisomes [1]. Their spores do penetrate and infect eukaryotic cells in various invertebrate and vertebrate organisms.

The literature reports epidemiology, causes, diagnosis, and digestive disorders related to microsporidiosis among HIV-patients [2–7].

In Kinshasa region, The capital city of The Democratic Republic of Congo (DRC), we detected *E. bieneusi* infection in HIV patients using only light microscopy and Fungi Fluor [8] as well as conventional polymerase chain reaction (PCR) method [9]. We could confirm the sensitivity of the diagnosis of *E. bieneusi* infection by a real-time PCR assay in comparison with traditional methods [10, 11].

E. bieneusi genotypes were also identified by PCR-restriction fragment length polymorphism (RFLP) analysis [12, 13].

Therefore, the objective of this study was to determine the prevalence and the genotypes of *E. bieneusi* in stool specimens among HIV patients by developing a rapid and efficient real-time PCR and PCR-RFLP approach.

2. Materials and Methods

2.1. Study Design. This study was designed as a descriptive cross-sectional approach between December 2009 and January 2012.

2.2. Ethical Considerations. The institutional review boards and the Committee of Ethics of the University of Kinshasa Faculty of Medicine approved the protocol of the study which was conducted in compliance with the principles of Helsinki Declaration. The procedures of the study were explained, and an informed consent sheet was signed by each participant or a designated literate substitute when necessary.

2.3. Study Setting. In the Kinshasa community, Democratic Republic of Congo, the Cliniques Universitaires de Kinshasa (CUK) as the teaching hospital at the south-western part of Kinshasa city, the general referral hospital of Kinshasa (HGRK) in the center of Kinshasa city, the general referral hospital of Kintambo (HGRKint) at the Northeastern Kinshasa city, and military referral hospital of Camp Kokolo (HMRK) at the western part of Kinshasa city were randomly selected.

2.4. Patients and Clinical Specimens. We included 242 consecutive HIV-infected patients. The clinical signs characteristic of HIV disease were collected among all participants.

2.5. Diagnosis of E. bieneusi Infection. We collected 242 fresh stool samples in pH 7.2 buffer stored at $+4°C$ before analysis. The stool specimens from all 242 patients were diluted at PBS solution for microscopic examination.

Microscopic examination and specific staining were done both in Kinshasa University Parasitology laboratories (CUK) and in the Pitié Salpêtrière Hospital (PSL) Parasitology Mycology Laboratory, Paris, France. Stool samples (one for each patient) were studied using optical microscopy (direct examination and trichrome specific staining as modified by Weber) for microsporidia detection [14].

The indirect immunofluorescence-monoclonal antibody (IFI-AcM) techniques were used for the identification of *E. bieneusi* and *E. intestinalis* [15, 16].

2.6. Genomic DNA Extraction. DNA extraction was performed by using the QIAamp DNA Mini Kit (Qiagen, Hilden, Germany) according to the supplier's protocol.

TABLE 1: Clinical signs of our HIV patients.

Clinical signs	N/242	%
Asthenia	88	36,3
Diarrhea	83	34,3
Pulmonary signs	52	21,4
Cutaneous signs	42	17,3
Anorexia	28	11,5
Fever	25	10,3
Emaciation	14	5,7
Anemia	5	2

2.7. Real-Time PCR. We carried out a real-time PCR for all samples at the Saint Louis Hospital Parasitology Mycology service in Paris, France, using a 7500 Real-Time PCR System (Applied Biosystems, Foster City, CA, USA) for all three species identification (*E. bieneusi* and *E. intestinalis*).

For *E. bieneusi*, the real-time PCR assay amplified a 102bp fragment of the small subunit ribosomal RNA gene, with FEB1 (5′-CGCTGTAGTTCCTGCAGTAAACTATGC-C-3′) and REB1 (5′-CTTGCGAGCGTACTATCCCCAGAG-3′) primers and a fluorescent TaqMan probe (5′-ACGTGG-GCGGGAGAAATCTTAGTGTTCGGG-3′), as previously described [10]. For *E. intestinalis*, the real-time PCR assay was performed by using FEI1 (5′-GCAAGGGAGGAATGG-AACAGAACAG-3′) and REI1 (5′-CACGTTCAGAAGCCC-ATTACACAGC-3′)-primers, with the following fluorescent TaqMan probe: 5′-FAM-CGGGCGGCACGCGCACTA-CGATA-TAMRA-3′, as previously described [10, 11].

2.8. PCR-RFLP for E. bieneusi Genotype Identification. The PCR-RFLP assay was performed on a 9700 PCR system (Applied Biosystems) as previously described [12, 13]. The RFLP analysis was performed on a 2% agarose gel by comparing the number and the length of the obtained PCR undigested and digested fragments by using Fnu4HI and NlaIII restriction enzymes.

2.9. Statistical Analysis. Data were expressed as proportions (%) for categorical variables and means with standard deviations for continuous variables. Differences were compared by the chi-square test for proportions and by the Student's t-test for continuous variables with results considered statistically significant for P value < 0.05. All analyses were performed by use of STATA (version 11) software package.

3. Results

3.1. Clinical Profile of Patients. Of 242 HIV/AIDS patients, 35.9% ($n = 87$) were males and 64.1% ($n = 155$) were females: sex ratio of 2 women: 1 man. The mean age of the participants was $39.2 ± 11.8$ years (range: 15–73).

Table 1 presents the clinical signs of the study population. Asthenia and diarrhea were the most frequent signs among the participants.

TABLE 2: Microsporidia (*E. bieneusi*, *E. intestinalis*, and genotypes).

N/19	IFI-AcM Eb,Ei	PCR RT Eb,Ei	Genotypes par RFLP
07	*Eb*	*Eb*	**Type 4**
08	No	*Eb*	**ND**
10	*Eb*	*Eb*	**Type 4**
12	No	*Eb*	**ND**
30	No	*Eb*	**Type 1**
34	*Eb*	*Eb*	**Type 4**
36	*Eb*	*Eb*	**ND**
37	*Eb*	*Eb*	**ND**
39	*Eb*	*Eb*	**ND**
40	*Eb*	*Eb*	**ND**
44	No	*Eb*	**Type 1**
49	No	*Eb*	**Type 1**
63	***Eb***	***Eb, Ei***	**Type 4**
89	No	*Eb*	**ND**
93	No	*Eb*	**ND**
105	No	*Eb*	**Type 4**
134	No	*Eb*	**ND**
183	No	*Eb*	**Type 1**
220	No	*Eb*	**Type 1**

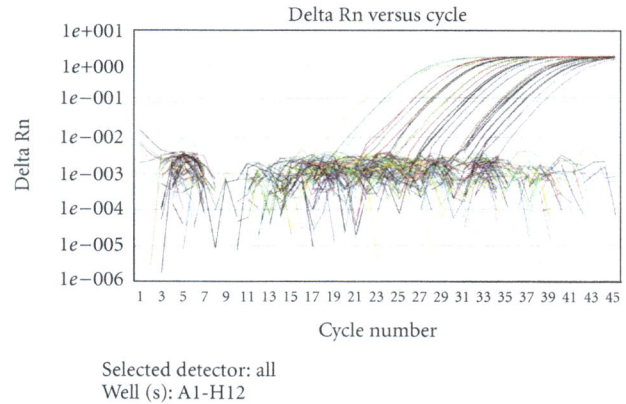

FIGURE 1: Amplification curves obtained with the *E. bieneusi*-specific real-time PCR assay.

Selected detector: all
Well (s): A1-H12

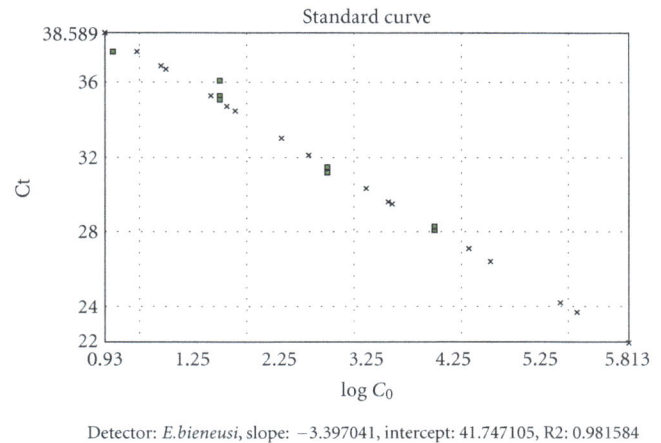

Detector: *E.bieneusi*, slope: −3.397041, intercept: 41.747105, R2: 0.981584

FIGURE 2: Standard curve representing the threshold cycle (Ct) values as a function of the decimal logarithms of *E. bieneusi* small subunit rRNA gene copy number per μl.

3.2. Molecular Evaluation and Prevalence. Out of 242 HIV-infected patients, using the real-time PCR, the prevalence of *E. bieneusi* was 7.9% ($n = 19$). Among the 19 *E. bieneusi*, one was coinfected with *E. intestinalis*.

Table 2 presents the findings from IFI-AcM, real-time PCR, and genotypes. The diagnosis efficiency of IFI-AcM was defined with comparison with the real-time PCR as follows: sensitivity of 42.1%, specificity of 100%, positive predictive value of 100%, and negative predictive value of 95%.

Figure 1 shows the function of the relative fluorescent signal (Delta Rn) according to the cycle number.

The sensitivity and reproducibility of real-time PCR was assessed by repeated testing of serial dilutions (Figure 2). The relation between Ct value and the decimal logarithm of *E. bieneusi* small subunit rRNA gene copy number per μl was as follows: slope = −3.397 and intercept = 41.747.

PCR-RFLP analysis of the amplification products of the ITS region was then performed on the 19 *E. bieneusi* stool isolates (Figure 3). We found two genetically unrelated lineages: type I strains without digestion of amplicons with Fnu 4HI, and type IV strains with digestion of amplicons with NlaIII and Fnu4HI.

4. Discussion

In the present study, we have used two real-time PCR assays and a PCR-RFLP assay for the quantitative detection of *E. bieneusi* DNA and strain genotyping from stool specimens.

Clinical features from the HIV-infected participants were similar to the frequency of diarrhea reported among other African patients [2–7].

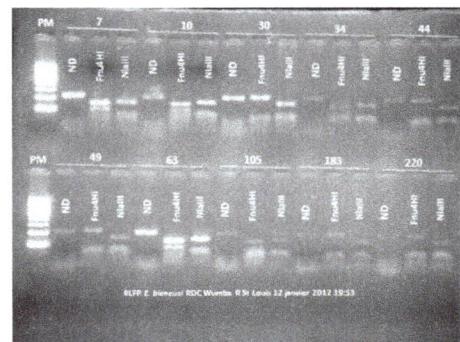

FIGURE 3: RFLP analysis of *E. bieneusi* PCR products after digestion with Fnu4HI and NlaIII enzymes. ND: not digested, PM: molecular weight marker.

The prevalence of *E. bieneusi* identified by PCR in these HIV Congolese patients was estimated at 8.2% (7.9% of *E. bieneusi*), which was higher than the prevalence of microsporidia found using similar PCR techniques in other African countries (less than 5%) [4, 17–25]. These low rates of microsporidiosis could be related to the location and availability of antiretroviral therapy (ART). Indeed, the prevalence of microsporidia including *E. bieneusi* in HIV-infected people has dramatically decreased in countries where ART is widely available [26, 27]. However, in most African countries including our Congolese study, few patients have access to ART [1, 8, 9], which could explain the higher prevalence found in our study and in some other African studies among HIV-infected individuals [8, 9, 28].

In this study, we used a rapid and efficient qPCR method combined with PCR-RFLP genotyping and IIF-MAb for determining intestinal microsporidiosis from stool specimens among HIV-infected patients. Thus, we confirmed the best diagnostic of *E. bieneusi* using more sensitive and specific real-time PCR than the diagnosis of *E. intestinalis* [10–13].

The literature reports that *E. bieneusi* is a relatively homogeneous entity with PCR-RFLP-based putative polymorphism of the ITS region of *E. bieneusi* [5]. This putative polymorphism of the ITS region of *E. bieneusi* had a genetic diversity of *E. bieneusi* [5].

Among the 19 *E. bieneusi* cases we studied, we identified 5 type I strains of *E. bieneusi* (26.3%) and 5 type IV strains. By contrast, HIV-infected patients in France were in majority infected with type I strains [12, 13]. Interestingly, type IV strains were also encountered in a previous study in Cameroon [18]. Furthermore, Tumwine et al. found a majority of genotype K strains, which correspond to type IV in our classification, in children from Uganda [29].

4.1. Findings and Current Understanding in the Field within the Field. The present work and the work by Liguory et al. [12, 13] were performed using the same PCR-RFLP developed by Liguory team. Our typing was based on DNA polymorphism of the ribosomal DNA internal transcribed spacer (ITS) region of *E. bieneusi*. PCR-RFLP generated with two restriction enzymes (Nla III and Fnu4HI) in PCR-amplified ITS products at classifying type I, type II, type III, and type IV [12, 13].

Santín et al. [30] were among the leaders to reduce confusion associated with the identification of genotypes within *E. bieneusi* after the meeting during IWOP-10. According to the consensus [30], previously, the correspondence for the nomenclature was as follows: genotype B belongs to type I, genotype C belongs to type II, genotype , undetermined genotype does not belong to type III, and genotype K belongs to genotype IV [13, 30, 31].

Despite the standard methods for determining the genotypes of *E. bieneusi* based on the DNA sequence of the internal transcribed spacer (ITS) region, the r-RNA gene in the publication of Santín et al. [30], the present work in Kinshasa (DRC), and the previous works in France [12, 13] and in Cameroun [31] showed a significant association

between HIV-infection and genotypes I and IV *E. bieneusi*. Genotype IV *E. bieneusi* was only present among HIV-patients from Nigeria [32], Uganda [29], Gabon [31], and Portugal [33]. The genotypes II and III *E. bieneusi* were not identified in the present study from Kinshasa (DRC) as they are not yet reported from Africa. However, genotypes II and III *E. bieneusi* are more frequent among HIV-negative people from Europe [12, 13]. Genotype I in HIV-patients is commoner and more frequent than genotype IV in Europe [12, 13, 34] than in HIV-patients from Central Africa including Democratic Republic of Congo with the present study and Cameroun [31].

In this study, the genotype I–genotype IV *E. bieneusi* ratio was 1 in HIVpatients and emerging: genotype I *E. bieneusi* in 5 cases of HIV/AIDS versus genotype IV *E. bieneusi* in 5 cases of HIV/AIDS. Possible rapid travels between France and francophone Central Africa may be a factor contributory to the emerging genotype I *E. bieneusi*.

4.2. Implications for Public Health. The significant diagnosis efficiency of PCR methods for *E. bieneusi* will have implications on management of HIV-related microsporidia.

The accurate identification and differentiation of microsporidian species by real-time PCR techniques will improve therapy, clinical manifestations, and prognosis [35–37].

Modes of transmission and sources of human infection by *E. bieneusi* or HIV and molecular analyses developed by real-time PCR and RFLP should be useful for epidemiological studies [1, 5, 8, 9, 35–39].

5. Conclusion

The prevalence of *E. bieneusi* is emerging. We used a sensitive, specific, rapid, and efficient approach for typing *E. bieneusi* obtained from stool specimens by real-time PCR and PCR-RFLP assays. Genotype I *E. bieneusi* is more prevalent among HIV-patients from Europe than the genotype I–genotype IV *E. bieneusi* estimated 1 in HIV-infected patients from the present study in Kinshasa, Democratic Republic of Congo.

Conflict of Interests

The authors have not received any funding or benefits from industry, agency of financing, or elsewhere to conduct this study.

Acknowledgments

The authors thank Professor Francis Derouin of the Hospital of Saint Louis, Paris, France, and the students of the University the Protestant of Congo (UPC), Kinshasa, DRC, for sample collection, Mrs. Annie Claude Guillo-Olczyk, Isabelle Jolly, Isabelle Meyer, and Liliane Ciceron for excellent technical assistance. We also thank Mr. Alain Gaulier and Mrs. Mireille Gaulier, Garenne Colombes, France, for excellent social assistance. they acknowledge with thanks also the Physician Directors and Physicians of The hospitals for permission to carry out this study.

References

[1] R. Weber, R. T. Bryan, D. A. Schwartz, and R. L. Owen, "Human microsporidial infections," *Clinical Microbiology Reviews*, vol. 7, no. 4, pp. 426–461, 1994.

[2] D. M. Asmuth, P. C. DeGirolami, M. Federman et al., "Clinical features of microsporidiosis in patients with AIDS," *Clinical Infectious Diseases*, vol. 18, no. 5, pp. 819–825, 1994.

[3] D. P. Kotler and J. M. Orenstein, "Clinical syndromes associated with microsporidiosis," in *The Microsporidia and Microsporidiosis*, M. Wittner and L. M. Weiss, Eds., pp. 258–292, ASM Press, Washington, DC, USA, 1999.

[4] I. Maiga, O. Doumbo, M. Dembele et al., "Human intestinal microsporidiosis in Bamako (Mali): the presence of *Enterocytozoon bieneusi* in HIV seropositive patients," *Cahiers Santé*, vol. 7, no. 4, pp. 257–262, 1997.

[5] O. Liguory, F. David, C. Sarfati et al., "Diagnosis of infections caused by *Enterocytozoon bieneusi* and *Encephalitozoon intestinalis* using polymerase chain reaction in stool specimens," *AIDS*, vol. 11, no. 6, pp. 723–726, 1997.

[6] E. S. Didier, J. M. Orenstein, A. Aldras, D. Bertucci, L. B. Rogers, and F. A. Janney, "Comparison of three staining methods for detecting microsporidia in fluids," *Journal of Clinical Microbiology*, vol. 33, no. 12, pp. 3138–3145, 1995.

[7] D. P. Kotler and J. M. Orenstein, "Prevalence of intestinal microsporidiosis in HIV-infected individuals referred for gastroenterological evaluation," *American Journal of Gastroenterology*, vol. 89, no. 11, pp. 1998–2002, 1994.

[8] R. Wumba, A. Enache-Angoulvant, M. Develoux et al., "Prévalence des infections opportunistes digestives parasitaires à kinshasa (république démocratique du congo), résultats d'une enquête préliminaire chez 50 patients au stade sida," *Revue Médecine Tropicale*, vol. 67, no. 2, pp. 145–148, 2007.

[9] R. Wumba, B. Longo-Mbenza, M. Mandina et al., "Intestinal parasites infections in hospitalized aids patients in kinshasa, democratic Republic Of Congo," *Parasite*, vol. 17, no. 4, pp. 321–328, 2010.

[10] J. Menotti, B. Cassinat, R. Porcher, C. Sarfati, F. Derouin, and J. M. Molina, "Development of a real-time polymerase-chain-reaction assay for quantitative detection of *Enterocytozoon bieneusi* DNA in stool specimens from immunocompromised patients with intestinal microsporidiosis," *Journal of Infectious Diseases*, vol. 187, no. 9, pp. 1469–1474, 2003.

[11] J. Menotti, B. Cassinat, C. Sarfati, O. Liguory, F. Derouin, and J. M. Molina, "Development of a real-time PCR assay for quantitative detection of *Encephalitozoon intestinalis* DNA," *Journal of Clinical Microbiology*, vol. 41, no. 4, pp. 1410–1413, 2003.

[12] O. Liguory, F. David, C. Sarfati, F. Derouin, and J. M. Molina, "Determination of types of *Enterocytozoon bieneusi* strains isolated from patients with intestinal microsporidiosis," *Journal of Clinical Microbiology*, vol. 36, no. 7, pp. 1882–1885, 1998.

[13] O. Liguory, C. Sarfati, F. Derouin, and J. M. Molina, "Evidence of different *Enterocytozoon bieneusi* genotypes in patients with and without human immunodeficiency virus infection," *Journal of Clinical Microbiology*, vol. 39, no. 7, pp. 2672–2674, 2001.

[14] R. Weber, R. T. Bryan, R. L. Owen, C. M. Wilcox, L. Gorelkin, and G. S. Visvesvara, "Improved light-microscopical detection of microsporidia spores in stool and duodenal aspirates," *The New England Journal of Medicine*, vol. 326, no. 3, pp. 161–166, 1991.

[15] O. A. Cisse, A. Ouattara, M. Thellier et al., "Evaluation of an immunofluorescent-antibody test using monoclonal antibodies directed against *Enterocytozoon bieneusi* and *Encephalitozoon intestinalis* for diagnosis of intestinal microsporidiosis in bamako (Mali)," *Journal of Clinical Microbiology*, vol. 40, no. 5, pp. 1715–1718, 2002.

[16] I. Accoceberry, M. Thellier, I. Desportes-Livage et al., "Production of monoclonal antibodies directed against the microsporidium *Enterocytozoon bieneusi*," *Journal of Clinical Microbiology*, vol. 37, no. 12, pp. 4107–4112, 1999.

[17] C. Sarfati, A. Bourgeois, J. Menotti et al., "Prevalence of intestinal parasites including microsporidia in human immunodeficiency virus-infected adults in Cameroon: a cross-sectional study," *The American Journal of Tropical Medicine and Hygiene*, vol. 74, no. 1, pp. 162–164, 2006.

[18] P. Kadende, T. Nkurunziza, J. J. Floch et al., "Infectious diarrhoea conconmitant with African AIDS. Review of 100 cases observed in Bujumbura (Burundi)," *Medecine Tropicale*, vol. 49, no. 2, pp. 129–213, 1989.

[19] F. Drobniewski, P. Kelly, A. Carew et al., "Human microsporidiosis in African AIDS patients with chronic diarrhea," *Journal of Infectious Diseases*, vol. 171, no. 2, pp. 515–516, 1995.

[20] P. Kelly, S. E. Davies, B. Mandanda et al., "Enteropathy in Zambians with HIV related diarrhoea: regression modelling of potential determinants of mucosal damage," *Gut*, vol. 41, no. 6, pp. 811–816, 1997.

[21] T. Gumbo, S. Sarbah, I. T. Gangaidzo et al., "Intestinal parasites in patients with diarrhea and human immunodeficiency virus infection in Zimbabwe," *AIDS*, vol. 13, no. 7, pp. 819–821, 1999.

[22] T. van Gool, E. Luderhoff, K. J. Nathoo, C. F. Kiire, J. Dankert, and P. R. Mason, "High prevalence of *Enterocytozoon bieneusi* infections among HIV-positive individuals with persistent diarrhoea in Harare, Zimbabwe," *Transactions of the Royal Society of Tropical Medicine and Hygiene*, vol. 89, no. 5, pp. 478–480, 1995.

[23] Y. Dieng, T. H. Dieng, G. Diouf, A. M. Coll-Seck, and S. Diallo, "Screening for microsporidian spores in AIDS patients at the fann University Hospital, Dakar (Senegal): preliminary results," *Medecine et Maladies Infectieuses*, vol. 28, no. 3, pp. 265–267, 1998.

[24] M. Lebbad, H. Norrgren, A. Nauclér, F. Dias, S. Andersson, and E. Linder, "Intestinal parasites in HIV-2 associated AIDS cases with chronic diarrhea in Guinea-Bissau," *Acta Tropica*, vol. 80, no. 1, pp. 45–49, 2001.

[25] A. Samé-Ekobo, J. Lohoué, and A. Mbassi, "Clinical and biological study of parasitic and fungal diarrhea in HIV patients in the urban and suburban areas of Yaounde," *Cahiers Sante*, vol. 7, no. 6, pp. 349–354, 1997.

[26] V. Le Moing, F. Bissuel, D. Costagliola et al., "Decreased prevalence of intestinal cryptosporidiosis in HIV-infected patients concomitant to the widespread use of protease inhibitors," *AIDS*, vol. 12, no. 11, pp. 1395–1397, 1998.

[27] A. Carr, D. Marriott, A. Field, E. Vasak, and D. A. Cooper, "Treatment of HIV-1-associated microsporidiosis and cryptosporidiosis with combination antiretroviral therapy," *The Lancet*, vol. 351, no. 9098, pp. 256–261, 1998.

[28] C. Mwachari, B. I. F. Batchelor, J. Paul, P. G. Waiyaki, and C. F. Gilks, "Chronic diarrhoea among HIV-infected adult patients

in Nairobi, Kenya," *Journal of Infection*, vol. 37, no. 1, pp. 48–53, 1998.

[29] J. K. Tumwine, A. Kekitiinwa, N. Nabukeera, D. E. Akiyoshi, M. A. Buckholt, and S. Tzipori, "*Enterocytozoon bieneusi* among children with diarrhea attending Mulago hospital in Uganda," *The American Journal of Tropical Medicine and Hygiene*, vol. 67, no. 3, pp. 299–303, 2002.

[30] M. Santín and R. Fayer, "*Enterocytozoon bieneusi* genotype nomenclature based on the internal transcribed spacer sequence: a consensus," *Journal of Eukaryotic Microbiology*, vol. 56, no. 1, pp. 34–38, 2009.

[31] J. Breton, E. Bart-Delabesse, S. Biligui et al., "New highly divergent rRNA sequence among biodiverse genotypes of *Enterocytozoon bieneusi* strains isolated from humans in Gabon and Cameroon," *Journal of Clinical Microbiology*, vol. 45, no. 8, pp. 2580–2589, 2007.

[32] F. O. Akinbo, C. E. Okaka, R. Omoregie, T. Dearen, E. T. Leon, and L. Xiao, "Molecular characterization of *Cryptosporidium* spp. in HIV-infected persons in Benin City, Edo State, Nigeria," *The The American Journal of Tropical Medicine and Hygiene*, vol. 86, no. 3, pp. 441–445, 2012.

[33] M. L. Lobo, L. Xiao, F. Antunes, and O. Matos, "Microsporidia as emerging pathogens and the implication for public health: a 10-year study on HIV-positive and -negative patients," *International Journal for Parasitology*, vol. 42, pp. 197–205, 2012.

[34] H. Rinder, S. Katzwinkel-Wladarsch, and T. Löscher, "Evidence for the existence of genetically distinct strains of *Enterocytozoon bieneusi*," *Parasitology Research*, vol. 83, no. 7, pp. 670–672, 1997.

[35] R. A. Hartskeerl, A. R. J. Schuitema, T. Van Gool, and W. J. Terpstra, "Genetic evidence for the occurrence of extraintestinal *Enterocytozoon bieneusi* infections," *Nucleic Acids Research*, vol. 21, no. 17, article 4150, 1993.

[36] J. M. Molina, J. Goguel, C. Sarfati et al., "Potential efficacy of fumagillin in intestinal microsporidiosis due to *Enterocytozoon bieneusi* in patients with HIV infection: results of a drug screening study," *AIDS*, vol. 11, no. 13, pp. 1603–1610, 1997.

[37] S. Pol, C. A. Romana, S. Richard et al., "Microsporidia infection in patients with the human immunodeficiency virus and unexplained cholangitis," *The New England Journal of Medicine*, vol. 328, no. 2, pp. 95–99, 1992.

[38] Y. Hutin, M. N. Sombardier, C. Sarfati, J. M. Decazes, J. Modai, and J. M. Molina, "Risk factors for intestinal microsporidiosis in patients infected with human immunodeficiency virus (HIV)," in *Proceedings of the 37th Interscience Conference on Antimicrobial Agents and Chemotherapy*, pp. 1–150, American Society for Microbiology, September 1997, Abstracts.

[39] B. Swaminathan and T. J. Barrett, "Amplification methods for epidemiologic investigations of infectious diseases," *Journal of Microbiological Methods*, vol. 23, no. 1, pp. 129–139, 1995.

11

Does Cattle Milieu Provide a Potential Point to Target Wild Exophilic *Anopheles arabiensis* (Diptera: Culicidae) with Entomopathogenic Fungus? A Bioinsecticide Zooprophylaxis Strategy for Vector Control

Issa N. Lyimo,[1] Kija R. Ng'habi,[1] Monica W. Mpingwa,[1] Ally A. Daraja,[1] Dickson D. Mwasheshe,[1] Nuru S. Nchimbi,[1] Dickson W. Lwetoijera,[1,2] and Ladslaus L. Mnyone[1]

[1] *Biomedical and Environmental Thematic Group, Ifakara Health Institute, P.O. Box 53, Off Mlabani, Ifakara, Morogoro, Tanzania*
[2] *Vector Group, Liverpool School of Tropical Medicine, Liverpool L3 5QA, UK*

Correspondence should be addressed to Issa N. Lyimo, ilyimo@ihi.or.tz

Academic Editor: Wej Choochote

Background. *Anopheles arabiensis* is increasingly dominating malaria transmission in Africa. The exophagy in mosquitoes threatens the effectiveness of indoor vector control strategies. This study aimed to evaluate the effectiveness of fungus against *An. arabiensis* when applied on cattle and their environments. *Methods.* Experiments were conducted under semi-field and small-scale field conditions within Kilombero valley. The semi-field reared females of 5–7 days old *An. arabiensis* were exposed to fungus-treated and untreated calf. Further, wild *An. arabiensis* were exposed to fungus-treated calves, mud-huts, and their controls. Mosquitoes were recaptured the next morning and proportion fed, infected, and survived were evaluated. Experiments were replicated three times using different individuals of calves. *Results.* A high proportion of *An. arabiensis* was fed on calves (>0.90) and become infected (0.94) while resting on fungus-treated mud walls than on other surfaces. However, fungus treatments reduced fecundity and survival of mosquitoes. *Conclusion.* This study demonstrates for the first time the potential of cattle and their milieu for controlling *An. arabiensis*. Most of *An. arabiensis* were fed and infected while resting on fungus-treated mud walls than on other surfaces. Fungus treatments reduced fecundity and survival of mosquitoes. These results suggest deployment of bioinsecticide zooprophylaxis against exophilic *An. arabiensis*.

1. Background

The feeding and resting behaviours of African malaria vectors are the key determinants of a high malaria transmission intensity [1]. Among African malaria vectors, *Anopheles gambiae* s.s and *An. funestus* are well adapted over many generations feeding and resting inside human houses (i.e. endophagy and endophily, resp. [1, 2]). In contrast, *Anopheles arabiensis* is opportunistically feeding on humans [3, 4] or cattle [1, 2, 5–7] and resting outside (exophily) [8] or inside (endophily) houses [3, 4], based on availability of their preferred host species. The endophilic vector species have been well controlled by insecticide-treated nets (ITNs)

[9, 10], long-lasting insecticide-treated nets (LLINs) [11], and indoor residual spraying (IRS) [12, 13] than exophilic population of *An. arabiensis* in most parts of Africa. These suggest that behaviours of malaria vectors are crucially important on designing effective control strategies.

Outdoor feeding behaviour of exophilic *An. arabiensis* minimizes the risk of being killed by ITN/LLIN and/or IRS since these measures are exclusively applied indoors. A recent study predicted that these interventions increase extrinsic mortality of endophilic, anthrophilic *An. gambiae* s.s, and *An. funestus* and consequently generate selection pressures for insecticide resistance [14]. These explain the phenomena of declining endophilic vectors, increase of exophilic

An. arabiensis in most parts of Africa [10, 11], and shift of *An. gambiae s.s* from endophagy to exophagy in some locations [15]. The widespread insecticide resistance in population of African malaria vectors [16, 17] suggests an urgent need for alternative strategies to complement the universal coverage of LLIN.

Entomopathogenic fungi of the group hyphomycetes, notably *Metarhizium anisopliae* and *Beauveria bassiana*, hold a great promise as complementary mosquito control bio-insecticides [18–23]. These are slow acting [19, 24] and non-repellent bio-insecticides to *Anopheles* and *Culex* mosquitoes [25]. These bio-insecticides kill mosquitoes between 4–10 d after exposure [18, 26–29], before malaria parasites become transmissible as such parasites require ≥12 d to develop within a mosquito [30]. Also fungus infection reduces blood feeding propensity, life-time fecundity, flight propensity, and flight stamina [19, 27, 31, 32]. Also fungus infection inhibits development of *Plasmodium* parasites within mosquitoes [20, 32] and kills both insecticide susceptible and resistant mosquitoes [22, 33, 34]. Based on these merits, fungi provide a potential candidate for bio-insecticide zooprophylaxis against exophilic *An. arabiensis*.

However, optimal methods for delivering fungi against outdoor feeding and resting malaria vectors are yet to be developed. An effective and practical fungus delivery method requires spores to be applied on sites whilst maximizing exposure, maintaining spore viability, and minimizing the required dose of conidia. Few point source delivery methods have been tested: (a) eave curtains, baffles to target host seeking mosquitoes [35], (b) odour-baited stations, clay pots, and cotton cloth attached on the ceiling roof to target resting mosquitoes [29, 36, 37]. Only two of these methods, however, achieved high infection rates (>75%) in a population of wild *An. arabiensis* through use of human sleepers [35], and synthetic human odours [29] to attract these exophilic, zoophilic mosquitoes. However, in rural settings, cattle, the naturally preferred hosts for exophilic *An. arabiensis*, are kept close to human houses but inside mud houses or wood posts shelters with thatched roof or palm fronds. Such environment (milieu) may favour spore viability and maintain their infectiousness against mosquitoes thereof. Equally important, this milieu may passively attract exophilic *An. arabiensis* to blood feed on their preferred cattle hosts and subsequently rest on mud walls and/or thatched roof. However, this cattle milieu has never been exploited as an option to apply fungi against exophilic *An. arabiensis*.

The purpose of the current study was to evaluate the efficacy of entomopathogenic fungus against the local population of exophilic *An. arabiensis* when applied on various delivery surfaces either individually or in combination: calf, mud walls, and cotton-cloth roofs.

2. Materials and Methods

2.1. Study Site. The study was conducted at the Ifakara Health Institute (IHI) in the Kilombero valley, south eastern Tanzania. The predominant malaria vectors in this valley include *An. gambiae s.s*, *An. funestus*, and *An. arabiensis* [38–41]. Recent field studies have indicated the decline of *An. gambiae s.s* population within Kilombero valley [11]. However, a population of exophilic *An. arabiensis* is increasing within this valley and other parts of Africa [10, 11]. The preferred hosts for this vector, cattle, are commonly kept in or near human houses within Kilombero valley [42].

2.2. Mosquitoes

2.2.1. Semifield Reared Anopheles arabiensis. Semifield experiments were conducted using female *An. arabiensis* reared under semifield conditions at the IHI. The colony of *An. arabiensis* was established with individuals from the village of Sagamaganga in 2007 and 2008 and is maintained at an ambient temperature varying from 25 to 32°C and a relative humidity of 51 to 90% within the semifield system [43, 44].

2.2.2. Wild Population of Anopheles arabiensis. Small-scale field trials were conducted against freely flying wild population of *An. gambiae s.l* at Lupiro village in Ulanga district, Kilombero valley (8.385°S and 36.670°E). The recent species identification using molecular biology techniques of polymerase chain reaction (PCR) demonstrated that 98% of *An. gambiae s.l* wild population in this village is composed of *An. arabiensis* [45]. This confirms that *An. arabiensis* is the most predominant malaria vector in this village. Generally, these mosquitoes are known to feed on cattle and rest outside human houses in most of villages within Kilombero valley [42].

2.3. Fungal Isolates, Formulation and Application. Two species of entomopathogenic fungi of the group hypho-mycetes were used: *Beauveria bassiana* isolate I93-825 and *Metarhizium anisopliae* isolate IP 46. The former and the latter were used in semifield and small-scale field experiments, respectively. These experiments were intended to demonstrate delivery methods of fungus and not a comparison between species. Therefore, before each experiment, conidia viability (>85% germination on Sabouraud dextrose agar) was confirmed.

Fungal conidia were suspended in a 1:1 mixture of highly refined Enerpar oil (Enerpar M002, BP Southern Africa Ltd) and Shellsol oil (Shellsol T, South Africa Chemicals). The test suspensions of conidia were prepared and applied to delivery surfaces based on procedures described by Mnyone et al. [18]. For semifield experiments, a calf was treated with 23 mL suspension of 2.3×10^{10} conidia (1×10^9 conidia mL^{-1}). The calf was sprayed with the suspension of conidia using a handheld pressure sprayer (Minijet SATA, Germany) at a constant pressure of 2 bars over its whole body including tail and legs. For small-scale field trials, hut walls and cotton-cloth roof were also sprayed with 23 mL conidia suspensions (5×10^{10} conidia m^{-2}). As in the semifield conditions, 23 mL suspension of 5×10^{10} conidia (2.2×10^9 mL^{-1}) was sprayed per calf in the small-scale field trials. The control hut and calves were treated with equal volumes of oil mixture alone. Treatments were done at the study site under tree shade to avoid the effect of intense sunlight on conidia. All surfaces, except cottoncloth, were treated 5 h prior to the experiments to allow proper drying. Cottoncloth was treated 24 h a prior. Calves were restrained and left to dry under tree shade.

2.4. Experimental Setup and Design

2.4.1. Net Huts. The rectangular net hut ($1.5 \times 1.8 \times 2.1$ m, Figure 1) was constructed from a regular bed net (Safi net) and placed within a netting enclosed tunnel ($100 \times 3.5 \times 2.70$ m) of the IHI semifield system. The rectangular shape was maintained by fixing wooden rectangular frame from inside the bed net. The rectangular net hut was partitioned into three chambers using pieces of white clothes: chamber 1 on the left-hand side, chamber 2 on the right-hand side (2 releasing chambers), and a middle chamber (a host chamber). Releasing chambers had a round opening with a sleeve through which mosquitoes were introduced. The top of the vertical white clothes partitioning releasing chambers from host chamber were slanted towards the middle chamber to form baffles with open eaves of 1.5 cm. These openings between top side of a net and the vertical white cloth mimic open eaves that allowed mosquitoes from the releasing chamber to enter the host chamber. Releasing chambers had two zipped slits. The first slit formed entrance into the releasing chamber from outside and the second slit allowed entering into the host chamber from releasing chamber. These slits also allowed introducing a calf into the host chamber. In this chamber, there was a small wooden cage ($1.10 \times 0.59 \times 1$ m) for restraining a calf not to damage the net hut. The floor of the net hut was made of nylon carpet for easy cleaning of calf urine and faeces and observing for dead mosquitoes. A strip of grease was kept on the nylon carpet surrounding the net hut from outside to ensure no ants enter the hut and eat (scavenging on) dead mosquitoes.

2.4.2. Mud Huts. Mud huts ($2.2 \times 1.6 \times 1.77$ m, Figure 2) were constructed in the same way people are building their local houses at Lupiro village. Hut walls were made from bamboo sticks and soil from the same village. The walls were plastered by mud. The roofs were made from thatches. The space between the roof and wall was 14 cm from outside. The baffles towards inside the hut were constructed to progressively reduce an eave of 14 cm to 3 cm. This tapered eave space allowed host-seeking mosquitoes to enter the hut but preventing them from exiting through hut. Therefore, wild population of *An. arabiensis* could be attracted to feed on a calf inside the hut and rest on mud plastered walls and/or cotton-cloth attached on thatched roofs.

2.4.3. Experimental Procedures. The semifield experiments were conducted following 2×2 Latin square design (LSD) in two rectangular net huts constructed within a unique Ifakara tunnel system. Two calves were randomly selected from Ifakara communities. One calf was treated with fungal conidia and the second calf remained untreated (control). One net hut contained fungal-treated calf (treatment) and

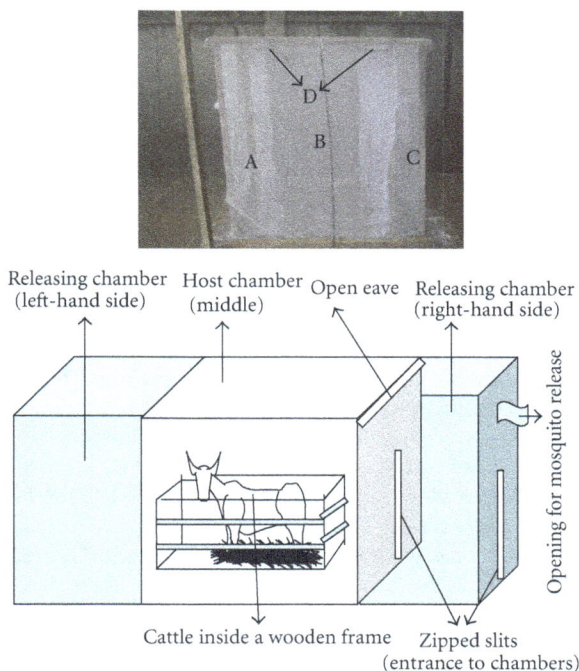

FIGURE 1: Picture of a rectangular net hut and its schematic representation. The sections of net hut are (A) left-hand side releasing chamber, (B) host or middle chamber, (C) right-hand side releasing chamber, and (D) open eaves with baffles to allow host-seeking mosquitoes to enter host chamber.

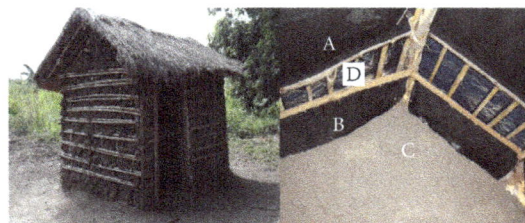

FIGURE 2: Mud hut at Lupiro village with fixed (A) cotton-cloth roof treated with fungal conidia, (B) baffles to reduce exit of mosquitoes, (C) mud walls either treated with fungal conidia or untreated for resting mosquitoes, and (D) open eave to allow host seeking mosquitoes to enter inside the hut.

the other hut contained untreated calf. These calves were introduced inside the host chamber in the evening (6:30 pm). Then 150–200 female *An. arabiensis* were introduced into each releasing chamber and left to forage overnight on the calf by entering through open eaves as they do in the natural environment. The next morning all mosquitoes were recaptured from releasing and host chambers. The recaptured mosquitoes were identified whether fed or unfed and then held in the semifield insectary to monitor for their subsequent fitness traits including longer-term survival and fecundity. The number of eggs and days survived were counted and recorded. Dead mosquitoes were put onto moist filter paper in petri dishes, sealed with parafilm, and incubated inside a humid chamber for 3–5 d, after which they were examined for fungal sporulation. The efficacy of fungus over time (persistence) was also preliminarily assessed by exposing mosquitoes to fungal-treated calves at 3 d after treatment. These experiments were replicated four times using four different individual calves.

The small-scale experiments were conducted following 3 × 3 LSD in three mud huts constructed at the periphery of the Lupiro village. Two mud huts were treated with fungal conidia (one was treated on the walls and the other on the roof) and one hut was left untreated. Three calves were randomly selected from the local communities. Only one of these calves was treated with fungal conidia and introduced inside untreated hut, while the other two untreated calves were placed inside wall- and roof-treated huts, respectively. These calves were introduced into the mud huts at evening time (6:00 pm) and left there overnight. The next morning, all mosquitoes were recaptured from inside each mud hut. The mosquitoes were identified whether fed or unfed. All these mosquitoes were individually transferred into a paper cup and held in the field insectary to monitor their survival. Also all blood fed mosquitoes were provided with wet filter paper in the paper cup for them to lay eggs at 4 d after blood feeding on calf (i.e. *Anopheles* mosquitoes develop eggs within 3-4 days after blood feeding [46]). The number of eggs and days survived was recorded. Also mosquitoes were processed for sporulation as in the semifield system. The calves were rotated between huts to control for the variation between huts. Therefore, the calves spent three days in the same hut before shifting into another hut. These experiments were replicated three times using three different individual calves to control the variation between their in attractiveness.

2.5. Statistical Analyses. Statistical analysis was conducted to evaluate effects of three fungus delivery methods (cattle, mud walls, and cotton-cloth roofs) against a population of exophilic *An. arabiensis*. Three key parameters were analysed: proportion of fed mosquitoes (feeding success) and showing fungal outgrowth (sporulation), number of eggs (fecundity), and postexposure survival of mosquitoes (number of days survived). The first two parameters are binomial response variable, whereas the last two are continuous response variables.

The binomial and continuous response variables were analysed using generalized linear mixed effect models using an appropriate link function in the *R* statistical package [47], with "treatments" as fixed effects and "replicate" as a random effect. A base model including only the random effect of "replicate" was constructed, to which the main effect of "treatment" was added to form a full model. The significance of additional fixed effects of treatments was tested by sequentially adding this term to a base model and applying likelihood-ratio test (LRTs) to test if they led to a significant improvement ($P < 0.05$). For semifield experiments where fungus-treated was compared with untreated calf, the chi-square (χ^2) generated from model comparisons was used to test for significant differences between control and treatment. Whereas for small-scale experiments, more than two treatments were compared and when treatment was

identified as being statistically significant, then Tukey's post hoc test (adjusting for multiple comparisons) was used to identify significant two-way differences between control and treatments and within treatments using z-values. All z-values reported small-scale experiments are those generated from multiple comparisons in *R* statistical software [47].

The continuous response variable of survival data (only infected mosquitoes noninfected were excluded from analysis) are rarely normally distributed and thus were appropriately analysed using Cox proportional hazards model (coxph) [47, 48]. This model tested whether survival of the mosquitoes varies between delivery methods (calf, house wall, and roof) and days after treatment. A frailty function was used to incorporate the random effect of "replicate", and "treatments" were fit as main effect in *R* statistical software [21]. The coxph model compared the survival curves of different treatments and gave statistical significant differences in overall mortality rates in hazard ratio (HR) values, which indicate the daily risk of dying [48]. These hazard ratios have now replaced direct comparison of mortality rates after a specific point in time using *t*-tests [49–51]. An HR value of 1 indicates equal mortality rates between treatments and control, an HR value > 1 indicates significantly greater mortality rates in treatment than control, and HR < 1 indicates significantly lower overall mortality rates in treatment than control.

3. Results

A total of 1,690 and 547 female *An. arabiensis* were attracted to calf and being recaptured from, respectively inside net huts in the semifield and mud huts in the field experiments. Almost all recaptured mosquitoes were blood fed (a proportion of 0.90 to 0.99) under both semifield and field conditions. This study evaluated the effects of fungus delivery options (calf, mud wall, and cotton-cloth roofs) on infection rates (sporulation), fecundity, and survival of blood fed, exophilic *An. arabiensis*.

3.1. Semifield Experiments

(i) Proportion of Fed Mosquitoes. The proportion of *An. arabiensis* fed on calf treated with *B. bassiana* was slightly higher than on untreated calf ($\chi_1^2 = 7.64$, $P = 0.01$, Figure 3(a)). However, the magnitude of proportion of fed mosquitoes was >0.93 in both cases (Figure 3(a)).

(ii) Proportion of Fungus-Infected Mosquitoes. A freshly fungal-treated calf infected a significantly higher proportion of fed *An. arabiensis* than 3-days posttreated calf ($\chi_2^2 = 101.53$, $P < 0.001$, Figure 3(b)). Calves infected a proportion of 0.90 of fed *An. arabiensis* immediately after treatment, and ~0.70, 3 d after treatment (Figure 3(b)).

(iii) Effect of Fungus on Mosquito Survival and Fecundity. The infection of semifield *An. arabiensis* with *B. bassiana* significantly reduced their subsequent number of eggs and survival (Figures 3(c) and 3(d)). The number of eggs laid

by *An. arabiensis* fed on fungal-treated calf was 17 eggs less than on untreated calves ($\chi_1^2 = 5.17$, $P = 0.02$, Figure 3(c)). Similarly, the postfeeding survival of fed *An. arabiensis* varied significantly between fungus-treated and untreated calves ($\chi_1^2 = 19.3$, $P < 0.001$, Figure 3(d)). The risk of death (hazard ratio-HR) of fed *An. arabiensis* on fungus-treated calves was almost twice that on untreated calves (HR = 1.63 (1.31–2.03), Figure 3(d)).

3.2. Small-Scale Field Experiments

(i) Proportion of Fed Wild Mosquitoes. The proportion wild *An. arabiensis* fed on their natural preferred host (calf) ranged from 0.81 to 1 under field conditions. The proportion of fed wild *An. arabiensis* from control was similar to all other treatments ($P > 0.05$, Figure 4(a)). Similarly, within treatments, two-way comparisons revealed that no significant differences in proportion of fed wild *An. arabiensis* between treatments ($P > 0.05$, Figure 4(a)).

(ii) Proportion of Fungus-Infected Mosquitoes. The proportion of wild *An. arabiensis* infected with *M. anisopliae* isolate IP 46 varied significantly between delivery methods ($\chi_5^2 = 228.05$, $P < 0.001$, Figure 4(b)). The multiple comparisons revealed that the proportion of infected mosquitoes observed in control was significantly lower than all treatments: treated (cloth + calf), TCTca ($z = 7.18$, $P < 0.001$, Figure 4(b)), treated cloth + untreated calf, TCUca ($z = 7.78$, $P < 0.001$, Figure 4(b)), treated (wall + calf), TwTca ($z = 5.73$, $P < 0.001$, Figure 4(b)), treated wall + untreated calf, TwUca ($z = 8.72$, $P < 0.001$, Figure 4(b)), and untreated wall + treated calf, UwTca ($z = 6.68$, $P < 0.001$, Figure 4(b)). Within fungal treatments, the proportion of infected mosquitoes was observed to differ significantly between TCUca and TwUca ($z = 3.25$, $P = 0.01$, Figure 4(b)).

(iii) Effect of Fungus on Mosquito Survival and Fecundity. The infection of mosquitoes with *M. anisopliae* significantly reduced the number of eggs laid by wild *An. arabiensis* ($\chi_2^2 = 13.61$, $P = 0.001$, Figure 4(c)). From pairwise multiple comparisons, mosquitoes from the control hut laid significantly more number of eggs than those from treated cottoncloth + untreated calf, TCUca, ($z = 4.98$, $P < 0.001$, Figure 4(c)) and treated walls + untreated calf, TwUca ($z = -2.43$, $P = 0.04$, Figure 4(c)). However, there was no significant differences between the number of eggs laid by mosquitoes from the hut with TCUca and TwUca ($z = 1.54$, $P = 0.27$, Figure 4(c)).

The infection of *An. arabiensis* with *M. anisopliae* had also significantly affected their longer-term survival ($\chi_5^2 = 83$, $P < 0.001$, Figure 4(d), Table 1). The risks of death (hazard ratio) of these mosquitoes were significantly lower on untreated surfaces than on fungus-treated surfaces (TCTca: $\chi_1^2 = 22.2$, $P < 0.001$, TCUca: $\chi_1^2 = 29.9$, $P < 0.001$, TwTca: $\chi_1^2 = 22.0$, $P < 0.001$, TwUca: $\chi_1^2 = 73$, $P < 0.001$, UwTca: $\chi_1^2 = 13.6$, $P < 0.001$, Figure 4(d), Table 1). However, no significant differences of the hazard ratio of fed *An. arabiensis*

TABLE 1: Hazard ratio of *An. arabiensis* after exposure on different fungal-treated surfaces and cattle. The numbers in brackets are 95% confidence intervals. The treatments are abbreviated as TCTca for (treated cloth roof + calf), TCUca for treated cloth + untreated calf, TwTca for (treated mud wall + treated calf), TwUca for (treated mud wall + untreated calf), and UwTca for (untreated mud wall + treated calf).

Fungal treatments	Hazard ratio (HR) Relative to control
Treated (cloth + calf)	2.56 (1.73–3.78)
Treated cloth + untreated calf	2.36 (1.73–3.21)
Treated (wall + calf)	4.05 (2.26–7.26)
Treated wall + untreated calf	4.13 (2.99–5.71)
Untreated wall + Treated calf	2.30 (1.48–3.58)

were observed across fungal treated surfaces ($P > 0.05$, Figure 4(d)).

4. Discussion

This study demonstrates for the first time the potential of applying fungus to cattle and their milieu (mud walls and roofs) for controlling wild population of exophilic *An. arabiensis*. A high proportion of these mosquitoes was strongly attracted and fed on cattle (>0.90) and became infected with fungus treatments. Notably, fungus-treated mud walls infected a higher proportion of mosquitoes (~0.94) than treated calf and cotton-cloth roof or their combinations under field conditions. The fecundity of fungus-infected mosquitoes on calf, mud wall and cotton-cloth roof was, respectively, 17, 40, and 27 eggs less than those on control. Surprisingly, the effects of fungus on survival of *An. arabiensis* were similar between delivery methods and/or their combinations. However, the magnitude of relative risk of death of mosquito on fungus-treated mud walls was 2 times more than that on treated calf and cotton-cloth roof. Therefore, cattle milieu, mostly mud walls of their houses, could be the best field delivery method of fungus against a population of exophilic, zoophilic *An. arabiensis*, suggesting a bio-insecticide zooprophylaxis.

Entomopathogenic fungi, notably *Metarhizium anisopliae* and *Beauveria bassiana,* are promising potential bio-insecticide against malaria vectors [23, 37, 50]. The mud walls treated with *M. anisopliae* IP 46 infected a higher proportion of wild *An. arabiensis* (0.94) than all other treatments or their combinations. This exophilic *An. arabiensis* is generally known to feed on cattle or on humans if available [5, 7]. Therefore, cattle attracted a high proportion (>0.90) of these mosquitoes that became fed and rest on fungus-treated mud walls because the oil formulation of fungal conidia is nonrepellent [25]. Besides, the four sides of the hut provided a bigger surface area of fungus exposure than a calf, and therefore, the likelihood of mosquitoes picking up more conidia on mud walls than on a calf. Furthermore, mud walls provided a natural medium (soil) for better fungal conidia attachment and viability. These results support the previous

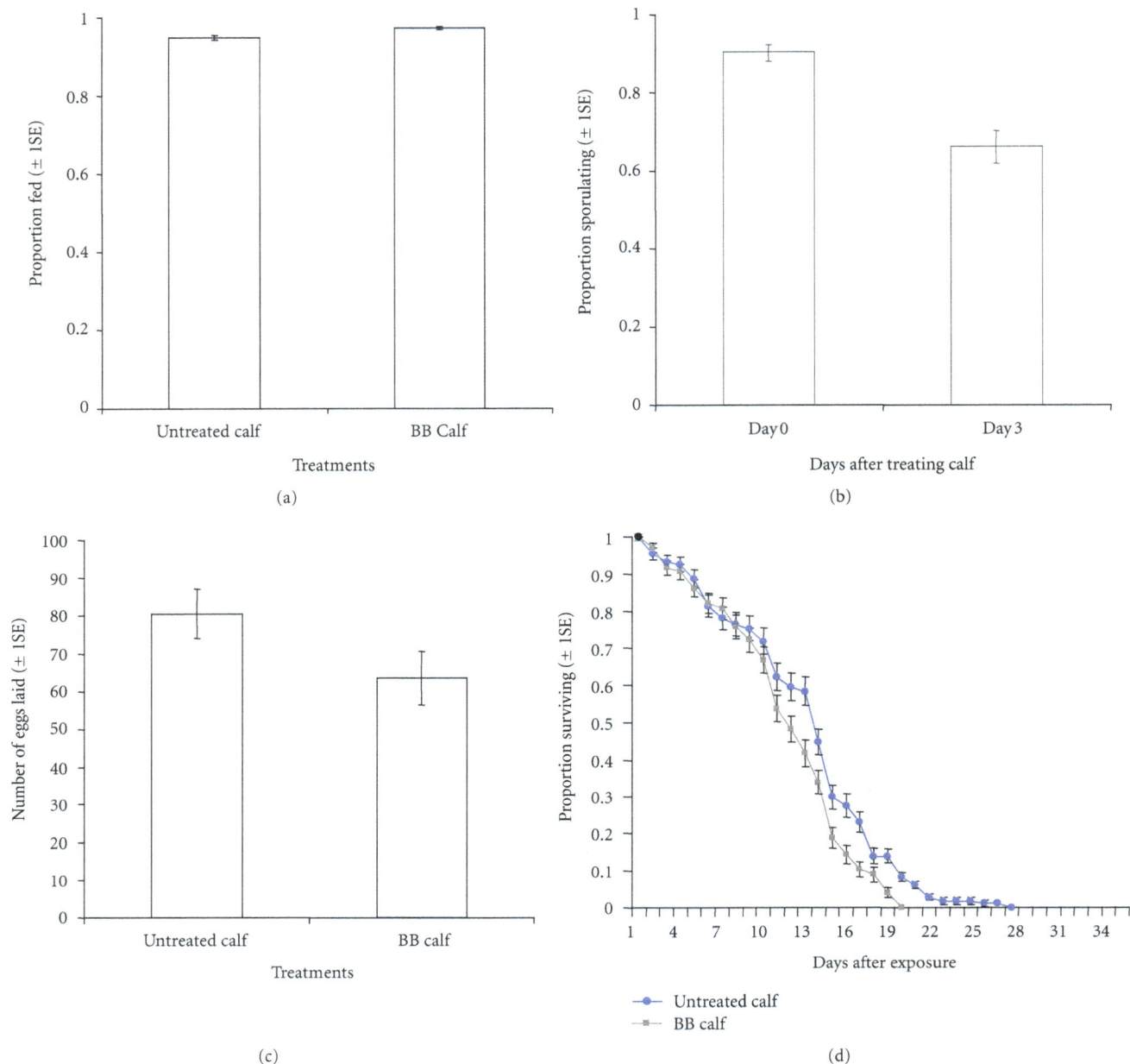

FIGURE 3: Effects of *B. bassiana* against semifield reared exophilic *An. arabiensis* populations: (a) Estimated proportion (± 1s.e) of fed after exposure to fungus-treated and untreated calf, BB calf indicates a calf sprayed with conidia suspension of *Beauvaria bassiana*. (b) Estimated proportion (± 1s.e) of infected mosquitoes after exposure to fungus treated calf on 0 d and 3 d. (c) Estimates (± 1s.e) of the mean number of eggs laid by mosquitoes after exposure to fungus-treated and untreated calf. (d) Survival of mosquitoes after exposure to fungus-treated and untreated calf, the lines represent the survival function as estimated from fitting Cox proportional hazard model (controlling for random variation between individual calves).

studies that demonstrate when fungal conidia applied on soil medium (e.g., mud panels [52], clay pots/tiles [19, 36]) the infection rates are higher relative to other substrates. One of these studies demonstrated that a humid and cool environment in a clay pot potentiates high infection rates (>92%) in *An. gambiae* s.s and *An. funestus* in the laboratory and suggested that a human synthetic odour should be incorporated to maximize the number of mosquitoes exposed

to fungal conidia [36]. The present study demonstrated that natural preferred hosts (cattle) could attract a high proportion of exophilic *An. arabiensis* that become infected with fungus applied on mud walls of cattle house

Unexpectedly, a significantly low proportion (~0.04) of fed wild *An. arabiensis* was infected by fungus treatments. The possible explanation for this observation could be that fed mosquitoes from fungus-treated huts flew into control

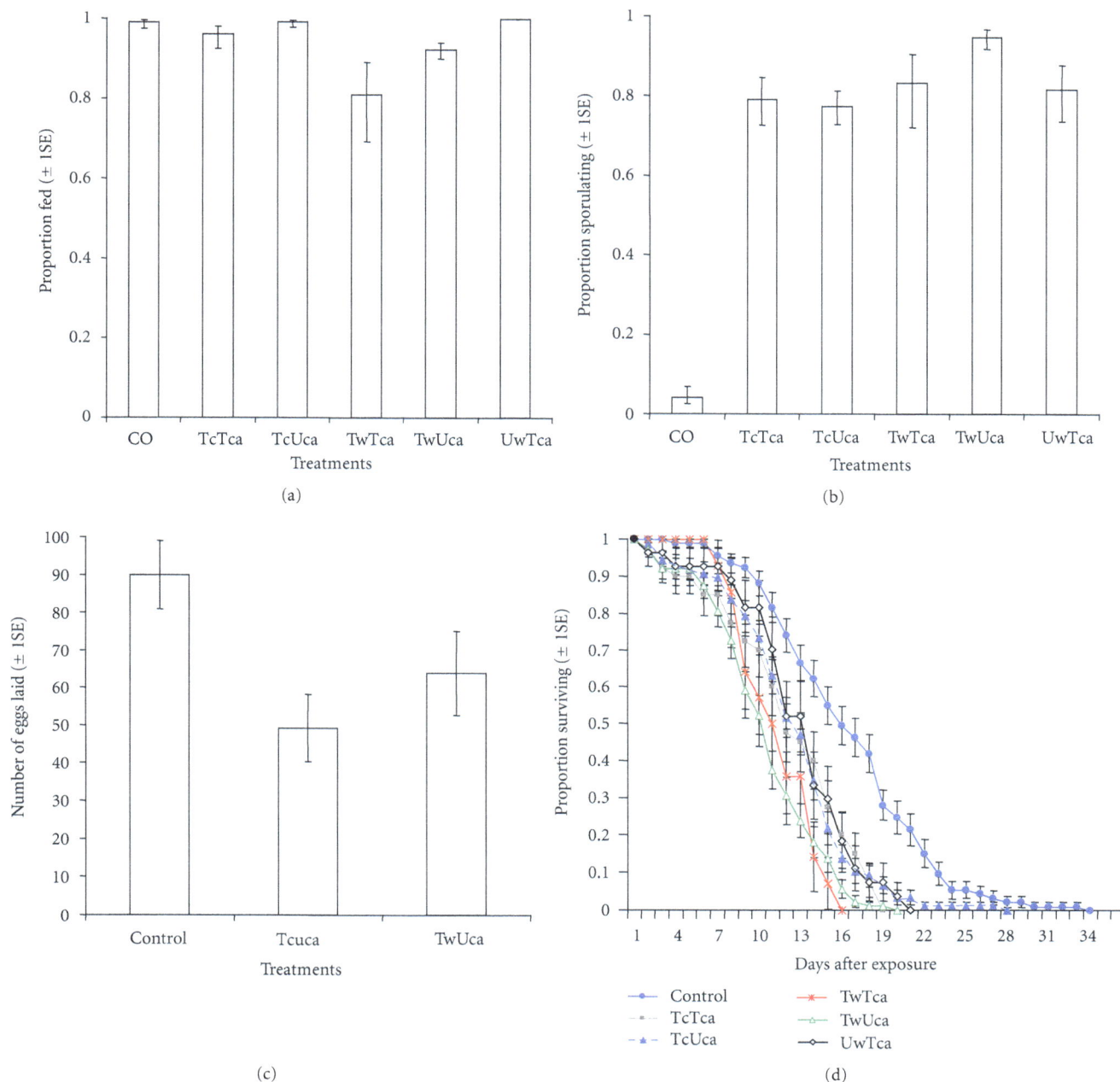

FIGURE 4: Effects of *M. anisopliae* IP 46 against wild exophilic *An. arabiensis*. (a) Estimated proportion (± 1s.e) of fed mosquitoes after exposure to fungus-treated and untreated calf inside experimental hut. (b) Estimated proportion (± 1s.e) of infected mosquitoes after exposure to fungus-treated and untreated surfaces. (c) Estimates (± 1s.e) of the mean number of eggs laid by mosquitoes after exposure to fungus-treated and untreated surfaces. (d) Survival of mosquitoes after exposure to fungus-treated and untreated surfaces. The lines represent the survival function as estimated from the fitting Cox proportional hazard model (controlling for random variation between individual calves).

hut while seeking for the resting places after blood feeding. Alternatively, mosquitoes could have been drifted by wind from fungus-treated huts to control hut.

The fungus is a nonrepellent [25], slow acting bioinsecticide [24], that allows mosquitoes to feed, rest on treated surfaces while developing eggs (e.g., for *Anopheles* mosquitoes take 3-4 days to develop eggs after blood

feeding, [46]), and subsequently lay eggs before being killed. This study demonstrated that fungus-infected *An. arabiensis* laid 17–40 eggs less than those mosquitoes in the control group. The possible explanation could be that fungus and mosquitoes compete on the same protein resources and therefore fed mosquitoes that became infected allocated these resources on their survival than egg development. The observations in this study are consistent with those reported elsewhere [19, 31]. Although a laboratory study revealed that no statistical significant differences between the numbers of eggs laid by fungus-infected and noninfected mosquitoes [53], this is largely linked with similar blood meal size between these two groups such that whatever depleted by fungus left these mosquitoes with threshold volume of blood meal required for egg development [54, 55].

Evolutionary forces act upon reproduction and survival of malaria vectors. The opportunity of surviving fungus infection to lay at least few eggs suggests that they are less likely to generate selection pressures for resistance [24]. The daily survival rates of An. arabiensis were significantly reduced by fungus-treated surfaces by >2 times than on control in both semifield and field conditions. However, the magnitude of the relative risk of death of mosquitoes on fungus-treated mud wall was consistently 4 times higher than on control. These findings are consistent with previous studies that have shown that fungus-treated surfaces reduce daily survival rates of An. gambiae s.l more than untreated surfaces [29, 37]. Similarly, the observations that the efficacy of fungus against mosquitoes was high when applied on mud walls are consistent with those reported under laboratory conditions on mud tiles [19] and mud panels [52]. Most importantly, the present study demonstrated that fungus-treated mud walls killed ~70% of the wild population at day 11 after exposure, suggesting that this option has potential of interrupting transmission of Plasmodium falciparum that requires 14 days to develop to transmissible sporozoites within mosquitoes [30].

Fungus kills mosquitoes in 4–10 days depending on exposure dose, viability and virulence of the fungal species/strain, and physiological status of mosquitoes [19, 49, 56]. Therefore, the results reported here have limitations before being compared with other studies or translated into application. First, the present study reported the effect of fungus infection on daily survival rates of blood fed mosquitoes (>95% blood fed mosquitoes) and therefore cannot be directly compared with nonblood fed mosquitoes in other studies. Previous studies have shown that blood resources improve daily survival rates of mosquitoes [57–59], and fungus infection kills nonblood fed mosquitoes much faster than blood fed ones [49]. Second, we briefly measured persistence of fungal conidia on calf under semifield conditions. This study found that ~70% of semifield reared An. arabiensis could be infected by a 3-days-post treated-calf. However, detailed experiments on persistence of fungal conidia on cattle and mud walls are now planned to be conducted under field conditions: (1) to test for the effect of sunlight (UV) and rainfall on persistence of fungal conidia in grazing cattle, (2) to test for the effect of wet and dry season or smoke on the persistence of fungal conidia on mud walls. Third, this is the first study to apply fungus on cattle for the control of malaria vectors, and therefore we could have underdose mosquitoes. We sprayed the whole body of a calf with a fungal conidia dose of $1.0 - 2.2 \times 10^9$ conidia mL^{-1} which was slightly higher than the dose tested (1×10^8 conidia mL^{-1}) on cattle to control ticks [60, 61]. However, ticks are sticking on cattle body for days while feeding until become engorged whereas mosquitoes fly and land on cattle body to feed for few minutes (temporary ectoparasites). This suggests that mosquitoes might have been underdosed. The full-field experiments have been designed to test different doses of fungal conidia applied on either mud walls or calf against wild An. arabiensis. Fourth, the use of different fungus species in these experiments: B bassiana for semifield experiments and M. anisopliae IP 46 for small-scale field experiments. The intention was not to compare their efficacy but was because of the availability of conidia. Although the two fungus species showed similar trend in the semifield and field conditions, now experiments are underway using M. anisopliae IP 46 and B. bassiana at the same time in the field to confirm if they act on the same direction against malaria vectors in the natural environments.

This study demonstrates for the first time the potential of applying fungus on cattle and their milieu (mud walls and cotton-cloth roof) for the control of exophilic wild An. arabiensis. A high proportion of exophilic An. arabiensis was attracted to both fungus-treated and untreated cattle where they fed and became infected by different fungus treatments. The fungus-treated mud walls infected a higher proportion of An. arabiensis than all other treated surfaces and their combinations. Surprisingly, fecundity and survival of infected mosquitoes were similar between treatments and their combinations, but varied from their controls. Although not significantly but the magnitude of risk of death of these mosquitoes on fungus-treated mud walls was 2 times more than on treated calf and cottoncloth roof. These results suggest that a combination of fungus-treated mud walls and untreated cattle in their milieu could be acceptable, cheap, and easy to apply in rural settings, thus making a perfect bio-insecticide zooprophylaxis that may compliment universal coverage of LLIN.

Conflict of Interests

The authors declare that they have no competing interests.

Authors Contribution

I. N. Lyimo and L. L. Mnyone designed and supervised execution of experiments. I. N. Lyimo analysed the data and drafted the paper. L. L. Mnyone, K. R. Ng'habi, and D. W. Lwetoijera critically reviewed the paper. M. W. Mpingwa, A. A. Daraja, N. S. Nchimbi, and D. D. Mwasheshe performed experiments. All authors read and approved the final version of the manuscript.

Acknowledgments

We would like to thank Nina and Thomas (Penn State University, USA) for providing the B. bassiana isolate I93-825 and Christian Luz (Instituto de Patologia, Tropical e Saude Publica, Universidade, Brazil) for providing M. anisopliae isolate IP 46.

References

[1] A. Kiszewski, A. Mellinger, A. Spielman, P. Malaney, S. E. Sachs, and J. Sachs, "A global index representing the stability of malaria transmission," *American Journal of Tropical Medicine and Hygiene*, vol. 70, no. 5, pp. 486–498, 2004.

[2] M. E. Sinka, M. J. Bangs, S. Manguin et al., "The dominant Anopheles vectors of human malaria in Africa, Europe and the Middle East: occurrence data, distribution maps and bionomic précis," *Parasites and Vectors*, vol. 3, no. 1, article 117, 2010.

[3] C. M. Fornadel and D. E. Norris, "Increased endophily by the malaria vector *Anopheles arabiensis* in southern Zambia and identification of digested blood meals," *American Journal of Tropical Medicine and Hygiene*, vol. 79, no. 6, pp. 876–880, 2008.

[4] M. Hadis, M. Lulu, Y. Makonnen, and T. Asfaw, "Host choice by indoor-resting *Anopheles arabiensis* in Ethiopia," *Transactions of the Royal Society of Tropical Medicine and Hygiene*, vol. 91, no. 4, pp. 376–378, 1997.

[5] I. Tirados, C. Costantini, G. Gibson, and S. J. Torr, "Blood-feeding behaviour of the malarial mosquito *Anopheles arabiensis*: implications for vector control," *Medical and Veterinary Entomology*, vol. 20, no. 4, pp. 425–437, 2006.

[6] R. J. Kent, P. E. Thuma, S. Mharakurwa, and D. E. Norris, "Seasonality, blood feeding behavior, and transmission of *Plasmodium falciparum* by *Anopheles arabiensis* after an extended drought in southern Zambia," *American Journal of Tropical Medicine and Hygiene*, vol. 76, no. 2, pp. 267–274, 2007.

[7] S. M. Muriu, E. J. Muturi, J. I. Shililu et al., "Host choice and multiple blood feeding behaviour of malaria vectors and other anophelines in Mwea rice scheme, Kenya," *Malaria Journal*, vol. 7, no. 43, pp. 1–20, 2008.

[8] G. B. White, S. A. Magayuka, and P. F. L. Boreham, "Comparative studies on sibling species of the *Anopheles gambiae* Giles complex (Diptera:Culicidae), bionomics and vectorial activity of species A and species B at Segera, Tanzania," *Bulletin of Entomological Research*, vol. 62, pp. 295–317, 1972.

[9] K. A. Lindblade, J. E. Gimnig, L. Kamau et al., "Impact of sustained use of insecticide-treated bednets on malaria vector species distribution and culicine mosquitoes," *Journal of Medical Entomology*, vol. 43, no. 2, pp. 428–432, 2006.

[10] M. N. Bayoh, D. K. Mathias, M. R. Odiere et al., "*Anopheles gambiae*: historical population decline associated with regional distribution of insecticide-treated bed nets in western Nyanza Province, Kenya," *Malaria Journal*, vol. 9, no. 1, article 62, 2010.

[11] T. L. Russell, D. W. Lwetoijera, D. Maliti et al., "Impact of promoting longer-lasting insecticide treatment of bed nets upon malaria transmission in a rural Tanzanian setting with pre-existing high coverage of untreated nets," *Malaria Journal*, vol. 9, no. 1, article 187, 2010.

[12] M. T. Gillies and A. Smith, "The effect of a residual house-spraying campaign in East Africa on species balance in the *Anopheles funestus* group. The replacement of *A. funestus* Giles by *A. rivulorum* Leeson," *Bulletin of Entomological Research*, vol. 51, no. 2, pp. 243–252, 1960.

[13] B. Pluess, F. C. Tanser, C. Lengeler, and B. L. Sharp, "Indoor residual spraying for preventing malaria," *Cochrane Database of Systematic Reviews*, vol. 4, article CD006657, pp. 1–46, 2010.

[14] H. Ferguson, N. Maire, W. Takken et al., "Selection of mosquito life-histories: a hidden weapon against malaria?" *Malaria Journal*, vol. 11, no. 1, p. 106, 2012.

[15] M. R. Reddy, H. J. Overgaard, S. Abaga et al., "Outdoor host seeking behaviour of *Anopheles gambiae* mosquitoes following initiation of malaria vector control on Bioko Island, Equatorial Guinea," *Malaria Journal*, vol. 10, article 184, 2011.

[16] H. Ranson, H. Abdallah, A. Badolo et al., "Insecticide resistance in *Anopheles gambiae*: data from the first year of a multi-country study highlight the extent of the problem," *Malaria Journal*, vol. 8, no. 1, article 299, 2009.

[17] H. Ranson, R. N'Guessan, J. Lines, N. Moiroux, Z. Nkuni, and V. Corbel, "Pyrethroid resistance in African anopheline mosquitoes: what are the implications for malaria control?" *Trends in Parasitology*, vol. 27, no. 2, pp. 91–98, 2011.

[18] L. L. Mnyone, M. J. Kirby, D. W. Lwetoijera et al., "Infection of the malaria mosquito, *Anopheles gambiae*, with two species of entomopathogenic fungi: effects of concentration, co-formulation, exposure time and persistence," *Malaria Journal*, vol. 8, no. 1, article 309, pp. 1–12, 2009.

[19] S. Blanford, W. Shi, R. Christian et al., "Lethal and pre-lethal effects of a fungal biopesticide contribute to substantial and rapid control of Malaria vectors," *PLoS ONE*, vol. 6, no. 8, Article ID e23591, 2011.

[20] W. Fang, J. Vega-Rodríguez, A. K. Ghosh, M. Jacobs-Lorena, A. Kang, and R. J. S. Leger, "Development of transgenic fungi that kill human malaria parasites in mosquitoes," *Science*, vol. 331, no. 6020, pp. 1074–1077, 2011.

[21] M. Farenhorst, A. Hilhorst, M. B. Thomas, and B. G. J. Knols, "Development of fungal applications on netting substrates for malaria vector control," *Journal of Medical Entomology*, vol. 48, no. 2, pp. 305–313, 2011.

[22] M. Farenhorst, B. G. J. Knols, M. B. Thomas et al., "Synergy in efficacy of fungal entomopathogens and permethrin against west african insecticide-resistant *Anopheles gambiae* mosquitoes," *PLoS ONE*, vol. 5, no. 8, Article ID e12081, 2010.

[23] E. J. Scholte, J. K. Bart, A. S. Robert, and T. Willem, "Entomopathogenic fungi for mosquito control: a review," *Journal of Insect Science*, vol. 4, no. 19, pp. 1–24, 2004.

[24] A. F. Read, P. A. Lynch, and M. B. Thomas, "How to make evolution-proof insecticides for malaria control," *PLoS Biology*, vol. 7, no. 4, pp. 1–10, 2009.

[25] L. L. Mnyone, C. J. Koenraadt, I. N. Lyimo, M. W. Mpingwa, W. Takken, and T. L. Russell, "Anopheline and culicine mosquitoes are not repelled by surfaces treated with the entomopathogenic fungi *Metarhizium anisopliae* and *Beauveria bassiana*," *Parasites and Vectors*, vol. 3, no. 1, article 80, 2010.

[26] E. Scholte, W. Takken, and B. Knols, "Pathogenicity of five east African entomopathogenic fungi against adult *Anopheles gambiae* s.s. mosquitoes (Diptera, Culicidae)," in *Proceedings of the Section Experimental and Applied Entomology of the Netherlands Entomological Society*, vol. 14, pp. 25–29, 2003.

[27] J. George, S. Blanford, M. J. Domingue, M. B. Thomas, F. R. Andrew, and C. B. Thomas, "Reduction in host-finding behaviour in fungus-infected mosquitoes is correlated with reduction in olfactory receptor neuron responsiveness," *Malaria Journal*, vol. 10, no. 219, pp. 1–13, 2011.

[28] L. L. Mnyone, T. L. Russell, I. N. Lyimo, D. W. Lwetoijera, M. J. Kirby, and C. Luz, "First report of *Metarhizium anisopliae* IP 46 pathogenicity in adult *Anopheles gambiae* s.s. and *An. arabiensis* (Diptera; Culicidae)," *Parasites and Vectors*, vol. 2, no. 1, article 59, pp. 1–4, 2009.

[29] D. W. Lwetoijera, R. D. Sumaye, E. P. Madumla et al., "An extra-domiciliary method of delivering entomopathogenic fungus, *Metharizium anisopliae* IP 46 for controlling adult

populations of the malaria vector, *Anopheles arabiensis*," *Parasites and Vectors*, vol. 3, no. 1, article 18, pp. 1–6, 2010.

[30] J. C. Beier, "Malaria parasite development in mosquitoes," *Annual Review of Entomology*, vol. 43, pp. 519–543, 1998.

[31] E. J. Scholte, B. G. J. Knols, and W. Takken, "Infection of the malaria mosquito *Anopheles gambiae* with the entomopathogenic fungus *Metarhizium anisopliae* reduces blood feeding and fecundity," *Journal of Invertebrate Pathology*, vol. 91, no. 1, pp. 43–49, 2006.

[32] S. Blanford, B. H. K. Chan, N. Jenkins et al., "Fungal pathogen reduces potential for malaria transmission," *Science*, vol. 308, no. 5728, pp. 1638–1641, 2005.

[33] M. Farenhorst, J. C. Mouatcho, C. K. Kikankie et al., "Fungal infection counters insecticide resistance in African malaria mosquitoes," *Proceedings of the National Academy of Sciences of the United States of America*, vol. 106, no. 41, pp. 17443–17447, 2009.

[34] A. F. V. Howard, R. N'Guessan, C. J. M. Koenraadt et al., "First report of the infection of insecticide-resistant malaria vector mosquitoes with an entomopathogenic fungus under field conditions," *Malaria Journal*, vol. 10, article 24, 2011.

[35] L. Mnyone, I. Lyimo, D. Lwetoijera et al., "Exploiting the behaviour of wild malaria vectors to achieve high infection with fungal biocontrol agents," *Malaria Journal*, vol. 11, no. 1, p. 87, 2012.

[36] M. Farenhorst, D. Farina, E. J. Scholte et al., "African water storage pots for the delivery of the entomopathogenic fungus *Metarhizium anisopliae* to the malaria vectors *Anopheles gambiae* s.s. and *Anopheles funestus*," *American Journal of Tropical Medicine and Hygiene*, vol. 78, no. 6, pp. 910–916, 2008.

[37] E. J. Scholte, K. Ng'Habi, J. Kihonda et al., "An entomopathogenic fungus for control of adult African malaria mosquitoes," *Science*, vol. 308, no. 5728, pp. 1641–1642, 2005.

[38] T. Smith, J. D. Charlwood, J. Kihonda et al., "Absence of seasonal variation in malaria parasitaemia in an area of intense seasonal transmission," *Acta Tropica*, vol. 54, no. 1, pp. 55–72, 1993.

[39] J. D. Charlwood, J. Kihonda, S. Sama et al., "The rise and fall of *Anopheles arabiensis* (Diptera: Culicidae) in a Tanzanian village," *Bulletin of Entomological Research*, vol. 85, no. 1, pp. 37–44, 1995.

[40] J. D. Charlwood, T. Smith, E. Lyimo et al., "Incidence of *Plasmodium falciparum* infection in infants in relation to exposure to sporozoite-infected anophelines," *American Journal of Tropical Medicine and Hygiene*, vol. 59, no. 2, pp. 243–251, 1998.

[41] J. D. Charlwood, T. Smith, F. F. Billingsley, W. Takken, E. O. K. Lyimo, and J. H. E. T. Meuwissen, "Survival and infection probabilities of anthropophagic anophelines from an area of high prevalence of *Plasmodium falciparum* in humans," *Bulletin of Entomological Research*, vol. 87, no. 5, pp. 445–455, 1997.

[42] V. Mayagaya, *The Impact of Livestock on Ecology of Malaria Vectors and Malaria Transmission in the Kilombero Valley*, University of Dar es Salaam, Dar es Salaam, Tanzania, 2010.

[43] I. Lyimo, *Ecological and Evolutionary Determinants of Anopheline Host Species Choice and Its Implications for Malaria Transmission*, University of Glasgow, Glasgow, UK, 2010.

[44] H. M. Ferguson, K. R. Ng'habi, T. Walder et al., "Establishment of a large semi-field system for experimental study of African malaria vector ecology and control in Tanzania," *Malaria Journal*, vol. 7, article 158, 2008.

[45] F. O. Okumu, G. F. Killeen, S. Ogoma et al., "Development and field evaluation of a synthetic mosquito lure that is more attractive than humans," *PloS ONE*, vol. 5, no. 1, article e8951, 2010.

[46] M. T. Gillies, "The duration of the gonotrophic cycle in *Anopheles gambiae* and *Anopheles funestus*, with a note on the efficiency of hand catching," *East African Medical Journal*, vol. 30, no. 4, pp. 129–135, 1953.

[47] M. J. Crawley, *The R Book*, John Wiley & Sons, 2007.

[48] D. R. Cox, "Regression of models and life tables," *Journal of the Royal Statistical Societ*, vol. 34, no. 2, pp. 187–220, 1972.

[49] L. L. Mnyone, M. J. Kirby, M. W. Mpingwa et al., "Infection of *Anopheles gambiae* mosquitoes with entomopathogenic fungi: effect of host age and blood-feeding status," *Parasitology Research*, vol. 108, no. 2, pp. 317–322, 2011.

[50] C. K. Kikankie, B. D. Brooke, B. G. Knols et al., "The infectivity of the entomopathogenic fungus *Beauveria bassiana* to insecticide-resistant and susceptible *Anopheles arabiensis* mosquitoes at two different temperatures," *Malaria Journal*, vol. 9, no. 1, article 71, 2010.

[51] M. Farenhorst and B. G. Knols, "A novel method for standardized application of fungal spore coatings for mosquito exposure bioassays," *Malaria Journal*, vol. 9, no. 1, article 27, 2010.

[52] L. L. Mnyone, M. J. Kirby, D. W. Lwetoijera et al., "Tools for delivering entomopathogenic fungi to malaria mosquitoes: effects of delivery surfaces on fungal efficacy and persistence," *Malaria Journal*, vol. 9, no. 1, article 246, 2010.

[53] S. N. Ondiaka, S. T. Bukhari, M. Farenhorst, W. Takken, and B. G. J. Knols, "Effects of fungal infection on the host-seeking behaviour and fecundity of the malaria mosquito *Anopheles gambiae* Giles," in *Proceedings of the Netherlands Entomological Society Meeting*, vol. 19, pp. 121–128, 2008.

[54] A. O. Lea, H. Briegel, and H. M. Lea, "Arrest, resorption, or maturation of oocytes in *Aedes aegypti*: dependence on the quantity of blood and the interval between blood meals," *Physiological Entomology*, vol. 3, no. 4, pp. 309–316, 1978.

[55] B. D. Roitberg and I. Gordon, "Does the Anopheles blood meal—fecundity curve, curve?" *Journal of Vector Ecology*, vol. 30, no. 1, pp. 83–86, 2005.

[56] A. S. Bell, S. Blanford, N. Jenkins, M. B. Thomas, and A. F. Read, "Real-time quantitative PCR for analysis of candidate fungal biopesticides against malaria: technique validation and first applications," *Journal of Invertebrate Pathology*, vol. 100, no. 3, pp. 160–168, 2009.

[57] I. N. Lyimo and H. M. Ferguson, "Ecological and evolutionary determinants of host species choice in mosquito vectors," *Trends in Parasitology*, vol. 25, no. 4, pp. 189–196, 2009.

[58] L. C. Harrington, J. D. Edman, and T. W. Scott, "Why do female *Aedes aegypti* (Diptera: Culicidae) feed preferentially and frequently on human blood?" *Journal of Medical Entomology*, vol. 38, no. 3, pp. 411–422, 2001.

[59] I. N. Lyimo, S. P. Keegan, L. C. Ranford-Cartwright, and H. M. Ferguson, "The impact of uniform and mixed species blood meals on the fitness of the mosquito vector *Anopheles gambiae* s.s: does a specialist pay for diversifying its host species diet?" *Journal of Evolutionary Biology*, vol. 25, no. 3, pp. 452–460, 2012.

[60] G. P. Kaaya, M. Samish, M. Hedimbi, G. Gindin, and I. Glazer, "Control of tick populations by spraying *Metarhizium anisopliae* conidia on cattle under field conditions," *Experimental and Applied Acarology*, vol. 55, no. 3, pp. 273–281, 2011.

[61] P. Polar, D. Moore, M. T. K. Kairo, and A. Ramsubhag, "Topically applied myco-acaricides for the control of cattle ticks: overcoming the challenges," *Experimental and Applied Acarology*, vol. 46, no. 1, pp. 119–148, 2008.

12

Barriers to Testing and Treatment for Chagas Disease among Latino Immigrants in Georgia

Rebecca M. Minneman,[1] Monique M. Hennink,[1] Andrea Nicholls,[1]
Sahar S. Salek,[1] Francisco S. Palomeque,[1] Amina Khawja,[1] Lauren C. Albor,[2]
Chester C. Pennock,[2] and Juan S. Leon[1]

[1] *Rollins School of Public Health, Emory University, Atlanta, GA 30322, USA*
[2] *Emory College, Emory University, Atlanta, GA 30322, USA*

Correspondence should be addressed to Rebecca M. Minneman, rebecca.minneman@gmail.com

Academic Editor: Vitaliano A. Cama

Background. The lack of testing and treatment of Chagas disease (CD), caused by Trypanosoma cruzi, amongst infected immigrants in the USA increases the risk of serious health complications and transmission (congenital or via blood transfusions). *Goal.* Our goal was to identify the barriers to testing and treatment of CD and understand the process of seeking healthcare amongst Latino immigrants in Georgia. *Methods.* In this qualitative study, eleven focus group discussions were conducted with 82 Latino immigrants, including migrant farm workers. Grounded theory was used to collect and analyze the data to develop an inductive conceptual framework to explain the context and process of seeking healthcare for CD amongst this at-risk population. *Results.* Participants were not aware of CD. Three healthcare seeking behaviors were identified: delaying treatment, using traditional remedies, and using either mainstream or alternative health providers. Behaviors and motivations differed by gender, and the use of licensed medical providers was considered a last resort due to the cost of healthcare, loss of earnings while seeking care, and fear of diagnosis with fatal illness. *Discussion.* Providing free or low cost services, mobile clinics, and education regarding CD is critical to increase testing and treatment of CD in the US.

1. Introduction

In the United States, the prevalence and burden of Chagas disease are increasing [1]. In the USA, the population most at-risk for developing Chagas disease, caused by the parasite *Trypanosoma cruzi*, are Latin American immigrants who became infected with *T. cruzi* in their home countries through vector borne transmission from triatomines, infants born of *T. cruzi* infected mothers (i.e., congenital transmission), and, rarely, individuals infected through autochthonous transmission (reviewed in [2–4]). The USA is the country with the highest number of Latin American immigrants [5]. In various immigrant populations, including migrant farm workers, seroprevalence rates ranging from 0.4% to 4.9% have been documented [6–8]. Further, it is estimated that approximately 300,167 Latin American immigrants in the USA are infected with *T. cruzi* [9]. If left untreated, *T. cruzi* infection may result in a dilated cardiomyopathy, gastrointestinal pathology, CNS abnormalities, and death [10]. In the USA, *T. cruzi* infection can be treated with two drugs, benznidazole and nifurtimox, which are exclusively controlled and distributed by the Centers for Disease Control and Prevention (CDC). Concerns with blood-borne transmission of *T. cruzi* led to screening of 65% of the US blood supply for anti-*T. cruzi* antibodies. From 2007 to 2010, this screening resulted in recognition of 1,113 cases of *T. cruzi* infection. Despite recommendation to seek medical treatment, only about 11% of these new *T. cruzi* seropositive blood donors or their physicians have contacted CDC for treatment consultation [1]. The low percentage of treatment consultations indicates a need to understand why so few seropositive donors have sought treatment. One hypothesis may be that these seropositive donors are marginalized Latin American immigrants who have difficulty accessing healthcare.

Testing and treatment for Chagas disease may be difficult among Latinos in the US because they are a marginalized, underserved population who are less likely to seek healthcare [11]. Hispanics are more likely to be poor and uninsured than any other race/ethnicity [12]. As a group, they are less likely than other races to favorably rate their quality of treatment or receive appropriate and timely treatment [11]. The reasons for these disparities include the cost of treatment, legal status, lack of insurance, and lack of interpreters (reviewed in [13]). In addition to the aforementioned structural barriers, many beliefs that are prevalent in the Latino community affect their access to healthcare. Often, Latinos fear the cost of health care, long waiting times, potential for deportation, and discrimination that they associate with accessing healthcare [11]. The combination of structural barriers and beliefs causes Latinos in the USA to use a variety of alternatives to mainstream healthcare, including self-medication and having friends or family members send medicines from their home country in order to avoid seeing a doctor in the USA [11].

In summary, due to the rapid growth of the Latino population [14] and their low levels of use of mainstream healthcare, the risk and burden of Chagas disease in the USA are increasing. The goal of this study is to understand the awareness of Chagas disease and identify common healthcare-seeking behaviors and barriers to prevention and treatment of Chagas disease among Latino immigrants in Georgia in order to create effective interventions to prevent and treat Chagas disease. The state of Georgia was chosen as the study area because the Latino population of Georgia is the third fastest growing population in any state, making it the tenth largest Latino population in the USA (8.8% of the state's population) [12, 14] roughly half being foreign born [1]. Focus group discussions with Latino immigrants in Georgia were analyzed to explore when and how Georgia Latinos used mainstream healthcare and their awareness and perceptions of Chagas disease.

2. Methodology

This study was conducted in both urban and rural areas of Georgia. The urban study site was Atlanta, which is one of the largest urban centers in the southeastern US and has a high concentration of Latino residents (approximately 8%), approximately half of whom are foreign born [12]. The rural study site comprised two farms in Moultrie, Georgia, an agricultural area relying heavily on manual farm labor where approximately 13% of the population is Latino [14]. Study participants were male and female immigrants, 18 years or older, who were born and lived in Latin America for at least ten years prior to the study, spoke Spanish, and were currently residing in Georgia.

Women's perspectives on health-seeking strategies for Chagas disease are important because of their traditional role in healthcare decision making and accessing health services. Additionally, women's awareness of Chagas disease

may influence the risk of congenital transmission of the disease. Latina women mainly reside in urban areas of Georgia and were therefore only recruited from the city of Atlanta. As female participants generally did not work outside the home, they were purposively recruited through formal and informal community networks as informed by the study's Community Advisory Board (CAB). The CAB consisted of representatives from the Latin American Association and the Hispanic Health Coalition of Georgia. These organizations were identified by the research team as representative of and influential in the Latino community in Georgia. The CAB advised on the cultural sensitivity of all aspects of the study including the discussion guide, participant recruitment, and dissemination of the findings to the Latino community. Focus group discussions explored when and how Latinos access health care and their awareness of Chagas disease. Formal networks included senior health classes and prenatal classes. The informal network was a group of friends who met to exercise together. The leader of each network was informed about the study and invited researchers to attend a regular meeting to describe the study and invite participation. A total of 45 female participants were recruited for six focus group discussions.

Men's perspectives about health seeking for Chagas disease are equally important given their limited access to health services. The lack of health insurance and low wages common among Latino men may also influence health-seeking strategies and the ability to pay for healthcare. Latino men were recruited from both urban and rural areas; they typically work in urban areas as day laborers in the informal job market or as unskilled labor in low-wage jobs and in rural areas in the agricultural sector (i.e., migrant farm workers). Day laborers often seek work in public locations; therefore, venue-based recruitment was used in locations where men congregated looking for work (e.g., close to hardware stores, specific gas stations). The study was explained, and eligible, immigrants were invited to participate. Only one focus group with six-day laborers was held due to their priority to continue touting for work. In order to include urban male immigrants who were not day laborers, one focus group was held with eight male immigrants recruited from a church. In the rural study site, men (migrant farm workers) were recruited at a mobile health clinic providing free, on-site care after working hours to immigrants on two farms. Men were approached (after their health care visit) and told about the study and invited to participate. Twenty-three male farmworkers were recruited for three focus group discussions. A total of 37 male participants were recruited for five focus group discussions.

Focus group discussions helped to identify a range of perceptions and experiences on participants' health seeking behaviors. Discussion groups were stratified by gender (five male groups, six female groups) for homogeneity so that participants felt comfortable sharing their perspectives and to allow for analytic comparisons by gender. Data were collected in two rounds: first from May to August of 2011 and second in February of 2012 until saturation of information was reached [15]. All focus group discussions were conducted in Spanish by Spanish-speaking moderators, gender matched to participants, and trained in human subjects research. A note taker was also present at each group discussion. Discussions lasted 60–75 minutes including the completion of the demographic survey. All group discussions were digitally recorded. Focus groups in the urban study site were held in private homes, churches, restaurants, and nongovernmental organization offices. In rural areas, group discussions were held outdoors, under a tent adjacent to the mobile health clinic. In the first round of data collection, participants were given refreshments as compensation for their time; during the second round of collection participants were also compensated with a travel stipend.

A semistructured discussion guide was developed, translated into Spanish, back-translated, pilot-tested, and refined with the help of the study's CAB before data collection. Additional demographic data were collected from each participant via a brief survey. The survey was also translated into Spanish, pilot-tested, and refined. The survey collected data from each participant on their country of origin, time spent in Latin America, education, and income. Study staff administered the surveys at the end of each group discussion in oral and written forms depending on participants' literacy levels.

This study was approved by the Emory University Institutional Review Board (IRB 00018964). Oral informed consent was sought from each participant in Spanish. To ensure clarity, participants were asked to summarize their understanding of what participation entailed prior to the discussion. After the discussion, each participant was given a written copy of the oral consent in Spanish.

All focus group discussions were transcribed verbatim in Spanish, checked for accuracy, deidentified, and uploaded into MAXQDA2007 (Marburg, Germany). The grounded theory approach was used to collect and analyze data, following a cyclical process in order to confirm findings and gain a richer understanding of context of issues that arose [16]. Data were read and annotated to identify core themes from which inductive and deductive codes were developed. All data were then coded by the research team. Intercoder reliability was assessed using an independent researcher who recoded a selection of text with the research team's codebook. Coding discrepancies were identified and addressed. Data were searched by topical themes, and a description encompassing the context, depth, and breadth of core themes in the data was developed. Themes were compared by gender and by urban and rural residence for males to identify differences in health seeking behavior. Four types of health-seeking behaviors were identified: using traditional remedies, delaying health seeking, using mainstream health services, and using alternative health services (e.g., herbalists, physicians in their home country). A conceptual framework of the pathways between these strategies was developed and validated with the data. Core differences in health-seeking pathways were identified between men and women.

3. Results

Focus group discussions were held to understand the awareness of Chagas disease and identify common healthcare-seeking behaviors and barriers to prevention and treatment of Chagas disease among Latino immigrants in Georgia. To understand why Latino immigrants do not seek testing and treatment for Chagas disease, we analyzed data describing when Latino immigrants seek healthcare, their process of seeking healthcare, including the types of behaviors used in seeking care, the motivation for choosing those behaviors, and the pathway between behaviors. Participants were unaware of Chagas disease; thus the moderator asked how participants would seek care if they had symptoms of Chagas disease (e.g., severe pain in their chest or stomach). Participants had not experienced these symptoms; therefore, they could not describe their actual health-seeking strategy; instead, participants described hypothetically how they would seek heath care if these symptoms arose. For example, they described their experience in seeking care for a range of health issues, from minor conditions like colds to severe and chronic conditions like breast cancer.

Approximately two-thirds of participants (66%) were from Mexico (Table 1), and participants had a median age of 40.5 years (range 19 to 80). Almost half (49%) reported a household income of less than $12,000 per annum ($n = 29$), although there was a high rate of nonresponse for this question (28%). 25% reported their highest level of education completed as primary school or less (Table 1).

3.1. Awareness of Chagas Disease. Before describing the process of seeking healthcare by Georgia Latino immigrants, it is important to describe the participants' perceptions of Chagas disease as it affects their motivation to be tested or treated for Chagas disease. All participants, but one, were unaware of Chagas disease. The only participant who had heard of Chagas disease learned of it through her daughter who had a Ph.D. in microbiology. All other participants were unaware of Chagas disease, but the majority were able to report basic knowledge of other vector-borne diseases (e.g., dengue fever and malaria).

When shown photos or a specimen of the triatomine, the vector of Chagas disease, recognition by its commonly known name *"chinche"* was more frequent among male participants. However, the local term *"chinche"* is used to refer to many different insects across Latin America that resembles the triatomine and does not necessarily refer specifically to the Chagas disease vector. Participants debated the name, origin, environment, ecology, and health effects of the bug that they referred to as *"chinche"* suggesting that few participants had specific knowledge of the vector. Both men and women commonly confused the triatomine with bed bugs and attributed a range of health problems to them such as diarrhea and rashes, which are not associated with triatomines.

3.2. Process of Seeking Healthcare. Due to the minimal awareness of Chagas disease amongst participants, the study team sought to identify common healthcare-seeking behaviors and barriers to prevention and treatment of Chagas disease by understanding the process of seeking healthcare for a range of health issues, by Latino immigrants.

Table 1: Demographic characteristics of focus group participants.

	% (n)
Country of origin (n = 74)	
Mexico	66% (49)
Central America	20% (15)
South America	14% (10)
Education (n = 72)	
Primary	25% (18)
Secondary	39% (28)
University	36% (26)
Yearly income (n = 59)	
< $12,000	49% (29)
$12,000–$24,000	37% (22)
> $24,000	14% (8)

Participants descriptions of the process of seeking healthcare for a range of health issues, from minor conditions like colds to severe and chronic conditions like breast cancer, highlighted three distinct stages: phase 1 involved the use of traditional remedies, phase 2 involved a period of stoicism, whereby no healthcare was sought, and phase 3 involved consulting a health provider, either a mainstream (i.e., licensed clinics, hospitals, and medical providers) or alternative provider (Figure 1). The pathway through these stages differed by gender and location of residence (rural men differed from urban men). Each stage is described in the following for each type of participant.

3.2.1. Women. Female participants indicated that their first response to illness was to use traditional remedies (Phase 1, Figure 1(a)), when confronted with familiar and less severe health problems such as colds or stomach aches. They described a variety of traditional remedies. Some were widely perceived as safe and effective, such as brewing various types of tea (e.g., lemon and cinnamon), using herbal compresses, or making remedies out of insects. Participants found other traditional practices used by Latinos to be outlandish including the use of tarot cards to treat heart disease and necklaces of tomatoes to treat sore throats. In general, traditional remedies were used exclusively as the first recourse for illness and were not used in conjunction with over-the-counter medicine from pharmacies. The following quote illustrates the common use of traditional remedies.

For the most part, we treat ourselves with natural or homemade medicines, but if it doesn't cure us, one goes to the hospital, to the doctor, but first we treat with what we know. (Female, Group 4, P1)

Women recognized that traditional remedies were not always effective and caused them to delay seeking formal care but continued to use them as the first recourse for illness. Women described using traditional remedies, despite their risk, because it was their cultural tradition and their lower cost compared to using mainstream health providers. Family and friends who were still in Latin America were

considered more knowledgeable regarding traditional remedies due to the long history of their use in Latin America. Therefore, women sought advice from family members in their home countries on how to use traditional remedies and had remedies unavailable in the USA mailed to them. The following extract demonstrates how female participants valued the advice of family members in their home countries about traditional remedies.

Because they [family in country of origin] almost always cure themselves with natural medicines like an herb or something. They make them into tea. They know more, a little more than one [here]. (Female, Group 4, P2)

In addition to the custom of using traditional remedies, women also preferred to use them because of the lower cost compared to mainstream healthcare providers. Women knew that if the traditional remedies cured their maladies, they saved money, but if they failed it was both dangerous and often more expensive. The following quote demonstrates their preference for traditional remedies due to their lower cost.

You call [your family] on the phone and you say to them, "my stomach hurts, what can I take? [They say] take this, take this." If it works you save [the money for] the doctor, but if it doesn't work, that's the bad part. It doesn't save you [money]. (Female, Group 5, P6)

If traditional remedies were ineffective or unavailable, women described a second phase of health treatment as "aguantando" (waiting stoically) (Phase 2, Figure 1(a)), which involved postponing caring for their illnesses, with the hope that the illness would cure itself. Women described four reasons for not seeking healthcare and waiting stoically. First, the lack of economic resources to seek medical care led women to wait for the symptoms to disappear on their own. For example,

P7: "We wait. When the pain is strong and we can't [wait] anymore, then we [go to the doctor]." Moderator: "Why do you wait?" P8: "Because of the same [reason], because there is no money." (Female, Group 5, P7-P8)

Second, women reported the fear of being diagnosed with a fatal or expensive illness as reason for waiting to seek healthcare. Women described how some Latinos prefer not to know if they were ill, even though they suffer and could potentially stop the spread of infection, because they believe they will die anyway due to a lack of resources. For example,

Because many times, we, I have friends that I tell [them] "let's go get a checkup" and they say "no, I won't because if I have some problem, I am scared and prefer not to know...because if I have something severe, I don't have any money and will die sooner, before my time." (Female, Group 4, P1)

Third, some women stated that they waited stoically because they lacked trust in mainstream healthcare providers

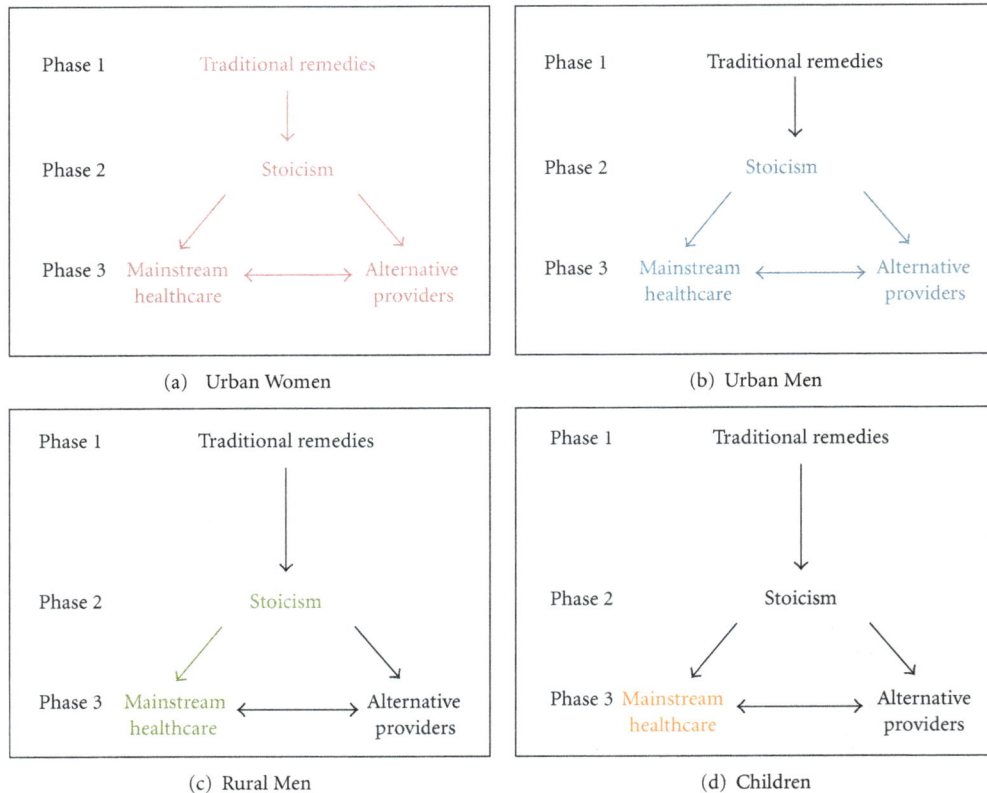

FIGURE 1: Process of seeking healthcare among focus group participants (Latino immigrants). "Phase" represents the strategy summarizing the healthcare-seeking behavior expressed by the participants. Arrows represent the change in strategy. The process applicable (colored font and arrow) and not applicable (gray font and arrow) to each population of participants is represented. (a) urban women, (b) urban men, (c) rural men, and (d) children.

in the USA. These women described how some providers prioritize earning money over caring for patients and cited examples of providers requiring excessive tests, being unwilling to use test results from other clinics, and charging for consultations without providing treatment. The following extract demonstrates these perceptions.

> When you go to the doctor, they say "I have to do an exam for this," but sometimes you say "this is just to get my money." Yes, because sometimes you already went to another clinic and they did a blood test, but in the other clinic [the new one] they repeat the same test to get information that you knew from the beginning. When they do it and you ask why, [they say] "I can't prescribe anything if I don't do the exam to know what you have." (Female, Group 5, P6)

The fourth reason for women waiting rather than seeking healthcare was that they felt their health was not a priority for them. They generalized Latinos as "descuidados" or careless with their health. Women portrayed Latinos as too lazy or irresponsible to seek healthcare, even if they had the resources to do so. When the moderator described Chagas disease and asked why Latinos would not seek treatment

if they knew they were infected, one woman responded as follows.

> There are many people who are lazy that have the help and that have the means and say "ay no." They prefer to sleep or they prefer to do other things. (Female, Group 3, P6)

If symptoms became insufferable, women entered into the third phase of the process of seeking healthcare (Phase 3, Figure 1(a)), which involved either consulting alternative health providers or mainstream healthcare providers depending on their available resources. Alternative providers included stores that provide prescription medications (e.g., antibiotics) without a prescription, natural doctors/herbalists who had no formal medical licenses, and doctors in their country of origin who would post prescriptions to the USA. Typically, women who did not have the financial resources to pay for mainstream healthcare used these alternative providers. Women also used alternative providers when they knew what medicines they needed or when they had strong relationships with physicians in their home countries. Some women explained

> I buy [prescription medicines] because they are cheaper than [going to a] doctor. Let's say, for

example that you have a sore throat, an infection, that won't go away with pharmacy cough syrup. It won't go away with pills because the infection is very strong. What do we do? We find antibiotics in the Latin stores. We find an injection of penicillin and we inject ourselves with penicillin, amoxicillin, pills. And with this, two shots of penicillin it's over. Sixty dollars and we will cure ourselves. (Female, Group 5, P8)

I suffered a lot from my ears. . . and he [a doctor] charged me about $200. . .and from then on when we had a problem, we had a doctor in Guatemala where I used to go and I would call him and tell him "I feel like this and he would tell me what to do and my family would try to send it [medicine] to me." (Female, Group 4, P3)

Some women chose to use mainstream healthcare rather than alternative providers when waiting became intolerable (Phase 3, Figure 1(a)). Women primarily chose to use mainstream providers if they had sufficient financial resources to pay for healthcare, or if they had very grave illnesses. Yet, within the category of mainstream healthcare, various factors were used to select providers, the most important, across all focus groups, was cost. For those eligible, acceptance of Medicare/Medicaid was the first requirement when searching for doctors. Those who were ineligible for government aid asked for recommendations from acquaintances for the most economic option among mainstream providers. Other women selected providers who offered payment plans and sliding payment scales. One woman described how she selected her provider based on affordability

It depends on your situation, the situation, what you feel, the symptoms. . . according to me I had problems in my breast, this breast, and I was thinking, it took me almost two months thinking "where, where, where will it be the cheapest?" (Female, Group 3, P4)

If mainstream healthcare failed to cure those women who first chose mainstream healthcare over alternative providers (Phase 3, Figure 1(a)), they described turning to alternative providers. Women cited experience (either personal or from acquaintances) where the doctors failed to cure their illnesses either by being too expensive or ineffective. They turned to alternative providers such as "natural doctors" or herbalists or having their doctor from their country of origin send a prescription through the mail (Phase 3, Figure 1(a)). One woman described how her friend turned to alternative providers.

There is a woman who went to different doctors, even American doctors, and all to see if [her illness] changed but no, [it stayed] the same. Her daughter went to a natural doctor here in Atlanta. . .she is taking some herbs, some plants that he recommended and is better and can move her leg and arm. (Female, Group 2, P3)

3.2.2. Children. Women considered the health of their children to be a priority, therefore, when a child was ill, women would directly seek mainstream healthcare, thereby skipping Phases 1 (traditional remedies) and 2 (waiting stoically) and moving straight to Phase 3 (only mainstream providers) in Figure 1(d). One woman cited that even though she knew what was wrong with her child, she would take her child to the doctor.

Even though [it is just a fever] one goes to the doctor if they have a fever or something and the only thing they [the doctor] tells you is to give them aspirin and it's over. And if you have to pay [no Medicaid] they charge you for the consult and everything. (Female, Group 4, P2)

Often children had Medicaid so mainstream healthcare was less costly, but women would find the time and the money to take children to mainstream providers, even if they were not eligible for Medicaid.

3.2.3. Men. In contrast to women, both urban and rural men described their first action when ill to be "*aguantando*" or waiting stoically, rather than attempting to treat their illnesses with traditional remedies. Therefore men entered the process of seeking healthcare at Phase 2 in Figures 1(b) and 1(c) by waiting for the illness to cure itself. Men cited the same reasons as women for waiting rather than seeking healthcare, including a lack of financial resources to pay for healthcare and fear of being diagnosed with a fatal or expensive illness. The following are examples of how Latino men dreaded going to the doctor.

I think getting sick is a problem for Latinos here in the United States because it costs us a lot. This is the biggest problem there is for [Latinos]. If you get gravely ill, they [doctors] won't see you because it is so expensive for [Latinos]. It [healthcare] is mostly for people from the United States. (Male, Group Rural Men 1)

We [Latino men] are much more careless with ourselves. Going to the doctor does not matter to us, we dread going to the doctor, we are always afraid of that they will tell us that we have this thing or the other. (Male, Group Urban Men, P4)

Men, similar to women, criticized Latinos for being irresponsible with their health, but the men attributed this negligence to a culture of machismo, in which men felt that they should not acknowledge illness in order to maintain their masculinity. Participants described how machismo prevented men from seeking healthcare.

We are careless [with our health]. Why? Because of machismo, how can we acknowledge it [that we are ill]? [Men say] How could it be? The doctor is mistaken.

When symptoms became intolerable, men entered Phase 3 in the process of seeking healthcare (Figures 1(b) and

1(c)), but Phase 3 varied by the area where the men lived (urban versus rural). Urban men used both mainstream and alternative providers, while rural men only used mainstream providers, as described in the following.

3.2.4. Urban Men.

When waiting became intolerable, urban male participants either sought care with mainstream healthcare providers (e.g., hospital emergency rooms and clinics) or with alternative providers (phase 3, Figure 1(b)). Despite being more expensive, mainstream healthcare was preferred over alternative providers and those who had sufficient funds would seek mainstream healthcare.

Urban men who did not have the financial resources to pay for mainstream healthcare used alternative providers (Phase 3, Figure 1(b)). These included Latino or Asian stores selling prescription drugs without prescriptions and unlicensed dentists and doctors in Georgia who would treat them at a lower cost. Although urban men recognized the risks of using these alternative providers and expressed concern over their effectiveness and safety, affordability was the priority. If alternative health providers failed to cure them, urban men described using mainstream healthcare providers. The following demonstrates their lack of confidence and their acknowledgment in the risk of delaying care.

> The people will tell you anything. They are not prescriptions of doctors; rather they are prescriptions of the people. You go to the pharmacy, Chinese or Hispanic; in the Hispanic stores around here they will give you a "mejoral," an aspirin, all that, and if it doesn't work you have to go to the doctor, but your condition is already serious. (Male, Group Urban Men, P2)

There were two differences in Phase 3 between urban men (Figure 1(b)) and women (Figure 1(a)). First, the type of alternative providers used by those urban men who could not afford mainstream healthcare differed from those used by women. Women described seeking care from physicians in their home countries while urban men described using unlicensed providers in Georgia. Second, unlike women, after mainstream providers failed to cure them, no urban men described turning to alternative providers; they only sought care with other mainstream providers.

3.2.5. Rural Men.

Rural men (i.e., migrant farm workers), similarly to urban men, began the process of seeking healthcare directly in Phase 2 of Figure 1(c) by waiting stoically. Yet their reason for waiting was a reluctance to take time off from work rather than fear, dread, carelessness, or lack of funds to pay for care. Rural male participants came to the USA to earn money and were hired on a temporary basis. They were unwilling to seek healthcare unless it was absolutely urgent because taking time off from work meant lost earnings. They preferred to wait and then seek care when they returned to Mexico in order to avoid losing time at work and to ensure continuity of care. When asked why he did not go to a clinic when ill, one participant responded as follows.

> In order not to lose a day [of work], to maintain a relationship with the boss, you have to work. Economically, economically you have to work. This is what we know. You have to take care of yourself, that's how it is. (Male, Group Rural Men 1)

When waiting became intolerable, rural men entered Phase 3 of Figure 1(c) but only sought care with mainstream health providers. Rural men made no mention of using any type of alternative providers. When selecting mainstream providers, rural men described their choice was based on convenience, in order to avoid losing more time at work.

3.3. Potential Healthcare Seeking for Chagas Disease

3.3.1. Perceptions of Chagas Disease.

In order to understand how the process of seeking healthcare would change if the participants were infected with Chagas disease, the study team explained the symptoms and transmission of Chagas disease. After the explanation of Chagas disease, participants expressed concern over being unknowingly infected. This concern stemmed from learning about the lack of definitive symptoms and the long asymptomatic period associated with Chagas disease. The lack of definitive symptoms, the length of the asymptomatic period, and the potential gravity of the disease were identified by participants to be the most important points to communicate regarding Chagas disease because of the difficulties in ascertaining whether or not testing was needed.

Women expressed more interest in all aspects of Chagas disease compared to men. Only women expressed concern about the possibility and mechanisms of congenital transmission. For example when asked about what they would do if they found they were infected with Chagas disease, one woman responded as follows.

> I am more worried about my daughter because you say that it can be transmitted to children [during pregnancy] as well. (Female, Group 4, P1)

In addition to women's apprehension regarding vertical transmission, participants highlighted the same barriers in seeking testing and treatment for Chagas disease as seeking care for other health problems. Economic limitations were particularly important. When asked what they would do if they were infected with Chagas disease women responded as follows.

> I would look for medical help.
>
> Depending on the cost.
>
> Yes because if it costs $22,000 let us say, something really expensive, I will never be able to pay. (Females, Group 3, P5 and P6)

Another woman emphasized the cost barrier to testing and treatment for Chagas disease when asked why Latinos wouldnot seek testing.

"The price. Right? It's not that one does not want to go to the doctor... It's because Hispanics generally don't have very much income, and even less than what we have [is available] because generally only the husbands work and everything has to come out of [their salary]." (Female, Group 2, P2)

In order to identify whether the length or side effects of treatment would be a barrier, the study team explained that the medications for treating Chagas disease had to be taken for three months and often had side effects including tingling, numbness, and rashes. Participants stated that this would not affect their decision to treat their disease, even likening it to the use of contraceptives.

We know that we take care of ourselves with contraceptives and it has consequences anyways we still take them despite the consequences.

Because if we don't then there are other consequences. (Females, Group 4, P1 and P2)

3.3.2. Modifications to the Process of Seeking Care. As previously mentioned, participants described their process of seeking healthcare for a range of issues because they had no experience seeking care for symptoms similar to those of Chagas disease (severe chest or stomach pain). When asked how this process would differ if they did have symptoms similar to Chagas disease, there was only one difference. Women said that if they and their family or friends were unaware of remedies for severe chest pain then they would skip Phase 1, Figure 1(a), and move directly into stoicism (Phase 2, Figure 1(a)). Additionally, both women and men said that they would be stoic as long as possible, but if the pain was intolerable they would move into Phase 3.

While participants criticized Latinos for being careless with their health, they excluded themselves from these reproaches by claiming that they prioritized their own health. When asked what they would do if they found out they were infected with Chagas disease, most said that they would find a way to cure it because they valued their health. The following are responses of participants when asked what they would do if they were infected with Chagas disease.

[I would take] the treatment [for Chagas disease]. I know it is expensive, but I would find the money, if I didn't have it to do the treatment. Because to leave it [untreated] is like not loving yourself. Because if you love yourself and you love your body, logically you will do what's right for it, right? But, many say, "no, it will go away, it will go away" but it doesn't. (Male, Group Urban Men, P1) (Medications to treat Chagas disease are provided at no cost in the United States, but at the time of this quote the group had not been informed of that, and the participant may have assumed that the medication would be expensive or may have been referring to other costs associated with missing work, paying for medical consults, etc.)

4. Discussion

The goal of this study was to understand awareness of Chagas disease and identify common healthcare-seeking behaviors and barriers to prevention and treatment of Chagas disease among Latino immigrants in Georgia from which to develop effective prevention and treatment interventions. Four barriers to being tested and treated for Chagas disease were reported by study participants including (1) a lack of awareness of the disease signifying a lack of motivation to be tested, (2) fear of being diagnosed with a fatal disease or an illness requiring expensive treatment, (3) reservations about the quality of care from mainstream healthcare providers among women and urban men, and (4) economic limitations to accessing mainstream healthcare. Finally, women described seeking mainstream healthcare immediately if their children became ill; thus no barriers were found to treating Chagas disease among children.

4.1. Lack of Awareness Regarding Chagas Disease. One of the prominent findings of this study was that the lack of awareness of Chagas disease was a significant barrier to testing. Currently, the authors are unaware of any studies measuring awareness of Chagas disease among Latino immigrants in the USA. However, a qualitative study in Spain among at-risk immigrants also demonstrated a lack of knowledge, fears, and false beliefs about Chagas disease [17]. The lack of awareness exhibited by participants in this study may be due to the low prevalence of Chagas disease in the USA, as few immigrants have been diagnosed with Chagas disease (previous studies have documented seroprevalence rates ranging from 0.4% to 4.9%) [6–8]. Additionally, awareness of Chagas disease is limited among medical providers in the USA; 68.2% of American College of Obstetricians and Gynecologists surveyed described their knowledge level about Chagas disease as "very limited" [18]. Yet, in areas where the disease is endemic, such as Argentina, there have been high rates of public participation in vector control campaigns which indicate an awareness of risk of Chagas disease [19].

The lack of awareness of Chagas disease documented in this study suggests a need for increased education about risk factors for contracting the disease and the need for testing due to a lack of definitive symptoms and an extended asymptomatic period. In the USA, there is only one educational campaign aimed at educating Latino immigrants in California about Chagas disease, in which health promoters educate them about their risk and encourage testing [20]. In Spain, the number of consultations for Chagas disease doubled after instituting a community-based educational program using leaflets. Finally new efforts have been made in the USA; the CDC has created a training module to educate medical providers and brochures printed in Spanish to inform Latin American immigrants about Chagas disease [21]. Participants in our study and others described their fears of being diagnosed with a fatal disease or illness that is expensive to treat [22, 23]. Increased education concerning the availability of free treatment may allay these fears and may also encourage testing.

4.2. Economic Limitations. In addition, our study, similar to others [11, 24], highlighted that mainstream healthcare was not sought because of the high cost and lack of health insurance. Among urban Latinos, our study and others found that the use of alternative providers (e.g., folk healers, homeopathic practitioners, and nonphysician prescribed medications) to overcome these structural barriers associated with mainstream care is common [11, 22, 25]. Increasing low cost and free services as well as other forms of support such as medical insurance for immigrants has been shown to be an effective way to increase access to healthcare among uninsured Latino immigrants [26]. Among migrant farm workers, our study and others [27] found that the primary structural barrier for seeking care was not cost as described by urban participants, but losing time at work and subsequent earnings. In rural areas, offering screenings, education, and preventive care in mobile clinics after work hours has been shown to increase access to mainstream healthcare for rural male immigrants [28–30].

4.3. Quality of Care. A further barrier to testing and treatment seeking for Chagas disease was reservation about the quality of care they would receive from mainstream healthcare providers. Often, Latinos fear the cost, potential for deportation, long waiting time, and discrimination that they associate with accessing healthcare [11] and consistently rate their experiences with medical staff more poorly than non-Hispanic whites [31]. This indicates a need for educating mainstream providers about ways to increase cultural sensitivity within their practices. Some educational resources for healthcare providers are already available and have been shown to increase overall knowledge and confidence in Latino cultural beliefs related to healthcare [32].

4.4. Use of Alternative Providers. Due to their inability to pay for care, urban participants described using a variety of alternative providers due to the lower cost associated with these providers. Some of these alternative providers practiced forms of complementary and alternative medicines (homeopathic healers and herbalists) which are increasingly being incorporated into mainstream healthcare [33–35]. Existing educational modules for improving cultural sensitivity amongst health providers incorporate awareness of alternative medicines. However, forming partnerships with some of the alternative providers identified (e.g., stores that sell prescription medications without prescriptions and physicians without licenses) is not feasible due to the unsafe and illegal nature of their practices.

No rural men described using alternative providers in order to bypass the barriers of seeking mainstream care. All the rural male participants were in Georgia on H-2A visas, which are temporary visas in which farmers hire, transport and house immigrants during the agricultural growing season, after which they are returned to their home countries [36]. Given the temporary and restricted nature of the farm workers' time in Georgia, it is possible that their lack of transportation may have limited their access to alternative providers. Additionally, farm workers might not have been aware of alternative providers, because of their limited time in Georgia and the isolated nature of being housed on farms. The dependence of rural men on mainstream healthcare providers when they are ill is an opportunity to increase preventive care for this population.

Finally, we found that mothers reported immediately taking their ill children to mainstream healthcare providers regardless of cost or other barriers. This practice is in contrast to previous studies which show that very few Latina mothers seek care for their children in a timely and appropriate manner [37, 38]. This process of immediately seeking mainstream healthcare for children may be a new opportunity to increase access to healthcare for women and their families. Although the authors are currently unaware of any targeted efforts to integrate care for Latina mothers and their children in the USA, integration of medical services has been shown to be effective in diverse settings and for a variety of preventive practices [39–42].

4.5. Potential Healthcare Seeking for Chagas Disease. A conceptual framework (Figure 1) was developed to understand the process of seeking healthcare by Latino immigrants and to understand the context of the barriers to testing and treatment for Chagas disease that were identified by participants. Some of the types of behaviors identified in this framework have been previously documented [11, 22, 24, 25], but the authors are unaware of any existing frameworks describing the pathways between behaviors and the applicable contexts of seeking healthcare among Latino immigrants. While participants were unable to describe their process of seeking healthcare specifically for Chagas disease, the model was consistent across groups and for all health problems mentioned by participants; thus, this model should also be applicable to Chagas disease.

Latino immigrants described seeking mainstream healthcare only when they could no longer tolerate symptoms of illnesses. Therefore, the lack of definitive symptoms and long asymptomatic period associated with Chagas disease means that Latino immigrants would not seek care for Chagas disease until they enter the clinically evident stage of the disease (e.g., cardiomyopathy, achalasia, and megacolon). Based on this model, upon entering the clinically evident stage of Chagas disease, a female Latina immigrant in Georgia would attempt to treat herself with traditional medicine. When traditional remedies failed, she would wait until the symptoms were unbearable and then seek care from either mainstream providers or alternative providers depending on her financial resources. Urban and rural male Latino immigrants would not use traditional remedies to treat themselves; rather they would wait until symptoms became unbearable and then seek care. Urban males would seek care with either mainstream or alternative providers depending on their financial resources, but rural males would only seek care from their closest mainstream provider. Finally, if a Latino child exhibited symptoms of Chagas disease, women would immediately take them to a mainstream provider.

4.5.1. Strengths and Limitations. The qualitative approach and inclusion of difficult to reach populations (e.g., Latino day laborers and migrant farm workers) in this study allowed a more comprehensive understanding of the perceptions of Chagas disease and the process of seeking healthcare among Latino immigrants. The lack of awareness of Chagas disease and experience of symptoms similar to that of Chagas disease (e.g., severe chest or stomach pains) among participants limited their ability to describe specific healthcare seeking for Chagas disease. Future research should explore how the gravity and familiarity with symptoms of any illness affect the process of seeking healthcare in order to confirm the applicability of the conceptual framework for different types of illnesses.

5. Conclusion

This study provides an in-depth understanding of the barriers regarding testing and treatment for Chagas disease among Latino immigrants by describing their processes of seeking healthcare, the motivations for using these processes, and the pathways between these processes. Latino immigrants tend to wait until they are gravely ill before seeking care due to beliefs and economic limitations. Financial resources primarily determine the process that Latinos use to seek health care. Educational campaigns on Chagas disease need to be implemented based on these treatment seeking pathways (e.g., combining education with Latino children medical visits) in order to increase testing and treatment. In addition, more low cost and free services or health insurance need to be made available for Latino immigrants in order to increase access to preventive care.

Acknowledgments

The authors thank all the participants for their willingness to take part in this project. They also thank colleagues with valuable contributions in study recruitment and cultural sensitivity: Dr. Judith L. Wold of the Lillian Carter Center for Global Health & Social Responsibility and Cynthia Hernandez, Director of the Ellenton Clinic, for facilitating data collection within the Farm Worker Family Health Program; and Heidy Guzman of the Hispanic Health Coalition of Georgia and Nolly Dyste of the Latin American Association for their valuable input throughout the project. This work was supported by Grant 1086 from the Healthcare Georgia Foundation. This work was also partially supported by grant 1K01AI087724-01 from the NIAID at the NIH (to J. S. Leon), Grant 2010–85212-20608 from the National Institute of Food and Agriculture at the U.S. Department of Agriculture (to J. S. Leon), and a grant from the Emory University Global Health Institute (to J. S. Leon, R. M. Minneman, C. C. Pennock, and S. I.). The authors' also acknowledge financial support from the Emory University Scholarly Inquiry and Research at Emory Program and Summer Undergraduate Research Program at Emory (to L. C. Albor.).

References

[1] S. Montgomery, *Chagas Disease in Your Backyard: Blood Donors and Trypanosoma Cruzi*, Atlanta, Ga, USA, 2009.

[2] J. H. F. Remme, P. Feenstra, P. R. Lever et al., *Tropical Diseases Targeted For Elimination: Chagas Disease, Lymphatic Filariasis, Onchocerciasis, and Leprosy*, 2006.

[3] A. Moncayo, F. Guhl, and C. Stein, *Global Burden of Chagas' Disease in the Year*, 2000.

[4] G. A. Schmunis and Z. E. Yadon, "Chagas disease: a Latin American health problem becoming a world health problem," *Acta Tropica*, vol. 115, no. 1-2, pp. 14–21, 2010.

[5] G. A. Schmunis, "Epidemiology of Chagas disease in non-endemic countries: the role of international migration," *Memorias do Instituto Oswaldo Cruz*, vol. 102, supplement 1, pp. 75–85, 2007.

[6] L. V. Kirchhoff, A. A. Gam, and F. C. Gilliam, "American trypanosomiasis (Chagas' disease) in Central American immigrants," *American Journal of Medicine*, vol. 82, no. 5, pp. 915–920, 1987.

[7] R. Arena, C. E. Mathews, A. Y. Kim, T. E. Lenz, and P. M. Southern, "Prevalence of antibody to Trypanosoma cruzi in Hispanic-surnamed patients seen at Parkland Health & Hospital System, Dallas, Texas," *BMC Research Notes*, vol. 4, article 132, 2011.

[8] S. Ciesielski, J. R. Seed, J. Estrada, and E. Wrenn, "The seroprevalence of cysticercosis, malaria, and Trypanosoma cruzi among North Carolina migrant farmworkers," *Public Health Reports*, vol. 108, no. 6, pp. 736–741, 1993.

[9] C. Bern and S. P. Montgomery, "An estimate of the burden of chagas disease in the United States," *Clinical Infectious Diseases*, vol. 49, no. 5, pp. e52–e54, 2009.

[10] J. S. Leon and D. M. Engman, "The contribution of autoimmunity to Chagas disease," in *World Class Parasites: American Trypanosomiasis*, K. M. Tyler and M. A. Miles, Eds., Kluer Academic, Boston, Mass, USA, 2003.

[11] H. E. Ransford, F. R. Carrillo, and Y. Rivera, "Health care-seeking among Latino immigrants: blocked access, use of traditional medicine, and the role of religion," *Journal of Health Care for the Poor and Underserved*, vol. 21, no. 3, pp. 862–878, 2010.

[12] National Council de la Raza, *Latinos in Georgia: A Closer Look*, 2005.

[13] M. K. Askim-Lovseth and A. Aldana, "Looking beyond "affordable" health care: cultural understanding and sensitivity-necessities in addressing the health care disparities of the u.s. hispanic population," *Health Marketing Quarterly*, vol. 27, no. 4, pp. 354–387, 2010.

[14] U. S. Census Bureau, *American FactFinder*, 2010.

[15] M. Hennink, I. Hutter, and A. Bailey, *Qualitative Research Methods*, SAGE Publications, London, UK, 2011.

[16] B. Glaser and A. Stauss, *The Discovery of Grounded Theory: Strategies for Qualitative Research*, Aldine de Gruyter, New York, NY, USA, 1967.

[17] M. Navarro, A. Perez-Ayala Guionnet et al., "Targeted screening and health education for Chagas disease tailored to at-risk migrants in Spain, 2007 to 2010," *Eurosurveillance*, vol. 16, no. 38, 2011.

[18] J. R. Verani, S. P. Montgomery, J. Schulkin, B. Anderson, and J. L. Jones, "Survey of obstetrician-gynecologists in the United States about Chagas disease," *American Journal of Tropical Medicine and Hygiene*, vol. 83, no. 4, pp. 891–895, 2010.

[19] I. Llovet, G. Dinardi, and F. G. de Maio, "Mitigating social and health inequities: community participation and Chagas

disease in rural Argentina," *Global Public Health*, vol. 6, no. 4, pp. 371–384, 2011.

[20] A. Petherick, "Campaigning for Chagas disease," *Nature*, vol. 465, no. 7301, pp. S21–S22, 2010.

[21] Centers for Disease Control and Prevention, *Chagas Disease: What U.S. Clinicians Need to Know*, 2010.

[22] I. C. Garcés, I. C. Scarinci, and L. Harrison, "An examination of sociocultural factors associated with health and health care seeking among latina immigrants," *Journal of Immigrant and Minority Health*, vol. 8, no. 4, pp. 377–385, 2006.

[23] S. Moreland, K. Engelman, K. A. Greiner, and M. S. Mayo, "Papanicolaou testing among native American and Hispanic populations," *Ethnicity and Disease*, vol. 16, no. 1, pp. 223–227, 2006.

[24] J. L. Hargraves and J. Hadley, "The contribution of insurance coverage and community resources to reducing racial/ethnic disparities in access to care," *Health Services Research*, vol. 38, no. 3, pp. 809–829, 2003.

[25] J. H. Lee, M. S. Goldstein, E. R. Brown, and R. Ballard-Barbash, "How does acculturation affect the use of complementary and alternative medicine providers among Mexican- and Asian-Americans?" *Journal of Immigrant and Minority Health*, vol. 12, no. 3, pp. 302–309, 2010.

[26] S. Khan, V. Velazquez, C. O'Connor et al., "Health care access, utilization, and needs in a predominantly latino immigrant community in providence," *Rhode Island*, vol. 94, no. 10, pp. 284–287, 2011.

[27] J. S. Meister, "The health of migrant farm workers," *Occupational Medicine*, vol. 6, no. 3, pp. 503–518, 1991.

[28] L. M. Stein, "Health care delivery to farmworkers in the Southwest: an innovative nursing clinic," *Journal of the American Academy of Nurse Practitioners*, vol. 5, no. 3, pp. 119–124, 1993.

[29] L. Heuer, C. W. Hess, and M. G. Klug, "Meeting the health care needs of a rural hispanic migrant population with diabetes," *Journal of Rural Health*, vol. 20, no. 3, pp. 265–270, 2004.

[30] A. Connor, L. Layne, and K. Thomisee, "Providing care for migrant farm worker families in their unique sociocultural context and environment," *Journal of Transcultural Nursing*, vol. 21, no. 2, pp. 159–166, 2010.

[31] L. S. Morales, W. E. Cunningham, J. A. Brown, H. Liu, and R. D. Hays, "Are Latinos less satisfied with communication by health care providers?" *Journal of General Internal Medicine*, vol. 14, no. 7, pp. 409–417, 1999.

[32] A. A. McGuire, I. C. Garces-Palacio, and I. C. Scarinci, "A successful guide in understanding latino immigrant patients: an aid for health care professionals," *Family & Community Health*, vol. 35, no. 1, pp. 76–84, 2012.

[33] D. P. Rakel, M. P. Guerrera, B. P. Bayles, G. J. Desai, and E. Ferrara, "CAM education: promoting a salutogenic focus in health care," *Journal of Alternative and Complementary Medicine*, vol. 14, no. 1, pp. 87–93, 2008.

[34] M. S. Wetzel, D. M. Eisenberg, and T. J. Kaptchuk, "Courses involving complementary and alternative medicine at US medical schools," *Journal of the American Medical Association*, vol. 280, no. 9, pp. 784–787, 1998.

[35] M. Frenkel and L. Cohen, "Incorporating complementary and integrative medicine in a comprehensive cancer center," *Hematology/Oncology Clinics of North America*, vol. 22, no. 4, pp. 727–736, 2008.

[36] "Commission for Labor Cooperation Guide to the H-2A Visa Program in the United States".

[37] R. E. Zambrana, K. Ell, C. Dorrington, L. Wachsman, and D. Hodge, "The relationship between psychosocial status of immigrant Latino mothers and use of emergency pediatric services," *Health & Social Work*, vol. 19, no. 2, pp. 93–102, 1994.

[38] G. Flores, M. Abreu, M. A. Olivar, and B. Kastner, "Access barriers to health care for Latino children," *Archives of Pediatrics and Adolescent Medicine*, vol. 152, no. 11, pp. 1119–1125, 1998.

[39] K. E. Mark, J. Meinzen-Derr, R. Stephenson et al., "Contraception among HIV concordant and discordant couples in Zambia: a randomized controlled trial," *Journal of Women's Health*, vol. 16, no. 8, pp. 1200–1210, 2007.

[40] D. S. Pope, A. N. Deluca, P. Kali et al., "A cluster-randomized trial of provider-initiated (opt-out) HIV counseling and testing of tuberculosis patients in South Africa," *Journal of Acquired Immune Deficiency Syndromes*, vol. 48, no. 2, pp. 190–195, 2008.

[41] S. Dodds, N. T. Blaney, E. M. Nuehring et al., "Integrating mental health services into primary care for HIV-infected pregnant and non-pregnant women: whole life—a theoretically derived model for clinical care and outcomes assessment," *General Hospital Psychiatry*, vol. 22, no. 4, pp. 251–260, 2000.

[42] D. Huntington and A. Aplogan, "The integration of family planning and childhood immunization services in Togo," *Studies in Family Planning*, vol. 25, no. 3, pp. 176–183, 1994.

13

Detection of Parasites and Parasitic Infections of Free-Ranging Wildlife on a Game Ranch in Zambia: A Challenge for Disease Control

Hetron Mweemba Munang'andu,[1] Victor M. Siamudaala,[2] Musso Munyeme,[3] and King Shimumbo Nalubamba[4]

[1] *Section of Aquatic Medicine and Nutrition, Department of Basic Sciences and Aquatic Medicine, Norwegian School of Veterinary Sciences, Ullevålsveien 72, P.O. Box 8146 Dep, 0033, Oslo, Norway*
[2] *Kavango Zambezi Transfrontier Conservation Area Secretariat, Kasane 821, Gaborone, Botswana*
[3] *Department of Disease Control, School of Veterinary Medicine, University of Zambia, P.O. Box 32379, Lusaka 10101, Zambia*
[4] *Department of Clinical Studies, School of Veterinary Medicine, University of Zambia, P.O. Box 32379, Lusaka 10101, Zambia*

Correspondence should be addressed to Hetron Mweemba Munang'andu, hetron2002@yahoo.co.uk

Academic Editor: Francisca Mutapi

Ex-situ conservancies are expanding alternatives to livestock production in Zambia albeit the lack of information on circulating infectious parasites from wildlife. Therefore, 12 wildlife species were examined on a game ranch were all species were found to be infected by *Rhipecephalus* spp. Haemoparasite infections were estimated at 7.37% ($n = 95$) with *Babesia* spp. detected in bushbuck (*Tragelaphus scriptus*); *Anaplasma marginale* in impala (*Aepyceros melampus*) and puku (*Kobus vardonii*) for the first time in Zambia. The majority of worm species isolated from bovids were not detected in equids and, *vice versa*. Our findings intimate ecological and behavioural patterns of some animals as deterministic to exposure. Kafue lechwe (*Kobus leche kafuensis*) had the widest range of worm species with more infected organs than other animals suggesting their semi aquatic nature contributory to prolonged worm exposure compared to other animals. On the other hand, Kafue lechwe had the least tick infections attributable more to shorter attachment periods as they spend prolonged periods submerged in water. Our findings indicate the vital role that wildlife plays in the epidemiology of parasitic diseases. To reduce the infection burden, control measures should be focused on reducing transmission to highly susceptible animal species as described herein.

1. Introduction

Ex-situ conservation is expanding in Zambia with the aim of promoting wildlife utilization alongside livestock production. The industry has turned out to be an alternative to cattle ranching given that the latter has been ravaged by tick-borne diseases that have caused a significant decline on the cattle population in Zambia [1, 2]. The shift from cattle ranching to game ranching reduces economic losses incurred in livestock production due to continuous prophylactic treatment of cattle unlike wildlife species that are resistant to tick-borne diseases [3]. Overall, game ranching promotes preservation of different wildlife species by protecting animals from poaching which is rare on the game ranches but common on state-owned national parks. In addition, the involvement of game ranches in stocking endangered species such as the kafue lechwe (*Kafue leche kafuensis*) and Black lechwe currently on the International Union Conservation of Nature (IUCN) red list of threatened species is a good conservation strategy which aims at serving these species from extinction [4]. Besides, the translocation of animals from different ecosystems into one habitat leads to stocking of animals that would, otherwise, have not shared a habitat under natural conditions. The mixing of animals from different ecosystems into one habitat is likely to be a proponent of introducing diseases sourced from different ecosystems into a new habitat thereby exposing animals to parasitic infections they would otherwise have never been exposed to. Hence, there is need to develop trace-back systems that track diseases obtained from different ecosystems. It has become paramount to investigate parasitic diseases of wildlife with a view of generating baseline data for use in trace-back systems during disease outbreaks and the translocation of animals from one ecosystem to the other. In the present study, we investigated the presence of endo, and ectoparasites of different wildlife species reared on a game ranch in central Zambia in order to obtain baseline data on the nature of parasitic infections obtained from wildlife in this part of the country. We also wanted to find out the prevalence levels of different parasitic infections on different wildlife species as a way of identifying control strategies that could be used to reduce parasitic burden on game ranches by advising game ranchers to use control strategies likely to reduce the prolonged survival of vector species engaged in transmission of different diseases. The challenge of developing effective disease control strategies for the control of parasitic infections in wildlife medicine is herein discussed. Although this study is based on survey data obtained from central Zambia, it brings into perspective challenges faced by veterinarians in the control of parasitic diseases given the expansion of game ranching across Africa.

2. Materials and Methods

2.1. Study Area. The study was carried out on a game ranch located approximately 45 km northeast of Lusaka. The ranch covers a total area of 4,500 km^2 and is located at an altitude of 1,100 meters. The mean annual rainfall was about 950 mm while summer temperatures varied between 20°C–32°C in the months of October to March. Winter temperatures varied between 10°C–26°C in the months of June to August. Relative humidity was below 40% throughout the year. Vegetation on the ranch comprised of miombo and acacias woodlands with open savannah grasslands. The ranch encompasses three periannual large dams that provide adequate water for the survival of various species including the semiaquatic kafue lechwe (*Kobus leche kafuensis*) and sitatunga (*Tragelaphus spekii*). Tick and worm infections were controlled by rotational burning of grass in the dry season and use of antihelminths and acaricides administered using Duncan applicators [5–7]. The ranch was surrounded by a 2.5 m fence with a 10-meter fire guard surrounding the entire game ranch.

2.2. Animals. The ranch is endowed with several mammalian species comprised of wild ungulates (Table 1), reptiles, and birds. In October 2005 blood samples and smears were collected from a total of 39 animals from six wildlife species captured for translocation (Table 1). The animals were immobilized using M99 (etorphine hydrochloride, Norvatis, Ltd., Johannesburg, South Africa) at standard doses and were later revived using M5050 (revivon, Norvatis, Ltd., Johannesburg, South Africa). In August to October 2004 and July to August 2005, 56 animals from 10 species were

sacrificed (killed using a rifle) for parasite infestation and disease surveys (Table 1). Only sacrificed animals ($n = 56$) were used for helminth surveys while all animals ($n = 95$), that is, both the sacrificed and immobilized, were used for blood parasite and tick infestation surveys.

2.3. Sampling of Ticks and Blood Smears. Ticks were collected and stored in bottles and transported to the School of Veterinary Medicine at the University of Zambia in Lusaka for identification. Thin blood smears were made from ear veins on site from all animals captured for translocation and those sacrificed for disease investigations (Table 1). Second sets of blood smears were made from buffy coats from blood collected in EDTA soon after arrival at the laboratory at the School of Veterinary Medicine, University of Zambia in Lusaka. For sacrificed animals, impression smears were also made from the prescapular and parotid lymph nodes on site. All slides were observed under the light microscope (×100) after staining with 20% Giemsa stain.

2.4. Sampling of Helminths. Components of the digestive system were separately ligated. From each segment, 180 ml of the contents was placed in a bottle and 20 mL of formalin was added to each bottle. Thereafter, the contents were emptied into sedimentation jars at the School of Veterinary Medicine at the University of Zambia in Lusaka. After sedimentation, worms in the supernatant were picked and stored in 10% formalin bottles. The mesentery was separated from the viscera and was carefully inspected for the presence of *Schistosoma* spp. while the veins were cut and squeezed to release the worms. The worms were picked and stored in 10% formalin. The liver and bile ducts were incised as described by Hansen and Perry [8] to check for flukes. The liver was sliced into small pieces and squeezed to let the flukes drop in water containers followed by draining the water through a 500 μm sieve (Endecotts Ltd., England). All flukes were collected and stored in 10% formalin. Other organs inspected were the trachea, lungs, heart, tongue, and skeletal muscles. Recovered worms were put in petri dishes containing lactophenol overnight. Thereafter, worms were identified using standard keys after mounting on glass slides [9].

3. Results

3.1. Blood Parasites. Identification of parasites was based on standard keys [9–11]. Figure 1 shows *Trypanosoma congolense* detected in greater kudu, while Figure 2 shows infection of *Babesia* spp. detected from bushbuck. *Anaplasma marginale* appeared as dense intraerythrocytic rounded bodies located on the edges of red blood cells ranging from 3.21–9.78 μm ($n = 52$) which is in line with observations made elsewhere [9, 12]. As shown in Figure 2, *Babesia* spp. were characterized by pairs of merozoites in blood smears which is in line with observations made by Homer et al. [10] and Schuster [11] who pointed out that detection of pairs or tetrads of merozites also known as "Maltose cross" in stained red blood cells is characteristic of *Babesia* spp. infection.

Detection of Parasites and Parasitic Infections of Free-Ranging Wildlife on a Game Ranch in Zambia: A Challenge for Disease Control

119

TABLE 1: Totals on the game ranch and number examined.

Species	Total on Game Ranch (2005)	Animals examined (n)			
		Sacrificed 2004	Sacrificed 2005	Immobilized 2005	Total examined
Bushbuck (*Tragelaphus scriptus*)	57	—	—	4	4
Defassa waterbuck (*Kobus ellipsiprymnus*)	63	3	5	—	8
Greater kudu (*Tragelaphus strepsiceros*)	25	3	3	5	11
Impala (*Aepyceros melampus*)	509	4	2	—	6
Kafue lechwe (*kobus leche kafuensis*)	380	4	4	14	22
Puku (*Kobus vardoni*)	252	4	2	10	16
Reedbuck (*Redunca redunca*)	72	—	—	4	4
Sable antelope (*Hippotragus niger*)	41	2	—	2	4
Tsessebe (*Damaliscus lunatus*)	42	2	—	—	2
Warthog (*Phacochoerus aethiopicus*)	205	4	—	—	4
Wildebeest (*Connochaetes taurinus*)	60	4	2	—	6
Zebra (*Equus burchelli*)	80	5	3	—	8
Totals	1,786	35	21	39	95

TABLE 2: Haemoparasites detected from blood smears.

Wildlife species	Total examined	Number of animals infected with				Totals
		Theileria piroplasms	*Babesia* species	*Anaplasma marginale*	*Trypanosoma congolense*	
Bushbuck (*Tragelaphus scriptus*)	4	—	1	—	—	1
Defassa waterbuck (*Kobus ellipsiprymnus*)	8	—	—	—	—	
Greater kudu (*Tragelaphus strepsiceros*)	11	1	—	—	2	3
Impala (*Aepyceros melampus*)	6	—	—	1	—	1
Kafue lechwe (*Kobus leche kafuensis*)	22	—	—	—	—	
Puku (*Kobus vardoni*)	16	—	1	1	—	2
Sable antelope (*Hippotragus niger*)	4	—	—	—	—	
Reedbuck (*Redunca redunca*)	4	—	—	—	—	
Tsessebe (*Damaliscus lunatus*)	2	—	—	—	—	
Warthog (*Phacochoerus aethiopicus*)	4	—	—	—	—	
Wildebeest (*Connochaetes taurinus*)	6	—	—	—	—	
Zebra (*Equus burchelli*)	8	—	—	—	—	
Totals	95	1	2	2	2	7

Overall, our findings show a low prevalence of blood parasite infections on the game ranch. As shown in Table 2, there were only seven animals having blood parasites giving an overall infection rate of 7.37% ($n = 95$). Prevalence rates for individual species of blood parasites were estimates at 2.11% ($n = 95$) for *Anaplasma marginale*, *Babesia* species, and *Trypanosoma congolense* while for Theileria piroplasms the infection rate was estimated at 1.05% ($n = 95$). *Trypanosoma congolence* was only detected in greater kudu at an infection rate of 18.18% ($n = 11$).

3.2. Ticks. Infection rates of different tick species for the different animals examined on the game ranch were generally high (Table 3). Some animal species were infected by different tick species while others were only infected by one species. *Rhipicephalus* spp. were the most prevalent tick species infesting all animal species examined (Table 3).

Amblyomma variegatum was collected from six species while *Hyaloma trancutum* together with other *Hyaloma* spp. were collected from five animal species. Bushbuck, defassa water-buck (*Kobus ellipsiprymnus*), and wildebeest (*Connochaetes taurinus*) were infested by a wider range of tick species unlike impala, kafue lechwe, reedbuck (*Redunca redunca*), and tsessebe (*Damaliscus lunatus*) which were only infested by *Rhipicephalus appendiculatus*. Kafue lechwe and impala had the least infection rates of 22.7% ($n = 22$) and 33.3% ($n = 6$), respectively. Only two tick control measures were used on the game ranch, namely, the use of Duncan applicators [13] and rotational burning. For rotational burning, the game ranch was divided into four sections and only one section was burnt at a time allowing the animals to graze on the unburnt areas. Duncan applicators were used for the control of ticks by administering acaricide pour-ons on animals. The efficacy

TABLE 3: Ticks collected from different wildlife species on game ranch.

Wildlife species	(n)	Tick species identified (*)
Bushbuck (*Tragelaphus scriptus*)	4	*Rhipiciphelus appendiculatus* (4), *Rhipiciphelus* species (4), *Amblyoma variegatum* (3), *Hyaloma* species (2).
Defassa waterbuck (*Kobus ellipsiprymnus*)	8	*Rhipiciphelus appendiculatus* (8), *Rhipiciphelus* species (6), *Amblyoma variegatum* (7), *Hyaloma trancutum* (3), *Boophilus decoloratus* (4).
Greater kudu (*Tragelaphus strepsiceros*)	11	*Rhipiciphelus appendiculatus* (11), *Rhipiciphelus* species (9), *Amblyoma variegatum* (8), *Amblyoma* species (2), *Hyaloma Trancutum (4), Hyaloma* species (2).
Impala (*Aepyceros melampus*)	6	*Rhipiciphelus appendiculatus* (2).
Kafue lechwe (*Kobus leche kafuensis*)	22	*Rhipiciphelus appendiculatus* (5),
Puku (*Kobus vardoni*)	16	*Rhipiciphelus appendiculatus* (8), *Rhipiciphelus* species (6), *Amblyoma variegatum* (7), *Hyaloma* species (4).
Sable antelope (*Hippotragus niger*)	4	*Rhipiciphelus appendiculatus* (4), *Rhipiciphelus* species (2).
Reedbuck (*Redunca redunca*)	4	*Rhipiciphelus appendiculatus* (4).
Tsessebe (*Damaliscus lunatus*)	2	*Rhipiciphelus appendiculatus* (2).
Warthog (*Phacochoerus aethiopicus*)	4	*Amblyoma variegatum* (4), *Rhicicephalus* spp. (4).
Wildebeest (*Connochaetes taurinus*)	6	*Amblyoma variegatum* (5), *Rhipiciphelus appendiculatus* (6), *Rhicicephalus* spp. (4), *Hyalomma* species (4),
Zebra (*Equus burchelli*)	8	*Rhipiciphelus appendiculatus* (6), *Rhicicephalus* spp. (4)

(*n*) = total of animals examined, (*) = number of infested animals.

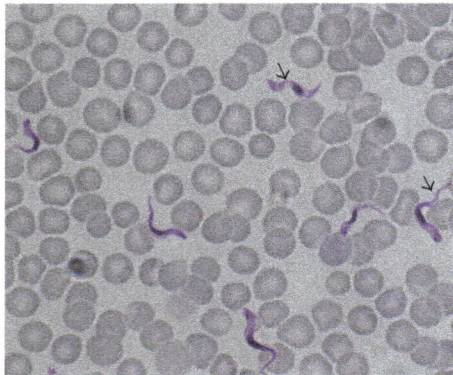

FIGURE 1: Show detection of *Trypanosoma congolense* (arrow) in greater kudu (*Tragelaphus strepsiceros*).

of these control measures was not evaluated in the present study.

3.3. Helminths. Table 4 shows a list of helminthes detected from 10 animal species examined on the game ranch. Generally, infection rates were high for most animal species (Table 4). Kafue lechwe and Burcelli's zebra (*Equus burchelli*) were infected by a wide range of worm species than other animal species. This can be attributed to the fact that there were more animals examined from these species than other animal species (Table 4). On contrast, defassa waterbuck, tsessebe, and greater kudu were only infected by one helminth species. *Schistosoma* species were only reported from kafue lechwe while *Gastrodiscus aegyptiacus, Gastrophilus meridionatis, Strongylus equines*, and *Strongylus vulgaris* were only reported from zebra (Table 4). *Oesophagosotomum* spp. were the most common worm species infecting both bovids and equids while the amphistomes and paramphistomes

were only recorded in bovids. Kafue lechwe had the widest organ distribution of worm infections being infected in seven different organs followed by Burcelli's zebra which had infections in three different organs. It is interesting to note that most worm species were specialized to specific organs despite infecting multiple hosts. For example, *Stillesia hepatica* was only found in the liver of infected kafue lechwe, puku, and greater kudu, while *Gaigeria panchyselis* was only found in the small intestines of puku, kafue lechwe, tsessebe, and impala. The only control measure used was the administering of antihelminthes using Dancun applicators.

4. Discussion

Prevalence for haemoparasite infections was generally low despite high-tick infection rates observed on the animals. It is interesting to note that all major tick-borne diseases infecting livestock diseases in Zambia [1, 14] were detected on the game ranch. The low infection rates observed in the present study could be attributed to the detection method used considering that the use of blood smears does not detect previous exposure and that low infection rates can easily be missed using this technique. Hence, it is likely that if we had used more robust diagnostic tools such as molecular-biology-based techniques that are more sensitive, higher infection rates would have been determined. On the other hand, the use of serological assays such as the enzyme linked immunosorbent assay (ELISA) would have determined the seroprevalence for animals previously exposed to haemoparasite infections. We did not find clinical cases at the times of the surveys although we did not analyze the blood samples to determine whether infections by hemoparasites caused changes in blood parameters. Besides, the sample size of the animals examined and the number of animals infected by different blood parasites obtained in the present study were

Detection of Parasites and Parasitic Infections of Free-Ranging Wildlife on a Game Ranch in Zambia: A Challenge for Disease Control

121

TABLE 4: Helminthes isolated from different wildlife species.

Table Species	Animals	Organ examined	Helminth Species	No infected	Percentage
Defassa waterbuck (*Kobus ellipsiprymnus*)	8	Large intestines	*Oesophagosotomum* spp.	6	75.0%
Geater kudu (*Tragelaphus strepsiceros*)	6	liver	*Stillesia hepatica*	4	66.7%
Impala (*Aepyceros melampus*)	6	Small intestines	*Gaigeria panchyselis,*	5	83.3%
		Large intestines	*Oesophagostomun* species,	2	33.3%
Kafue lechwe (*Kobus leche kafuensis*)	8	Liver	*Fasciola gigantica*	8	100.0%
		Mesentery	*Schistosoma* spp.	5	62.5%
		Peritoneum	*Setaria* species	7	87.5%
		Rumen	*Amphistoma* spp.,	7	87.5%
			Parampistoma spp.	7	87.5%
		Abomasum	*Amphistoma* spp.,	7	87.5%
		Small intestines	*Gaigeria panchyselis,*	4	50.0%
			Borrostomum trignocephalum	3	37.5
		Large intestines	*Trichuris* spp.,	5	62.5%
			Oesophagostomum species	7	87.5%
Puku (*Kobus vardoni*)	6	liver	*Stillesia hepatica*	4	66.7%
		Large intestines	*Oesophagostomum* species	4	66.7%
Sable antelope (*Hippotragus niger*)	2	Small intestines	*Gaigera pachyscelis*	2	100.0%
Tsessebe (*Damaliscus lunatus*)	2	Small intestines	*Gaigera pachyscelis*	2	100.0%
Warthogs (*Phacochoerus aethiopicus*)	4	Lage intestines	*Oesophagosotomum* spp.,	2	50.0%
			Trichuris species,	3	75.0%
			Trichostrongylus species	3	75.0%
Wildebeest (*Connochaetes taurinus*)	6	Rumen	*Paramphystomes*	4	66.7%
Zebra (*Equus burchelli*)	8	Ceacum,	*Gastrodiscus aegyptiacus,*	5	62.5%
			Stelizia species	3	37.5%
			Gastrophilus meridionatis	4	50.0%
		Large intestines	*Oesophagostomum* spp.	7	87.5%
			Strongylus equinus	4	50.0%
			Strongylus vulgaris	4	50.0%
		Small intestines	*Anaplocephala perfoliata*	6	75.0%
Totals	56				

inadequate to carry out analysis on the impact of haemoparasites on blood parameters. However, these findings are consistent with other studies that have shown that detection of blood parasites in wildlife is often incidental. This is supported by observations made by other scientists that wildlife are resistant to haemoparasite infections and that clinical disease is often stress related. Besides, Munang'andu et al. [15, 16] recently reported *Babesia* spp. infections in free ranging pukus and *Trypanosoma brucei* in free ranging greater kudu without clinical disease on game ranches in Zambia. To our knowledge, this is the first report of *Babesia* spp. infections in bushbuck and *Anaplasma marginale* in puku and impala in Zambia. Overall, these findings point to the fact that wildlife could play an important role in the epidemiology of haemoparasites in Zambia. This implies that while tick control using acaricides could be reducing the occurrence of tick-borne diseases in livestock, the expansion of game ranching could have a long-term adverse effect of expanding the reservoir host occupancy range of tick-borne diseases whose spillover into cattle ranching would impact negatively on livestock production. This poses a significant challenge for control of tick-borne diseases especially in interface areas where concurrent expansion of wildlife and livestock production is taking places. However, there is need for detailed epidemiology studies to determine the role of different wildlife species in the epidemiology of these diseases in countries where game ranching is expanding.

Generally, *Rhipicephalus appendiculatus* and *Rhipicephalus* species were the most common tick-species infecting multiple host species. *Amblyomma variegatum* and *Hyalomma trancutum* were collected from fewer animal hosts than *Rhipephelus appendiculatus*. It is not known whether this difference was based on host preference or the relative abundance of the different tick species on the game ranch. Moreover, some animal species like tsessebe, reedbuck, and impala were only infected by *Rhipicephalus appendiculatus*

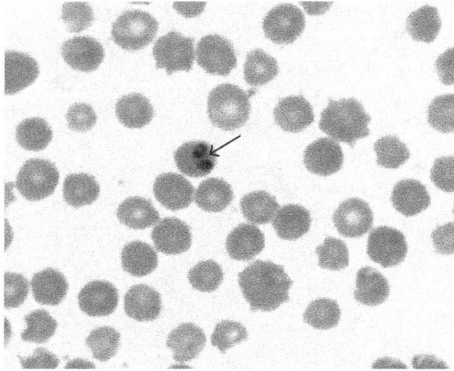

FIGURE 2: Shows detection *Babesia* spp. in bushbuck (*Tragelaphus scriptus*).

further showing that *Rhipicephalus appendiculatus* was the most common tick species infecting both the bovids and equids on the game ranch. Kafue lechwe are semiaquatic medium sized antelopes often submerged up to the shoulders sometimes leaving only the nostrils when frightened [17, 18]. Hence, the only foreseeable reason why kafue lechwe had low infection rates is that because of its semiaquatic nature, ticks infecting this animal species are likely to drop-off when the animals are submerged in water thereby reducing the attachment time on the host. However, we observed low infection rates on impala which is purely an on-land species, and not semiaquatic like kafue lechwe, although the sample size obtained in this study was low (n = 4, Table 3). We did not establish whether impala is a less favored host for tick infection while species such as the deffasa waterbuck, bushbuck, puku, and greater kudu were not only infected by a wide range of tick species but also had high infection rates for most tick species (Table 2). However, there is need for detailed experimental studies to determine the host preference of tick infections in wildlife and to establish reasons why some animal species are less infested than others. Information obtained from such studies would help in selecting wildlife species for culling especially in situation where population reduction of selected wildlife species is aimed at reducing the tick burden. For example, when tick burden is high, animal species that are more vulnerable to infestation can be reduced by culling or safari hunting while the less infested species are left to increase.

Kafue lechwe and Burchelli's zebra were the most infected by helminthes. In addition, kafue lechwe had the widest organ distribution of infections than other animal species. Elsewhere [8, 19], it has been shown that gastrointestinal worm infection rates are dependent on a number of factors which include the number of infective larvae ingested by the host, which in turn is influenced by climatic factors, vegetation, and animal density. Dry open areas prone to excessive heat are hostile for the survival of infective larvae while moist areas near water sources are conducive for the survival infective larvae. This would account for reasons why Ng'ang'a et al. [19] consistently recovered infective larvae around watering points throughout their study period

unlike semiarid open areas that had no infective larvae during the dry season. In their conclusion [19], they noted that watering points were an important source of infection for animals, especially during the dry season when other pastures were noninfective. On moist herbage, larvae of different nematodes migrate up and down the blades of grass which facilitate the uptake of infective larvae by grazing animals. During the dry season, areas around water sources attract more animals for grazing thereby increasing the animal population density. As pointed out by Chingwena et al. [20], animals that aggregate in these places are likely to get infected by infective larvae. This would account for the reason why kafue lechwe had a wide infection rate of different worm species in different organs as a result of constant exposure to infection by grazing on moist pastures that harbor high infection rates of infective larvae around water sources close to their habitats. Moreover, the timing of the current surveys which was in the dry season between August and October when there was scarcity of pasture and water on the game ranch, moist conditions prevailing at water sources indicate that these areas had the highest levels of infective larvae leading to transmission of these larvae to the semiaquatic kafue lechwe that graze around the water sources close to their habitat unlike other animal species found in open dry pastures that are less infective during the dry season. By being definitive host, kafue lechwe are likely to serve as critical determinants of infection to other animals as they contaminate the pastures around the water sources with fecal droppings containing infective larvae. Besides, infective larvae deposited in water by defecating lechwes during times when they are submerged in water are likely to infect other animals that come to drink water. By maintaining an active transmission cycle, kafue lechwe is likely to save as a continuous source of infection to other animal species. These observations indicate that reducing infection to kafue lechwe at water sources is likely to reduce the source of infection to other animals. Hence, these is a need for innovative disease control strategies that would reduce the cycle of transmission between infected pastures at water sources and kafue lechwe in order to reduce worm burden infections of wildlife species reared on game ranches.

Phiri et al. [21] pointed out that snail intermediate of worm species like *Schistosoma* spp. are often concentrated in marshy areas or marginal shallow water areas of ox-bow lakes, lagoons, and rivers. Animals that aggregate in these places increase the contact between miracidia and snail intermediate hosts. Hamond [22] pointed out that the higher the number of final hosts and snails are found together at one site, the more the likelihood that worm infection will propagate and be transmitted to other species. Hence, kafue lechwe which are final hosts and predominantly occupy marshy areas are likely to maintain a high transmission cycle of *Schistosoma* spp. with snail intermediate hosts found on the edges of water sources on the game ranch. This would account for reasons why kafue lechwe were the only species infected by *Schistoma* spp. on the game ranch.

It is interesting to note that most worm species identified in kafue lechwe in the Kafue basin were also detected in the present study [23]. It is likely that these worms could have been introduced on the game ranch by the first breeding stock that was translocated from the kafue basin. We, therefore, advocate that treatment of animals against parasitic infections and use of pour-on acaricides and antihelminths should be carried out prior to or during translocation to reduce the transmission of parasites from one ecosystem to the other. Some helminthes were isolated from several wildlife hosts while others were limited to single hosts. For example, *Gaigera panchyscelis* was isolated from kafue lechwe, impala, and sable antelopes while *Borrostomum trignocephalum* was only isolated from kafue lechwe. In addition, some worm species were only isolated from the bovids and not the equids. For example, *Stillesia hepatica* was only isolated from greater kudu and puku which are bovids while *Strongylus equinus*, *Strongylus vugaris*, and *Anaplocephala perfoliata* were only isolated from Burchelli's zebra which is an equid. These findings suggest that there is host preference for some worm species. On the contrary, some species were collected from a wide host range suggesting that there is interspecies transmission between different animal hosts. For example, *Oesophagostomum* spp. were found to infect both the bovids and equids.

Although different approaches have been used for control of tick and worm infections in wildlife [5, 6, 24], there has been no comprehensive study that assessed the efficacy of these techniques in Zambia. McGranahan [25] assessed the perceptions of game ranchers on the use of rotational burning as a tick control strategy and observed that there was a low attitude generally as most game ranchers did not understand the effectiveness of this technique. Hence, there is need for a quantitative assessment to determine the efficacy of rotational burning as a tick control strategy in game ranching. The major limiting factor to use of Duncan applicators as a tick control strategy for wildlife is that not all animals on game ranches get in contact with the applicators, and that this technique works better for animals kept in captivity under closed confinements. For free-ranging animals, it is unlikely that all animal will rub contacts with Dancun applicators for animals to get a pouring of the acaricide on their body surfaces. In some cases, the use of livestock as a tick control strategy has been suggested in situations where cattle are allowed to graze on the game ranch and latter dipped in acaricide dip-tanks to get rid of the ticks. Doing this a number of times is expected to reduce the tick-burden as cattles are used to sweep off the tick-population on the game ranch. However, the danger with this technique is the transmission of animal diseases between cattle and wildlife which could spark unexpected disease outbreaks. Observations made from this study clearly show that much as control of parasites and parasitic diseases in livestock and other domestic animal species have reached advanced stages, control measures in wildlife medicine are still in their infancy. Hence, there still remains the challenge of finding the most effective way of controlling tick infection and other parasitic infections of wildlife. Given the rapid expansion of the wildlife industry in Southern African, there is urgent need for more effective innovations that would help reduce disease transmission of various parasitic diseases in wildlife.

Authors' Contribution

All authors were involved in sample collection and analysis of data. H. M. Munang'andu prepared the manuscript, all authors read and approved the contents of the manuscript.

Acknowledgments

The authors are grateful to the School of Veterinary Medicine at the University of Zambia for using their facilities to carry out laboratory examination of the study materials.

References

[1] L. H. Makala, P. Mangani, K. Fujisaki, and H. Nagasawa, "The current status of major tick borne diseases in Zambia," *Veterinary Research*, vol. 34, no. 1, pp. 27–45, 2003.

[2] A. Nambota, K. Samui, C. Sugimoto, T. Kakuta, and M. Onuma, "Theileriosis in Zambia: etiology, epidemiology and control measures," *Japanese Journal of Veterinary Research*, vol. 42, no. 1, pp. 1–18, 1994.

[3] R. D. Fyumagwa, V. Runyoro, I. G. Horak, and R. Hoare, "Ecology and control of ticks as disease vectors in wildlife of the Ngorongoro Crater, Tanzania," *South African Journal of Wildlife Research*, vol. 37, no. 1, pp. 79–90, 2007.

[4] International Union for Conservation of Nature (IUCN), IUCN Red List of Threatened Species, Version 2010.1, 2010.

[5] M. J. Burridge, L. A. Simmons, E. H. Ahrens, S. A. J. Naudé, and F. S. Malan, "Development of a novel self-medicating applicator for control of internal and external parasites of wild and domestic animals," *Onderstepoort Journal of Veterinary Research*, vol. 71, no. 1, pp. 41–51, 2004.

[6] I. M. Duncan, "Tick control on cattle with flumethrin pour-on through a duncan applicator," *Journal of the South African Veterinary Association*, vol. 63, no. 3, pp. 125–127, 1992.

[7] J. M. Pound, J. A. Miller, J. E. George, and C. A. Lemeilleur, "The '4-poster' passive topical treatment device to apply acaricide for controlling ticks (Acari: Ixodidae) feeding on white-tailed deer," *Journal of Medical Entomology*, vol. 37, no. 4, pp. 588–594, 2000.

[8] H. Hansen and B. D. Perry, *The Epidemiology, Diagnosis and Control of Helminth Parasites of Ruminants*, ILRAD, Nairobi, Kenya, 1994.

[9] E. Soulsby, *Helminths, Arthropods and Protozoa of Domesticated Animals*, Bailliere Tindall, London, UK, 6th edition, 2011.

[10] M. J. Homer, I. Aguilar-Delfin, S. R. Telford, P. J. Krause, and D. H. Persing, "Babesiosis," *Clinical Microbiology Reviews*, vol. 13, no. 3, pp. 451–469, 2000.

[11] F. L. Schuster, "Cultivation of Babesia and Babesia-like blood parasites: agents of an emerging zoonotic disease," *Clinical Microbiology Reviews*, vol. 15, no. 3, pp. 365–373, 2002.

[12] Office International des Epizooties (OIE), "Bovine anaplasmosis," in *Terresterial Manual of Standards*, pp. 599–610, Office International des Epizooties (OIE), Paris, France, 2008.

[13] I. M. Duncan, "The use of flumethrin pour-on for de-ticking black rhinoceros (*Diceros bicornis*) prior to translocation in Zimbabwe," *Journal of the South African Veterinary Association*, vol. 60, no. 4, pp. 195–197, 1989.

[14] F. Jongejan, B. D. Perry, P. D. S. Moorhouse, F. L. Musisi, R. G. Pegram, and M. Snacken, "Epidemiology of bovine babesiosis

and anaplasmosis in Zambia," *Tropical Animal Health and Production*, vol. 20, no. 4, pp. 234–242, 1988.

[15] H. M. Munang'andu, V. Siamudaala, A. Nambota, M. Unyeme, and K. S. Nalubamba, "Detection of Babesis species in free ranging puku, Kobus vordonii, on a game ranch in Zambia," *Korean Journal of Parasitology*, vol. 166, no. 1-2, pp. 163–166, 2009.

[16] H. M. Munang'andu, V. Siamudaala, M. Munyeme, A. Nambota, S. Mutoloki, and W. Matandiko, "Trypanosoma brucei infection in asymptomatic greater kudus (*Tragelaphus strepsiceros*) on a game ranch in Zambia," *Korean Journal of Parasitology*, vol. 48, no. 1, pp. 67–69, 2010.

[17] R. J. C. Nefdt and S. J. Thirgood, "Lekking, resource defense, and harassment in two subspecies of lechwe antelope," *Behavioral Ecology*, vol. 8, no. 1, pp. 1–9, 1997.

[18] W. A. Rees, "Ecology of Kafue Lechwe—as affected by Kafue gorge hydroelectric scheme," *Journal of Applied Ecology*, vol. 15, pp. 205–217, 1978.

[19] C. J. Ng'ang'a, N. Maingi, P. W. N. Kanyari, and W. K. Munyua, "Development, survival and availability of gastrointestinal nematodes of sheep and pastures in a semi-arid area of Kajiado District of Kenya," *Veterinary Research Communications*, vol. 28, no. 6, pp. 491–501, 2004.

[20] G. Chingwena, S. Mukaratirwa, T. K. Kristensen, and M. Chimbari, "Larval trematode infections in freshwater snails from the highveld and lowveld areas of Zimbabwe," *Journal of Helminthology*, vol. 76, no. 4, pp. 283–293, 2002.

[21] A. M. Phiri, A. Chota, and I. K. Phiri, "Seasonal pattern of bovine amphistomosis in traditionally reared cattle in the Kafue and Zambezi catchment areas of Zambia," *Tropical Animal Health and Production*, vol. 39, no. 2, pp. 97–102, 2007.

[22] J. A. Hamond, *Studies on fascioliasis with special reference to Fasciola gigantica in East Africa*, Ph.D. thesis, University of Edinburg, Edinburgh, UK, 1970.

[23] A. M. Phiri, A. Chota, J. B. Muma, M. Munyeme, and C. S. Sikasunge, "Helminth parasites of the Kafue lechwe antelope (*Kobus leche kafuensis*): a potential source of infection to domestic animals in the Kafue wetlands of Zambia," *Journal of helminthology*, vol. 85, no. 1, pp. 20–27, 2011.

[24] I. M. Duncan and N. Monks, "Tick control on eland (*Taurotragus oryx*) and buffalo (*Syncerus caffer*) with flumethrin 1% pour-on through a Duncan Applicator," *Journal of the South African Veterinary Association*, vol. 63, no. 1, pp. 7–10, 1992.

[25] D. A. McGranahan, "Managing private, commercial rangelands for agricultural production and wildlife diversity in Namibia and Zambia," *Biodiversity and Conservation*, vol. 17, no. 8, pp. 1965–1977, 2008.

14

The Effect of Antihelminthic Treatment on Subjects with Asthma from an Endemic Area of Schistosomiasis: A Randomized, Double-Blinded, and Placebo-Controlled Trial

**Maria Cecilia F. Almeida,[1] Givaneide S. Lima,[1] Luciana S. Cardoso,[1,2,3]
Robson P. de Souza,[1] Régis A. Campos,[1] Alvaro A. Cruz,[4] Joanemile P. Figueiredo,[1]
Ricardo R. Oliveira,[1,3,5] Edgar M. Carvalho,[1,3,6] and Maria Ilma Araujo[1,3,6]**

[1] *Serviço de Imunologia, Universidade Federal da Bahia (UFBA), 40110-160 Salvador, BA, Brazil*
[2] *Departamento de Ciências da Vida, Universidade do Estado da Bahia UNEB, 41.150-000 Salvador, BA, Brazil*
[3] *Instituto Nacional de Ciências e Tecnologia em Doenças Tropicais (INCT-DT/CNPq-MCT), Brazil*
[4] *ProAR, Núcleo de Excelência em Asma, Universidade Federal da Bahia 40110-160 Salvador, BA, Brazil*
[5] *Faculdade de Farmácia, UFBA 40110-160 Salvador, BA, Brazil*
[6] *Escola Bahiana de Medicina e Saúde Pública, EBMSP, Salvador, Bahia, Brazil*

Correspondence should be addressed to Maria Ilma Araujo, mia@ufba.br

Academic Editor: Elena Pinelli

This is a prospective, double-blinded, and placebo-controlled trial evaluating the influence of antihelminthic treatments on asthma severity in individuals living in an endemic area of schistosomiasis. Patients from group 1 received placebo of Albendazole or of Praziquantel and from group 2 received Albendazole and Praziquantel. Asthma severity was assessed by clinical scores and by pulmonary function test. There was no significant difference in the asthma scores from D0 to D1–D7 after Albendazole or Praziquantel and from D0 to D30–90 after Albendazole or Praziquantel in both, group 1 and 2. It was observed, however, a clinical worsening of the overall studied population after 6 months and 12 months of antihelminthic treatments. Additionally, we observed increased frequency of forced expiratory volume in 1 second (FEV1) < 80% on 12 and 18 months after treatment. The worsening of asthma severity after repeated antihelminthic treatments is consistent with the hypothesis of the protective role conferred by helminths in atopic diseases.

1. Introduction

Helminthic infections and allergic diseases are highly prevalent in many parts of the world, and both lead to type 2 immune response with secretion of IL-4, IL-5, and IL-13, with a consequent increase in the production of IgE and eosinophilia. Lynch et al. [1] studying the skin prick test (SPT) response in children from an ascariasis endemic area observed a decreased reactivity to the test and subsequently an increased response after treatment of intestinal helminths [2].

Studies conducted by the Immunology Service (SIM) of the Federal University of Bahia, Brazil, demonstrated a negative association between the cutaneous immediate hypersensitivity response to the skin prick test to aeroallergens and *Schistosoma mansoni* parasite load, measured by a quantitative assessment of the number of eggs per gram of stool [3]. Over a year-long follow-up study in the state of Bahia, a group of researchers from SIM compared three populations from impoverished areas of the state, including one group from an endemic area of *S. mansoni*, one group from a rural nonendemic area, and a third group from a slum area in Salvador, the capital of the state. In the study, it was observed that asthma severity indicators were lower in individuals living in the endemic areas of schistosomiasis when compared to the other two areas in which there were no recorded cases of *S. mansoni* transmission [4].

Several factors may explain the lower frequency of positive SPT and lower asthma severity in populations infected with helminths. The possible hypothesis includes high production of polyclonal IgE and reduced levels of allergen-specific IgE [1, 2], high concentrations of antigen-specific IgG4 [5], activation of regulatory cells, and regulatory cytokine production [6]. For instance, IL-10 can promote a decrease in the release of histamine and other mast cell mediators [7]. Since infection by *S. mansoni* induces high production of IL-10, it is possible that this is the main mechanism by which the allergic response is suppressed in infected individuals. This hypothesis is reinforced by studies of van den Biggelaar et al. [6], which showed that reduced reactivity to SPT in African children infected with *S. haematobium* was associated with an increase in IL-10 production *in vitro* by cells of these individuals. Our hypothesis for the current study is that treatment of helminth infections, including a drug to treat schistosomiasis, alters the immune response leading to worsening in asthma outcomes. The aims of this study were, firstly, evaluating early events associated to asthma severity resulting from treatment of helminth infections, and secondly, determining the degree of interference of antihelminthic treatment in the clinical course of asthma in a randomized, double-blinded, and placebo-controlled trial.

2. Materials and Methods

2.1. Subjects and Study Design. This is a randomized, double-blinded, and placebo-controlled trial with two groups of asthmatics living in a *S. mansoni* endemic area. The study was carried out in Água Preta, a small village near the city of Gandu in the State of Bahia in Brazil. Gandu is located 280 km south of Salvador, the capital of the State of Bahia. Água Preta is composed of a residential area in the center of the village and surrounding farms. There are approximately 800 people living in that community. They live in poor sanitary conditions and agriculture is their predominant occupation. The Água Preta community was identified by our group as a community with a frequent infection of *Schistosoma mansoni* (49.5% in a survey of 427 residents in 2006) and helminthic infection in general (prevalence of *Ascaris lumbricoides*, *Trichuris trichiura*, and hookworm at that time was 24.2%, 33.8%, and 22.2%, resp.). Six hundred one individuals from Água Preta agreed to participate in this study. They were screened for asthma through a direct questionnaire on the basis of the International Study of Asthma and Allergies in Childhood (ISAAC) [8]. In addition, questions concerning concomitant illnesses, socioeconomic status, and living conditions were asked in a complementary questionnaire. Patients were selected as having asthma if their responses to the ISAAC questionnaire were considered by a physician to be indicative of a personal history of asthma in the past 12 months, and if they were 5 to 50 years old. Children under 5 years old were not included due to difficulties in performing the pulmonary function test. Subjects over 50 years old were not included due to increased rates of chronic obstructive pulmonary disease in this age group. Only fifty asthmatic individuals from

the community met the inclusion criteria and they were randomized either as group 1 (who would receive initially placebo of antihelminthic treatments; $n = 25$) or group 2 (who would receive Albendazole in a split dose of 400 mg, followed by Praziquantel 50 mg/Kg of body weight a week later; $n = 25$). Five individuals from group 1 left the study in the first week after the beginning of the study for personal reasons, remaining 20 individuals in this group. The loss of these individuals did not alter significantly demographic features of group 1, such as the age and gender distribution. At each evaluation, every patient underwent a physical examination, always performed by the same physician (blind to the type of treatment). At the evaluations, common asthma symptoms and signs were checked, such as cough, dyspnea, and wheezing. In addition, a questionnaire elicited information on asthma (i.e., presence and frequency of asthma attacks, type of treatment received during the attack at home, emergency department, or hospital) and the use of prophylactic or symptomatic antiasthma drugs (e.g., antihistamine, inhaled, or oral beta-2-agonist, and inhaled, oral, or parenteral corticosteroid) since the previous evaluation. These parameters were scored as follows: physical examination as 0 (normal examination) and 1 (at least one abnormal finding); asthma exacerbations as 0 (no), 1 (yes, if treated at home), and 2 (yes, if treated at an emergency department or at a hospital); use of antiasthma drugs as 0 (no), 1 (yes, except for oral or parenteral corticosteroid use), and 2 (yes, if oral or parenteral corticosteroid had been used) as reported in a previous study [4].

Human subject study guidelines of the US Department of Health and Human Service were followed in the conduction of this study. The study was approved by the Ethics Committee of Professor Edgard Santos Hospital. Informed consent was obtained from all patients or their legal guardians.

In the initial assessment, patients responded to a questionnaire to assess the clinical score of asthma [4] underwent a physical examination, chest radiograph, pulmonary function test, and blood sampling for evaluation of immune response (IL-5, IL-10, and IFN-γ in supernatants of cultures stimulated with SWAP and Der p1). Stool exams for parasites were also performed on each patient. The two groups were then treated with antihelminthics or placebo and reevaluated after the completion of treatment as shown in the study design flowchart (Figure 1). Chest radiographs and pulmonary function tests were performed at the time of enrolment and at day 7 after treatment. Spirometry test was repeated 30 days after treatment. Immunological evaluation was performed at enrolment and repeated at 7 and 90 days after treatment. After 90 days of enrolment, both placebo and treatment groups were treated with Albendazole and Praziquantel. From there, the two groups were assessed monthly by clinical examination, questionnaire, and pulmonary function test for a total period of 18 months according to the study design (Figure 1).

2.2. Pulmonary Function Tests, Chest X-Ray, and Skin Prick Test to Aeroallergens (SPT). Pulmonary function tests were performed on all subjects at enrollment and at each visit thereafter. The parameter used was Forced Expiratory

The Effect of Antihelminthic Treatment on Subjects with Asthma from an Endemic Area of Schistosomiasis:
A Randomized, Double-Blinded, and Placebo-Controlled Trial

127

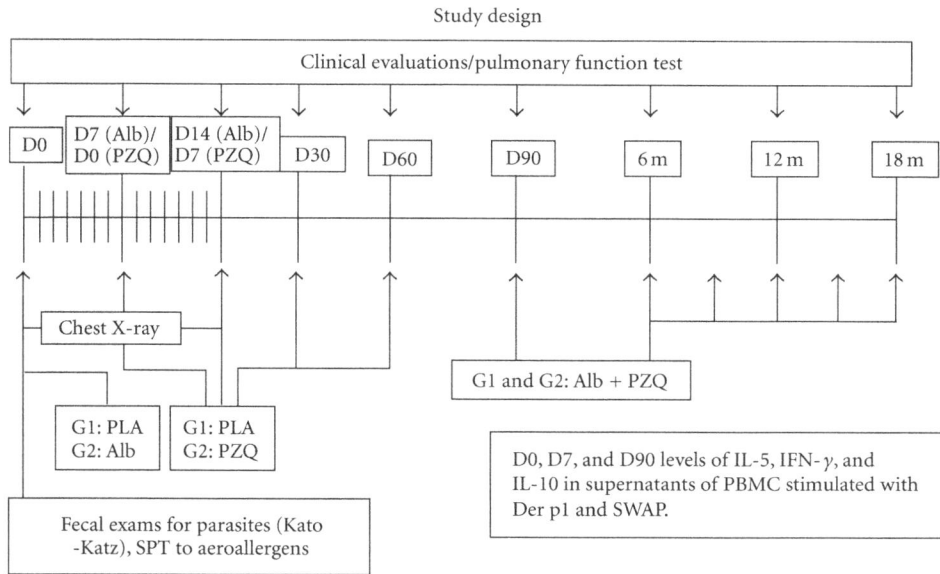

FIGURE 1: Flowchart of the study design.

Volume in 1 second (FEV1). Results were considered as normal when FEV1 value was $\geq 80\%$ [9].

Chest X-ray was performed in a specialized clinic in Gandu, at baseline and at D7 after Praziquantel treatment. SPTs were performed on the right forearms of all individuals at enrolment using *Dermatophagoides pteronyssinus* (Der p), *D. farinae* (Der f), *Blomia tropicalis* (Blo t), *Periplaneta americana* (Per a), and *Blattella germanica* (Bla g) glycerinated allergen extracts (FDA Allergenic). Histamine (1 : 1000) and glycerinated saline were used as positive and negative controls, respectively. A positive skin reaction was defined as formation of a wheal with a mean diameter greater than 3 mm. The SPT results were read 20 minutes after application, and a SPT response was considered as positive if there was at least one positive test of the five tested allergens.

2.3. Immune Response—Cell Culture and Cytokine Measurements. All enrolled individuals had their blood taken for immunological studies. Peripheral blood mononuclear cells (PBMC) from the blood samples of the two groups were analyzed for *in vitro* immune response, which included measurement of IL-10, IL-5, and IFN-γ production by PBMCs in response to soluble *S. mansoni* adult worm antigen (SWAP, kindly provided by Dr. Alfredo Góes from the Federal University of Minas Gerais, Brazil) and to the antigen 1 from *D. pteronyssinus* (Der p1 extract, cosmo Bio Co., Ltd.).

PBMCs from individuals of the study were obtained through the Ficoll-Hypaque gradient and adjusted to the concentration of 3×10^6 cells/mL in complete RPMI medium (Life technologies GIBCO BRL, Gaithersburg, MD). Cells were cultured *in vitro* with the antigens Der p1 (25 μg/mL) and SWAP (10 μg/mL). The mitogen phytohemagglutinin (PHA) in the final concentration of 2 μg/mL was also used in the cultures. The cultures were incubated

for 72 hours at 37°C and 5% CO_2 and the supernatants were collected for cytokine measurements. Levels of IL-5, IL-10, and IFN-γ were determined by ELISA sandwich technique, using commercially available kits (R&D Systems) and the results were expressed in picograms per milliliter based on a standard curve.

2.4. Fecal Examinations for Parasites. Three stool samples from each individual were examined using the Hoffman sedimentation method to identify helminths and enteric protozoa, and the Kato-Katz method was used to estimate parasite load [10].

2.5. Sample Size and Statistical Analysis. Only fifty asthmatic individuals from the village where the study was carried out met the inclusion criteria. The power of the study was calculated taking into account the results of a previous study from our group [4] which demonstrated that the frequency of asthma symptoms in asthmatic individuals living in an endemic area of schistosomiasis is 18.6%, whereas the prevalence of symptoms in a worm-free population is 58.7%. Based on these data, a sample size of 20 in the placebo group and 25 in the treated group has 76% power to detect a difference between proportions with a significance alpha level of 0.05 (two-tailed).

Statistical analyses were performed using the software Statistical Package for Social Science (version 9.0 for Windows; SPSS). Fisher's exact test was used to compare proportions. The Mann-Whitney U test was used to compare levels of cytokines between groups, and the Wilcoxon matched-pairs signed rank test was used to compare the levels of cytokines intragroup before and after antihelminthic treatments. Statistical significance was established at the 95% confidence interval.

3. Results

3.1. Features of the Studied Subjects. The study included 45 asthmatic patients. They were divided into two groups: one group received placebo of antihelminthic treatments (placebo of Albendazole and placebo of Praziquantel (group 1 or placebo group), while the other group received Albendazole to treat geohelminths and Praziquantel to treat *S. mansoni* infection (group 2 or Praziquantel group). The demographic data of subjects enrolled in the study are shown in Table 1. *S. mansoni* parasite burden and infection with other helminths as well as the frequency of positive skin prick test to aeroallergens are also shown in Table 1.

Gender distribution did not differ significantly between groups who received placebos (G1; 50% male) and those who received antihelminthic treatment (G2; 32% male, $P > 0.05$; Table 1).

The mean age of patients included in the study was 17.6 ± 13.1 years. There was no significant difference in the mean age between groups (21.3 ± 15.4 and 14.7 ± 10.3 years in Group 1 and Group 2, resp.) with age ranges from 6 to 20 years found in 65% of G1 and in 72% of G2 ($P > 0.05$) and 21 to 50 years in 35% and 28% of G1 and G2, respectively ($P > 0.05$; Table 1).

The frequency of rhinitis did not differ significantly between patients from group 1 and 2, being 50% and 72%, respectively ($P > 0.05$; Table 1). The frequency of active smokers was similar between the placebo and Praziquantel groups (30% and 28%, resp.; $P > 0.05$), while the frequency of second-hand smoking exposure was higher in patients from group 1 (90%) then in group 2 (36%; $P < 0.0005$; Table 1).

There was no significant difference in the frequency of FEV1 $\leq 80\%$ between the placebo and Praziquantel groups at baseline (5% and 12%, resp.; $P > 0.05$). A positive response to the skin prick test to aeroallergens was found in 36% of patients from the Praziquantel group compared to 45% of placebo group at baseline ($P > 0.05$; Table 1). The frequency of positive response to the different aeroallergens such as Der p, Der f, Blo t, Per a and Bla g also did not differ between the two groups of patients ($P > 0.05$). A positive response to histamine was found in 100% of patients from group 1 and 90.9% of those from group 2 ($P > 0.05$; Table 1).

The frequency of *S. mansoni* infection did not differ between groups ($P > 0.05$). The *S. mansoni* parasite burden also did not differ significantly between group 1 and group 2 (115 ± 49 and 427 ± 228 eggs/g feces, resp.; $P > 0.05$). Patients were also infected with other helminths, such as *Ascaris lumbricoides*, *Trichuris trichiura*, and hookworm. However, there was no significant difference in the frequency of other helminthic infection between groups (Table 1; $P > 0.05$). Co-infection with *S. mansoni* and one or more other helminths were observed in 90% and 100% of patients from G1 and G2, respectively ($P > 0.05$; Table 1).

Clinical scores for asthma were initially evaluated at baseline (day 0 or pretreatment) and for seven consecutive days during the first week after treatment with placebo or Albendazole (D1 to D7). At day 7 after Albendazole treatment, patients from group 1 and group 2 were treated

with placebo of Praziquantel and Praziquantel, respectively. They were clinically evaluated and had their asthma score recorded during the following seven days. The results of clinical scores of asthma after treatment with placebo of Albendazole and placebo of Praziquantel are shown in Figure 2(a). There was no significant difference in frequency of asthma scores at D1 to D7 compared to D0 ($P > 0.05$) in patients from group 1 who received placebos of Albendazole and Praziquantel (Figure 2(a)).

There was also no significant difference in the frequency of asthma scores from D0 to D1–D7 after Albendazol and Praziquantel in group 2 ($P > 0.05$; Figure 2(b)).

The mid-term effect of Praziquantel treatment on the asthma severity was evaluated during the first 90 days of posttreatment. Three monthly consecutive evaluations (D30 to D90) were performed in each patient in this time period. The mean frequencies of asthma clinical scores are shown in Figure 3(a). There was no significant difference in the frequency of clinical scores 0, 1, 2, 3, or 4 between D0 (80%, 5%, 15%, 0%, 0%) and D30 to D90 ($73 \pm 8.8\%$, $6.7 \pm 6.2\%$, $19 \pm 5.7\%$, 0%, 0%) in the placebo group (Figure 3(a)). In the group treated with Praziquantel, the frequency of different scores (0, 1, 2, 3, and 4) also did not differ from D0 (55%, 25%, 20%, 0%, 0%) to D30–D90 ($68 \pm 5\%$, $15.7 \pm 6.9\%$, $16 \pm 5.4\%$, 0%, 0%; $P > 0.05$; Figure 3(b)).

The long-term effect of antihelminthic treatment on asthma severity was evaluated in three different time periods, 6, 12, and 18 months after-treatments. The placebo group was treated with Albendazole and Praziquantel at day 90, when the Praziquantel group received the second treatment with these two drugs. Since, after day 90, both groups were treated with antihelminthics, the groups were combined into one group of treated patients thereafter. The frequency of score zero in the studied population was lower after 6 months of treatment (58%) compared to D0 (73%; $P < 0.05$; Figure 3(c)), being the frequency of score one higher (20%) compared to baseline (6%; $P < 0.05$). Likewise, the frequency of score 2 was higher after 12 months of treatment (30%) compared to D0 (15%; $P < 0.05$, Figure 3(c)).

There was, however, no significant difference in the frequency of different asthma clinical scores at 18 months after treatments (65%, 15%, 15%, 3%, 0%, to scores 1, 2, 3, and 4, resp.) compared to baseline (73%, 6%, 15%, 2% and 2% to scores 1, 2, 3, and 4, resp.; $P > 0.05$; Figure 3(c)).

The severity of asthma was also evaluated through pulmonary function test (PFT). The result of FEV1 in patients from group one and two are shown in Figure 4. There was no significant difference in the frequency of FEV1 <80% either in Praziquantel or in the placebo groups when the baseline values were compared to D7 post Albendazole, D7 after Praziquantel and also to D90 after Praziquantel ($P > 0.05$, Figure 4(a)). On the other hand, the frequency of FEV1 <80% among all the subjects with asthma ($n = 45$) was higher at 12 months (22.2%) and subsequently at 18 months after treatment (34.8%) compared to the frequency at baseline (9%; $P < 0.05$, Figure 4(b)).

TABLE 1: Baseline characteristics of studied subjects.

	Group 1 placebo $n = 20$	Group 2 Praziquantel $n = 25$	P value
Gender n			
Male n (%)	10 (50)	8 (32)	>0.05
Female n (%)	10 (50)	17 (68)	
Age group n (%)			
Child/teenagers (6–20 years old)	12 (65)	18 (72)	>0.05
Adult (21–50 years old)	8 (35)	7 (28)	>0.05
Rhinitis n (%)	10 (50)	18 (72)	>0.05
Smoking n (%)			
Active	6 (30)	7 (28)	>0.05
Passive	18 (90)	9 (36)	<0.005
Positive SPT response n (%)			
Dermatophagoides pteronyssinus	2 (10)	5 (20)	>0.05
D. farinae	2 (10)	5 (20)	>0.05
Blomia tropicalis	4 (20)	7 (28)	>0.05
Periplaneta americana	1 (5)	3 (12)	>0.05
Blatella germanica	2 (10)	3 (12)	>0.05
Positive total	9 (45)	9 (36)	>0.05
Current helminth infections n (%)			
Schistosoma mansoni	16 (80)	19 (76)	>0.05
Ascaris lumbricoides	10 (50)	14 (56)	>0.05
Hookworm	7 (35)	10 (40)	>0.05
Trichuris trichiura	10 (50)	13 (52)	>0.05
Coinfection (*S. mansoni* /1 or + helminths)	14 (90)	19 (100)	>0.05

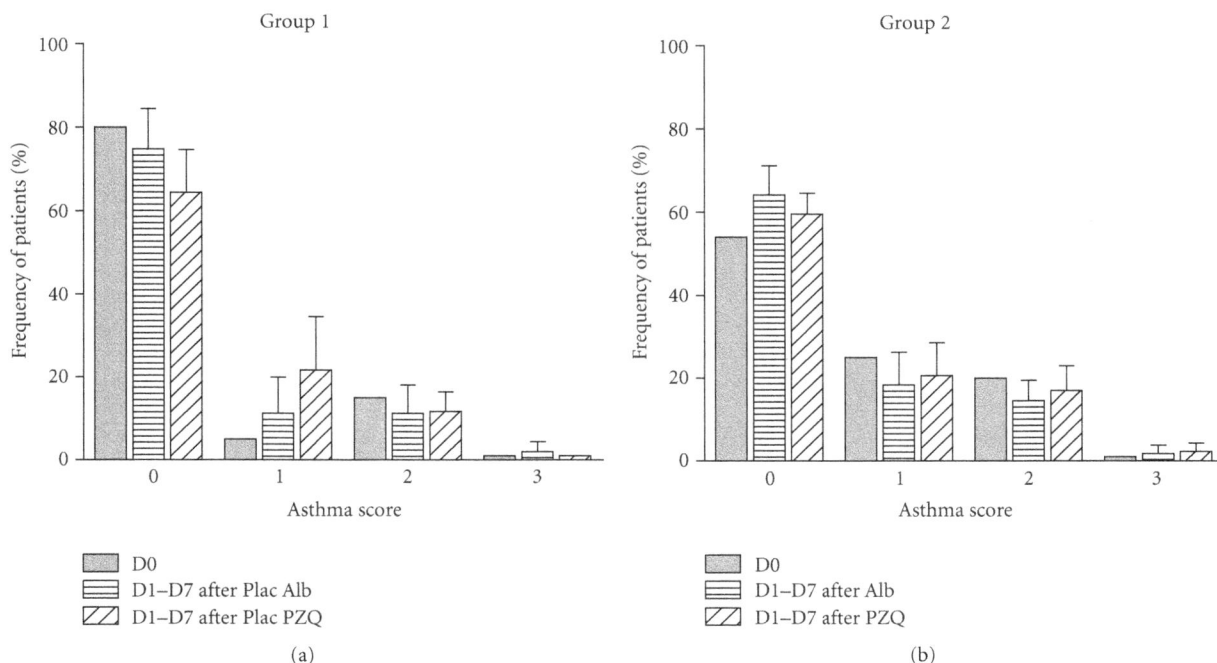

FIGURE 2: Frequency of clinical scores of asthma in patients from Group 1 at day zero (D0) and one to seven days (D1–D7) after treatment with placebo (Plac) of Albendazole (Alb) and placebo of Praziquantel (PZQ) (a); and at day zero (D0) and one to seven days (D1–D7) after treatment with Albendazole or Praziquantel (Group 2) (b). Praziquantel was given seven days after the treatment with Albendazole in the Group 2. Data were represented as mean ± SD. There was no significant difference in the frequency of asthma scores between D0 versus D1–D7 in none of groups ($P > 0.05$; Chi-square Test for Independence).

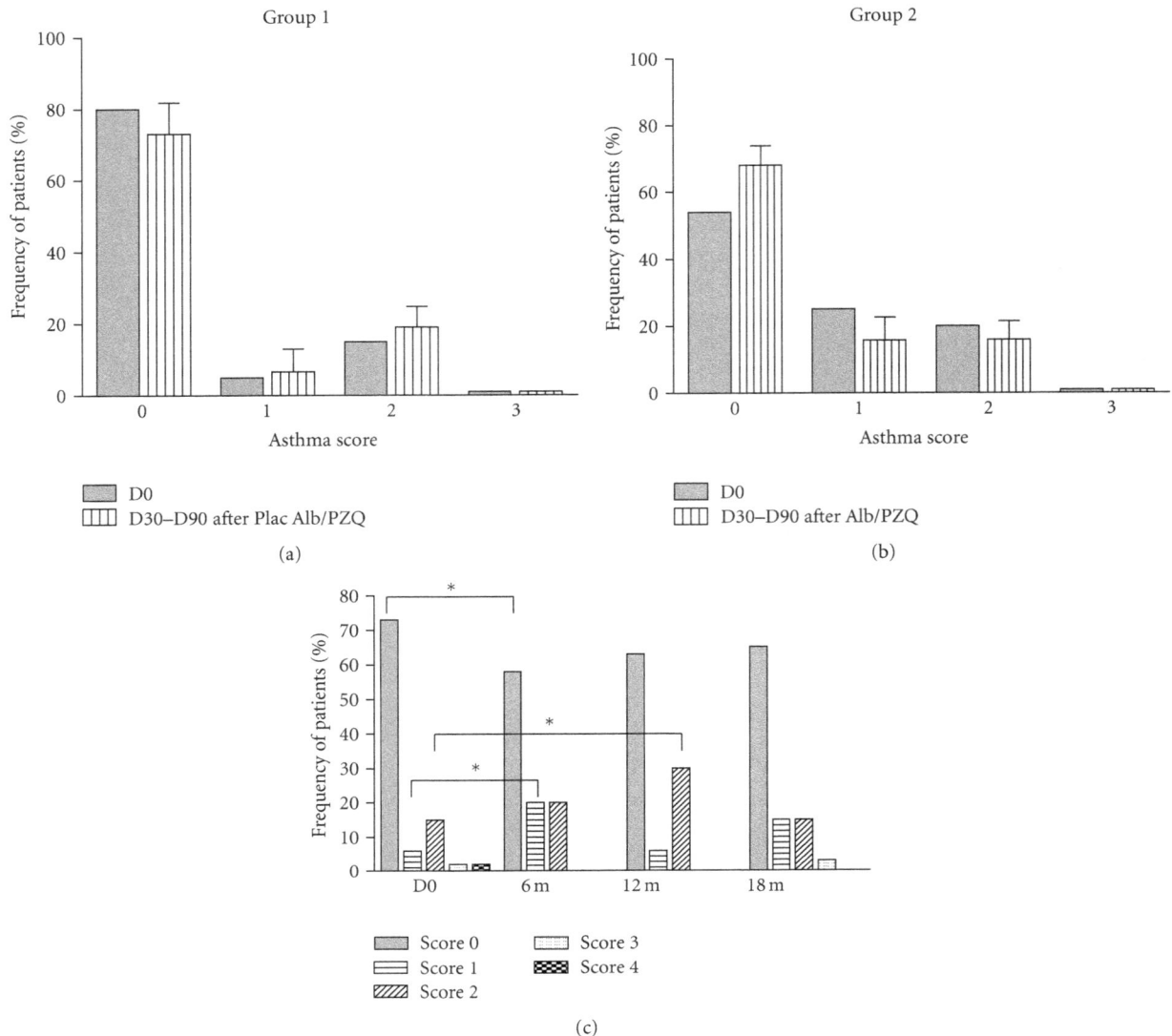

FIGURE 3: Frequency of clinical scores of asthma at day zero (D0) and day 30 to day 90 (D30–D90) after treatment with placebo (Plac) of Albendazole (Alb) and placebo of Praziquantel (PZQ) (Group 1; (a)); and frequency of clinical scores of asthma at day zero (D0) and day 30 to 90 (D30–D90) after treatment with Albendazole and Praziquantel (Group 2; (b)). Data were represented as mean ± SD. (c) shows the frequency of clinical scores of asthma (0, 1, 2, 3, 4) at day zero (D0) and at 6 months, 12 months and 18 months after treatment with Praziquantel (Figure 3(c)). Patients were evaluated monthly from day 30 to day 90 (D30–D90). After 90 days of enrollment, both placebo and treatment groups were treated with Albendazole and Praziquantel. From there clinical scores of asthma were assessed each 6 months for a total period of 18 months. *D0 versus 6 m or D0 versus 12 m, $P < 0.05$; Fisher's exact test.

3.2. Cytokine Profile Induced by S. mansoni and D. pteronyssinus Antigens in PBMCs of the Studied Population.

We measured the cytokines IFN-γ, IL-5, and IL-10 in the supernatants of PBMC cultures stimulated with the *S. mansoni* soluble adult worm antigen (SWAP) and the antigen 1 of the *D. pteronyssinus* (Der p1). There was no significant difference in the levels of IFN-γ between D0 and D7 or D0 and D90 in response to SWAP and Der p1 ($P > 0.05$, Table 2). The baseline of IL-5 mean levels in response to Der p1 in cultures was lower in the placebo group (150 pg/mL) than in Praziquantel group (463 pg/mL; $P < 0.05$). There was, however, no significant difference between the baseline mean values and those found at D7 or D90 after treatment in response to SWAP or Der p1 in either group ($P > 0.05$; Table 2). On the other hand, in the group of Praziquantel, the mean levels of IL-10 in response to SWAP decreased from 642 pg/mL at D0 to 175.6 pg/mL at D90 after treatment ($P < 0.05$; Table 2). There was no significant difference in the mean level of IL-10 in response to Der p1 comparing baseline values with those from D7 and D90 in both placebo and Praziquantel groups (Table 2). There was a high production of IL-10, IL-5, and IFN-γ in cultures stimulated with the mitogen PHA compared to nonstimulated cultures ($P < 0.05$) in both groups of patients (data not shown).

The Effect of Antihelminthic Treatment on Subjects with Asthma from an Endemic Area of Schistosomiasis: A Randomized, Double-Blinded, and Placebo-Controlled Trial

131

TABLE 2: Levels of cytokines in supernatants of PBMC cultures in the studied subjects.

Cytokine	Antigen	D0		D7		D90	
		G1/placebo	G2/PZQ	G1/placebo	G2/PZQ	G1/placebo	G2/PZQ
IFN-γ	Without stimulus	32.7 (31.2–276.0)	31.6 (31.2–91.0)	31.2 (31.2–46.7)	39.7 (31.2–583.6)	31.2 (31.2–130.7)	32.7 (31.2–2563.0)
	SWAP	149.5 (42.0–10337.0)	88.0 (31.2–2092.0)	56.0 (43.8–1415.0)	137.8 (37.4–15543.0)	70.0 (15.6–3329.0)	130.7 (31.2–2563.0)
	Der p1	338.5 (102.4–7475.0)	710.0 (149.0–2568.0)	116.7 (42.0–420.2)	333.9 (32.7–9669.0)	201.0 (31.2–4818.0)	149.4 (15.6–1275.0)
IL-5	Without stimulus	15.6 (15.6–112.0)	15.6 (15.6–15.6)	15.6 (15.6–15.6)	15.6 (15.6–40.7)	15.6 (15.6–25.0)	15.6 (15.6–635.0)
	SWAP	468.5 (15.6–5156.0)	310.2 (15.6–5904.0)	542.0 (15.6–4444.0)	1863.0 (15.6–4606.0)	1670.0 (21.5–4444.0)	2850 (136.0–4444.0)
	Der p1	150.0 (15.6–630.0)[a]	463.0 (207.0–4606.0)[a]	28.8 (15.6–407.4)	60.0 (15.6–4606.0)	28.7 (15.6–1720.0)	352.0 (15.6–3509.0)
IL-10	Without stimulus	27.5 (15.6–475.0)	28.0 (15.6–102.0)	57.0 (16.8–589.0)	23.1 (15.6–1144.0)	36.0 (15.6–446.9)	15.6 (15.6–127.0)
	SWAP	264.5 (26.0–1688.0)	642.0 (125.0–1132.0)[b]	288.0 (29.4–641.0)	667.3 (15.6–2447.0)	240.0 (21.1–638.4)	175.6 (60.3–1062.0)[b]
	Der p1	451.0 (56.0–1060.0)	349.0 (177.0–1539.0)	924.0 (297.1–2063.0)	288.0 (15.6–2086.0)	390.0 (54.3–1744.0)	204.0 (60.0–638.0)

Data showed as median (minimum and maximum values) pg/ml;
[a]Placebo D0 versus PZQ D0 $P < 0.05$ (Mann-Whitney t-test);
[b]PZQ D0 versus PZQ D90 $P < 0.05$ (Wilcoxon matched-pairs signs ranks test).

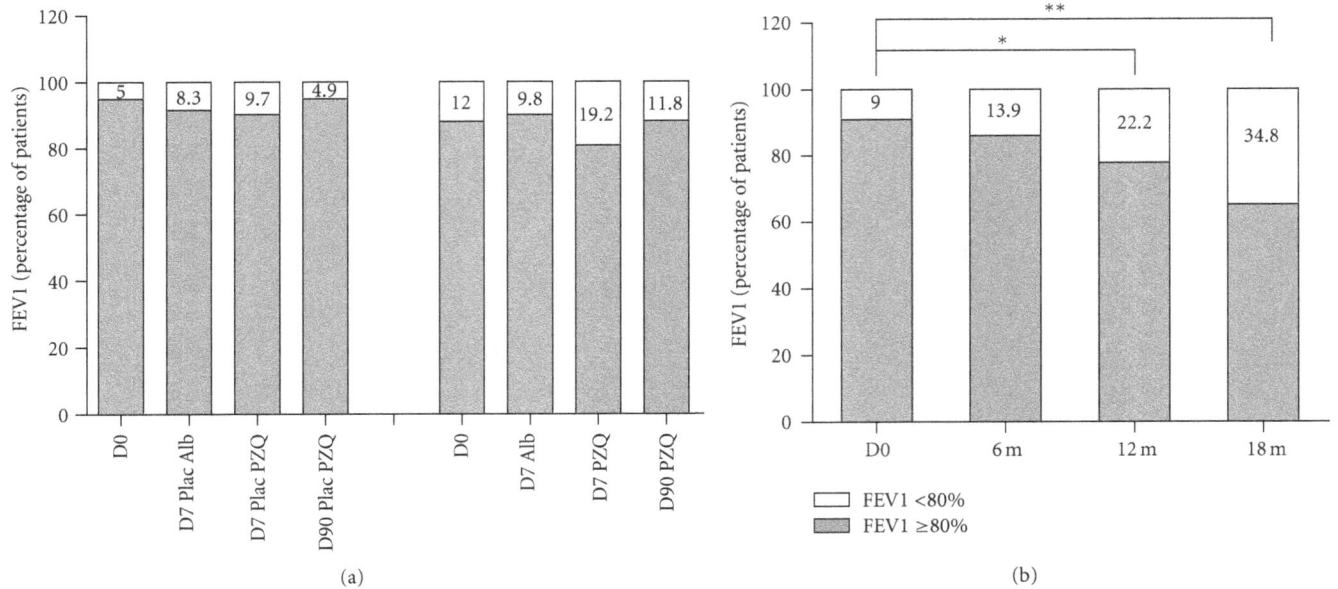

FIGURE 4: Frequency of Forced Expiratory Volume in 1 second (FEV1) <80% at D0, D7 after placebo of Albendazole treatment and D7 and D90 after placebo of Praziquantel treatments in the Group 1 and after treatments with Albendazole and Praziquantel in the Group 2 (Figure 4(a)). Frequency of FEV1 <80% at day zero (D0) and 6, 12 and 18 months after treatment with Albendazole and Praziquantel (b). Results were considered as normal when FEV1 value was ≥80%. Numbers in the top of the bars represent the percentage of patients who had FEV1<80%. *D0 versus 12 m, $P < 0.05$; **D0 versus 18 m, $P = 0.0001$; Fisher's exact test.

4. Discussion

The aim of this study was to determine in a randomized, double-blinded, and placebo-controlled trial whether anti-helminthic treatment would interfere with the clinical course of asthma in individuals living in a *S. mansoni* endemic area. Additionally, we evaluated the *in vitro* cytokine profile in response to an aeroallergen and to *S. mansoni* antigen before and after antihelminthic treatment. There was no change in the asthma score during the first weeks after treatment in both groups, neither was there any variation after 30 to 90 days. At day 90, group 1 and group 2 were combined into one group who received Albendazole and Praziquantel each three months thereafter. When comparing the baseline score (D0) of this group, we found a higher frequency of score one and two at 6 and 12 months after treatments. The most likely straightforward explanation for these findings is that early after treatment the antigens released during the parasite killing maintain the protective effect on asthma severity. The effect of the antihelminthic treatments on the clinical score of asthma was perceived, however, only six to 12 months after repeated treatments. Considering that patients remained living in the same area using the contaminated water in their activities, we believe that only after sequential treatments, the loss of the protective effect of parasite infection over asthma symptoms could be observed. Although the asthma clinical score scale used in this study is not validated, against other measures of asthma severity or control, it gives an idea of the clinical course of the disease, when obtained repeatedly. It considers the use of antiasthmatic drugs and corticosteroids, as well as the need for the patient to visit emergency rooms

or the hospital due to asthma attacks. The relevance of this score categorization was supported by the results of pulmonary function test which showed a significantly higher frequency of patients with abnormal lung function 12 and 18 months after antihelminthic treatments. For medical and ethical reasons, the group of placebo could not be left without antihelminthic treatment for more than 90 days. Hence, the two groups of patients received Albendazole and Praziquantel thereafter. One may ask if the adverse effect of long-term treatment on asthma severity we observed in this study would be due some seasonal differences during the follow-up study. We are aware that it is a limitation in this type of study design; however, as the study was conducted in a region with a tropical climate with no defined seasons and low weather variations during the year, this factor may not affect significantly the asthma severity.

Although our initial hypothesis in this study was that *S. mansoni* infection protects against asthma, the observation herein does not allow us to rule out the effect of other helminth infections in this protection, as have been proposed by other authors in experimental models of OVA-induced asthma [11–13]. In a polyhelminth endemic area, it is difficult to establish which parasite is protecting the host against a harmfully Th2-mediated pathological process. Previous systematic reviews and meta-analysis studying the effect of geohelminth infections on the risk of asthma showed that these parasites in general do not protect against asthma, but hookworm was shown to reduce the risk of the disease [14, 15]. A clinical trial using *Trichuris suis* ova resulted in no significant changes in symptom scores of allergic rhinitis [16]. Furthermore, treatment of hookworm infection, which leads to a reduction in the worm burden, increased the risk of allergen skin sensitization but did not interfere with the symptoms of allergic diseases [17], symptom scores of allergic rhinitis [16] and asthma [18]. In a systematic review and meta-analysis of epidemiological studies that researched the association between intestinal parasite infection and the presence of atopy, the authors found a consistent protective effect on allergic sensitization in patients with *Ascaris lumbricoides*, *Trichuris trichiura*, hookworm, or *Schistosoma* sp, infection [18]. There is evidence that in *S. mansoni* infection, the large number of regulatory cells and high levels of modulators molecules in the host system lead to a downmodulation not only of the parasite immune response, but also the immune response to bystander antigens as revised by Dunne and Cooke [19].

A possible explanation for these conflicting observations may include the time of exposure to the worm (if acute or chronic) and also the helminthic species. For instance, it was demonstrated in a randomized, double-blinded and placebo controlled trial that Albendazole and Praziquantel treatment of infected women during pregnancy was associated with increased risk of eczema and wheeze in the offspring [20]. New data presented here reinforce the idea that deworming may be associated with asthma worsenings.

Regarding the effect of antihelminthic treatments on cytokine production, we were able to measure IFN-γ, IL-5, and IL-10 in response to Der p1 and to SWAP in supernatants of PBMC cultures before treatment and 7 and 90 days after treatment. Although Th2 molecules such as IL-4, IL-5, and IL-13 are key cytokines involved in the inflammatory response in asthma, IFN-γ has also been associated with asthma severity [21, 22]. The baseline levels of IL-5 in response to Der p1 in the present study were higher in the Praziquantel group compared to the placebo group. However, there was no significant difference in the pre- versus posttreatment levels of this cytokine in the two studied groups. On the other hand, the production of IL-10 in response to SWAP decreased 90 days after antihelminthic treatment in the Praziquantel group. This result suggests that the regulatory mechanisms exerted by the parasites begin to diminish early after treatment.

In the past few years, it has been shown that chronic helminth infections or parasite products induce the production of T-regulatory cells and molecules such as IL-10. This response has been associated with a down-modulation of allergic inflammatory mediators, such as Th2-cytokines, eosinophils, and histamine in murine models of allergic asthma [7, 12, 23]. We have characterized the *in vitro* immune response of asthmatic patients to some *S. mansoni* antigens and found a high production of *S. mansoni* antigen-specific IL-10 not only in cells of *S. mansoni* infected individuals but also in cells of noninfected asthmatic individuals [24]. There was also a significantly higher production of IL-10 and lower production of Th2-cytokine IL-5 in response to Der p1 in PBMC cultures of asthmatic patients infected with *S. mansoni* compared to non-infected asthmatic patients [25]. Our findings of decreased levels of IL-10 production after antihelminthic treatment in the present study, therefore,

are in agreement with our's [25] and other authors, [26] previous studies.

Other regulatory mechanisms may contribute to the suppression of allergic inflammation induced by helminths. For instance, Pacífico et al. showed that T CD4$^+$CD25$^+$ cells protect mice against allergen induced airway inflammation through an IL-10 independent mechanism [12]. This finding differs from other studies, which have demonstrated that IL-10 is a key cytokine that suppresses the inflammatory response in OVA-induced asthma in mice infected with helminths. In mice infected with *H. polygyrus*, for example, the reduction in the number of eosinophils and in the levels of IL-5 was associated with IL-10 production and migration of regulatory cells to the draining lymph nodes [27].

It has also been demonstrated that cytotoxic T-lymphocyte antigen 4 (CTLA-4), a molecule rapidly upregulated after T-cell activation that provides negative feedback signaling, and limits the immune response as reviewed by Deurloo et al. [28], is involved in the suppression of allergic response in asthma [29]. We previously demonstrated that the lower levels of Th2-cytokines in asthmatics infected with *S. mansoni* compared to non-infected asthmatics were associated with a higher frequency of CD4$^+$ T-cells expressing CTLA-4 [30]. Based on these findings, it is likely that the mechanisms underlying the regulation of inflammatory responses in asthma by *S. mansoni* antigens involve IL-10 [31], T-regulatory cells [12], and other mechanisms such as the expression of CTLA-4 [29, 30].

In the present study, where the number of asthmatic individuals who filled out the inclusion criteria and who agreed to participate was small, there was a significant worsening in the clinical scores of asthma as well as in the pulmonary function after repeated antihelminthic treatments. These findings are consistent with the hypothesis of the protective role of helminths on atopic diseases. Changes in clinical scores of asthma and in the pulmonary function tests were observed later after sequential antihelminthic treatments. Based on these results, we argue that the antigens released by the parasite when it dies can maintain protection against atopic diseases. After the clearance of antigens and repeated antihelminthic treatments to maintain a low parasite load, a loss of protection of parasite infection in the clinical course of asthma is observed. The data suggests that IL-10 could be implied in this protection. However, a better understanding of immune events following antihelminthic treatment is still required and may have practical consequences in the development of future therapies of allergic diseases.

Acknowledgments

We are very grateful to all volunteers from the community of Água Preta - Gandu who agreed to participate in this study and to the local health agent Irene Jesus for assistance with the field work. We thank Dr. Irismá Souza for performing the Chest-X-ray evaluation and Cristina Toledo for the review of the manuscript. This work was supported by the Brazilian National Research Council (CNPq). MIA, AAC and EMC are investigators supported by CNPq.

References

[1] N. R. Lynch, I. A. Hagel, M. E. Palenque et al., "Relationship between helminthic infection and IgE response in atopic and nonatopic children in a tropical environment," *Journal of Allergy and Clinical Immunology*, vol. 101, no. 2 I, pp. 217–221, 1998.

[2] I. Hagel, N. R. Lynch, M. Perez, M. C. di Prisco, R. Lopez, and E. Rojas, "Modulation of the allergic reactivity of slum children by helminthic infection," *Parasite Immunology*, vol. 15, no. 6, pp. 311–315, 1993.

[3] M. I. Araujo, A. A. Lopes, M. Medeiros et al., "Inverse association between skin response to aeroallergens and *Schistosoma mansoni* infection," *International Archives of Allergy and Immunology*, vol. 123, no. 2, pp. 145–148, 2000.

[4] M. Medeiros Jr., J. P. Figueiredo, M. C. Almeida et al., "*Schistosoma mansoni* infection is associated with a reduced course of asthma," *Journal of Allergy and Clinical Immunology*, vol. 111, no. 5, pp. 947–951, 2003.

[5] R. Hussain, R. W. Poindexter, and E. A. Ottesen, "Control of allergic reactivity in human filariasis: Predominant localization of blocking antibody to the IgG4 subclass," *Journal of Immunology*, vol. 148, no. 9, pp. 2731–2737, 1992.

[6] A. H. J. van den Biggelaar, R. van Ree, L. C. Rodrigues et al., "Decreased atopy in children infected with *Schistosoma haematobium*: a role for parasite-induced interleukin-10," *The Lancet*, vol. 356, no. 9243, pp. 1723–1727, 2000.

[7] B. Royer, S. Varadaradjalou, P. Saas, J. J. Guillosson, J. P. Kantelip, and M. Arock, "Inhibition of IgE-induced activation of human mast cells by IL-10," *Clinical and Experimental Allergy*, vol. 31, no. 5, pp. 694–704, 2001.

[8] M. I. Asher and S. K. Weiland, "The international study of asthma and allergies in childhood (ISAAC)," *Clinical and Experimental Allergy, Supplement*, vol. 28, no. 5, pp. 52–66, 1998.

[9] S. Stanojevic, A. Wade, J. Stocks et al., "Reference ranges for spirometry across all ages: a new approach," *American Journal of Respiratory and Critical Care Medicine*, vol. 177, no. 3, pp. 253–260, 2008.

[10] N. Katz, P. M. Coelho, and J. Pellegrino, "Evaluation of Kato's quantitative method through the recovery of *Schistosoma mansoni* eggs added to human feces," *Journal of Parasitology*, vol. 56, no. 5, pp. 1032–1033, 1970.

[11] D. E. Elliott, J. Li, A. Blum et al., "Exposure to schistosome eggs protects mice from TNBS-induced colitis," *American Journal of Physiology*, vol. 284, no. 3, pp. G385–G391, 2003.

[12] L. G. G. Pacífico, F. A. V. Marinho, C. T. Fonseca et al., "*Schistosoma mansoni* antigens modulate experimental allergic asthma in a murine model: a major role for CD4+ CD25+ Foxp3+ T cells independent of interleukin-10," *Infection and Immunity*, vol. 77, no. 1, pp. 98–107, 2009.

[13] H. M. Mo, J. H. Lei, Z. W. Jiang et al., "*Schistosoma japonicum* infection modulates the development of allergen-induced airway inflammation in mice," *Parasitology Research*, vol. 103, no. 5, pp. 1183–1189, 2008.

[14] J. Leonardi-Bee, D. Pritchard, and J. Britton, "Asthma and current intestinal parasite infection: systematic review and meta-analysis," *American Journal of Respiratory and Critical Care Medicine*, vol. 174, no. 5, pp. 514–523, 2006.

[15] C. Flohr, S. K. Weiland, G. Weinmayr et al., "The role of atopic sensitization in flexural eczema: findings from the international study of asthma and allergies in childhood phase two," *Journal of Allergy and Clinical Immunology*, vol. 121, no. 1, pp. 141–147, 2008.

[16] P. Bager, J. Arnved, S. Rønborg et al., "*Trichuris suis* ova therapy for allergic rhinitis: a randomized, double-blind, placebo-controlled clinical trial," *Journal of Allergy and Clinical Immunology*, vol. 125, no. 1–3, pp. 123–130, 2010.

[17] C. Flohr, L. N. Tuyen, R. J. Quinnell et al., "Reduced helminth burden increases allergen skin sensitization but not clinical allergy: a randomized, double-blind, placebo-controlled trial in Vietnam," *Clinical and Experimental Allergy*, vol. 40, no. 1, pp. 131–142, 2010.

[18] J. Feary, J. Britton, and J. Leonardi-Bee, "Atopy and current intestinal parasite infection: a systematic review and meta-analysis," *Allergy*, vol. 66, no. 4, pp. 569–578, 2011.

[19] D. W. Dunne and A. Cooke, "A worm's eye view of the immune system: consequences for evolution of human autoimmune disease," *Nature Reviews Immunology*, vol. 5, no. 5, pp. 420–426, 2005.

[20] H. Mpairwe, E. L. Webb, L. Muhangi et al., "Anthelminthic treatment during pregnancy is associated with increased risk of infantile eczema: randomised-controlled trial results," *Pediatric Allergy and Immunology*, vol. 22, no. 3, pp. 305–312, 2011.

[21] S. H. Cho, L. A. Stanciu, S. T. Holgate, and S. L. Johnston, "Increased interleukin-4, interleukin-5, and interferon-γ in airway CD4+ and CD8+ T cells in atopic asthma," *American Journal of Respiratory and Critical Care Medicine*, vol. 171, no. 3, pp. 224–230, 2005.

[22] R. Stephens, D. A. Randolph, G. Huang, M. J. Holtzman, and D. D. Chaplin, "Antigen-nonspecific recruitment of Th2 cells to the lung as a mechanism for viral infection-induced allergic asthma," *Journal of Immunology*, vol. 169, no. 10, pp. 5458–5467, 2002.

[23] L. S. Cardoso, S. C. Oliveira, A. M. Góes et al., "*Schistosoma mansoni* antigens modulate the allergic response in a murine model of ovalbumin-induced airway inflammation," *Clinical and Experimental Immunology*, vol. 160, no. 2, pp. 266–274, 2010.

[24] L. S. Cardoso, S. C. Oliveira, A. M. Góes et al., "*Schistosoma mansoni* antigens modulate the allergic response in a murine model of ovalbumin-induced airway inflammation," *Clinical and Experimental Immunology*, vol. 160, no. 2, pp. 266–274, 2010.

[25] M. I. A. S. Araujo, B. Hoppe, M. Medeiros Jr. et al., "Impaired T helper 2 response to aeroallergen in helminth-infected patient with asthma," *Journal of Infectious Diseases*, vol. 190, no. 10, pp. 1797–1803, 2004.

[26] L. E. P. M. van der Vlugt, L. A. Labuda, A. Ozir-Fazalalikhan et al., "Schistosomes induce regulatory features in human and mouse CD1dhi B cells: inhibition of allergic inflammation by IL-10 and regulatory T cells," *Plos One*, vol. 7, no. 2, Article ID e30883, 2012.

[27] K. Kitagaki, T. R. Businga, D. Racila, D. E. Elliott, J. V. Weinstock, and J. N. Kline, "Intestinal helminths protect in a murine model of asthma," *Journal of Immunology*, vol. 177, no. 3, pp. 1628–1635, 2006.

[28] D. T. Deurloo, B. C. A. M. van Esch, C. L. Hofstra, F. P. Nijkamp, and A. J. M. van Oosterhout, "CTLA4-IgG reverses asthma manifestations in a mild but not in a more "severe" ongoing murine model," *American Journal of Respiratory Cell and Molecular Biology*, vol. 25, no. 6, pp. 751–760, 2001.

[29] S. Tsuyuki, J. Tsuyuki, K. Einsle, M. Kopf, and A. J. Coyle, "Costimulation through B7-2 (CD86) is required for the induction of a lung mucosal T helper cell 2 (Th2) immune

response and altered airway responsiveness," *Journal of Experimental Medicine*, vol. 185, no. 9, pp. 1671–1679, 1997.

[30] R. R. Oliveira, K. J. Gollob, J. P. Figueiredo et al., "*Schistosoma mansoni* infection alters co-stimulatory molecule expression and cell activation in asthma," *Microbes and Infection*, vol. 11, no. 2, pp. 223–229, 2009.

[31] L. M. Araujo, J. Lefort, M. A. Nahori et al., "Exacerbated Th2-mediated airway inflammation and hyperresponsiveness in autoimmune diabetes-prone NOD mice: a critical role for CD1d-dependent NKT cells," *European Journal of Immunology*, vol. 34, no. 2, pp. 327–335, 2004.

15

Genetic Variation and Population Genetics of *Taenia saginata* in North and Northeast Thailand in relation to *Taenia asiatica*

Malinee Anantaphruti, Urusa Thaenkham, Teera Kusolsuk, Wanna Maipanich, Surapol Saguankiat, Somjit Pubampen, and Orawan Phuphisut

Department of Helminthology, Faculty of Tropical Medicine, Mahidol University, 420/6 Ratchawithi Road, Bangkok 10400, Thailand

Correspondence should be addressed to Malinee Anantaphruti; tmmtr@mahidol.ac.th

Academic Editor: Wej Choochote

Taenia saginata is the most common human *Taenia* in Thailand. By *cox1* sequences, 73 isolates from four localities in north and northeast were differentiated into 14 haplotypes, 11 variation sites and haplotype diversity of 0.683. Among 14 haplotypes, haplotype A was the major (52.1%), followed by haplotype B (21.9%). Clustering diagram of Thai and GenBank sequences indicated mixed phylogeny among localities. By MJ analysis, haplotype clustering relationships showed paired-stars-like network, having two main cores surrounded by minor haplotypes. Tajima's *D* values were significantly negative in *T. saginata* world population, suggesting population expansion. Significant Fu's F_s values in Thai, as well as world population, also indicate that population is expanding and may be hitchhiking as part of selective sweep. Haplotype B and its dispersion were only found in populations from Thailand. Haplotype B may evolve and ultimately become an ancestor of future populations in Thailand. Haplotype A seems to be dispersion haplotype, not just in Thailand, but worldwide. High genetic *T. saginata* intraspecies divergence was found, in contrast to its sister species, *T. asiatica*; among 30 samples from seven countries, its haplotype diversity was 0.067, while only 2 haplotypes were revealed. This extremely low intraspecific variation suggests that *T. asiatica* could be an endangered species.

1. Introduction

Human taeniasis occurs worldwide. It is caused by *Taenia saginata* and *T. solium*. A third species, *T. asiatica*, is an additional source of intestinal infection in a number of Asian countries. Cattle are the most common source of *T. saginata* infection, while the most common cause of both *T. solium* and *T. asiatica* infection is swine. *T. solium* metacestodes, called *Cysticercus cellulosae*, reside in the animal's muscle, whereas *T. asiatica* metacestodes, called *C. viscerotropica*, parasitize the liver. In its evolution, speciation of *Taenia* appears to be linked primarily to host switching among carnivore definitive hosts [1]. This association between *Taenia* and humans is thought to have developed about 10,000 years ago, coincidental with the development of agriculture and the domestication of food animals, like cattle and pigs [2]. Contraction of *Taenia* tapeworms by humans happened independently twice, both times by host switching from carnivore definitive hosts to primate definitive hosts [3]. One

is an ancestor of *T. saginata* + *T. asiatica*, and the other is *T. solium*. Geographical distribution has been extensively modified by European exploration and colonization since the 1500s and by ongoing globalization of agriculture and the changing patterns of human migration [4].

Understanding the genetic population structure of parasites helps to elucidate parasite transmission patterns and develop control measures [5]. The population structure and genetic variation of *T. solium* revealed two separate groups: Asian and African/Latin American genotypes [6]. Intraspecies strain variation among *T. solium* has been found to be minimal [7, 8]. Livestock farming of intermediate swine hosts may be reducing the possibility of genetic variation in *T. solium*. On the other hand, cattle, an intermediate host of *T. saginata*, are found in herds in a wide range of pastures. With different farming methods of this intermediate host, the population structure of *T. saginata* requires investigation.

Cytochrome *c* oxidase subunit I (*cox1*) genes of mitochondrial DNA have been commonly used for studying

FIGURE 1: Map of Thailand showing study areas. N, Nan; U, Ubon Ratchathani; K, Khon Kaen.

phylogenetic relationships among taeniid cestodes. Distinct intraspecific variations have been detected among various species, for example, *Echinococcus granulosus* [9], *E. multilocularis* [10], and *T. taeniaeformis* [11]. However, little is known about *cox1* genetic variation within a human *Taenia* species. A minor variation was observed in one isolate of *T. saginata* from Kenya and Poland [12], which compared mitochondrial *cox1* and nuclear rDNA 28S sequences. In this study, we focused on the genetic variation of *T. saginata* among samples collected from various geographical localities in north and northeast Thailand, where this parasite is highly prevalent among local inhabitants [13, 14]. In-depth studies of the genetic divergence of *T. saginata* specimens in Thailand have never been conducted. Indeed, the genetic structure among populations of this species and its evolution in Thailand and throughout the world remains limited. Our aim, using partial sequencing of the mitochondrial *cox1* gene, was to examine intraspecific variations and the population genetics of *T. saginata* in Thailand. The genetic divergence of the sister species, *T. asiatica*, from Thailand was also considered.

2. Materials and Methods

2.1. Studied Host Population and Parasites. Parasites were collected during the years 2009-2010 from four sites in north and northeast Thailand. The two sites in the north—one in the lowlands, the other in the upland hill tribe communities—are both in the Thung Chang district, Nan province, an area on the northern border with Lao PDR. The two sites in the northeastern region are in two different provinces: Ubon Ratchathani and Khon Kaen. Ubon Ratchathani shares borders with both southern Lao PDR and northern Cambodia. Khon Kaen lies closer to the center in the upper half of Thailand (Figure 1).

T. saginata was collected from both *Taenia* egg-positive persons and from persons who spontaneously discharged

gravid proglottids. The worms were identified morphologically as *T. saginata* by scolex and/or gravid proglottids. Male infection rates were almost double females, at a ratio of 43 : 23. The ages of the infected individuals ranged between 12 and 83 years. The worms were fixed in 70% ethanol for molecular analysis. This study was approved by the Ethics Committees of the Faculty of Tropical Medicine, Mahidol University, and the Ministry of Public Health, Thailand. Informed consent was obtained prior to subject participation.

Twelve *T. asiatica*, previously collected from Kanchanaburi Province [15], were coprocessed with *T. saginata*. The *cox1* sequences of 33 *T. saginata* and 18 *T. asiatica* from various different countries in the world, deposited in the GenBank database (Table 1), were also analyzed.

2.2. Molecular Studies

2.2.1. DNA Analysis. Partial proglottid fragments of individual strobila were separated and washed with distilled water to remove any ethanol remaining from the fixation process. The genomic DNA of each worm was extracted using a Genomic DNA Mini Kit (Geneaid, Sijhih City, Taiwan) per the manufacturer's instructions. DNA was resuspended in 50 μL elution buffer (provided as part of the kit). The PCR amplicons were amplified using two oligonucleotide primers: *cox1* (forward), 5′-CATGGAATAATAATGATTTTC-3′, and *cox1* (reverse), 5′-ACAGTACACACAATTTTAAC-3′. These primers were designed from the alignment of *T. saginata* and *T. asiatica* mitochondrial *cox1* genes (AB533171 and AB533175, resp.). PCR amplicons were produced in 50 μL of reaction mixture, containing: 10 ng genomic DNA; 0.5 μM of each primer; and 1x TopTaq Master Mix Kit (comprising TopTaq DNA Polymerase, PCR Buffer with 1.5 mM $MgCl_2$, and 200 μM each dNTP) (QIAGEN, Germany). Amplification conditions were as follows: initial heating at 94°C for 3 min, followed by 30 amplification cycles, consisting of denaturation at 95°C for 30 sec, annealing at 53°C for 30 sec, and elongation at 72°C for 50 sec. PCR products were run into 1.2% agarose gel and visualized with a UV illuminator. The PCR amplicons were purified and sequenced by dideoxytermination method, using an ABI3730XL sequencer and BigDye v 3.1 (Applied Biosystems, Foster City, CA, USA) at Macrogen Inc. (Geumcheon-gu, Seoul, Republic of Korea). DNA sequences were aligned using the BioEdit program, version 7.0 [16]. There is no conflict of interests with the commercial identities in this paper.

2.2.2. Population Genetic Analysis. The 924 bp *cox1* gene population genetics of *T. saginata* samples from four different localities was analyzed. Genetic diversity values, including polymorphic sites between populations (*S*), haplotype numbers (*h*), haplotype diversity (*Hd*), nucleotide diversity (π), theta-w (θw), and theta-π ($\theta \pi$) estimators to measure DNA polymorphism [17–19], were calculated using DnaSP version 4.0 [20] and the Arlequin computer program, version 3.1 [21, 22]. These programs were also used to evaluate the genetic structure of the parasites under the population expansion effect, via Tajima's *D* test and Fu's F_s test [23].

TABLE 1: Accession numbers of 73 *T. saginata cox1* sequences in this study, 33 from 11 different geographical countries, and 30 *T. asiatica* isolates from 7 different countries deposited in GenBank.

Species	Number of samples	Locality (country)	Accession numbers
T. saginata	5	China	AB107239, AB107247, AB533168, AB533169, and AB533171
	1	Korea	AB465246
	1	Japan	AB465244
	2	Indonesia	AB107240, AB465240
	1	Cambodia	AB465241
	1	Nepal	AB107243
	2	Ecuador	AB107238, AB465243
	3	Brazil	AB107237, AB107246, and AB465238
	3	Ethiopia	AB107241, AB465237, and AB465245
	1	Belgium	AB107242
	13	Thailand	AB107244, AB107245, AB465231, AB465232, AB465233, AB465234, AB465235, AB465236, AB465239, AB465242, AB465247, AB465248, and AB533173
	73	Thailand, this study	JN986646–JN986718
T. asiatica	4	China	AB465211, AB465212, AB465213, and AB465227
	1	Taiwan	AB465230
	2	Korea	AB465224, AB465225
	3	Japan	AB608736, AB608739, and AB608742
	1	Philippines	AB465229
	3	Indonesia	AB465215, AB465216, and AB465228
	4	Thailand	AB533174, AB533175, AB465222, and AB465223
	12	Thailand, this study	JQ517298–JQ517309

2.2.3. Clustering Diagram and Haplotype Network Analysis. The *cox1* sequences were aligned by ClustalX version 2.0 [24], and the haplotypes then distinguished. A neighbor-joining (NJ) phylogram was constructed under *p*-distance model by MEGA version 5.0 [25]. Bootstrap analyses were conducted using 1,000 replicates. A median-joining (MJ) network of *cox1* haplotypes was illustrated by Network 4.5.1.6 Software (Fluxus Technology Ltd. (http://www.fluxus-engineering .com/)). The *T. saginata* world populations (Table 1) were also tested for genetic differentiation without regional separation by global AMOVA.

3. Results

3.1. Parasites and Infections. A total of 73 *Taenia saginata* isolates were collected from 66 cases across the four study sites in north (Nan lowland, NL; Nan highland, NH) and northeast (Ubon Ratchathani, UB; Khon Kaen, KK) Thailand (Figure 1). The samples studied were 16 and 32 isolates from lowland and highland Nan Province in the north, and 9 and 16 isolates from Ubon Ratchathani and Khon Kaen provinces, in the northeast.

3.2. Mitochondrial cox1 DNA Sequence Analyses. Total DNA was extracted from 73 *Taenia* samples from four different geographical localities and then processed for sequencing. The partial cytochrome *c* oxidase subunit 1 (*cox1*) sequences

confirmed that they were all *T. saginata* (GenBank accession nos. JN986646 to JN986718). The 924 bp *cox1* sequences of these samples were divided into 14 discrete groups, represented as haplotypes A–N, and revealed 11 segregation (polymorphic) sites (S). Percentage intraspecific variation was 1.2%, with 1–5 nucleotide substitutions (Tables 2 and 3). Among these diverse haplotypes, two main ones had the highest ratio. Haplotype A was the most dominant haplotype (38/73, 52.1% of samples), in total and across all four localities. Haplotype B was the second most dominant (16/73 isolates, 21.9%) and was also detected in all localities. Between the two main haplotypes, A and B, there were two nucleotide substitutions (0.2%). The other haplotypes (haplotypes C–N) were detected in only 1–3 isolates (Table 2).

3.3. Population Genetics. The genetic diversity value of the 73 samples taken from the four localities, determined by haplotype diversity (Hd), was 0.683 ± 0.05; the nucleotide diversity (π) was 0.00146 ± 0.00017. θw was greater than $\theta \pi$. Tajima's D test of neutrality showed no significant value (-1.102, $P = 0.112$) in these samples. A significant Fu's F_s value was, however, revealed (-7.565, $P = 0.001$) (Table 3). The *cox1* gene sequences of 33 *T. saginata* in different geographical areas from the GenBank database (Table 1) were included in the analysis. From 106 samples, 23 haplotypes (h) and 23 polymorphic sites (S) were revealed. Entire intraspecific variation was 2.5%. Mean haplotype diversity (Hd) was

TABLE 2: Haplotype, nucleotide variation sites of partial *coxI* gene (924 bp length), and frequency of 73 *Taenia saginata* samples in this study (THA*, haplotype A–N) and 33 from 11 different geographical countries** (Haplotype O–W).

Position of nucleotide change

Number	Haplotype	039	063	174	186	219	231	237	300	382	490	571	629	630	661	729	767	781	793	809	866	900	912	924
1	A	C	G	G	A	C	A	C	T	A	A	A	G	A	C	T	T	A	T	C	A	C	G	T
2	B	.	.	.	.	.	.	.	.	.	.	.	.	.	.	.	.	T	.	.	.	.	.	.
3	C	.	.	.	.	.	.	.	.	.	.	.	.	.	.	.	.	T	.	.	.	.	.	.
4	D	.	.	.	.	.	.	.	G	.	.	.	.	.	.	.	.	.	.	.	.	.	.	.
5	E	.	.	.	.	.	.	G	.	.	.	.	.	.	.	.	.	.	.	.	.	.	.	.
6	F	.	.	.	.	.	.	.	.	.	.	.	.	.	.	.	.	T	.	.	.	.	.	.
7	G	.	.	.	.	.	.	.	.	.	.	.	.	.	.	.	.	T	.	.	.	.	.	.
8	H	.	.	.	.	.	.	.	.	.	.	.	.	.	C	.	.	.	.	.	.	.	.	.
9	I	.	.	.	.	.	.	.	.	.	.	.	.	.	.	T	.	.	.	.	.	.	.	.
10	J	.	.	.	.	.	.	.	.	.	.	.	.	.	.	.	C	.	.	.	.	.	.	.
11	K	.	.	.	.	.	.	.	.	.	.	.	.	.	C	.	.	.	.	.	.	.	.	.
12	L	.	.	.	.	.	.	.	.	.	.	.	.	.	.	.	.	T	.	.	.	.	.	.
13	M	.	.	.	.	.	.	.	.	.	.	.	.	G	.	.	.	.	.	.	.	.	.	.
14	N	.	.	.	.	.	.	.	.	.	.	T	.	.	.	.	.	T	.	.	.	.	.	.
15	O	.	.	.	.	.	.	.	.	.	.	.	.	.	.	.	.	T	.	T	.	C	.	.
16	P	.	.	.	.	.	.	.	.	.	.	.	.	.	.	.	.	T	A	T	.	A	.	.
17	Q	.	.	.	.	.	.	G	.	.	.	.	.	.	.	.	.	.	.	.	.	.	.	.
18	R	A	.	.	.	.	.	.	.	.	.	.	.	.	.	.	.	.	.	.	.	.	.	.
19	S	.	A	.	G	.	A	.	.	.	.	.	.	.	.	.	.	.	.	.	.	.	.	.
20	T	.	.	T	.	.	.	.	.	.	.	.	.	.	.	.	.	.	.	.	.	.	.	.
21	U	.	.	.	.	.	.	.	.	T	.	.	.	.	.	.	.	.	.	.	.	.	.	.
22	V	T	.	T	.	.	.	.	.	.	.	.	.	.	.	.	.	.	C	.	.	.	T	.
23	W	.	.	.	C	.	.	.	.	.	.	.	.	.	.	.	.	.	.	.	A	.	.	.

Number of individuals per population

Number	Haplotype	NL	NH	UB	KK	Subtotal THA*	THA	CHN	KOR	JPN	IND	CAM	NEP	ECU	BRA	ETH	BEL	Total
1	A	10	17	5	6	38	7	2	1	1	2	1	1	1	2	1	1	57
2	B	1	6	3	6	16	3	—	—	—	—	—	—	—	—	—	—	19
3	C	1	2	—	—	3	—	—	—	—	—	—	—	—	—	—	—	3
4	D	1	2	—	—	3	—	—	—	—	—	—	—	—	—	—	—	3
5	E	1	1	—	—	2	—	—	—	—	—	—	—	—	—	—	—	2
6	F	—	2	—	—	2	—	—	—	—	—	—	—	—	—	—	—	2
7	G	—	1	1	—	2	—	—	—	—	—	—	—	—	—	—	—	2
8	H	—	—	—	1	1	—	—	—	—	—	—	—	—	—	—	—	1
9	I	—	—	1	—	1	—	—	—	—	—	—	—	—	—	—	—	1
10	J	1	—	—	—	1	—	—	—	—	—	—	—	—	—	—	—	1
11	K	1	—	—	—	1	—	—	—	—	—	—	—	—	—	—	—	1
12	L	—	—	—	1	1	—	—	—	—	—	—	—	—	—	—	—	1
13	M	—	1	—	—	1	—	—	—	—	—	—	—	—	—	—	—	1
14	N	—	—	—	1	1	—	—	—	—	—	—	—	—	—	—	—	1
15	O	—	—	—	—	—	1	—	—	—	—	—	—	—	—	—	—	1
16	P	—	—	—	—	—	1	—	—	—	—	—	—	—	—	—	—	1
17	Q	—	—	—	—	—	1	—	—	—	—	—	—	—	—	—	—	1
18	R	—	—	—	—	—	1	—	—	—	—	1	—	—	—	—	—	2
19	S	—	—	—	—	—	—	1	—	—	—	—	—	—	—	—	—	1
20	T	—	—	—	—	—	—	1	—	—	—	—	—	—	—	—	—	1
21	U	—	—	—	—	—	—	—	—	—	—	—	—	1	—	—	—	1
22	V	—	—	—	—	—	—	—	—	—	—	—	—	—	1	—	—	1
23	W	—	—	—	—	—	—	—	—	—	—	—	—	—	—	2	—	2

Dots reprepresent homology with haplotype A sequence.

NL: Nan lowland, NH: Nan highland, UB: Ubon Ratchathani, KK: Khon Kaen.

**11 countries including THA: Thailand, CHN: China, KOR: Korea, JPN: Japan, IND: Indonesia, CAM: Cambodia, NEP: Nepal, ECU: Ecuador, BRA: Brazil, ETH: Ethiopia, and BEL: Belgium.

The numbering of nucleotide position of 1–924 referred to position 400–1324 of the complete mtDNA sequence (1620 bp) of *T. saginata* (GenBank acc. no. AB066495).

0.686, and mean intrapopulation nucleotide diversity (π) was 0.00155. θw was greater than $\theta \pi$ (Table 3). The nucleotide sequence of Haplotype A in our study (38) was identical to *T. saginata* in GenBank data for Thailand(7) and the other countries analyzed (China(2), Korea(1), Japan(1), Indonesia(2), Nepal(1), Ecuador(1), Brazil(2), Ethiopia(1), and Belgium(1)). The nucleotide sequence of Haplotype B was unique among the Thailand isolates, that is, 16 samples in this study and 3 from GenBank. The frequency of Haplotype A was 53.8% (57/106 samples); the frequency of Haplotype B was 17.9% (19 from 106 samples). Our sample sequences of Haplotype C–N and the GenBank sequences of Haplotype O–Q were unique among the Thailand isolates (Table 2). Some samples from China, Cambodia, Ecuador, Brazil, and Ethiopia showed nonidentical sequences (Haplotype R–W) (Table 2). In the GenBank data samples analyzed, noteworthy Tajima's *D* value and Fu's F_s value (−1.878, −18.798) were observed (Table 3).

In looking at the sister species, *T. asiatica*, significant differences between *T. asiatica* and *T. saginata* were discovered. The 924 bp *cox1* gene sequences of the 12 isolates from Kanchanaburi Province, Thailand, were all identical. Among 30 samples from seven different countries—China (4), Taiwan (1), Korea (2), Japan (3), the Philippines (1), Indonesia (3), and Thailand (16)—only two haplotypes of the *cox1* gene were found, where the major haplotype comprised 29 samples. Only one sample from China had other haplotypes, and only one polymorphic site was found. Haplotype diversity *Hd* was 0.067. However, the value of nucleotide diversity π of each of the two species was very low (Table 3).

3.4. Clustering Diagram and Haplotype Network. The clustering diagram of *T. saginata cox1* gene indicated no significant genetic differentiation among the populations from the four different localities studied. *T. saginata* could not be separated by geographical locality, that is, upper north, east of, northeast, and central northeast Thailand (Figure 2). Similarly, *T. saginata* from the different countries (see Table 1) was not geographically discriminated. The median-joining (MJ) clustering relationship was constructed as a haplotype network (Figure 3). All haplotypes appeared to be separated into two groups which each cored at the two main haplotypes, A and B. Connectivity between them was seen with a number of shared divergent haplotypes. In addition, unique sequences of samples from different geographical countries, Haplotypes Q–W, showed connectivity to Haplotype A. Among these haplotypes, 1–4 nucleotide substitutions were revealed (Table 2). Additionally, of samples from Thailand only in the GenBank database, three had Haplotype B sequences; two unique sequence haplotypes (O and P) were connected to Haplotype B (Figure 3).

4. Discussion

In this study, we investigated the population variation of *Taenia saginata* in Thailand. The results showed 11 haplotypes where distribution was not related to geographical locality. Samples in our study from the north (*n* = 48), northeast (*n* = 25), and west (*n* = 8) (AB465231-3, AB465242, AB465247-8,

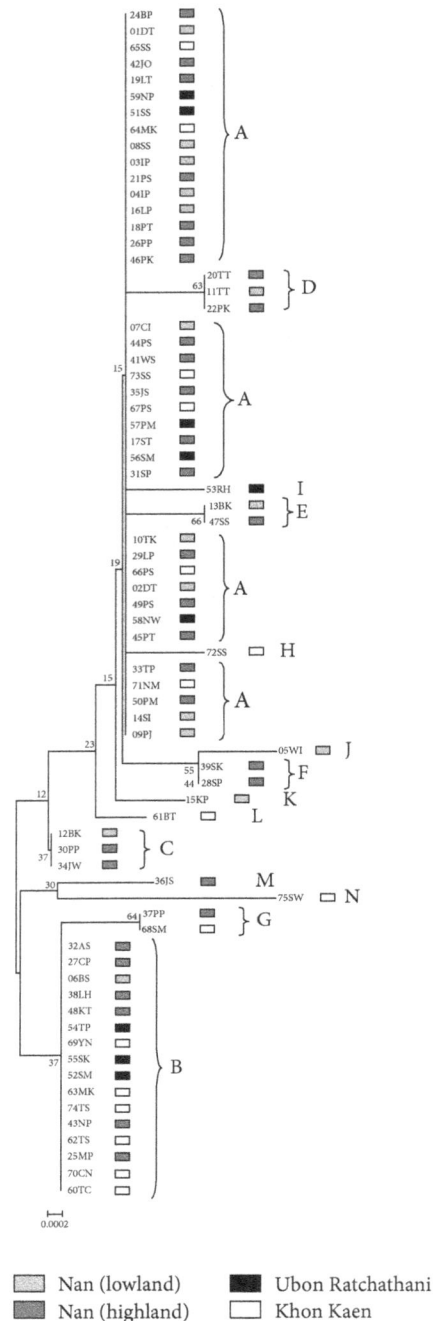

FIGURE 2: A neighbor-joining phylogram of mitochondrial *cox1* gene of 73 *T. saginata* from 4 localities of Thailand. A to N indicates haplotype.

AB465236, and AB465239) of Thailand were mixed together. Intraspecific variations in *cox1* genes have been reported among *T. saginata* samples from a number of different localities, including China, Ethiopia, France, Indonesia, Korea, Lao PDR, the Philippines, Taiwan, Thailand, and Switzerland. Global genetic divergence was 1.2–1.8% in the nucleotide variant positions of the total 1620 bp length [26, 27]. In

TABLE 3: Genetic diversity and test of *T. saginata* population, 73 from 4 different geographical regions of Thailand, 33 from 11 different countries and of *T. asiatica*, 12 from Kanchanaburi province, and 18 from 7 different countries.

Species	Population	Number of samples	h	S (%)	Genetic diversity				Neutrality tests	
					Hd	π	Theta-w	Theta-π	Tajima's D (P value)	Fu's F_s (P value)
T. saginata	Thai population	73	14	11 (1.2)	0.683 ± 0.050	0.00146 ± 0.000	2.263 ± 0.875	1.350 ± 0.093	-1.102 (0.112)	-7.565^* (0.001)
	World population	106	23	23 (2.5)	0.686 ± 0.045	0.0155 ± 0.0001	4.194 ± 1.319	1.487 ± 1.002	-1.878^* (0.004)	-18.798^* (0.000)
T. asiatica	Thai population	12	0	0	0	0	ND	ND	ND	ND
	Asian population	30	2	1	0.067	0.000	ND	ND	-1.15	ND

h: haplotype numbers, S: number of segregation sites, Hd: haplotype diversity, π: nucleotide diversity, Theta-w: Watterson's theta based on S, Theta-π: the theta based on π. * Significance ($P < 0.05$).

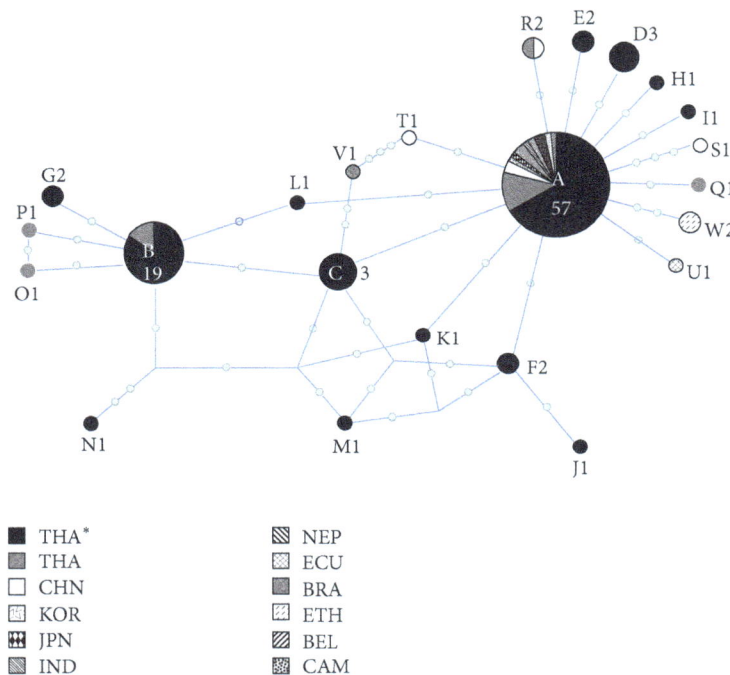

FIGURE 3: A median-joining network of *T. saginata* from Thailand (THA*, $n = 73$) and other 11 countries ($n = 33$). Haplotype codes (A-W are shown inside/adjacent to the circles. The size of circle denotes that a haplotype is proportional to the number of isolates of each haplotype shown inside/adjacent to the circle. Small circles indicate the number of nucleotide substitutions. THA: Thailand, CHN: China, KOR: Korea, JPN: Japan, IND: Indonesia, NEP: Nepal, ECU: Ecuador, BRA: Brazil, ETH: Ethiopia, BEL: Belgium, and CAM: Cambodia.

terms of nucleotide divergence of large-scale samples from Thailand, 11S, 1.2% displayed the results of isolates from global infection. However, among the worldwide *T. saginata* populations ($n = 106$), the Thai isolates ($n = 73$) and GenBank isolates ($n = 33$) used in this analysis, a high level of genetic variation (23S, 2.5%) was found. The genetic diversity values, haplotype diversity (Hd), and nucleotide diversity (π) found in populations in this study and in other combined populations were similar.

By MJ network, Haplotype B showed connectivity to Haplotype A in the *T. saginata* world population. In the *T. saginata* sequences, 19 of 33 (57.6%) from GenBank were identical to Haplotype A. Consequently, the star-like expansion in the MJ network of the major haplotype confirms

Haplotype A as an ancestor among the *T. saginata* world population. It also suggests that the subpopulation of minor haplotypes recently experienced a significant increase from its ancestors. Haplotype B and its star-like expansion network were unique to the Thailand isolates. Indeed, it is possible for Haplotype B diverged genetically to be the recent common ancestor of *T. saginata* in Thailand. However, the major Haplotype A contained half ($n = 57$) of the total population we analyzed. This means that *T. saginata* Haplotype A may share its genetic ancestry with populations from a variety of different geographical areas, in Asia as well as those in other continents. $\theta\pi$ less than θw indicates purifying selection, which results in selective removal of the deleterious allele in the population [18]. Tajima's D value is significantly negative in the world population but not in the Thailand population. The significant Fu's F_s value revealed in isolates from both Thailand and the world population, however, suggests that the population is growing and is hitchhiking due to population expansion and selective sweep [23]. The samples analyzed from each country were too small to be able to estimate this expansion and the genetic structure of *T. saginata* populations worldwide.

T. asiatica is the third *Taenia* tapeworm of humans and is reported only in Asian countries. It is distributed in specific areas across several countries, including Taiwan [28], Korea [29], China [30], Vietnam [31], Indonesia [32], and Thailand [33]. It is estimated that *T. saginata* and *T. asiatica* diverged from other human tapeworms about one million years ago (Mya), 0.414–1.616 Mya. *T. saginata* lineages emerged at an earlier period than *T. asiatica*. Lineages of *T. saginata* emerged at 238,000 years, while that of *T. asiatica* were at 41,000 years [34]. Very low intraspecies diversity of *T. asiatica* has been observed. Identical partial *cox1* gene nucleotide sequences have been found in 5 isolates (366 bp) from unspecified areas of Taiwan [12]; 17 isolates (337 bp) have been found across different localities in Korea [27]; and 12 isolates (924 bp) have been found in Kanchanaburi Province of Thailand. Of the total 1620 bp sequence length, only two variant nucleotide positions (0.1%) were detected in 5 isolates from China (2), the Philippines (1), and Korea (2) [26]. The low genetic variation of *T. asiatica* suggests populations of *T. asiatica* tapeworm to be small. The prevalence of *T. asiatica* is low when compared to *T. saginata* infections in most countries [35–37]. Also, Tajima's *D* value revealed no gene flow in *T. asiatica*, a result which indicates obvious differences in the population structures of *T. saginata* and *T. asiatica*. To date, since the emergence of *T. saginata* and *T. asiatica*, cattle have been known to be the intermediate hosts for *T. saginata* and swine for *T. asiatica*. The livestock management of these two intermediate host species has been different, and this may be an impact factor for these parasites. Pigs are raised feeding in restricted shed areas; gene flow among parasite populations in pigs, therefore, is diminished. In cattle farming, especially in Thailand, the animals are herded by grazing on naturally grown pastures, particularly pastures of postharvest rice which often cover wide distances. Furthermore, such cattle are frequently untransported, while daily herds are moved to a main city slaughterhouse over a relatively long period of time. The chances of cattle coming into contact with contaminated *Taenia* eggs, whether from human carriers during grazing or whilst drinking stream water from place to place, remain. This supports no specific locality in *T. saginata* populations for each genotype. It may even suggest that *T. saginata* tapeworm populations migrated during host cattle farming. Gene flow among *T. saginata* may have been influenced by host population migration. The difference in population genetics found in *T. saginata* suggests that intraspecies populations are growing. On the other hand, its sister species, *T. asiatica*, reveals very low genetic diversity. Such low divergence may indicate a loss of potential adaptive alleles for surviving in a changing environment, which could lead to the overall reduction of *T. asiatica* populations.

Cattle also act as definitive hosts for the liver fluke *Fasciola* spp. Ichikawa et al. [38] investigated the 535 bp partial nucleotide sequences of the *nad1* gene in 88 adult *Fasciola* flukes from three localities in Myanmar and found 27 substitution sites that yielded 20 haplotypes. A major haplotype revealed 54.6% (48/88 flukes) frequency and was seen in all three areas regardless of locality. The intraspecies genetic variation in *Fasciola* spp. is thought to have been introduced to Myanmar through ancient anthropogenetic movements of domestic ruminants. This seems to be the main factor determining mixing of the parasite population. Likewise, cattle host movements suggest intraspecific genetic variations of *T. saginata* populations in our study. Despite the fact that cattle serve as the definitive host for *Fasciola* spp. but the intermediate hosts for *T. saginata*, this status of intermediate host and definitive host does not influence different genetic variation of parasite species.

Our work shows that *T. saginata* adult worm isolates from humans, from two locations in the north and two provinces in northeast Thailand, exploited intraspecific genetic variability, without correlation with the geographical region of origin. The phylogenetic network of *cox1* sequences revealed 14 haplotypes from 73 samples. Thirty-three sequences from GenBank were added, and 23 haplotypes were exploited among the 106 samples. The genetic divergence of world *T. saginata* populations was 2.5%. Two main haplotypes, A and B, showed connectivity between them. Haplotype A seems to be an ancestor of *T. saginata* in the world population. Haplotype B and its dispersion are unique to the Thailand population. Intensive studies and a greater number of samples from different geographical areas are required to clarify the population genetics of *T. saginata* both in Thailand and worldwide.

Acknowledgments

Special thanks are extended to staff of the various health centers at each study site and particularly to Mr. Somjet Yosalai at Thung Chang Hospital in Nan Province. The authors would also like to thank the Director of the Northeastern Region Hydro Power Plant of the Electricity Generating Authority of Thailand and its staff for aiding their cooperation with local health centers and communities and for providing accommodation at Sirindhorn Dam and Ubolratana Dam during their stay in Ubon Ratchathani and Khon Kaen. The authors sincerely thank Dr. Yukifumi Nawa for his comments and suggestions for improving this Paper. This study was supported by an internal research grant from Mahidol University.

References

[1] E. P. Hoberg, A. Jones, R. L. Rausch, K. S. Eom, and S. L. Gardner, "A phylogenetic hypothesis for species of the genus *Taenia* (Eucestoda: Taeniidae)," *Journal of Parasitology*, vol. 86, no. 1, pp. 89–98, 2000.

[2] E. P. Hoberg, N. L. Alkire, A. De Queiroz, and A. Jones, "Out of Africa: origins of the *Taenia* tapeworms in humans," *Proceedings*

of the Royal Society B: Biological Sciences, vol. 268, no. 1469, pp. 781–787, 2001.

[3] A. De Queiroz and N. L. Alkire, "The phylogenetic placement of Taenia cestodes that parasitize humans," Journal of Parasitology, vol. 84, no. 2, pp. 379–383, 1998.

[4] E. P. Hoberg, "Taenia tapeworms: their biology, evolution and socioeconomic significance," Microbes and Infection, vol. 4, no. 8, pp. 859–866, 2002.

[5] G. Campbell, H. H. Garcia, M. Nakao, A. Ito, and P. S. Craig, "Genetic variation in Taenia solium," Parasitology International, vol. 55, supplement, pp. S121–S126, 2006.

[6] M. Nakao, M. Okamoto, Y. Sako, H. Yamasaki, K. Nakaya, and A. Ito, "A phylogenetic hypothesis for the distribution of two genotypes of the pig tapeworm Taenia solium worldwide," Parasitology, vol. 124, no. 6, pp. 657–662, 2002.

[7] K. Hancock, D. E. Broughel, I. N. S. Moura et al., "Sequence variation in the cytochrome oxidase I, internal transcribed spacer 1, and Ts14 diagnostic antigen sequences of Taenia solium isolates from South and Central America, India, and Asia," International Journal for Parasitology, vol. 31, no. 14, pp. 1601–1607, 2001.

[8] M. Okamoto, M. Nakao, Y. Sako, and A. Ito, "Molecular variation of Taenia solium in the world," Southeast Asian Journal of Tropical Medicine and Public Health, vol. 32, no. 2, pp. 90–93, 2001.

[9] R. B. Gasser and N. B. Chilton, "Characterisation of taeniid cestode species by PCR-RFLP of ITS2 ribosomal DNA," Acta Tropica, vol. 59, no. 1, pp. 31–40, 1995.

[10] M. Okamoto, Y. Bessho, M. Kamiya, T. Kurosawa, and T. Horii, "Phylogenetic relationships within Taenia taeniaeformis variants and other taeniid cestodes inferred from the nucleotide sequence of the cytochrome c oxidase subunit I gene," Parasitology Research, vol. 81, no. 6, pp. 451–458, 1995.

[11] H. Azuma, M. Okamoto, Y. Oku, and M. Kamiya, "Intraspecific variation of Taenia taeniaeformis as determined by various criteria," Parasitology Research, vol. 81, no. 2, pp. 103–108, 1995.

[12] J. Bowles and D. P. McManus, "Genetic characterization of the Asian Taenia, a newly described taeniid cestode of humans," American Journal of Tropical Medicine and Hygiene, vol. 50, no. 1, pp. 33–44, 1994.

[13] P. Radomyos, B. Radomyos, and A. Tungtrongchitr, "Multiinfection with helminths in adults from northeast Thailand as determined by post-treatment fecal examination of adult worms," Tropical Medicine and Parasitology, vol. 45, no. 2, pp. 133–135, 1994.

[14] B. Radomyos, T. Wongsaroj, P. Wilairatana et al., "Opisthorchiasis and intestinal fluke infections in Northern Thailand," Southeast Asian Journal of Tropical Medicine and Public Health, vol. 29, no. 1, pp. 123–127, 1998.

[15] M. T. Anantaphruti, U. Thaenkham, D. Watthanakulpanich et al., "Genetic diversity of Taenia asiatica from Thailand and other geographical locations as revealed by cytochrome c oxidase subunit 1 sequences," Korean Journal of Parasitology, vol. 50, no. 1, pp. 55–59, 2013.

[16] T. A. Hall, "BioEdit: a user-friendly biological sequence alignment editor and analysis program for Windows 95/98/NT," Nucleic Acids Symposium Series, no. 41, pp. 95–98, 1999.

[17] M. Nei, Molecular Evolutionary Genetics, Columbia University Press, New York, NY, USA, 1987.

[18] F. Tajima, "The effect of change in population size on DNA polymorphism," Genetics, vol. 123, no. 3, pp. 597–601, 1989.

[19] K. Misawa and F. Tajima, "Estimation of the amount of DNA polymorphism when the neutral mutation rate varies among sites," Genetics, vol. 147, no. 4, pp. 1959–1964, 1997.

[20] J. Rozas, J. C. Sánchez-DelBarrio, X. Messeguer, and R. Rozas, "DnaSP, DNA polymorphism analyses by the coalescent and other methods," Bioinformatics, vol. 19, no. 18, pp. 2496–2497, 2003.

[21] L. Excoffier, P. E. Smouse, and J. M. Quattro, "Analysis of molecular variance inferred from metric distances among DNA haplotypes: application to human mitochondrial DNA restriction data," Genetics, vol. 131, no. 2, pp. 479–491, 1992.

[22] L. Excofer, G. Laval, and S. Schneider, "Arlequin (version3.0): an integrated software package for population genetics data analysis," Evolutionary Bioinformatics Online, vol. 1, pp. 47–50, 2005.

[23] Y.-X. Fu, "Statistical tests of neutrality of mutations against population growth, hitchhiking and background selection," Genetics, vol. 147, no. 2, pp. 915–925, 1997.

[24] J. D. Thompson, T. J. Gibson, F. Plewniak, F. Jeanmougin, and D. G. Higgins, "The CLUSTAL X windows interface: flexible strategies for multiple sequence alignment aided by quality analysis tools," Nucleic Acids Research, vol. 25, no. 24, pp. 4876–4882, 1997.

[25] K. Tamura, D. Peterson, N. Peterson, G. Stecher, M. Nei, and S. Kumar, "MEGA5: molecular evolutionary genetics analysis using maximum likelihood, evolutionary distance, and maximum parsimony methods," Molecular Biology and Evolution, vol. 28, no. 10, pp. 2731–2739, 2011.

[26] H. K. Jeon and K. S. Eom, "Taenia asiatica and Taenia saginata: genetic divergence estimated from their mitochondrial genomes," Experimental Parasitology, vol. 113, no. 1, pp. 58–61, 2006.

[27] H.-K. Jeon, J.-Y. Chai, Y. Kong et al., "Differential diagnosis of Taenia asiatica using multiplex PCR," Experimental Parasitology, vol. 121, no. 2, pp. 151–156, 2009.

[28] P. C. Fan, C. Y. Lin, C. C. Chen, and W. C. Chung, "Morphological description of Taenia saginata asiatica (Cyclophyllidea: Taeniidae) from man in Asia," Journal of Helminthology, vol. 69, no. 4, pp. 299–303, 1995.

[29] K. S. Eom and H. J. Rim, "Morphologic descriptions of Taenia asiatica sp. n.," Korean Journal of Parasitology, vol. 31, no. 1, pp. 1–6, 1993.

[30] L. Zhang, H. Tao, B. Zhang et al., "First discovery of Taenia saginata asiatica infection in Yunnan province," Zhongguo Ji Sheng Chong Xue Yu Ji Sheng Chong Bing Za Zhi, vol. 17, no. 2, pp. 95–96, 1999.

[31] N. V. De, L. T. Hoa, N. Q. Doanh, N. B. Ngoc, and L. D. Cong, "Report on a new species of Taenia (Taenia asiatica) in Hanoi, Vietnam," Journal of Malaria and Parasitic Diseases Control, vol. 3, pp. 80–85, 2001.

[32] T. Wandra, A. A. Depary, P. Sutisna et al., "Taeniasis and cysticercosis in Bali and North Sumatra, Indonesia," Parasitology International, vol. 55, pp. S155–S160, 2006.

[33] M. T. Anantaphruti, H. Yamasaki, M. Nakao et al., "Sympatric occurrence of Taenia solium, T. saginata, and T. asiatica, Thailand," Emerging Infectious Diseases, vol. 13, no. 9, pp. 1413–1416, 2007.

[34] L. Michelet and C. Dauga, "Molecular evidence of host influences on the evolution and spread of human tapeworms," Biological Reviews, vol. 87, no. 3, pp. 731–741, 2012.

[35] T. Li, P. S. Craig, A. Ito et al., "Taeniasis/cysticercosis in a Tibetan population in Sichuan Province, China," *Acta Tropica*, vol. 100, no. 3, pp. 223–231, 2006.

[36] T. Wandra, P. Sutisna, N. S. Dharmawan et al., "High prevalence of *Taenia saginata* taeniasis and status of *Taenia solium* cysticercosis in Bali, Indonesia, 2002–2004," *Transactions of the Royal Society of Tropical Medicine and Hygiene*, vol. 100, no. 4, pp. 346–353, 2006.

[37] M. T. Anantaphruti, J. Waikagul, H. Yamasaki, and A. Ito, "Cysticercosis and taeniasis in Thailand," *Southeast Asain Journal of Tropical Medicine and Public Health*, vol. 38, no. 1, supplement, pp. 151–158, 2007.

[38] M. Ichikawa, S. Bawn, N. N. Maw et al., "Characterization of *Fasciola* spp. in Myanmar on the basis of spermatogenesis status and nuclear and mitochondrial DNA markers," *Parasitology International*, vol. 60, no. 4, pp. 474–479, 2011.

16

Analysis of Circulating Haemocytes from *Biomphalaria glabrata* following *Angiostrongylus vasorum* Infection Using Flow Cytometry

Thales A. Barçante,[1,2] **Joziana M. P. Barçante,**[2] **Ricardo T. Fujiwara,**[3] **and Walter S. Lima**[3]

[1] *Departamento de Medicina Veterinária, Pontifícia Universidade Católica de Minas Gerais (PUC Minas), 30535-901 Belo Horizonte, MG, Brazil*
[2] *Departamento de Medicina Veterinária, Preventiva, Universidade Federal de Lavras, 37200-000 Lavras, MG, Brazil*
[3] *Departamento de Parasitologia, Universidade Federal de Minas Gerais, 31270-901 Belo Horizonte, MG, Brazil*

Correspondence should be addressed to Joziana M. P. Barçante, joziana@dmv.ufla.br

Academic Editor: C. Genchi

Angiostrongylus vasorum is an emerging parasite of dogs and related to carnivores that have an indirect life cycle, with a wide range of terrestrial and aquatic gastropods as the obligatory intermediate host. Unfortunately, the relationship between *A. vasorum* and their snail hosts remains poorly understood. Circulating haemocytes are the main line of cellular defence involved in the destruction of helminths in snails. Aiming to further characterize the haemocyte subsets in *Biomphalaria* snails, we have performed a flow cytometric analysis of whole haemolymph cellular components using a multiparametric dual colour labelling procedure. Our findings demonstrated that *B. glabrata* infected with *A. vasorum* have two major circulating haemocyte subsets, referred to as small and large haemocytes. Differences in the cell proportion occurred over time. The development of better invertebrate infection control strategies would certainly result in the better control of human diseases caused by other species of the genus *Angiostrongylus*. Such knowledge will assist in the establishment of novel control strategies aimed at parasites that use molluscs as intermediate hosts and clarify new aspects of the parasite-host relationship regarding cell recognition and activation mechanisms, which are also found in the innate response of vertebrates.

1. Introduction

Angiostrongylus vasorum is a nematode belonging to the superfamily Metastrongyloidea that parasitizes domestic dogs (*Canis familiaris*) and wild carnivores. *Angiostrongylus vasorum* is distributed throughout Europe, the Americas, and Africa [1]. Isolated endemic foci are generally observed, although reports of infections outside of these focal areas have increasingly been documented [2]. A determinant factor in the pathology of canine angiostrongylosis seems to be related to the location of the parasite in the definitive host. The presence of the parasite inside the arteries and arterial branches of the host promotes a mechanical and metabolic action on the vessel walls, which may alter homeostasis [3, 4], resulting in pneumonia, loss of racing performance, coughing, and anaemia [5]. Severely infected dogs may develop

heart failure and pulmonary fibrosis, followed by weight loss, haemorrhagic diatheses, and death [6, 7].

Angiostrongylus vasorum has an indirect life cycle, with a wide range of terrestrial and aquatic gastropods as the obligatory intermediate host [8–13]. Some terrestrial molluscs are natural hosts of *A. vasorum*. In the laboratory, this nematode can be maintained in some species of planorbids. *Biomphalaria glabrata* is a fresh water planorbid of considerable importance as experimental intermediate host of *Angiostrongylus* sp. Species of *Biomphalaria* are capable of maintaining their integrity mainly through the activity of their internal defence system, which is made up of both cellular and soluble components. Circulating haemocytes are the main line of cellular defence involved in the destruction of helminths in snails that serve as the intermediate host [14–16]. These cells can bind to infectious agents by phagocytosing the syncytial tegument and/or releasing cyto-toxic compounds [17].

Circulating haemocytes in *B. glabrata* are composed of at least two cell populations: hyalinocytes and granulocytes [16, 18–21]. These cell populations differ substantially in morphological, biochemical, and functional aspects [22, 23]. The successful elimination of potential infective agents requires granulocytes to engulf particles and further eliminate living pathogens through enzymatic or oxidative degradation. In contrast, hyalinocytes are thought to be responsible primarily for wound repair [24, 25], requiring aggregation at an injury site.

During infection by helminths, the internal defence system of species of *Biomphalaria* is capable of recognising parasite molecules through a system similar to (soluble or membrane-binding) pattern-recognition receptors and activating phagocytes, which leads to parasite encapsulation and destruction [26]. Unfortunately, the relationship between *A. vasorum* and their snail hosts remains poorly understood. The development of better invertebrate infection control strategies would certainly result in the better control human diseases caused by other species of the genus *Angiostrongylus*.

The aim of the present study was to describe the profile of circulating haemocytes in *B. glabrata* following *A. vasorum* infection using flow cytometry analysis.

2. Materials and Methods

2.1. Parasite Source.
The strain of *A. vasorum* used in the experiments was isolated from the faeces of a domestic dog. The nematode was maintained by serial infections of domestic dogs and aquatic snails (*B. glabrata*) at the Laboratory of Veterinary Helminthology, Department of Parasitology, *Universidade Federal de Minas Gerais* (UFMG, Brazil).

The experimental protocols were carried out in compliance with the Ethical Principles in Animal Experimentation adopted by the UFMG Ethics Committee in Animal Experimentation and were approved under process number 060/03.

2.2. Parasitic Infection.
First-stage larvae of *A. vasorum* (L1) were isolated from the faeces of an experimentally infected dog using a Baermann apparatus, followed by filtration using filter paper, as described by Barçante et al. [27]. The number of viable L1 larvae was estimated under a stereomicroscope (magnification: 40x). *Biomphalaria glabrata* snails were individually exposed to 1000 L1 larvae of *A. vasorum* for 24 h in 2 mL of dechlorinated water [28]. Groups of 10 individuals were kept in containers throughout the trial, maintained at room temperature (25 to 27°C), and fed on lettuce.

2.3. Haemolymph Collection.
The whole haemolymph was collected from the snails at different times during *A. vasorum* infection: 30 min, 1, 2, 3, 4, 5, 6, 12, 24, 48, and 72 h after infection as well as 10, 20, 30, and 60 days after infection. The haemolymph was collected using the method described by Martins-Souza et al. [16]. Each snail shell was cleaned with 70% alcohol, dried with absorbent tissue paper, and the haemolymph was collected by a cardiac puncture using a 21-gauge needle. To avoid cell agglutination, the whole haemolymph was collected and diluted [1 : 1 (v/v)] in Chernin's balanced salt solution (CBSS) (47.7 mM of NaCl, 2.0 mM of KCl, 0.49 mM of anhydrous Na_2HPO_4, 1.8 mM of $MgSO_4 \cdot 7H_2O$, 3.6 mM of $CaCl_2 \cdot 2H_2O$, 0.59 mM of $NaHCO_3$, 5.5 mM of glucose and 3 mM of trehalose) containing citrate/EDTA buffer (50 mM of sodium citrate,10 mM of EDTA, and 25 mM of sucrose) at pH 7.2.

2.4. Haemocyte Count.
The haemolymph from three snails of each time group was divided in triplicate (1 mL) in microtubes and tested. After debris sedimentation for two min, the whole haemolymph was transferred to another microtube (1 mL). Total haemocyte counts were performed using 10 μL of whole haemolymph diluted (1/10) in CBSS buffer containing 0.5% neutral red solution [neutral red retention assay (NRRA)]. Haemocytes that stained red were considered granulocytes and those that did not were considered hyalinocytes.

2.5. Flow Cytometry Analysis.
Flow cytometry analysis was performed as described by Martins-Souza et al. [16]. After individual collection, the haemolymphs from three snails from the same experimental group were pooled and three separate pools were prepared and tested for each experimental group.

Haemolymph incubation with propidium iodide (PI) and acridine orange (Ao) solution allowed the separation of the viable circulating haemocytes (Ao positive/Eb negative cells) from the dead cells (Ao negative/Eb positive cells) and small fragments (Ao and Eb negative events). A total of 150 μL of the pooled whole haemolymph was mixed with an equal volume of propidium iodide (25 mg of PI/mL 95% alcohol) and acridine orange (7.5 mg of Ao/mL 95% alcohol) diluted (1 : 1000) in CBSS citrate/EDTA.

Haemocyte suspension was incubated for one h on ice in the absence of light. The suspension was then analysed using a FACScan flow cytometer (BD—Becton Dickinson, USA). Flow cytometry analysis of the cellular components of the whole haemolymph was performed using instrument settings to capture the fluorescence signals from propidium iodide (FL2) and acridine orange (FL3), using log amplification scales. A total of 15,000 events were analysed for each pooled haemolymph sample.

Data analysis was performed using the Cell Quest program (Becton Dickinson, USA). FL2 versus FL3 dot plot distribution graphs were constructed to differentiate live cells from dead cells and debris. Live haemocytes were selected based on size versus internal complexity (laser forward scatter (FSC) and side scatter (SSC), resp.). Two major haemocyte subpopulations were selected based on FSC. Percentages of haemocyte subsets obtained from the flow cytometry analysis were further converted into absolute counts taking into account the total viable haemocyte counts (NRRA) performed in a Neubauer chamber obtained with the same haemolymph sample.

2.6. Statistical Analysis.
Data referring to the numbers of circulating haemocytes within each cell subset are reported as mean and standard deviation and analysed using one-way

Analysis of Circulating Haemocytes from Biomphalaria glabrata following Angiostrongylus vasorum
Infection Using Flow Cytometry

147

(a)

(b)

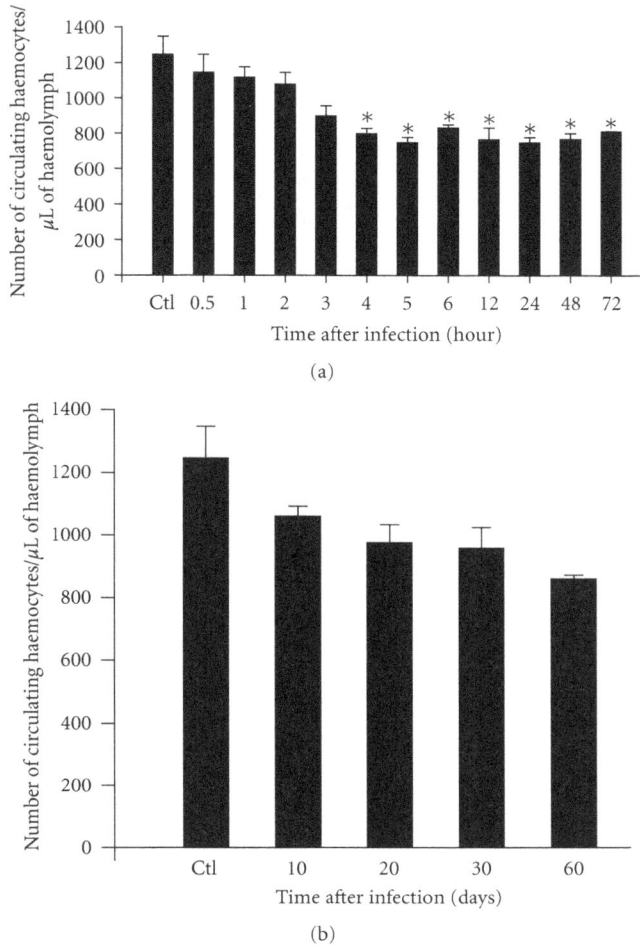

FIGURE 1: Mean number of circulating haemocytes/μL of haemolymph in three snails from each experimental group. Ctl—Biomphalaria glabrata free from infection with A. vasorum and B. glabrata exposed to 1,000 first-stage larvae of A. vasorum at different time of infection. Data are presented as mean number ± standard deviation of circulating haemocyte subpopulations. (*) Represents significant differences ($P < 0.05$) in the number of haemocytes of each point of infection to control group.

analysis of variance (ANOVA) followed by the Bonferroni post test.

3. Results

During the course of infection, a significant reduction occurred in the number of circulating haemocytes from four to 72 hours after exposure to A. vasorum L1 larvae ($P < 0.05$) (Figure 1(a)). This reduction was gradual through five hours after infection. From six to 72 hours after infection, the number of circulating haemocytes in the snails remained without much alteration, although at a significantly lower number than that in the control group ($P < 0.05$). From 10 to 60 days after infection, normal patterns of total haemocytes were reestablished in the haemolymphs, with no significant differences between groups ($P > 0.05$) (Figure 1(b)).

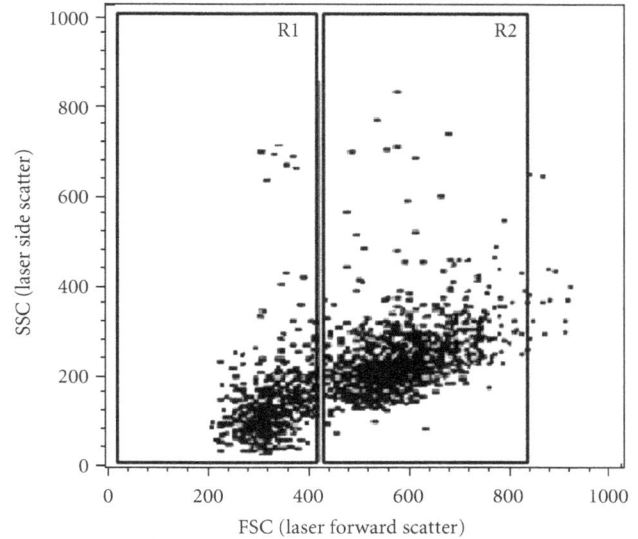

FIGURE 2: Profile of circulating haemocyte population in Biomphalaria glabrata snails. Two haemocyte subpopulations (R1 = small—FSC between 200–400, R2 = large—FSC between 440–840 can be identified by flow cytometric dot plot distributions based on their laser forward scatter (FSC) versus laser side scatter properties (SSC).

Haemolymph incubation with propidium iodide and acridine orange solution allowed the separation of viable circulating haemocytes from dead haemocytes (Figure 2). Viable circulating haemocytes from Biomphalaria snails were separated into two major cell subpopulations based mainly on size (small (FSC channels between 200 and 440) and large (FSC channels between 440 and 840)) and granularity. The haemocyte subpopulations were denominated small (R1) and large (R2).

In the control, 41% of the haemocytes were small and 59% were large (Figure 3). Thirty minutes after infection, there was a significant reduction in the percentage of small circulating haemocytes. This phenotype profile remained constant until four hours after infection (Figure 3(a)). In the subsequent hours, the number of small cells demonstrated a tendency toward a reduction until reaching significantly lower values than those of the control at 48 and 72 hours ($P < 0.05$). From 10 to 60 days after infection, the number of small haemocytes recovered in the infected snails remained below that found in the control group ($P < 0.05$) (Figure 3(b)).

4. Discussion

Although invertebrates account for the vast majority of living beings, publications addressing defence mechanisms against pathogens have been restricted mainly to interactions with pathogens of vertebrate animals [15]. Since the description of the participation of molluscs as intermediate hosts in the evolutionary cycle of A. vasorum by Guilhon [29], very little has been revealed regarding the infection biology and parasite-host interaction, especially in relation to the intermediate host. Studies demonstrate differences in the

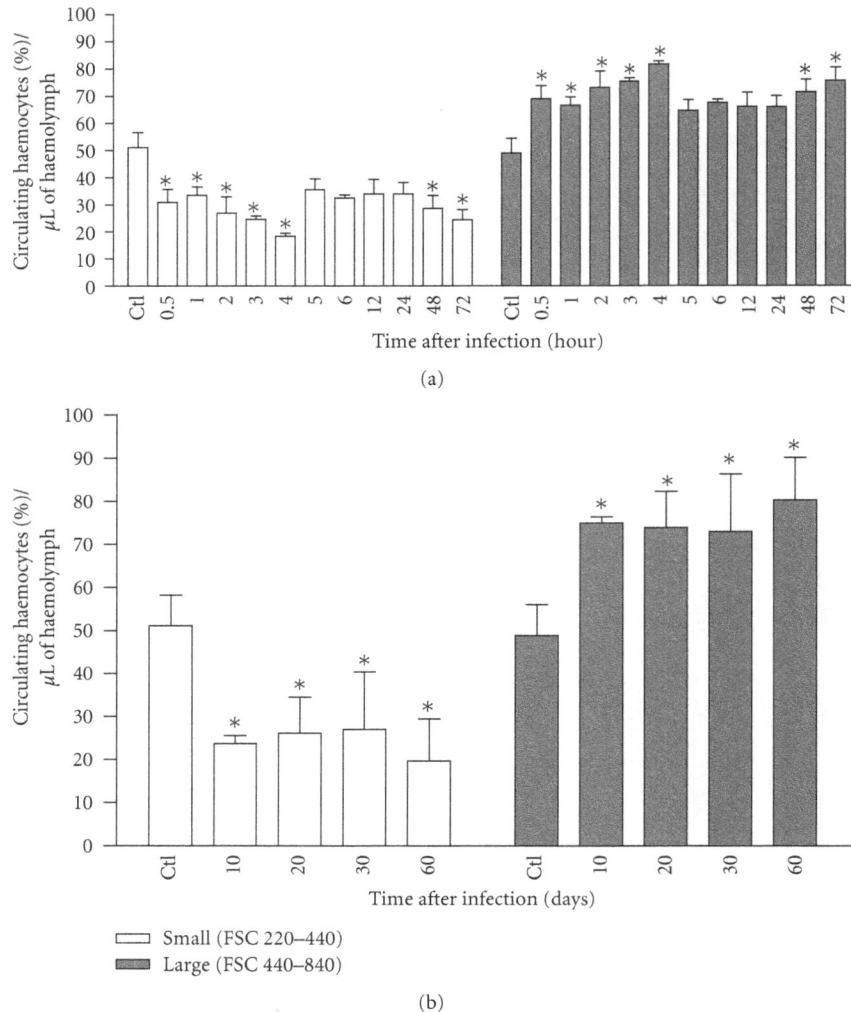

FIGURE 3: Percentual number of small (R1) and large (R2) haemocyte subpopulations in 1 μm of haemolymph of *Biomphalaria glabrata* free from infection with *Angiostrongylus vasorum* (Ctl) and in 1 μm of haemolymph of *B. glabrata* exposed to 1,000 first-stage larvae of *A. vasorum* at different times of infection. Data are presented as mean number ± standard deviation of circulating haemocyte subpopulations. (*) Represents significant differences ($P < 0.05$) in the number of haemocytes at each point of infection for the control group.

susceptibility of some species of molluscs to infection by *A. vasorum* [13, 30].

The capacity of molluscs to respond strongly to stimuli depends on haemocyte viability and functional capacity. Studies carried out on *B. tenagophila* demonstrate that the temporary reduction in the number of circulating granulocytes results in increased susceptibility to infection by *Schistosoma mansoni* [23].

In the present study, the number of circulating haemocytes in the haemolymph of *B. glabrata* reduced significantly soon after infection by *A. vasorum* through to 72 hours after infection. As the circulatory system of molluscs is generally open, haemocytes can move freely in and out of the tissues [31]. In histological cuts of *B. glabrata* infected by *A. vasorum*, Barçante [32] found that a perilarval cellular infiltrate occurs in the tissues five hours after infection. Similarly,

Bezerra et al. [33] report a reduction in the population of circulating haemocytes in *B. glabrata* in the first five hours after infection by *S. mansoni*. These observations indicate that, in infections of *B. glabrata* by *A. vasorum*, with the onset of a perilarval tissue reaction and the formation of a typical granuloma, circulating cells migrate to the tissue, leading to a reduction in the amount of these cells in the haemolymph.

Over the course of the infection, the chemotaxis of cells to the site of larval infection ceases, with a consequent reestablishment of cell levels, as seen beginning 10 days after infection [32]. According to Barçante [32], there is a reduction in perilarval cellularity in the long term, which explains the stabilisation in the cell composition of the haemolymph beginning 10 days after infection. Similarly, Pereira et al. [23] report an increase in haemocyte activation with no corresponding increase in the number of circulating haemocytes

Analysis of Circulating Haemocytes from Biomphalaria glabrata following Angiostrongylus vasorum
Infection Using Flow Cytometry

149

in *B. tenagophila* infected by *A. vasorum* due to the migration of circulating cells to the infection site during the encapsulation process.

The haemocytes of other molluscs and invertebrates are generally rather easy to define morphologically and to distribute among categories of known functions. In contrast, the haemocytes in *B. glabrata* do not offer simple criteria for recognition, such as specific secretory granules [34]. Thus, flow cytometry analysis proved to be a useful tool for the analysis of circulating haemocytes. Specifically in *B. glabrata*, two subpopulations of cells were characterized in noninfected snails: small round cells ranging from 5 to 6 μm in diameter with little granularity and large cells ranging from 6.5 to 8 μm in diameter with a greater degree of granularity [22].

The present study found two subpopulations with similar characteristics to those described previously as well as changes in the cell profile during infection by *A. vasorum*. Differences in the cell proportion occurred over time. A number of studies demonstrate that hyalinocytes are precursors of granulocytes and that, in cytometric counts, small cells are hyalinocytes and small granulocytes, whereas large cells are granulocytes. Thus, with the onset of infection by *A. vasorum*, the proportion of small cells diminishes due to the conversion of hyalinocytes into granulocytes, with the subsequent migration of these granulocytes to the tissues after five hours (as seen in histological cuts), thereby explaining the increase in the proportion of small cells in the haemolymph at this point [32]. This finding corroborates the results described by Bezerra et al. [33], who explain the reduction in the number of granulocytes in the haemolymph by the fact that granulocytes are effecter cells that migrate to the infection site, as demonstrated by histological analysis. In the later stages of infection, the percentages of small cells are lower due to the fact that the mollusc is infected and the effecter cells are acting, even at a lesser intensity, as seen in histological cuts [32].

5. Conclusion

The study of the parasite-host relationship by evaluating of the infection of *B. glabrata* by *A. vasorum* has much to contribute toward knowledge on the mechanisms used by the internal defence system of these invertebrates. Such knowledge will assist in the establishment of novel control strategies aimed at parasites that use molluscs as intermediate hosts and clarify new aspects of the parasite-host relationship regarding cell recognition and activation mechanisms, which are also found in the innate response of vertebrates. Moreover, *A. vasorum* is a nematode of increasing veterinary importance for which the possibility of being a causal agent of zoonoses has not yet been discarded, making studies on the interaction between molluscs and parasites important to the understanding of the biology and epidemiology of angiostrongylosis.

References

[1] M. Schnyder, A. Fahrion, B. Riond et al., "Clinical, laboratory and pathological findings in dogs experimentally infected with *Angiostrongylus vasorum*," *Parasitology Research*, vol. 107, no. 6, pp. 1471–1480, 2010.

[2] J. Koch and J. L. Willesen, "Canine pulmonary angiostrongylosis: an update," *Veterinary Journal*, vol. 179, no. 3, pp. 348–359, 2009.

[3] C. G. Schelling, C. E. Greene, A. K. Prestwood, and V. C. Tsang, "Coagulation abnormalities associated with acute *Angiostrongylus vasorum* infection in dogs," *American Journal of Veterinary Research*, vol. 47, no. 12, pp. 2669–2673, 1986.

[4] J. O. Costa and W. L. Tafuri, "Estudo anátomo-patológico de cães infectados experimentalmente pelo *Angiostrongylus vasorum* (Baillet, 1866) Kamenski 1905," *Arquivos Brasileiros de Medicina Veterinária e Zootecnia*, vol. 49, pp. 389–407, 1997.

[5] G. W. Jones, C. Neal, and G. R. Turner, "*Angiostrongylus vasorum* infection in dogs in Cornwall," *Veterinary Record*, vol. 106, no. 4, p. 83, 1980.

[6] K. Dodd, "*Angiostrongylus vasorum* (Baillet, 1866) infestation in a greyhound kennels," *Veterinary Record*, vol. 92, no. 8, pp. 195–197, 1973.

[7] M. C. Cury and W. S. Lima, "Estudo anàtomo-patologico de caes infectados experimentalmente pelo *Angiostrongylus vasorum*," *Arquivo Brasileiro de Medicina Veterinaria e Zootecnia*, vol. 48, pp. 27–34, 1996.

[8] J. Guilhon and B. Cens, "*Angiostrongylus vasorum* (Baillet, 1866). Étude biologique et morfologique," *Annales de Parasitologie Humaine et Comparee*, vol. 48, no. 4, pp. 567–596, 1973.

[9] L. Rosen, L. R. Ash, and G. D. Wallace, "Life history of the canine lungworm *Angiostrongylus vasorum* (Baillet)," *American Journal of Veterinary Research*, vol. 31, no. 1, pp. 131–143, 1970.

[10] G. Bolt, J. Monrad, P. Henriksen et al., "The fox (*Vulpes vulpes*) as a reservoir for canine angiostrongylosis in Denmark. Field survey and experimental infections," *Acta veterinaria Scandinavica*, vol. 33, no. 4, pp. 357–362, 1992.

[11] A. C. Bourque, G. Conboy, L. M. Miller, and H. Whitney, "Pathological findings in dogs naturally infected with *Angiostrongylus vasorum* in Newfoundland and Labrador, Canada," *Journal of Veterinary Diagnostic Investigation*, vol. 20, no. 1, pp. 11–20, 2008.

[12] T. Ferdushy, C. M. O. Kapel, P. Webster, M. N. S. Al-Sabi, and J. Grnvold, "The occurrence of *Angiostrongylus vasorum* in terrestrial slugs from forests and parks in the Copenhagen area, Denmark," *Journal of Helminthology*, vol. 83, no. 4, pp. 379–383, 2009.

[13] W. S. Lima, L. R. Mozzer, L. C. Montresor, and T. H.D.A. Vidigal, "*Angiostrongylus vasorum*: experimental infection and larval development in *Omalonyx matheroni*," *Journal of Parasitology Research*, vol. 2011, Article ID 178748, 4 pages, 2011.

[14] T. P. Yoshino and G. R. Vasta, "Parasite-invertebrate host immune interactions," in *Invertebrate Immune Responses: Cell Activities and the Environment*, E. L. Cooper, Ed., pp. 125–167, Springer, Heidelberg, Germany, 1996.

[15] A. D. Negrão-Corrê, C. A. J. Pereira, F. M. Rosa, R. L. Martins-Souza, Z. A. Andrade, and P. M. Z. Coelho, "Molluscan response to parasite, *Biomphalaria* and *Schistosoma mansoni* interaction," *Invertebrate Survival Journal*, vol. 4, pp. 101–111, 2007.

[16] R. L. Martins-Souza, C. A. J. Pereira, P. M. Z. Coelho, O. A. Martins-Filho, and D. Negrão-Corrêa, "Flow cytometry analysis of the circulating haemocytes from *Biomphalaria glabrata* and Biomphalaria tenagophila following *Schistosoma mansoni* infection," *Parasitology*, vol. 136, pp. 67–76, 2009.

[17] K. K. Sapp and E. S. Loker, "A comparative study of mechanisms underlying digenean-snail specificity: in vitro interactions between hemocytes and digenean larvae," *Journal of Parasitology*, vol. 86, no. 5, pp. 1020–1029, 2000.

[18] K. R. Harris, "The fine structure of encapsulation in *Biomphalaria glabrata*," *Annals of the New York Academy of Sciences*, vol. 266, pp. 446–464, 1975.

[19] P. T. LoVerde, J. Gherson, and C. S. Richards, "Amebocytic accumulations in *Biomphalaria glabrata*: fine structure," *Developmental and Comparative Immunology*, vol. 6, no. 3, pp. 441–449, 1982.

[20] K. J. Lie, K. H. Jeong, and D. Heyneman, "Molluscan host reactions to helminthic infection," in *Protozoa, Arthropods and Invertebrates*, E. J. L. Soulsby, Ed., pp. 211–270, CRC-Press Inc., Boca Raton, Fla, USA, 1987.

[21] M. A. Barracco, A. A. Steil, and R. Gargioni, "Morphological characterization of the hemocytes of the pulmonate snail *Biomphalaria tenagophila*," *Memorias do Instituto Oswaldo Cruz*, vol. 88, no. 1, pp. 73–83, 1993.

[22] L. A. Johnston and T. P. Yoshino, "*Larval Schistosoma* mansoni excretory-secretory glycoproteins (ESPs) bind to hemocytes of *Biomphalaria glabrata* (gastropoda) via surface carbohydrate binding receptors," *Journal of Parasitology*, vol. 87, no. 4, pp. 786–793, 2001.

[23] C. A. J. Pereira, R. L. Martins-Souza, P. M. Z. Coelho, W. S. Lima, and D. Negrão-Corrêa, "Effect of *Angiostrongylus vasorum* infection on *Biomphalaria tenagophila* susceptibility to *Schistosoma mansoni*," *Acta Tropica*, vol. 98, no. 3, pp. 224–233, 2006.

[24] C. L. Ruddell, "The fine structure of oyster agranular amebocytes from regenerating mantle wounds in the Pacific oyster, *Crassostrea gigas*," *Journal of Invertebrate Pathology*, vol. 18, no. 2, pp. 260–268, 1971.

[25] A. K. Sparks, *Invertebrate Pathology: Noncommunicable Diseases*, Academic Press, New York, NY, USA, 1972.

[26] C. J. Bayne, U. K. Hahn, and R. C. Bender, "Mechanisms of molluscan host resistance and of parasite strategies for survival," *Parasitology*, vol. 123, pp. S159–S167, 2001.

[27] T. A. Barçante, J. M. P. Barçante, S. R. C. Dias, and W. S. Lima, "*Angiostrongylus vasorum* (Baillet, 1866) Kamensky, 1905: emergence of third-stage larvae from infected *Biomphalaria glabrata* snails," *Parasitology Research*, vol. 91, no. 6, pp. 471–475, 2003.

[28] J. M. P. Barçante, T. A. Barçante, V. M. Ribeiro et al., "Cytological and parasitological analysis of bronchoalveolar lavage fluid for the diagnosis of *Angiostrongylus vasorum* infection in dogs," *Veterinary Parasitology*, vol. 158, no. 1-2, pp. 93–102, 2008.

[29] J. Guilhon, "Role of the Limacidae in the evolutive cycle of *Angiostrongylus vasorum* (Baillet, 1866)," *Comptes Rendus de l'Académie des Sciences*, vol. 251, pp. 2252–2253, 1960.

[30] J. Guilhon and A. Afghahi, "Larval development of *Angiostrongylus vasorum* (Baillet, 1866) in the body of various species of terrestrial mollusks," *Comptes Rendus de l'Académie des Sciences*, vol. 268, no. 2, pp. 434–436, 1969.

[31] W. P. W. van der Knaap and E. S. Loker, "Immune mechanisms in trematode-snail interactions," *Parasitology Today*, vol. 6, no. 6, pp. 175–182, 1990.

[32] T. A. Barçante, *Aspectos do desenvolvimento de* Angiostrongylus vasorum *(Baillet, 1866) Kamensky, 1905 em* Biomphalaria glabrata *(Say, 1818)*, Doutorado Tese, Departamento de Parasitologia, Universidade Federal de Minas Gerais, 2006.

[33] F. S. D. M. Bezerra, J. A. Nogueira-Machado, M. M. Chaves, R. L. Martins, and P. M. Z. Coelho, "Quantification of the population and phagocytary activity of hemocytes of resistant and susceptible strains of *Biomphalaria glabrata* and *Biomphalaria tenagophila* infected with *Schistosoma mansoni*," *Revista do Instituto de Medicina Tropical de Sao Paulo*, vol. 39, no. 4, pp. 197–201, 1997.

[34] M. Matricon-Gondran and M. Letocart, "Internal defenses of the snail *Biomphalaria glabrata*: I. characterization of hemocytes and fixed phagocytes," *Journal of Invertebrate Pathology*, vol. 74, no. 3, pp. 224–234, 1999.

17

Helminths: Immunoregulation and Inflammatory Diseases—Which Side Are *Trichinella* spp. and *Toxocara* spp. on?

Carmen Aranzamendi,[1] Ljiljana Sofronic-Milosavljevic,[2] and Elena Pinelli[1]

[1] *Centre for Infectious Disease Control Netherlands, National Institute for Public Health and the Environment (RIVM), P.O. Box 1, 3720 BA Bilthoven, The Netherlands*
[2] *Institute for the Application of Nuclear Energy (INEP), University of Belgrade, Banatska 31b, 11080 Belgrade, Serbia*

Correspondence should be addressed to Elena Pinelli; elena.pinelli.ortiz@rivm.nl

Academic Editor: Fabrizio Bruschi

Macropathogens, such as multicellular helminths, are considered masters of immunoregulation due to their ability to escape host defense and establish chronic infections. Molecular crosstalk between the host and the parasite starts immediately after their encounter, which influences the course and development of both the innate and adaptive arms of the immune response. Helminths can modulate dendritic cells (DCs) function and induce immunosuppression which is mediated by a regulatory network that includes regulatory T (Treg) cells, regulatory B (Breg) cells, and alternatively activated macrophages (AAMs). In this way, helminths suppress and control both parasite-specific and unrelated immunopathology in the host such as Th1-mediated autoimmune and Th2-mediated allergic diseases. However, certain helminths favour the development or exacerbation of allergic responses. In this paper, the cell types that play an essential role in helminth-induced immunoregulation, the consequences for inflammatory diseases, and the contrasting effects of *Toxocara* and *Trichinella* infection on allergic manifestations are discussed.

1. Introduction

Immune responses induced by helminths are predominantly of the Th2 type involving cytokines such as interleukin-3 (IL-3), IL-4, IL-5, IL-9, IL-10, and IL-13. These cytokines mediate immune responses typically characterized by increased levels of circulating IgE antibodies, eosinophils, basophils, and mast cells [1]. During infection, the immune system is exposed to different helminth-derived molecules, including proteins, lipids, and glycoconjugates present either at the surface of the worms or in the excretory-secretory (ES) products [2]. Interaction of helminth-derived molecules with host cells can result in a shift of the immune response, from an inflammatory towards an anti-inflammatory type of response. Helminth-derived molecules can modify dendritic cells (DCs) function and downregulate adaptive immune responses, through the induction of a regulatory network that include regulatory T (Treg) cells, alternatively activated macrophages (AAMs), and regulatory B (Breg) cells. The induced immunosuppresive network, together with cytokines produced by diverse hematopoietic and non-hematopoietic cells as integral part of immunoregulatory pathways, appears to be essential for parasite survival and its effect can be extended to other inflammatory disorders such as allergies and autoimmune diseases [3, 4]. However, the association between helminth infections and allergy does not always have an unequivocal outcome. While certain helminth infections protect against allergic diseases (reviewed in [5]), other helminths can exacerbate this immunopathology (reviewed in [6]). Here, the role of DCs, Treg, and other regulatory cells in helminth-induced immunoregulation, the consequences for inflammatory diseases, and the contrasting effects of *Toxocara* and *Trichinella* infections on allergic manifestations are discussed.

2. Dendritic Cells

DCs are sentinels on alert for possible danger signals to immediately activate local innate immune responses and subsequently, after antigen presentation, initiate the proper adaptive immune responses. Interaction with DCs determines the function and cytokine production of lymphocytes. DCs play therefore an essential role in shaping the immune response and controlling the course of infection [7]. These cells are located throughout the body forming a complex network that allows them to communicate with different populations of lymphocytes. Different DC subsets may have distinct locations, where they acquire antigens to be transported to the draining lymph nodes for T-cell priming [8]. DCs as well as other innate immune cells possess various families of pattern recognition receptors (PRRs) such as Toll-like receptors (TLRs), NOD-like receptors, RIG-like receptors, and the C-type lectin receptors (CLRs) that allow them to recognize a great variety of pathogen-associated molecular patterns (PAMPs). After pathogen recognition via various PRR, DCs produce molecules that induce polarization of different types of responsiveness such as Th1-, Th2-, Th17-, or Treg-related. The response of DCs to pathogens is mediated in large part via TLR, with input from other PRR resulting in changes in gene expression that leads to DCs maturation. Maturation of these cells refers to a transition from a resting state into a more dynamic state in which the cells present antigen in the context of MHC, express costimulatory molecules such as CD40, CD80, and CD86, and secrete a broad spectrum of cytokines and chemokines [9]. TLR-mediated responses are controlled mainly by the MyD88-dependent pathway, which is used by all TLR except TLR3- and by the TIR-domain-containing adapter-inducing interferon-β-(TRIF)-dependent pathway, which is used by TLR3 and TLR4 [10]. TLRs have been implicated in the recognition of helminth products by DC. For instance, Lacto-N-fucopentaose III (LNFPIII) produced by trematode *Schistosoma mansoni,* and ES-62, a phosphorylcholine-containing protein secreted by the nematode *Acanthocheilonema viteae,* can condition DCs to induce Th2 responses through TLR4 [11]. Likewise, monoacetylated phosphatidyl serine lipids from schistosomes specifically instruct DCs to preferentially induce IL-10-producing Treg in a TLR2-dependent manner [12]. This was also demonstrated in TLR2-deficient mice that showed a reduced number of CD4+CD25+ Treg cells and immunopathology during schistosomiasis [13]. CLRs on DCs also play an important role in sensing helminth glycans. Studies using schistosomal antigens suggest that helminth glycans may be the conserved molecular pattern that instructs DCs via CLR to drive Th2-polarized responses [14]. Other recent studies demonstrate that host-like glycan antigens expressed by many helminths are recognized by DCs via lectin receptors [2]. Schabussova et al. reported on blood group-like glycans from *T. canis* that bind the lectin DC-SIGN (Dendritic Cell-Specific Intercellular adhesion molecule-3-Grabbing Nonintegrin) [15], which may enable the activation of signal transduction pathways involving Raf-1 and subsequent modulation of DC maturation resulting in skewing towards a Th2 responses [16]. Lewis X antigen, a host-like glycan expressed on the surface of schistosomes in all life stages which is also present in secreted products such as the soluble egg antigens (SEAs), also binds to DC-SIGN [17].

DC maturation is considered to be essential for DCs to induce T-cell responses. It has become clear that DCs responding to helminth products do not mature in the conventional way upon encountering parasitic antigens but acquire a semimatured status and are still capable of inducing T-cell polarization. Several studies support the findings that helminth products fail to directly activate DCs and other studies show that helminth products suppress DC maturation. For instance, SEA suppresses lipopolysaccharide-(LPS-) induced activation of immature murine DCs, as indicated by decreased MHC class II and costimulatory molecules expression in addition to IL-12 production. This resulted in increased LPS-induced production of IL-10 by DC after incubation by SEA [18]. Pretreatment of DCs and macrophages with ES-62 also inhibits their ability to produce IL-12p70 in response to LPS [19]. In another study, a mixture of high molecular weight components from *Ascaris suum* was found to reduce the LPS-induced expression of MHCII, CD80, CD86, and CD40 molecules on mouse CD11c+ DCs and to hampered T-cell proliferative responses *in vitro*. This inhibitory effect was abolished in IL-10-deficient mice [20]. *Fasciola hepatica* tegumental antigen alone did not induce cytokine production or cell surface marker expression on murine DCs; however, it significantly suppressed cytokine production and cell surface marker expression in DCs matured with a range of TLR and non-TLR ligands [21]. *In vitro* studies on the impact of *T. spiralis* excretory/secretory products (TspES) on mice DCs revealed that these parasitic antigens suppress DC maturation induced by LPS derived from different bacteria [22]. In this study, different TLR agonists were used showing that the suppressive effect of TspES on DC maturation is restricted to TLR4. These helminth products were also shown to interfere with the expression of several genes related to the TLR-mediated signal transduction pathways. For rat bone-marrow-derived DC it has been shown that after incubation with TspES these DCs acquire a semimatured status which is reflected in moderate upregulation of CD86, significant upregulation of ICAM1 (Intercellular Adhesion Molecule 1), and no upregulation of MHC II, accompanied by impaired production of IL-12 p70 [23].

3. Regulatory T Cells

Treg cells control peripheral immune responses and are likely to play a central role in autoimmune, infectious, and allergic diseases. Three phenotypes of Treg have been described to date, categorized according to their origin, function, and expression of cell surface markers: natural Treg cells (CD4+CD25+Foxp3+) and inducible Treg cells that include the IL-10-producing Tr1 cells and the Foxp3+ T cells induced in the periphery [24]. In spite of the complexity of regulatory cell types, CD4+CD25+Foxp3+ Treg cells are the most prominent population of immunoregulatory cells known so far to be induced during helminth infections [4].

Early studies had already suggested regulatory T-cell activity during chronic helminth infections in humans. Doetze et al. reported that IL-10 and transforming growth factor-β (TGF-β) production mediated the hyporesponsiveness observed in PBMC from individuals with generalized onchocercosis caused by the filarial nematode *Onchocerca volvulus* [25]. In a study with filariasis patients, lymphedema was associated with a deficiency in the expression of Foxp3, GITR (glucocorticoid-induced tumour-necrosis-factor-receptor-related protein), TGF-β, and CTLA-4 (cytotoxic T-lymphocyte antigen 4), known to be expressed by Treg cells [26], while in children infected with intestinal nematodes (*Ascaris lumbricoides* and *Trichuris trichiura*) high levels of IL-10 and TGF-β in addition to generalized T-cell hyporesponsiveness were found [27, 28]. Likewise, *Schistosome*-infected individuals in Kenya and Gabon had higher CD4+CD25+ and CD4+CD25+Foxp3+ T-cell levels compared with uninfected individuals [29]. One of the studies providing evidence on the suppressive effect of Treg cells from helminth-infected individual is the one reported by Wammes et al. [30]. In this study, carried out in Indonesia, Treg cells from geohelminth-infected individuals were more effective at suppressing proliferation and IFN-γ production by effector T cells in response to malaria antigens and BCG than Treg cells from healthy individuals. A filarial parasite of humans, *Brugia malayi*, was found to secrete TGH-2 (transforming growth factor homologue-2), a homologue of host TGF-β [31]. Since the recombinant TGH-2 can bind to the mammalian TGF-β receptor, it has been suggested that it can promote the generation of regulatory T cells, as it has been shown for mammalian TGF-β. In another study a significant increased expression of Foxp3 and regulatory effector molecules such as TGF-β, CTLA-4, PD-1 (programmed death 1) and ICOS (inducible costimulatory molecule) was found in filaria-infected compared to uninfected individuals in response to live infective-stage larvae or microfilariae of *Brugia malayi* [32].

Various studies on the role of Treg cells in helminth infections have used animal models. In mice, CD25+ Treg cells were shown to restrain the pathology to helminth eggs during schistosome infection [13] and to *Trichuris muris* in the gut [33]. Moreover, depletion of CD25+ Treg cells with combined antibodies to CD25 and GITR resulted in enhanced immunity to filarial nematode *Litomosoides sigmodontis* in mice [34]. Generation of Treg cells with elevated expression of Foxp3 during helminth infection has also been demonstrated. For instance, infection of BALB/c mice with *Brugia pahangi* third-stage larvae (L3) resulted in expansion of a population of CD4+CD25+ T cells which was highly enriched in Foxp3 and IL-10 gene expression [35]. Induction of Treg cells was demonstrated to be necessary to establish a chronic *L. sigmodontis* infection since depletion of Treg cells in susceptible mice prior to infection enhanced parasitic killing and cleared the infection [36]. In chronic infection with the gastrointestinal helminth *Heligmosomoides polygyrus*, it was established that levels of Foxp3 expression within the CD4+ T-cell population of mice mesenteric lymph nodes were significantly increased and that purified CD4+CD25+ Treg cells possess suppressive activity *in vitro* [37, 38].

The effect of TspES on T-cell activation *in vitro* was investigated using splenocytes derived from ovalbumin- (OVA)- TCR transgenic D011.10 mice that were incubated with TspES-pulsed DC+OVA. Results indicate that the presence of TspES resulted in expansion of CD4+CD25+ T cells that express high levels of Foxp3+. These Treg cells were shown to have suppressive activity and to produce TGF-β. Together these results indicate that *T. spiralis* secretion products can induce expansion of functional Treg cells *in vitro* [22].

In a rat model, the infection with *T. spiralis* is accompanied with the increase proportion of Foxp3+ Treg cells [23]. *In vitro* studies showed that DCs stimulated with TspES caused strong Th2 polarization, accompanied by elevated production of the regulatory cytokines IL-10 and TGF-β [23]. However, unlike the mouse model described previously, conditioned rat DCs generated no increase in the proportion of CD4+CD25+Foxp3+ T cells. *In vivo* T-cell priming with TspES stimulated DCs resulted in mixed Th1/Th2 cytokine response, with the dominance of the Th2 type and elevated levels of regulatory cytokines. Significant increase in the proportion of CD4+CD25+Foxp3+ cells was found in spleen cells of recipients that received TspES stimulated DCs compared to the control value obtained from rats that received DCs cultivated in medium only.

4. Other Regulatory Cells

Helminth infections may also lead to expansion of immunoregulatory cells other than Treg cells, including alternatively activated macrophages (AAMs) and regulatory B cells. Signals encountered during migration by developing macrophages determine their function at sites of inflammation or infection. Among these signals, cytokines are responsible for the development of highly divergent macrophage phenotypes: classically activated and AAMs [39]. Prieto-Lafuente et al. reported that the homologues of the mammalian cytokine macrophage migration inhibitory factor (MIF) expressed by *Brugia malayi* synergized with IL-4 to induce the development of suppressive AAMs *in vitro* [40]. One pathway for this effect may be through the MIF-mediated induction of IL-4R expression on macrophages, amplifying in this way the potency of IL-4 itself. Thus, in a Th2 environment, MIF may prevent the classical activation of macrophages. The suppressive effect of AAMs on the immune response is most likely dependent on the expression of arginase-1 (Arg-1) as indicated by studies in which mice macrophages lacking Arg-1 failed to suppress Th2 responses (reviewed in [41]).

B cells possess a variety of immune functions, including production of antibodies, presentation of antigens, and production of cytokines. IL-10-producing regulatory B cells have great potential to regulate T-cell-mediated inflammatory responses [5] and to downmodulate experimental autoimmune encephalomyelitis, collagen-induced arthritis, and inflammatory bowel disease [42]. In addition, in mouse models of chronic parasitic inflammation, such as chronic schistosomiasis, IL-10-producing B cells were also reported to be associated with protection against anaphylaxis [43]. Moreover, *H. polygyrus*-infected mice induced regulatory B cells that can downmodulate both allergy and autoimmunity in an IL-10 independent manner [44].

5. Helminth Infections and Inflammatory Diseases

The helminth-induced immunosuppresive network may not only be beneficial for the parasite, but it can also have beneficial outcomes for the host, reducing allergic and autoimmune diseases [41, 45]. Epidemiological, cross-sectional studies support an inverse correlation between allergic diseases and helminth infection [46, 47] including infections by nematode species like *A. lumbricoides* and *Necator americanus* [48]. An increased skin reactivity to house dust mites was found after antihelminthic treatment against infection with *A. lumbricoides* and *T. trichiura* [49]. Studies performed in animal model system have confirmed that helminth infection can protect against allergic disease and in particular lung-associated inflammation. For instance, *S. mansoni*-infected BALB/c mice were protected against OVA-induced experimental allergic airway inflammation (EAAI) as indicated by reduction of eosinophils in BAL, Th2 cytokine production, OVA-specific IgE levels and reduction of the number of inflammatory cells in lungs. Here, induction of CD4+CD25+Foxp3+ regulatory T cells was independent of IL-10 [50]. Dittrich et al. found that chronic infection with the filarial parasite *Litomosoides sigmodontis* suppressed all pathological features of the OVA-induced EAAI model [51]. Additionally, these authors observed significantly increased numbers of Treg cells in spleen and mediastinal lymph nodes in infected OVA-treated mice compared to OVA-controls animals. Suppression of EAAI during the course of *H. polygyrus* infection was shown to involve the induction of Treg cells [52]. Infection with the same parasite resulted also in the inhibition of allergic response to peanut extract [53].

Several epidemiological studies have investigated the protective effect of parasitic infections in different autoimmune diseases like multiple sclerosis and type 1 diabetes [54]. Studies indicate that persons infected with chronic parasitic worm infections have lower rates of inflammatory bowel disease (IBD) than persons without these infections [55]. Experiments carried out using animal models of human autoimmune diseases have shown that parasites can interfere with autoimmunity. *Schistosoma mansoni* infection has been shown to protect from type 1 diabetes [56] and reduces the severity of EAE [57] while infection with *H. polygyrus* suppresses the experimental colitis [58]. Infection with *L. sigmodontis* prevented diabetes in NOD mice. In this study, protection was associated with increased Th2 responses and Treg cell numbers [59].

The immunomodulatory effect of helminth-derived products has been extensively studied. Table 1 provides an overview of different helminthic antigens with immunoregulatory properties. Findings regarding the use of parasite antigens to suppress experimental inflammatory diseases are summarized in Table 2.

Although the majority of data suggest that infection with helminths is associated with a suppression of allergic and autoimmune responses, some examples provide the opposite view. Epidemiological studies indicate that infection with *Ascaris* spp, *Toxocara* spp, *Fasciola hepatica*, hookworms, or *Enterobius vermicularis* has no protective effects or even enhanced allergic responses (reviewed in [95]).

There are also experimental studies that show that infection with some helminths have a positive association with allergy. A study using a murine model has shown that *T. canis* infection results in exacerbation of EAAI [96]. Other animal experiments provided evidence that parasites like *Nippostrongylus brasiliensis* [97] and *B. malayi* [98] could also induce or exacerbate allergic responses. Exacerbation of anaphylaxis has been shown to occur during *T. spiralis* infection [99]. The links between infections and autoimmunity are complex and there is scarce evidence on the induction or exacerbation of autoimmune responses by helminths [100].

6. Contrasting Effect of *Toxocara* and *Trichinella* Infections on Inflammatory Diseases

Toxocara canis and *Toxocara cati* are roundworms of dogs and cats, respectively, that can also infect humans worldwide. After ingestion of the infectious *Toxocara* eggs, the larvae migrate to the intestine, liver, and lungs. While in dogs and cats under the age of 6 months, the larva migrate back to the intestine; in humans migration continues to other organs where they can persist for many years [101]. *Toxocara* infection results in the induction of Th2 cells that make cytokines such as IL-4, IL-5, and IL-13, which induce responses to the parasite such as increased IgE levels and eosinophilia (reviewed in [6]). *Trichinella spiralis* is also a roundworm that infects different mammals including humans and mice. After ingestion of *Trichinella* infected meat, the larvae migrate to the intestine and matures to the adult stage, the parasites mate, and finally the newborn larvae (NBL) migrate to striated muscle cells where they become encysted. Infection with *T. spiralis* is characterized by the induction of a Th1 type of response at the beginning of the intestinal phase. When the NBL disseminate, a dominant Th2 type of response develops which is essential for parasite expulsion [102]. Ingestion of both *Toxocara* spp. and *Trichinella* spp. commonly results in chronic infections. Interestingly these helminths have a contrasting effect on inflammatory diseases, while infections with *Trichinella* spp. can suppress (reviewed in [103]) *Toxocara* spp. exacerbate inflammatory diseases [6]. Studies using animal models for human autoimmune and allergic diseases indicate that *Trichinella* infection ameliorates these immune disorders (Table 3). Khan et al. showed that *T. spiralis* infection reduces the severity of dinitrobenzenesulphonic-acid- (DNBS-) induced colitis in C57BL/6 mice [88]. Motomura et al. demonstrated that in addition to the protection exerted by the actual infection, rectal submucosal administration of *T. spiralis* crude muscle larvae antigen can also protect [78]. *T. spiralis* infection also ameliorated autoimmune diabetes in NOD mice [89] and modulated severity of the disease in the experimental model of multiple sclerosis (MS), namely, experimental autoimmune encephalomyelitis (EAE) in Dark Agouti rats in a dose-dependent manner [90]. In this study severity of EAE as

TABLE 1: Helminth-derived antigens with immunoregulatory properties.

Helminth	Antigen	Immunoregulatory mechanism	References
Schistosoma mansoni	LNFPIII (lacto-N-fucopentaoseIII)/SEA (soluble egg antigen)	Interact with TLR4 to produce Th2 polarizing DCs	[60]
Schistosoma mansoni	Schistosome lysophosphatidylserine	Interact with TLR2 to induce Treg polarizing DCs	[61]
Acanthocheilonema viteae	ES-62	Exert immunomodulatory effects on macrophages and DCs by a TLR4-dependent mechanism with consequent Th2 polarisation	[62–64]
Nippostrongylus brasiliensis	Excretory-secretory antigen (NES)	Potently induce Th2 type of response via DC	[65]
Brugia malayi adult	Cystatins (cysteine protease inhibitors) CIP-2	Interfere with antigen processing in human cells and inhibits B cells	[66, 67]
	Cytokine homologue MIF-1/2	Alternatively activate macrophages	[68]
Brugia malayi microfilariae	Serpins (serine protease inhibitors) SPN-2	Block neutrophil protease and promote Th1 type of response	[69]
Brugia malayi L3 larvae	ALT-1/2 proteins	Inhibit macrophage resistance and present good filarial vaccine candidate	[70]
Toxocara canis	TES32—C type lectin (CTL)	Inhibit TLR responses on DC and compete with host lectins for ligands, thereby blocking host immunity	[71, 72]
Heligmosomoides polygyrus	Excretory-secretory antigen (HES)	Induce regulatory T cells through TGF-βR	[37]
Teladorsagia circumcincta	Excretory-secretory antigen	Induce generation of Foxp3+ regulatory T cells through TGF-β mimicking effect	[73]
Trichinella spiralis	Adult excretory-secretory antigen (AdES); newborn larvae antigen (NBL); crude muscle larvae antigen (MLCr)	All antigens from different life stages induce polarization towards mixed Th1/Th2 with predominance of Th2 response, via semimatured DC	[74]
	Excretory secretory muscle larvae antigen	Induce mixed Th1/Th2 response with the predominance of Th2 component and elicit regulatory arm of immune response	[75]
	Excretory secretory muscle larvae antigen	Interfere with LPS-induced DC maturation and induce expansion of Foxp3+ regulatory T cells	[22]
Fasciola hepatica	Thioredoxin peroxides	Alternatively activated macrophages	[76]

TABLE 2: Suppression of experimental inflammatory diseases by parasite-derived antigens.

Helminth	Antigen	Model	Reference
Trichinella spiralis	Soluble antigens of muscle larvae	EAE	[77]
		DNBS-induced colitis	[78]
Trichuris suis	Soluble antigens of adult worm	EAE	[77]
Ancylostoma ceylanicum	Soluble and excretory-secretory antigens of adult worm	DSS-induced colitis	[79]
Hymenolepis diminuta	Soluble antigens of adult worm	DNBS-induced colitis	[80]
Heligmosomoides polygyrus	Excretory-secretory antigens (HES) of adult worm	EAAI	[73]
Ancylostoma caninum	Excretory-secretory antigens of adult worm	TNBS-induced colitis	[81]
Acanthocheilonema vitae	ES-62	Collagen-induced arthritis	[82]
Schistosoma mansoni	Soluble antigens of adult worm	TNBS-induced colitis	[81]
	SEA and soluble adult worm antigen	T1D	[83]
	Recombinant proteins (Sm22·6, Sm29) and soluble adult worm fraction (PIII)	EAAI	[84]
Schistosoma japonicum	SEA	EAAI	[85]
Nippostrongylus brasiliensis	Excretory-secretory antigens (NES) of adult worm	EAAI	[86]
Ascaris suum	Soluble antigens of adult worm	EAAI	[87]

EAE: experimental autoimmune encephalomyelitis; DNBS: dinitrobenzene sulfonic acid; TNBS: trinitrobenzene sulfonic acid; DSS: dextran sodium sulfate; T1D: type 1 diabetes; EAAI: experimental allergic airway inflammation.

TABLE 3: Experimental models of Th2-mediated inflammatory diseases successfully treated by *Trichinella* infection or administration of *Trichinella* antigens.

Trichinella spp. or their products	Experimental disease model	Reference
T. spiralis	Exp. colitis	[88]
	T1D	[89]
	EAE	[90]
	EAE	[91]
	EAAI	[92]
	EAAI	[22]
T. pseudospiralis	EAE	[93]
	EAE	[94]
T. spiralis crude muscle larvae antigen	Exp. colitis	[78]
	EAE	[77]

Exp. colitis: experimental colitis; T1D: type 1 diabetes; EAE: experimental autoimmune encephalomyelitis; EAAI: experimental allergic airway inflammation.

judged by cumulative disease index, maximal clinical score, duration of illness, and the number of mononuclear cells infiltrating the spinal cord in *T. spiralis* infected animals were all reduced in comparison to the uninfected EAE-induced group. In a following study, these authors reported that alleviation of the disease in infected-EAE rats coincided with reduced IFN-γ and IL-17 production and increased IL-4, IL-10, and TGF-β production. They suggested that mechanisms underlying the observed beneficial effect include Th2 and regulatory responses provoked by the parasite. Transfer of T-cell-enriched spleen cells from *T. spiralis*-infected rats that contained a higher proportion of CD4+CD25+Foxp3+ regulatory T cells into rats in which EAE was induced caused amelioration of EAE, which indirectly points to the role of Treg in restraining inflammatory conditions [91]. Boles et al. have shown with another *Trichinella* species, namely, with *Trichinella pseudospiralis*, that infection results in suppression of MS in the rat [93]. These authors used this model to compare the anti-inflammatory effects of the intestinal and late migratory phases of *T. pseudospiralis* infection on development of myelin-basic-protein- (MBP-) induced MS-like debilitation. Findings from this study indicate that the late migratory phase of infection which occurred during the peak of MBP-induced debilitation significantly improved performance scores in mobility, coordination, and strength. Wu et al. also reported on amelioration of clinical severity and delayed onset of EAE after *T. pseudospiralis* infection. This effect was associated with suppression of Th17 and Th1 responses induced by infection [94]. *Trichinella pseudospiralis* is markedly different from *T. spiralis* in that it is smaller in size and that the muscle stage larvae are not surrounded by a capsule [104]. Whether the mechanisms involved in immunosuppression varies depending on *Trichinella* species remains to be investigated.

Infection with *T. spiralis* can also ameliorate EAAI [92]. In this study, the concentrations of IL-10 and TGF-β were significantly increased and the recruitment of Treg into draining lymph nodes was elevated as the result of *T. spiralis* infection. This protective effect has been recently shown to occur during acute as well as chronic phases of *Trichinella* infection [105]. Protection against EAAI to OVA was stronger during the chronic phase of infection and associated with increased numbers of splenic CD4+CD25+Foxp3+ Treg cells with suppressive activity. Adoptive transfer of CD4+ T cells from chronically infected mice with elevated numbers of Treg cells in the spleen induced partial protection against EAAI [105]. The possible mechanisms by which helminths or their products could inhibit allergic responses are depicted in Figure 1.

In contrast to the suppressive effect of *Trichinella* infections on allergic diseases experimental as well as epidemiological studies indicate that *Toxocara* infections are risk factors for allergies, including allergic asthma [6].

Studies using murine models for toxocariosis indicate that infection with *T. canis* leads to persistent pulmonary inflammation, eosinophilia, increase levels of circulating IgE, airway hyperreactivity, and production of Th-2 type cytokines. Pulmonary inflammation has been shown to develop as soon as 48 hours after infection, it occurs in a dose-dependent manner, and it can persist up to 2 or 3 months. Eosinophil counts also increase in the bronchoalveolar lavage (BAL) of *Toxocara*-infected mice [106, 107]. Relative quantification of cytokine expression in lungs of mice infected with different *T. canis* doses showed that while a proportional increased expression of the IL-4, IL-5, and IL-10 transcripts was observed, the expression of IFN-γ was not different from that of uninfected controls [107]. Results from this study indicate that infection of mice with *T. canis* results in a dominant Th2 type of immune response, independent of the inoculum size [108]. In addition, infection of BALB/c mice with 1,000 *T. canis* embryonated eggs results in hyperreactivity of the airways that persisted up to 30 days p.i. Evaluation of parasite burden revealed that few *T. canis* larvae were still present in the lungs of infected mice at 60 days p.i. which could explain the persistent pulmonary inflammation observed in these mice [107].

Common features between allergic asthma and toxocariosis are the induction of a Th2-cell mediated immune response including the production of high levels of IgE, and eosinophilia. In addition, infection with *Toxocara* spp. shares common clinical features with allergic asthma such as inflammation of the airways accompanied with wheezing, coughs, mucus hypersecretion, and bronchial hyperreactivity. In order to study the effect of *Toxocara* infection on allergic manifestations two murine models were combined, namely, the murine model for toxocariosis described above and a murine model for allergic airway inflammation. For this study BALB/c mice were infected with 500 embryonated *T. canis* eggs and exposed to OVA sensitization followed by OVA-challenge. Results indicate that infection with *T. canis* in combination with OVA treatment led to exacerbation of pulmonary inflammation; eosinophilia; airway hyperresponsiveness; increase of OVA specific and total IgE; increased expression of IL-4 compared to mice that were only *T. canis* infected or OVA treated. The observed exacerbation of EAAI was independent of the timing of infection in relation to

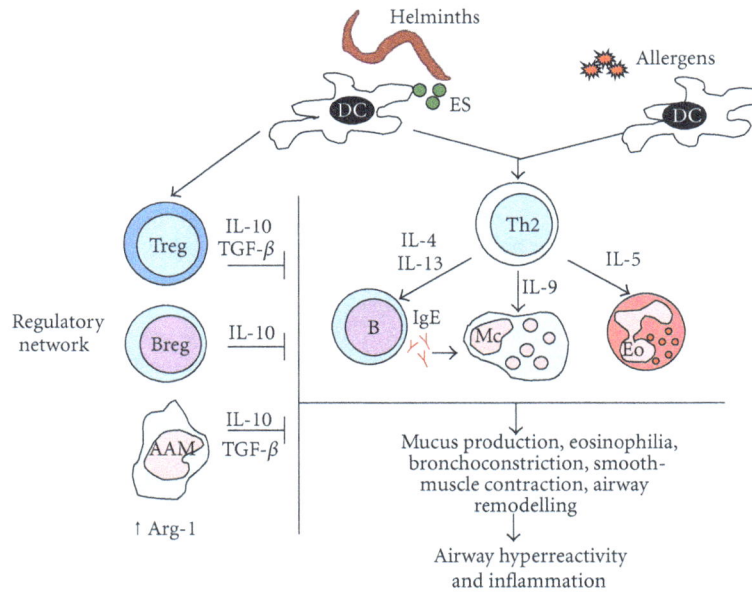

FIGURE 1: Mechanisms involved in immunosuppression induced by helminths and its effect on allergic responses. Helminths can modulate dendritic cells (DCs) function and induce regulatory T (Treg) cells. Other cells from the regulatory network include regulatory B (Breg) cells and alternatively activated macrophages (AAMs). These cells create an immunosuppressive (⊣) environment in which increased expression of arginase 1 (Arg 1) in AAMs and the production of the cytokines IL-10 and TGF-β play an essential role in reducing allergic effector mechanisms.

allergen exposure. In conclusion, infection with *T. canis* leads to exacerbation of EAAI [96].

Several factors may influence the differential effect of helminth infections on allergic diseases [6]. One of these factors is whether the host is definitive or accidental. The normal or definitive hosts for *T. canis* are dogs whereas humans are accidental hosts for this parasite. In an accidental host the parasite does not usually develop to the adult stage and in case of *Toxocara* spp. the continuous migration of the larvae through different organs including the lungs can cause more damage comparing to what happens in a definitive host where migration is transitory. *T. spiralis* can infect many different mammals including mice and humans in which the parasite completes its life cycle. The infected mammal is therefore a definitive host for this helminth. And although *T. spiralis* pass through the lung microvascular system on its way to the skeletal muscle, it is a rapid process in which the larvae are usually not trapped in the lungs [109]. It is likely that there are differences between parasites of humans such as *T. spiralis* that have evolved with their host and have developed strategies to survive without causing much damage compared to parasites such as *Toxocara* spp. for which humans are accidental host [6, 110].

In conclusion, helminths induce an anti-inflammatory response, which could ameliorate inflammatory diseases; however, this is not a universal property of all helminths and different factors such as the helminth species, and whether the host is definitive or accidental, the parasite load and acute versus chronic infections may all influence the overall effect of helminth infections on inflammatory diseases. Identification of the helminth molecules that induce immunosuppression

and elucidation of the mechanisms involved is essential for the development of alternative strategies for prevention and/or treatment of inflammatory diseases.

References

[1] J. E. Allen and R. M. Maizels, "Diversity and dialogue in immunity to helminths," *Nature Reviews Immunology*, vol. 11, no. 6, pp. 375–388, 2011.

[2] I. van Die and R. D. Cummings, "Glycan gimmickry by parasitic helminths: a strategy for modulating the host immune response?" *Glycobiology*, vol. 20, no. 1, pp. 2–12, 2010.

[3] R. Maizels and M. Yazdanbakhsh, "T-cell regulation in helminth parasite infections: implications for inflammatory diseases," *Chemical Immunology and Allergy*, vol. 94, pp. 112–123, 2008.

[4] M. D. Taylor, N. van der Werf, and R. M. Maizels, "T cells in helminth infection: the regulators and the regulated," *Trends in Immunology*, vol. 33, no. 4, pp. 181–189, 2012.

[5] H. H. Smits, B. Everts, F. C. Hartgers, and M. Yazdanbakhsh, "Chronic helminth infections protect against allergic diseases by active regulatory processes," *Current Allergy and Asthma Reports*, vol. 10, no. 1, pp. 3–12, 2010.

[6] E. Pinelli and C. Aranzamendi, "Toxocara infection and its association with allergic manifestations," *Endocrine Metabolic Immune & Disorders Drug Targets*, vol. 12, no. 1, pp. 33–44, 2012.

[7] H. Moll, "Dendritic cells and host resistance to infection," *Cellular Microbiology*, vol. 5, no. 8, pp. 493–500, 2003.

[8] G. T. Belz and S. L. Nutt, "Transcriptional programming of the dendritic cell network," *Nature Reviews*, vol. 12, no. 2, pp. 101–113, 2012.

[9] O. Joffre, M. A. Nolte, R. Spörri, and C. R. E. Sousa, "Inflammatory signals in dendritic cell activation and the induction of adaptive immunity," *Immunological Reviews*, vol. 227, no. 1, pp. 234–247, 2009.

[10] T. Kawai and S. Akira, "The role of pattern-recognition receptors in innate immunity: update on toll-like receptors," *Nature Immunology*, vol. 11, no. 5, pp. 373–384, 2010.

[11] L. Carvalho, J. Sun, C. Kane, F. Marshall, C. Krawczyk, and E. J. Pearce, "Review series on helminths, immune modulation and the hygiene hypothesis: mechanisms underlying helminth modulation of dendritic cell function," *Immunology*, vol. 126, no. 1, pp. 28–34, 2009.

[12] D. Van der Kleij, E. Latz, J. F. H. M. Brouwers et al., "A novel host-parasite lipid cross-talk. Schistosomal lysophosphatidylserine activates toll-like receptor 2 and affects immune polarization," *The Journal of Biological Chemistry*, vol. 277, no. 50, pp. 48122–48129, 2002.

[13] L. E. Layland, R. Rad, H. Wagner, and C. U. Prazeres da Costa, "Immunopathology in schistosomiasis is controlled by antigen-specific regulatory T cells primed in the presence of TLR2," *European Journal of Immunology*, vol. 37, no. 8, pp. 2174–2184, 2007.

[14] B. Everts, H. H. Smits, C. H. Hokke, and M. Yazdanbakhsh, "Helminths and dendritic cells: sensing and regulating via pattern recognition receptors, Th2 and Treg responses," *European Journal of Immunology*, vol. 40, no. 6, pp. 1525–1537, 2010.

[15] I. Schabussova, H. Amer, I. van Die, P. Kosma, and R. M. Maizels, "O-Methylated glycans from *Toxocara* are specific targets for antibody binding in human and animal infections," *International Journal for Parasitology*, vol. 37, no. 1, pp. 97–109, 2007.

[16] U. Švajger, M. Anderluh, M. Jeras, and N. Obermajer, "C-type lectin DC-SIGN: an adhesion, signalling and antigen-uptake molecule that guides dendritic cells in immunity," *Cellular Signalling*, vol. 22, no. 10, pp. 1397–1405, 2010.

[17] S. Meyer, E. Van Liempt, A. Imberty et al., "DC-SIGN mediates binding of dendritic cells to authentic pseudo-Lewis Y glycolipids of *Schistosoma mansoni* cercariae, the first parasite-specific ligand of DC-SIGN," *The Journal of Biological Chemistry*, vol. 280, no. 45, pp. 37349–37359, 2005.

[18] C. M. Kane, L. Cervi, J. Sun et al., "Helminth antigens modulate TLR-initiated dendritic cell activation," *Journal of Immunology*, vol. 173, no. 12, pp. 7454–7461, 2004.

[19] H. S. Goodridge, W. Harnett, F. Y. Liew, and M. M. Harnett, "Differential regulation of interleukin-12 p40 and p35 induction via Erk mitogen-activated protein kinase-dependent and -independent mechanisms and the implications for bioactive IL-12 and IL-23 responses," *Immunology*, vol. 109, no. 3, pp. 415–425, 2003.

[20] S. R. Silva, J. F. Jacysyn, M. S. Macedo, and E. L. Faquin-Mauro, "Immunosuppressive components of *Ascaris suum* down-regulate expression of costimulatory molecules and function of antigen-presenting cells via an IL-10-mediated mechanism," *European Journal of Immunology*, vol. 36, no. 12, pp. 3227–3237, 2006.

[21] C. M. Hamilton, D. J. Dowling, C. E. Loscher, R. M. Morphew, P. M. Brophy, and S. M. O'Neill, "The *Fasciola hepatica* tegumental antigen suppresses dendritic cell maturation and function," *Infection and Immunity*, vol. 77, no. 6, pp. 2488–2498, 2009.

[22] C. Aranzamendi, F. Fransen, M. Langelaar et al., "*Trichinella spiralis*-secreted products modulate DC functionality and expand regulatory T cells in vitro," *Parasite Immunology*, vol. 34, no. 4, pp. 210–223, 2012.

[23] N. Ilic, M. Colic, A. Gruden-Movsesijan, I. Majstorovic, S. Vasilev, and L. Sofronic-Milosavljevic, "Characterization of rat bone marrow dendritic cells initially primed by *Trichinella spiralis* antigens," *Parasite Immunology*, vol. 30, no. 9, pp. 491–495, 2008.

[24] Y. Belkaid and W. Chen, "Regulatory ripples," *Nature Immunology*, vol. 11, no. 12, pp. 1077–1078, 2010.

[25] A. Doetze, J. Satoguina, G. Burchard et al., "Antigen-specific cellular hyporesponsiveness in a chronic human helminth infection is mediated by T(h)3/T(r)1-type cytokines IL-10 and transforming growth factor-β but not by a T(h)1 to T(h)2 shift," *International Immunology*, vol. 12, no. 5, pp. 623–630, 2000.

[26] S. Babu, S. Q. Bhat, N. P. Kumar et al., "Filarial lymphedema is characterized by antigen-specific Th1 and Th17 proinflammatory responses and a lack of regulatory T cells," *PLoS Neglected Tropical Diseases*, vol. 3, no. 4, article no. e420, 2009.

[27] J. D. Turner, J. A. Jackson, H. Faulkner et al., "Intensity of intestinal infection with multiple worm species is related to regulatory cytokine output and immune hyporesponsiveness," *Journal of Infectious Diseases*, vol. 197, no. 8, pp. 1204–1212, 2008.

[28] C. A. Figueiredo, M. L. Barreto, L. C. Rodrigues et al., "Chronic intestinal helminth infections are associated with immune hyporesponsiveness and induction of a regulatory network," *Infection and Immunity*, vol. 78, no. 7, pp. 3160–3167, 2010.

[29] K. Watanabe, P. N. M. Mwinzi, C. L. Black et al., "T regulatory cell levels decrease in people infected with *Schistosoma mansoni* on effective treatment," *American Journal of Tropical Medicine and Hygiene*, vol. 77, no. 4, pp. 676–682, 2007.

[30] L. J. Wammes, F. Hamid, A. E. Wiria et al., "Regulatory T cells in human geohelminth infection suppress immune responses to BCG and *Plasmodium falciparum*," *European Journal of Immunology*, vol. 40, no. 2, pp. 437–442, 2010.

[31] N. Gomez-Escobar, W. F. Gregory, and R. M. Maizels, "Identification of tgh-2, a filarial nematode homolog of *Caenorhabditis elegans* daf-7 and human transforming growth factor β, expressed in microfilarial and adult stages of *Brugia malayi*," *Infection and Immunity*, vol. 68, no. 11, pp. 6402–6410, 2000.

[32] S. Babu, C. P. Blauvelt, V. Kumaraswami, and T. B. Nutman, "Regulatory networks induced by live parasites impair both Th1 and Th2 pathways in patent lymphatic filariasis: implications for parasite persistence," *Journal of Immunology*, vol. 176, no. 5, pp. 3248–3256, 2006.

[33] R. D'Elia, J. M. Behnke, J. E. Bradley, and K. J. Else, "Regulatory T cells: a role in the control of helminth-driven intestinal pathology and worm survival," *Journal of Immunology*, vol. 182, no. 4, pp. 2340–2348, 2009.

[34] M. D. Taylor, L. LeGoff, A. Harris, E. Malone, J. E. Allen, and R. M. Maizels, "Removal of regulatory T cell activity reverses hyporesponsiveness and leads to filarial parasite clearance in vivo," *Journal of Immunology*, vol. 174, no. 8, pp. 4924–4933, 2005.

[35] V. Gillan and E. Devaney, "Regulatory T cells modulate Th2 responses induced by *Brugia pahangi* third-stage larvae," *Infection and Immunity*, vol. 73, no. 7, pp. 4034–4042, 2005.

[36] M. D. Taylor, N. van der Werf, A. Harris et al., "Early recruitment of natural CD4+Foxp3+ Treg cells by infective larvae determines the outcome of filarial infection," *European Journal of Immunology*, vol. 39, no. 1, pp. 192–206, 2009.

[37] C. A. M. Finney, M. D. Taylor, M. S. Wilson, and R. M. Maizels, "Expansion and activation of CD4+CD25+ regulatory T cells in *Heligmosomoides polygyrus* infection," *European Journal of Immunology*, vol. 37, no. 7, pp. 1874–1886, 2007.

[38] S. Rausch, J. Huehn, D. Kirchhoff et al., "Functional analysis of effector and regulatory T cells in a parasitic nematode infection," *Infection and Immunity*, vol. 76, no. 5, pp. 1908–1919, 2008.

[39] S. Gordon, "Alternative activation of macrophages," *Nature Reviews*, vol. 3, no. 1, pp. 23–35, 2003.

[40] L. Prieto-Lafuente, W. F. Gregory, J. E. Allen, and R. M. Maizels, "MIF homologues from a filarial nematode parasite synergize with IL-4 to induce alternative activation of host macrophages," *Journal of Leukocyte Biology*, vol. 85, no. 5, pp. 844–854, 2009.

[41] R. M. Maizels, E. J. Pearce, D. Artis, M. Yazdanbakhsh, and T. A. Wynn, "Regulation of pathogenesis and immunity in helminth infections," *Journal of Experimental Medicine*, vol. 206, no. 10, pp. 2059–2066, 2009.

[42] S. Fillatreau, D. Gray, and S. M. Anderton, "Not always the bad guys: B cells as regulators of autoimmune pathology," *Nature Reviews Immunology*, vol. 8, no. 5, pp. 391–397, 2008.

[43] N. E. Mangan, R. E. Fallon, P. Smith, N. Van Rooijen, A. N. McKenzie, and P. G. Fallon, "Helminth infection protects mice from anaphylaxis via IL-10-producing B cells," *Journal of Immunology*, vol. 173, no. 10, pp. 6346–6356, 2004.

[44] M. S. Wilson, M. D. Taylor, M. T. O'Gorman et al., "Helminth-induced CD19+CD23hi B cells modulate experimental allergic and autoimmune inflammation," *European Journal of Immunology*, vol. 40, no. 6, pp. 1682–1696, 2010.

[45] Y. Osada and T. Kanazawa, "Parasitic helminths: new weapons against immunological disorders," *Journal of Biomedicine & Biotechnology*, vol. 2010, Article ID 743758, 9 pages, 2010.

[46] C. Flohr, R. J. Quinnell, and J. Britton, "Do helminth parasites protect against atopy and allergic disease?" *Clinical and Experimental Allergy*, vol. 39, no. 1, pp. 20–32, 2009.

[47] W. Harnett and M. M. Harnett, "Parasitic nematode modulation of allergic disease," *Current Allergy and Asthma Reports*, vol. 8, no. 5, pp. 392–397, 2008.

[48] F. G. Selassie, R. H. Stevens, P. Cullinan et al., "Total and specific IgE (house dust mite and intestinal helminths) in asthmatics and controls from Gondar, Ethiopia," *Clinical and Experimental Allergy*, vol. 30, no. 3, pp. 356–358, 2000.

[49] A. M. J. Van Den Biggelaar, L. C. Rodrigues, R. Van Ree et al., "Long-term treatment of intestinal helminths increases mite skin-test reactivity in Gabonese schoolchildren," *Journal of Infectious Diseases*, vol. 189, no. 5, pp. 892–900, 2004.

[50] L. G. G. Pacífico, F. A. V. Marinho, C. T. Fonseca et al., "*Schistosoma mansoni* antigens modulate experimental allergic asthma in a murine model: a major role for CD4+ CD25+ Foxp3 + T cells independent of interleukin-10," *Infection and Immunity*, vol. 77, no. 1, pp. 98–107, 2009.

[51] A. M. Dittrich, A. Erbacher, S. Specht et al., "Helminth infection with *Litomosoides sigmodontis* induces regulatory T cells and inhibits allergic sensitization, airway inflammation, and hyperreactivity in a murine asthma model," *Journal of Immunology*, vol. 180, no. 3, pp. 1792–1799, 2008.

[52] M. S. Wilson, M. D. Taylor, A. Balic, C. A. M. Finney, J. R. Lamb, and R. M. Maizels, "Suppression of allergic airway inflammation by helminth-induced regulatory T cells," *Journal of Experimental Medicine*, vol. 202, no. 9, pp. 1199–1212, 2005.

[53] M. E. H. Bashir, P. Andersen, I. J. Fuss, H. N. Shi, and C. Nagler-Anderson, "An enteric helminth infection protects against an allergic response to dietary antigen," *Journal of Immunology*, vol. 169, no. 6, pp. 3284–3292, 2002.

[54] H. Okada, C. Kuhn, H. Feillet, and J. F. Bach, "The 'hygiene hypothesis' for autoimmune and allergic diseases: an update," *Clinical and Experimental Immunology*, vol. 160, no. 1, pp. 1–9, 2010.

[55] J. V. Weinstock and D. E. Elliott, "Helminths and the IBD hygiene hypothesis," *Inflammatory Bowel Diseases*, vol. 15, no. 1, pp. 128–133, 2009.

[56] A. Cooke, P. Tonks, F. M. Jones et al., "Infection with *Schistosoma mansoni* prevents insulin dependent diabetes mellitus in non-obese diabetic mice," *Parasite Immunology*, vol. 21, no. 4, pp. 169–176, 1999.

[57] A. C. La Flamme, K. Ruddenklau, and B. T. Bäckström, "Schistosomiasis decreases central nervous system inflammation and alters the progression of experimental autoimmune encephalomyelitis," *Infection and Immunity*, vol. 71, no. 9, pp. 4996–5004, 2003.

[58] D. E. Elliott, T. Setiawan, A. Metwali, A. Blum, J. F. Urban, and J. V. Weinstock, "*Heligmosomoides polygyrus* inhibits established colitis in IL-10-deficient mice," *European Journal of Immunology*, vol. 34, no. 10, pp. 2690–2698, 2004.

[59] M. P. Hübner, J. Thomas Stocker, and E. Mitre, "Inhibition of type 1 diabetes in filaria-infected non-obese diabetic mice is associated with a T helper type 2 shift and induction of FoxP3+ regulatory T cells," *Immunology*, vol. 127, no. 4, pp. 512–522, 2009.

[60] P. G. Thomas, M. R. Carter, O. Atochina et al., "Maturation of dendritic cell 2 phenotype by a helminth glycan uses a toll-like receptor 4-dependent mechanism," *Journal of Immunology*, vol. 171, no. 11, pp. 5837–5841, 2003.

[61] D. Van der Kleij, E. Latz, J. F. H. M. Brouwers et al., "A novel host-parasite lipid cross-talk. Schistosomal lysophosphatidylserine activates toll-like receptor 2 and affects immune polarization," *The Journal of Biological Chemistry*, vol. 277, no. 50, pp. 48122–48129, 2002.

[62] H. S. Goodridge, E. H. Wilson, W. Harnett, C. C. Campbell, M. M. Harnett, and F. Y. Liew, "Modulation of macrophage cytokine production by ES-62, a secreted product of the filarial nematode *Acanthocheilonema viteae*," *Journal of Immunology*, vol. 167, no. 2, pp. 940–945, 2001.

[63] M. Whelan, M. M. Harnett, K. M. Houston, V. Patel, W. Harnett, and K. P. Rigley, "A filarial nematode-secreted product signals dendritic cells to acquire a phenotype that drives development of Th2 cells," *Journal of Immunology*, vol. 164, no. 12, pp. 6453–6460, 2000.

[64] H. S. Goodridge, F. A. Marshall, K. J. Else et al., "Immunomodulation via novel use of TLR4 by the filarial nematode phosphorylcholine-containing secreted product, ES-62," *Journal of Immunology*, vol. 174, no. 1, pp. 284–293, 2005.

[65] M. J. Holland, Y. M. Harcus, P. L. Riches et al., "Proteins secreted by the parasitic nematode *Nippostrongylus brasiliensis* act as adjuvants for Th2 responses," *European Journal of Immunology*, vol. 30, no. 7, pp. 1977–1987, 2000.

[66] W. F. Gregory and R. M. Maizels, "Cystatins from filarial parasites: evolution, adaptation and function in the host-parasite relationship," *International Journal of Biochemistry and Cell Biology*, vol. 40, no. 6-7, pp. 1389–1398, 2008.

[67] B. Manoury, W. F. Gregory, R. M. Maizels, and C. Watts, "Bm-CPI-2, a cystatin homolog secreted by the filarial parasite *Brugia*

malayi, inhibits class II MHC-restricted antigen processing," *Current Biology*, vol. 11, no. 6, pp. 447–451, 2001.

[68] F. H. Falcone, P. Loke, X. Zang, A. S. MacDonald, R. M. Maizels, and J. E. Allen, "A *Brugia malayi* homolog of macrophage migration inhibitory factor reveals an important link between macrophages and eosinophil recruitment during nematode infection," *Journal of Immunology*, vol. 167, no. 9, pp. 5348–5354, 2001.

[69] X. Zang and R. M. Maizels, "Serine proteinase inhibitors from nematodes and the arms race between host and pathogen," *Trends in Biochemical Sciences*, vol. 26, no. 3, pp. 191–197, 2001.

[70] W. F. Gregory, A. K. Atmadja, J. E. Allen, and R. M. Maizels, "The abundant larval transcript-1 and -2 genes of *Brugia malayi* encode stage-specific candidate vaccine antigens for filariasis," *Infection and Immunity*, vol. 68, no. 7, pp. 4174–4179, 2000.

[71] A. Loukas, A. Doedens, M. Hintz, and R. M. Maizels, "Identification of a new C-type lectin, TES-70, secreted by infective larvae of *Toxocara canis*, which binds to host ligands," *Parasitology*, vol. 121, part 5, pp. 545–554, 2000.

[72] A. Loukas and R. M. Maizels, "Helminth C-type lectins and host-parasite interactions," *Parasitology Today*, vol. 16, no. 8, pp. 333–339, 2000.

[73] J. R. Grainger, K. A. Smith, J. P. Hewitson et al., "Helminth secretions induce de novo T cell Foxp3 expression and regulatory function through the TGF-β pathway," *Journal of Experimental Medicine*, vol. 207, no. 11, pp. 2331–2341, 2010.

[74] N. Ilic, J. J. Worthington, A. Gruden-Movsesijan et al., "*Trichinella spiralis* antigens prime mixed Th1/Th2 response but do not induce de novo generation of Foxp3+ T cells in vitro," *Parasite Immunology*, vol. 33, no. 10, pp. 572–582, 2011.

[75] A. Gruden-Movsesijan, N. Ilic, M. Colic et al., "The impact of *Trichinella spiralis* excretory-secretory products on dendritic cells," *Comparative Immunology, Microbiology and Infectious Diseases*, vol. 34, no. 5, pp. 429–439, 2011.

[76] S. Donnelly, S. M. O'Neill, M. Sekiya, G. Mulcahy, and J. P. Dalton, "Thioredoxin peroxidase secreted by *Fasciola hepatica* induces the alternative activation of macrophages," *Infection and Immunity*, vol. 73, no. 1, pp. 166–173, 2005.

[77] L. M. Kuijk, E. J. Klaver, G. Kooij et al., "Soluble helminth products suppress clinical signs in murine experimental autoimmune encephalomyelitis and differentially modulate human dendritic cell activation," *Molecular Immunology*, vol. 51, no. 2, pp. 210–218, 2012.

[78] Y. Motomura, H. Wang, Y. Deng, R. T. El-Sharkawy, E. F. Verdu, and W. I. Khan, "Helminth antigen-based strategy to ameliorate inflammation in an experimental model of colitis," *Clinical and Experimental Immunology*, vol. 155, no. 1, pp. 88–95, 2009.

[79] G. G. Cancado, J. A. Fiuza, N. C. de Paiva et al., "Hookworm products ameliorate dextran sodium sulfate-induced colitis in BALB/c mice," *Inflammatory Bowel Diseases*, vol. 17, no. 11, pp. 2275–2286, 2011.

[80] M. J. G. Johnston, A. Wang, M. E. D. Catarino et al., "Extracts of the rat tapeworm, *Hymenolepis diminuta*, suppress macrophage activation in vitro and alleviate chemically induced colitis in mice," *Infection and Immunity*, vol. 78, no. 3, pp. 1364–1375, 2010.

[81] N. E. Ruyssers, B. Y. De Winter, J. G. De Man et al., "Therapeutic potential of helminth soluble proteins in TNBS induced colitis in mice," *Inflammatory Bowel Diseases*, vol. 15, no. 4, pp. 491–500, 2009.

[82] I. B. McInnes, B. P. Leung, M. Harnett, J. A. Gracie, F. Y. Liew, and W. Harnett, "A novel therapeutic approach targeting articular inflammation using the filarial nematode-derived phosphorylcholine-containing glycoprotein ES-62," *Journal of Immunology*, vol. 171, no. 4, pp. 2127–2133, 2003.

[83] P. Zaccone, Z. Feheérvári, F. M. Jones et al., "*Schistosoma mansoni* antigens modulate the activity of the innate immune response and prevent onset of type 1 diabetes," *European Journal of Immunology*, vol. 33, no. 5, pp. 1439–1449, 2003.

[84] L. S. Cardoso, S. C. Oliveira, A. M. Góes et al., "*Schistosoma mansoni* antigens modulate the allergic response in a murine model of ovalbumin-induced airway inflammation," *Clinical and Experimental Immunology*, vol. 160, no. 2, pp. 266–274, 2010.

[85] J. Yang, J. Zhao, Y. Yang et al., "*Schistosoma japonicum* egg antigens stimulate CD4+ CD25 + T cells and modulate airway inflammation in a murine model of asthma," *Immunology*, vol. 120, no. 1, pp. 8–18, 2006.

[86] C. M. Trujillo-Vargas, M. Werner-Klein, G. Wohlleben et al., "Helminth-derived products inhibit the development of allergic responses in mice," *American Journal of Respiratory and Critical Care Medicine*, vol. 175, no. 4, pp. 336–344, 2007.

[87] C. Lima, A. Perini, M. L. B. Garcia, M. A. Martins, M. M. Teixeira, and M. S. Macedo, "Eosinophilic inflammation and airway hyper-responsiveness are profoundly inhibited by a helminth (*Ascaris suum*) extract in a murine model of asthma," *Clinical and Experimental Allergy*, vol. 32, no. 11, pp. 1659–1666, 2002.

[88] W. I. Khan, P. A. Blennerhasset, A. K. Varghese et al., "Intestinal nematode infection ameliorates experimental colitis in mice," *Infection and Immunity*, vol. 70, no. 11, pp. 5931–5937, 2002.

[89] K. A. Saunders, T. Raine, A. Cooke, and C. E. Lawrence, "Inhibition of autoimmune type 1 diabetes by gastrointestinal helminth infection," *Infection and Immunity*, vol. 75, no. 1, pp. 397–407, 2007.

[90] A. Gruden-Movsesijan, N. Ilic, M. Mostarica-Stojkovic, S. Stosic-Grujicic, M. Milic, and L. Sofronic-Milosavljevic, "*Trichinella spiralis*: modulation of experimental autoimmune encephalomyelitis in DA rats," *Experimental Parasitology*, vol. 118, no. 4, pp. 641–647, 2008.

[91] A. Gruden-Movsesijan, N. Ilic, M. Mostarica-Stojkovic, S. Stosic-Grujicic, M. Milic, and L. Sofronic-Milosavljevic, "Mechanisms of modulation of experimental autoimmune encephalomyelitis by chronic *Trichinella spiralis* infection in Dark Agouti rats," *Parasite Immunology*, vol. 32, no. 6, pp. 450–459, 2010.

[92] H. K. Park, M. K. Cho, S. H. Choi, Y. S. Kim, and H. S. Yu, "*Trichinella spiralis*: infection reduces airway allergic inflammation in mice," *Experimental Parasitology*, vol. 127, no. 2, pp. 539–544, 2011.

[93] L. H. Boles, J. M. Montgomery, J. Morris, M. A. Mann, and G. L. Stewart, "Suppression of multiple sclerosis in the rat during infection with *Trichinella pseudospiralis*," *Journal of Parasitology*, vol. 86, no. 4, pp. 841–844, 2000.

[94] Z. Wu, I. Nagano, K. Asano, and Y. Takahashi, "Infection of non-encapsulated species of *Trichinella* ameliorates experimental autoimmune encephalomyelitis involving suppression of Th17 and Th1 response," *Parasitology Research*, vol. 107, no. 5, pp. 1173–1188, 2010.

[95] K. J. Erb, "Can helminths or helminth-derived products be used in humans to prevent or treat allergic diseases?" *Trends in Immunology*, vol. 30, no. 2, pp. 75–82, 2009.

[96] E. Pinelli, S. Brandes, J. Dormans, E. Gremmer, and H. Van Loveren, "Infection with the roundworm *Toxocara canis* leads

to exacerbation of experimental allergic airway inflammation," *Clinical and Experimental Allergy*, vol. 38, no. 4, pp. 649–658, 2008.

[97] A. J. Coyle, G. Kohler, S. Tsuyuki et al., "Eosinophils are not required to induce airway hyperresponsiveness after nematode infection," *European Journal of Immunology*, vol. 28, no. 9, pp. 2640–2647, 1998.

[98] L. R. Hall, R. K. Mehlotra, A. W. Higgins, M. A. Haxhiu, and E. Pearlman, "An essential role for interleukin-5 and eosinophils in helminth-induced airway hyperresponsiveness," *Infection and Immunity*, vol. 66, no. 9, pp. 4425–4430, 1998.

[99] R. T. Strait, S. C. Morris, K. Smiley, J. F. Urban, and F. D. Finkelman, "IL-4 exacerbates anaphylaxis," *Journal of Immunology*, vol. 170, no. 7, pp. 3835–3842, 2003.

[100] A. S. Gounni, K. Spanel-Borowski, M. Palacios, C. Heusser, S. Moncada, and E. Lobos, "Pulmonary inflammation induced by a recombinant *Brugia malayi* γ-glutamyl transpeptidase homolog: involvement of humoral autoimmune responses," *Molecular Medicine*, vol. 7, no. 5, pp. 344–354, 2001.

[101] P. A. M. Overgaauw, "Aspects of *Toxocara* epidemiology: toxocarosis in dogs and cats," *Critical Reviews in Microbiology*, vol. 23, no. 3, pp. 233–251, 1997.

[102] N. Ilic, A. Gruden-Movsesijan, and L. Sofronic-Milosavljevic, "*Trichinella spiralis*: shaping the immune response," *Immunologic Research*, vol. 52, no. 1-2, pp. 111–119, 2012.

[103] F. Bruschi and L. Chiumiento, "Immunomodulation in trichinellosis: does *Trichinella* really escape the host immune system?" *Endocrine Metabolic Immune & Disorders Drug Targets*, vol. 12, no. 1, pp. 4–15, 2012.

[104] D. Wakelin, P. K. Goyal, M. S. Dehlawi, and J. Hermanek, "Immune responses to *Trichinella spiralis* and *T. pseudospiralis* in mice," *Immunology*, vol. 81, no. 3, pp. 475–479, 1994.

[105] C. Aranzamendi, A. de Bruin, R. Kuiper et al., "Protection against allergic airway inflammation during the chronic and acute phases of *Trichinella spiralis* infection," *Clinical and Experimental Allergy*, vol. 43, no. 1, pp. 103–115, 2013.

[106] J. Buijs, W. H. Lokhorst, J. Robinson, and F. P. Nijkamp, "*Toxocara canis*-induced murine pulmonary inflammation: analysis of cells and proteins in lung tissue and bronchoalveolar lavage fluid," *Parasite Immunology*, vol. 16, no. 1, pp. 1–9, 1994.

[107] E. Pinelli, C. Withagen, M. Fonville et al., "Persistent airway hyper-responsiveness and inflammation in *Toxocara canis*-infected BALB/c mice," *Clinical and Experimental Allergy*, vol. 35, no. 6, pp. 826–832, 2005.

[108] E. Pinelli, S. Brandes, J. Dormans, M. Fonville, C. M. Hamilton, and J. V. der Giessen, "*Toxocara canis*: effect of inoculum size on pulmonary pathology and cytokine expression in BALB/c mice," *Experimental Parasitology*, vol. 115, no. 1, pp. 76–82, 2007.

[109] F. Bruschi, S. Solfanelli, and R. A. Binaghi, "*Trichinella spiralis*: modifications of the cuticle of the newborn larva during passage through the lung," *Experimental Parasitology*, vol. 75, no. 1, pp. 1–9, 1992.

[110] P. J. Cooper, "Interactions between helminth parasites and allergy," *Current Opinion in Allergy and Clinical Immunology*, vol. 9, no. 1, pp. 29–37, 2009.

18

Enhancing Schistosomiasis Control Strategy for Zimbabwe: Building on Past Experiences

Moses J. Chimbari

Okavango Research Institute, University of Botswana, P. Bag 285, Maun, Botswana

Correspondence should be addressed to Moses J. Chimbari, mjchimbari@gmail.com

Academic Editor: Francisca Mutapi

Schistosoma haematobium and *Schistosoma mansoni* are prevalent in Zimbabwe to levels that make schistosomiasis a public health problem. Following three national surveys to map the disease prevalence, a national policy on control of schistosomiasis and soil transmitted helminths is being developed. This paper reviews the experiences that Zimbabwe has in the area of schistosomiasis control with a view to influence policy. A case study approach to highlight key experiences and outcomes was adopted. The benefits derived from intersectoral collaboration that led to the development of a model irrigation scheme that incorporates schistosomiasis control measures are highlighted. Similarly, the benefits of using plant molluscicides and fish and duck biological agents (*Sargochromis codringtonii* and *Cairina moschata*) are highlighted. Emphasis was also placed on the importance of utilizing locally developed water and sanitation technologies and the critical human resource base in the area of schistosomiasis developed over years. After synthesis of the case studies presented, it was concluded that while there is a need to follow the WHO recommended guidelines for schistosomiasis control it is important to develop a control strategy that is informed by work already done in the country. The importance of having a policy and local guidelines for schistosomiasis control is emphasized.

1. Introduction

Schistosomiasis has, for many decades, been among the top ten causes of hospital admissions in Zimbabwe, an indication of its public health importance [1]. Before the advent of HIV and AIDS, the disease ranked second after malaria in terms of public health importance. *Schistosoma haematobium* and *S. mansoni* are prevalent countrywide and their epidemiology has been studied extensively [2–6] Apart from site specific prevalence and incidence studies [7–11], three national surveys have been conducted since 1982 [2, 3]. Ndhlovu et al. [3] reported that *S. haematobium* was more widely distributed in Matabeleland South province than previously reported by Taylor and Makura [4]. The authors [3] also reported presence of *S. mansoni* in areas where it was not previously reported. Ndhlovu et al. [3] attributed the observed differences in distribution and prevalence of schistosomiasis to increased dam projects in the provinces (Figure 1) and population movements following the country's independence from colonial rule as well a laxity in schistosomiasis control activities. The most recent survey [2] has shown that schistosomiasis is still prevalent in Zimbabwe with overall *S. haematobium* and *S. mansoni* prevalences of 20.8% and 9%, respectively.

Given the more thorough approach taken in the latest national survey [2], future efforts to control schistosomiasis in Zimbabwe should be informed by results of that survey. It should, however, be noted that because of the global shift from an integrated approach to schistosomiasis control that included control of intermediate host snails to a treatment based approach [12], the most recent survey did not include snail aspects. Nonetheless, a good overview of the distribution of the two intermediate hosts in Zimbabwe (*Bulinus globosus* and *Biomphalaria pfeifferi*) for schistosomiasis is known from previous studies [13].

On the basis of the most recent national schistosomiasis survey [2], a national control policy for Zimbabwe was drafted and will soon go through the necessary national structures responsible for policy formulation. At the policy formulation workshop, where the draft policy document was drafted, the evidence of successes made in controlling schistosomiasis through inclusion of other strategies apart

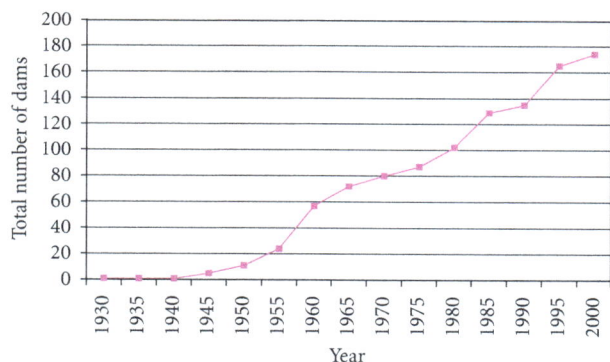

FIGURE 1: Cumulative number of small dams in Zimbabwe (1930–2000) Adapted from Senzanje and Chimbari (2002). Inventory of small dams in Africa: A case study for Zimbabwe.

from chemotherapy was highlighted and inclusion of such strategies in the policy was proposed.

This paper reviews schistosomiasis control activities that have been conducted in Zimbabwe over the years, highlighting key lessons that may be applied to develop a home grown strategy for controlling schistosomiasis in Zimbabwe.

2. Methodology

This paper is based on case studies on schistosomiasis control activities in Zimbabwe, and all the work presented has been published elsewhere or exists as grey literature mainly in project reports available at the National Institute of Health Research (formerly Blair Research Laboratory), Harare Zimbabwe. Although much work on schistosomiasis control has been done in Zimbabwe since 1960s, this paper focuses on key case studies that made significant impact on the prevalences of the two parasites and therefore should be used as lessons and should inform the proposed national schistosomiasis control strategy/policy. Some of the cases are in the form of research projects and intervention trials, while others are robust control interventions implemented over protracted periods of time. Figure 2 shows the locations of the case studies reviewed.

3. Case Studies on Zimbabwe Schistosomiasis Control Experiences

The case studies described in this paper highlight the experience that Zimbabwe has regarding alternative control strategies for schistosomiasis. The case studies are as follows: (i) Kariba Dam schistosomias is control programme, (ii) Mushandike schistosomiasis control programme, (iii) Hippo Valley Sugar Estates schistosomiasis control programme, (iv) Madziwa and Goromonzi schistosomiasis control programmes, and (v) Plant-based molluscicides for schistosomiasis control.

3.1. Kariba Dam Schistosomiasis Control Programme. The schistosomiasis control programme for Kariba was initiated in 1967 after cases of the disease attended to at local health

facilities increased significantly from 1963 when the lake filled for the first time. The control programme mainly focused on focal mollusciciding using niclosamid, and systematic screening and treatment of all residents. Shorelines were kept free of weeds particularly *Salvinia auriculata*, which was known to reintroduce snails in sprayed areas. The programme was funded by local companies and implemented by the Lake Kariba Area Coordinating Committee with technical backup from the Blair Research Laboratory, a disease control unit of the Ministry of Health. Routine snail surveys which informed what areas needed to be sprayed indicated that one area where the company had refused to participate in the control programme continued to harbor snails. Table 1 shows compiled results of surveys conducted between 1967 and 2001. Although the systematic control activities were terminated in the late 1980s, assessments done after year 2000 [7] showed lower prevalence on the Zimbabwean side of Lake Kariba compared to the Zambian side, and the differences were attributed to a long history of schistosomiasis control activities on the Zimbabwean side and better water and sanitation facilities than on the Zambian side [7].

3.2. Hippo Valley Sugar Estates Schistosomiasis Control Programme. The Hippo Valley Sugar Estates Schistosomiasis control programme was started in 1971 as a pilot project [17] that covered both the Hoppo Valley and Triangle Sugar Estates located in the south east lowveld region of Zimbabwe. The pilot project was later scaled up, and the programme placed greater emphasis on snail control using niclosamide and ducks as biological control agency. Alongside the snail control aspects, the programme had an annual chemotherapy component targeting school children and an intensive water and sanitation component. Assessments of efficacy of the control programme [9, 18] showed a significant decline in both prevalence and intensity over long periods and a sustained phase of prevalence below 10% (Figures 3 and 4). The success of the Hippo Valley story cannot be fully attributed to any one of the control strategies as each component made a significant contribution.

3.3. Mushandike Schistosomiasis Control Programme. The Mushandike project is a good example of a win-win intersectoral collaboration. The project was initiated in 1986 [19] with the objective to increase agricultural production of small-scale farmers through irrigation. Farmers were allocated farms ranging from 0.5 to 1.5 hectares, and irrigation was through siphoning water from tertiary canals onto the fields. The infield canals were fed by a 25 km main canal that made it necessary to have infield night storage ponds for smooth commanding of the fields.

At conceptualization and design of the irrigation scheme, there was consultation between health professionals interested in disease control and engineers responsible for designing the scheme. It was agreed that schistosomiasis was a potential health hazard that would impact negatively on crop production. Thus, the design was influenced such that the infield network of canals would all be lined in order to ensure fast movement of water to dislodge any snails present

FIGURE 2: Map of Zimbabwe showing locations of case studies reviewed *(produced by Mrs. A. Makati, GIS Laboratory, Okavango Research Institute)*.

in the system and to avoid unnecessary water seepage. The in-field canal system included special features designed to flush snails (drop structures with stilling basins, special off takes, and duck bill weirs). Toilets constructed and arranged in a matrix system that ensured that people in the fields were at all times closer to a toilet than to a bush [20]. Water management was designed in such a way that canals in some irrigation blocks would be completely dry when not under irrigation, and only water needed for irrigation was released thus limiting end of field flooding. This was made possible by making sure that each block had one crop and, therefore, water demand would be the same. While the night storage

ponds were undesirable, they could not be avoided from an engineering perspective but it was envisaged that proper operation of the night storage ponds would expose snails to predators during the draw down period. Furthermore the changing water levels would make the environment not conducive for snail colonization and establishment.

Monitoring of schistosomiasis conducted at Mushandike for a period of 5 years consistently showed higher levels of infection in the irrigation scheme where schistosomiasis control measures were not introduced (control farm) compared to the irrigation scheme where schistosomiasis control measures were introduced (intervention farms) [21].

TABLE 1: Results of schistosomiasis surveys conducted around Kariba town [15].

Year	Area	Population category	Prevalence of S. haematobium (%)	Prevalence of S. mansoni (%)
1967	Kariba Town	Adult workers	13.3	8.6
1979	Mahombekombe	School children	54.6	68.0
	Nyamhunga	School children	48.0	64.0
1984	All government departments and industries	Adult employees of all government departments and private sector	9.4	14.3
1985	All government departments and industries	Adult employees of all government departments and private sector	4.8	8.1
1986	All government departments and industries	Adult employees of all government departments and private sector	8.4	10.5
2001	Mahombekombe, Nyamhunga and Charara	School children	9.0	2.5
		Subsistence fishermen	7.3	12.5
		Commercial fishermen	0	26.3

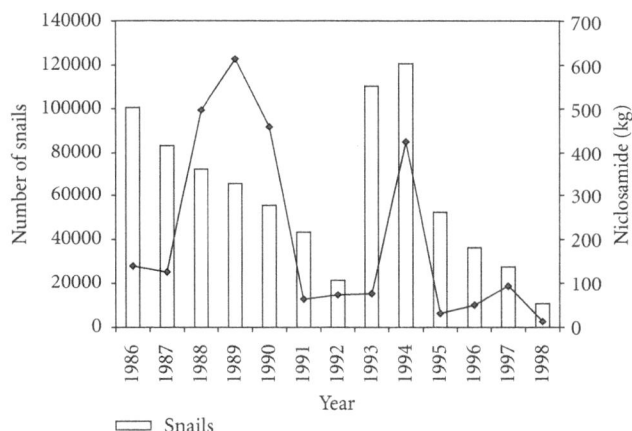

FIGURE 3: Snail population densities and quantities of niclosamide applied to reduce the snail numbers from 1986 to 1998 [9].

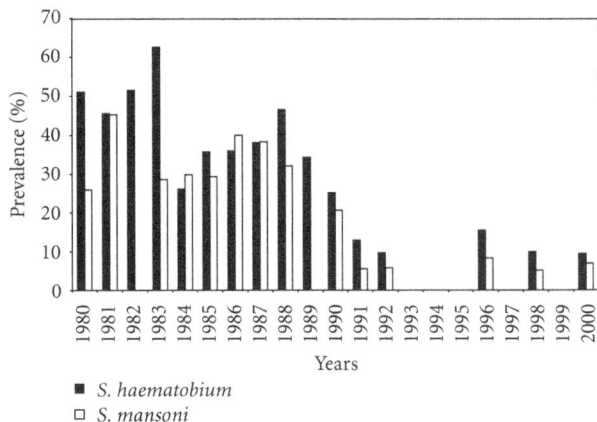

FIGURE 4: Prevalences of Schistosoma haematobium and S. mansoni among school children in Hippo Valley Estates for the period 1980–2000 [9].

Similar observations were made in a survey conducted 10 years after the project became operational (Table 2). Furthermore, a comparison of prevalences for 1989 to those obtained in 1999 showed that both S. haematobium and S. mansoni prevalences did not change significantly in villagers not attending school over a period of 10 years [16]. The prevalence of S. haematobium in the control farm was significantly higher than that of intervention farms implying that the engineering and environmental interventions may have contributed towards the difference.

Snail surveys conducted during the initial 5-year period and in 1991 also consistently showed high numbers of intermediate host snails present in the control farm than in the intervention farms. Furthermore, larger proportions of intermediate host snails collected in the control farm were infected with schistosome parasites compared to those collected in the intervention farms.

Although the costs involved in developing the "Mushandike Model" irrigation scheme are substantial, the Department of Irrigation in Zimbabwe adopted the model as the standard for all-small scale irrigation schemes. From a disease control perspective, the costs are justified as indicated by the incremental ratio of -$446 010.31 per 1% schistosomiasis prevalence, which meant that a saving of $446 010.31 per schistosomiasis prevalence of 1% was realized over a 10-year period [16].

3.4. Madziwa and Goromonzi Schistosomiasis Control Programmes. The Madziwa and Goromonzi schistosomiasis control projects were implemented in 1985–1989 [22] and 1994–1997 [23], respectively. Common to both projects were strong water and sanitation components, chemotherapy targeted to school children, and health education. The main differences in approaches used in the two intervention studies were that the Goromonzi health education component adopted the participatory health and hygiene education (PHHE) approach and mollusciciding was done once at the beginning of the project along the main stream in

TABLE 2: Showing the prevalence of *S. haematobium* in the study population [16].

| | Intervention villages | | | | | Control village |
	Village 12	Village13	Village 14	Village15	Total	Chikore
S. haematobium						
Number positive	5	5	4	8	22	10
Number negative	98	58	42	57	255	69
Prevalence	4.9%	7.9%	8.7%	12.3	7.9%	12.7%
S. mansoni						
Number positive	0	3	3	3	9	No data
Number negative	103	60	43	62	268	No data
Prevalence	0	4.8	6.5	4.6	3.4%	No data

Madziwa while for Goromonzi only monitoring of sites for intermediate host snails was done in all major rivers and streams.

A 60% to 20% reduction in prevalence of *S. haematobium* infections among children aged 7–15 years was achieved at Madziwa. Furthermore, a 95% reduction in heavy infections among the targeted age group (7–15 years) was also achieved. Heavy infections were defined as greater than 50 *S. haematobium* eggs per 10 mL of urine or greater than 100 *S. mansoni* eggs per gram of faeces. In Goromonzi, prevalence of schistosomiasis among children aged 6–15 years declined from 20% to less than 5% for *S. mansoni* and from 40% to 10% *S. haematobium*.

In both studies (Goromonzi and Madziwa), the differences in prevalence of infection between schools in intervention villages and schools in control villages were not significantly different and this was attributed to spill over of interventions largely because the villages were close to each other. Furthermore, infections increased to preintervention levels when chemotherapy was discontinued [23].

4. JICA Funded School Screening, Treatment and Education Programme

Upon a request by the Ministry of Health and Child Welfare (MOHCW), the Japanese International Cooperation Agency (JICA) partnered with the ministry to embark on a project with the following purposes: (i) to control specified infectious diseases such as schistosomiasis and malaria in the eight model districts and (ii) to formalize the National Schistosomiasis Control Policy based on the Project's experiences.

The project focused on the following 8 districts that became known as the model districts: Hurungwe, Mt Darwin, UMP, Lupane, Gokwe, Bulilimamangwe, Chipinge, and Mwenezi. School children in grade one (6 years) to grade 5 (10 years) were screened for schistosome infections and treated under a programme referred to as School Screening, Treatment and Education (SSTE) over a period of two years (1997–1999). Staff in 131 local health centres and at provincial and district level were trained on how to conduct SSTE. The trained staff with technical support from Blair Research Laboratory and JICA experts conducted SSTE in 497 out of

the 631 schools in the model districts resulting in 85 578 out of the 102 000 children enrolled in the schools being screened and 99.4% of those found infected treated.

Prevalence and intensity of infection was significantly reduced over the two-year period, and a study conducted in one of the model districts (Mt Darwin) showed improved knowledge on schistosomiasis by school children but not a change in behavior [24], and no correlation between level of knowledge and infection rates was established. The draft policy document on schistosomiasis was adopted by model districts and therefore formed the basis for development of the final policy document worked on following the latest national survey [2].

4.1. Plant-Based Molluscicides for Schistosomiasis Control. Two plant-based molluscides (*Phytollacca dodecandra* and *Jatropha curcas*) have been studied with a view to use them in preference to the WHO recommended molluscicide, niclosamide. *Phytolacca dodecandra* has been studied in sufficient detail to justify its application in selected areas [25–30]. *Jatropha curcas* studies in Zimbabwe were only done in the laboratory [31] where the potency of the plant berries was demonstrated, showing that the unripe stage (green) of the berries was more potent than the ripe (yellow) and overripe stages (brown). The advantage of *J. curcas* over *P. dodecandra* is that the former has multipurposes (including potential for bio-fuel) and hence presents an incentive for farmers to grow it for financial benefits. However, the potency of a water extract of *J. curcas* is much lower (75 ppm) compared to that of *P. dodecandra* (10 ppm) implying that larger quantities of Jatropha berries would be required to sustain snail control activities.

Contrary to *J. curcas*, *P. dodecandra* has been extensively studied [25–30]. The variety of the plant that produces the most potent berries under the Zimbabwean agro-conditions was identified [26], and trials conducted along two natural streams showed that sites at which the molluscicide was applied was kept free from snail infestation for 7 months [28]. It was demonstrated under laboratory conditions that sublethal doses (<10 ppm) could be used to stop miracidia from successfully penetrating snail intermediate host snails for schistosomiasis [29]. The extent to which communities could be empowered to grow, harvest, process, and apply the molluscicide with minimum technical support has been described [31]. The results showed low level community participation due to, among other reasons, poor leadership, low economic value of the plant, inaccessible fields, and lack of tangible benefits. Despite these challenges, some districts adopted the use of the plant in the control programmes.

4.2. Biological Control Trials. Studies on exploring the possibilities of controlling intermediate host snails for schistosomiasis using a variety of biological agents have been conducted in Zimbabwe. The most studied biological agents tested include ducks, fish (*Sargochromis codringtonii*), and competitor snails (*Bulinus tropicus*).

4.2.1. Ducks. Ducks were used in the Hippo Valley schistosomiasis control programme for many years as a supplement to application of niclosamide. The use of ducks was restricted to night storage ponds where a number of ducks would be allowed to swim around in one pond for 8 hours before being moved to another pond. While the ducks made significant impact in terms of reducing snail numbers in ponds there were several challenges faced with this intervention strategy. The costs associated with transportation of the ducks and looking after them to avoid poaching were high. Furthermore, the breeding and maintenance costs of the ducks were high as they were exotic species. In an effort to want to reduce costs of duck operations, semifield pond studies to investigate the potential of using indigenous ducks for snail control were conducted. The study concluded that there was potential for using indigenous ducks as biological snail control agents but further work needed to be done [32].

4.2.2. Fish (Sargochromis codringtonii). Inspired by the observations that overfishing of cichlid fish in Lake Malawi shorelines resulted in increased numbers of snails and increased cases of schistosomiasis [33] and studies conducted in Lake Kariba [34], comprehensive studies aimed at testing the potential of using an indigenous cichlid to Zimbabwe, *S. codringtoni,* as a biological agent for snail control were conducted. Laboratory studies [35, 36] demonstrated the snail eating tendencies of *S. codringtonii* and the interactions of snails (prey) and fish (predator) under aquaria conditions. Enclosure [37], and exclosure [38] showed the effects of snail predation on *S. codringtonii* under different treatments: with vegetation, in combination with fish herbivore (*Tilapia rendalli*); with a wider choice of snails (both pulmonates and prosobranchs). The results showed that pulmonates but not necessarily intermediate host snails were preferred by *S. codringronii* and that vegetation provided refugia for snails against the predator fish. However, the combination of *S. codringtonii* and *T. rendalli* was not desirable as the later was attacked by the former. Field studies conducted in night storage ponds [37] further demonstrated the potential use of *S. codringtonii* as a biological agent but the results were not conclusive as the monitoring period was short (Figure 5). It was, however, evident that *S. codringtoni,* which is mainly found in Lake Kariba, could acclimatize to small ponds (100 m × 100 × 1–1.5 m depth) in the Lowveld of Zimbabwe. The predator-prey interactions of *S. codringtonii* and snails have also been studied [14, 38–41].

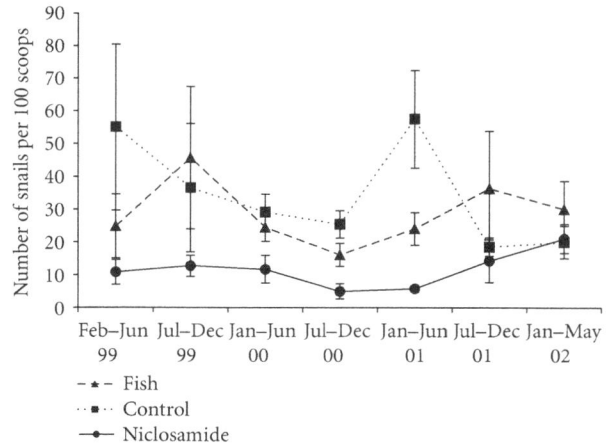

FIGURE 5: Average number of snails collected from night storage ponds from February 1999 to May 2002 [14].

The extensive studies on *S. codringtonii* as biological agent for controlling intermediate host snails for schistosomiasis provide convincing evidence to justify use of the fish in appropriate settings like night storage ponds in irrigation schemes where the fish could be a good source of protein and serve as a snail control agent.

4.2.3. Competitor Snails (Bulinus tropicus). Motivated by the observation that *B. globosus* (intermediate host snail for schistosomiasis) and *B. tropicus* (non-intermediate host snail for schistosomiasis) do not share the same niche although they share similar habitats, studies aimed at investing the potential of using *B. tropicus* as a competitor snail of *B. globosus* with the ultimate goal of controlling schistosomiasis were conducted [42, 43]. Laboratory and quasifield studies showed a significant reduction in reproductivity of *B. globosus* in the presence of *B. tropicus* and evidence of *B. tropicus* preying on *B. globosus* eggs [42]. However, further enclosure studies [43] did not show any significant effect of *B. tropicus* on *B. globosus* population density suggesting the competition between the two snail species was not important control of schistosomiasis.

5. Discussion

Understanding the life cycle of a parasite and the epidemiology of the disease caused by the parasite are fundamental to disease control. Following the first description of schistosomiasis in man by Theodor Bilharz in 1851, the life cycle was studied and described [44]. The transmission dynamics of the disease in Zimbabwe has been well documented [3–11, 17, 18, 22]. There are four broad interventions that can be made to disrupt the life cycle of the parasite and hence its transmission; (1) treatment of infected individuals to reduce, and remove morbidity, reduce mortality and reduce contamination of the environment with schistosome parasite eggs, (2) providing communities with adequate, appropriate sanitation to reduce environmental contamination and hence minimize the chances of miracidia finding and penetrating the intermediate host snails, (3) snail control to minimize the chances of miracidia finding an appropriate intermediate host and therefore significantly reducing the number of cercariae available for infecting people at water contact sites, and (4) provision of adequate and accessible safe water to reduce the chances of people getting in contact with water that may be infested with cercariae and hence limits the chances of cercaria locating the human host and infect them in its limited life span. All the aforementioned possible interventions have been studied in detail globally and at local level, and it is appreciated that simultaneous implementation of all the measures may not be cost effective. Hence, treatment has been prioritized as it reduces early and late life morbidity and mortality, and eventually reduces the force of transmission by reducing contamination.

The experiences in treatment of infected individuals reviewed in this paper for Zimbabwe clearly show that scaling up this strategy will not be a difficult task. The Hippo Valley and Mushandike [8, 18, 19] experiences can inform the treatment strategy for communities in both large- and small-scale irrigation schemes. The local level capacity developed during the SSTE JICA programme and during the 1992 and 2010 surveys [2, 3] is an asset in rolling out a national control programme that is school based in line with WHO guidelines [12]. Thus, with adequate government commitment to resource the control programme and donor/partner support particularly in the area of drug procurement, success in schistosomiasis control in Zimbabwe can be achieved.

While WHO guidelines [12] do not negate other key schistosomiasis control measures described in this paper, it is clear that greater emphasis is placed on treatment. However, at country level the experience gained in the other nontreatment measures cannot be ignored. The use of niclosamide for control of intermediate host snails is not practical for application in communal areas and other poorly resourced communities because of logistical, financial, and environmental reasons. However, this is a strategy that can continue to be promoted in the lowveld where it has been proven to be successful. Given the huge research investment made on *P. dodecandra* and the positive results obtained in field trials [26–31], there is justification in scaling up this intervention particularly in communal areas and small-scale irrigation schemes. However, the challenges associated with its application in communal areas [31] will need to be addressed and close monitoring of environmental impacts will need to be done. Given the current drive towards use of *J. curcas* for biofuel, there is a need to conduct further research on possible use of the plant as molluscicide as there are likely to be better incentives to grow *J. curcas* than to grow *P. dodecandra*. In general, the mollusciciding strategy should be focal in nature to minimize costs and environmental impacts with the exception of irrigation systems where there is a need to treat the complete canal network.

The challenges associated with biological control in general are known [45]. Predator-prey interactions will lead to some equilibrium and that equilibrium threshold may not be adequate for purposes of controlling the prey to the desired level. This is particularly so in the case where *B. tropicus* may needs to be used as a competitor for *B. globosus*. The inconclusive results as reviewed in this paper show that this may not be an area to make further investments. However, the potential use of indigenous ducks and *S. codringtonii* need to be seriously considered particularly in irrigation ponds but also in communal ponds or small dams. This is because these biological agents have broad spectrum diets, which will allow them to switch to another less preferred prey if the preferred one is absent. Furthermore, the agents may contribute significantly to community protein requirements. Ducks would be provided with alternative feeds, and a possibility for supplementing fish with inexpensive feeds can be explored once the snail numbers have reached too low numbers to maintain a reasonable population size of the fish. Dietary shift studies will need to be carried out to establish if there might not be a permanent shift of diet from snails to other food items.

Zimbabwe is better positioned to apply the ordinarily expensive interventions of water and sanitation in schistosomiasis control as home grown technologies have been developed and tested in the field [46]. Furthermore, promotion of water and sanitation interventions will impact on more than one neglected tropical disease [47] and will generally improve the quality of human life particularly in rural settings. The latter reason is likely to garner support of NGOs and other international organizations keen on improving rural community health and livelihoods.

In conclusion, I recommend that Zimbabwe should adopt the WHO recommended strategy for controlling schistosomiasis and in doing so should seriously consider some of the measures proven to be effective at local level but are less emphasized in the guidelines. Specifically, the snail aspects should be seriously considered to avoid a situation where the only safe efficacious drug for schistosomiasis, praziquantel, may one day be compromised by parasite resistance and result in an outbreak. Besides, snail control complement well the treatment strategy. In an effort to achieve the objective of controlling schistosomiasis there is a need to ensure that the policy for control is passed by parliament and guidelines to operationalize the policy are developed. The huge human resource base in the area of schistosomiasis developed since 1990 should be fully utilized to achieve control and move towards elimination. Since 1990, fifteen staff were trained to Ph.D. level and more than 20 technicians were trained most of them to M. S. level. While a small proportion have passed-on, the remaining "Zimbabwe Schistosomiasis Scientists" spread in the southern Africa region and overseas are very committed to the cause of control and are currently supporting in-country initiatives.

Acknowledgments

The author sincerely thanks management of the Zimbabwe Ministry of Health and Child Welfare and in particular the National Institute of Health Research for involving him in the schistosomiasis activities leading to the 2010 national survey, for allowing him to participate in the survey, and for inviting him to participate at the policy formulation workshop where the idea of writing this paper was muted. Writing this paper was made possible by utilization of materials developed and published by other scientists. Authors to all papers used in this paper and principal investigators of all case studies referred to are duly acknowledged. Lastly, he wants to acknowledge the encouragement from Dr. Mutapi to write this paper.

References

[1] Ministry of Health & Child Welfare, Epidemiology & Disease Control National Health Information & Surveillance Unit, *Top 10 causes of hospital admissions*, National Health Profile, 2008.

[2] N. Midzi, D. Sangweme, S. Zinyowera et al., *Report on the National Soil Transmitted Helminthiasis and Schistosomiasis Survey*, Ministry of Health and Child Welfare, Harare, Zimbabwe, 2011.

[3] P. Ndhlovu, M. Chimbari, J. Ndamba, and S. K. Chandiwana, *National Schistosomiasis Survey*, Ministry of Health and Child Welfare, Harare, Zimbabwe, 1992.

[4] P. Taylor and O. Makura, "Prevalence and distribution of schistosomiasis in Zimbabwe," *Annals of Tropical Medicine and Parasitology*, vol. 79, no. 3, pp. 287–299, 1985.

[5] S. K. Chandiwana and M. E. J. Woolhouse, "Heterogeneities in water contact patterns and the epidemiology of Schistosoma haematobium," *Parasitology*, vol. 103, no. 3, pp. 363–370, 1991.

[6] P. Hagan, S. Chandiwana, P. Ndhlovu, M. Woolhouse, and A. Dessein, "The epidemiology, immunology and morbidity of Schistosoma haematobium infections in diverse communities in Zimbabwe. I. The study design," *Tropical and Geographical Medicine*, vol. 46, no. 4, pp. 227–232, 1994.

[7] M. J. Chimbari, E. Dhlomo, E. Mwadiwa, and L. Mubila, "Transmission of schistosomiasis in Kariba, Zimbabwe, and a cross-sectional comparison of schistosomiasis prevalences and intensities in the town with those in Siavonga in Zambia," *Annals of Tropical Medicine and Parasitology*, vol. 97, no. 6, pp. 605–616, 2003.

[8] S. K. Chandiwana, P. Taylor, M. Chimbari et al., "Control of schistosomiasis transmission in a newly established small holder irrigation scheme," *Transactions of Royal of Tropical Medicine and Hygiene*, vol. 82, pp. 874–880, 1998.

[9] M. J. Chimbari and B. Ndlela, "Successful control of schistosomiasis in large sugar irrigation estates of Zimbabwe," *Central African Journal of Medicine*, vol. 47, no. 7, pp. 169–172, 2001.

[10] K. C. Brouwer, P. D. Ndhlovu, Y. Wagatsuma, A. Munatsi, and C. J. Shiff, "Urinary tract pathology attributed to Schistosoma haematobium: does parasite genetics play a role?" *The American Journal of Tropical Medicine and Hygiene*, vol. 68, no. 4, pp. 456–462, 2003.

[11] N. Midzi, D. Sangweme, S. Zinyowera et al., "The burden of polyparasitism among primary schoolchildren in rural and farming areas in Zimbabwe," *Transactions of the Royal Society of Tropical Medicine and Hygiene*, vol. 102, no. 10, pp. 1039–1045, 2008.

[12] World Health Organisation, *The Prevention and Control of Schistosomiasis and Soil Transmitted Helminthiasis*, WHO Technical Report Series No. 912., World Health Organisation, Geneva, Switzerland, 2002.

[13] O. Makura, "TK: national freshwater snail survey of Zimbabwe," in *Proceedings of the 10th International Malacological Congress*, pp. 227–232, Tübingen, Germany, 1989.

[14] M. J. Chimbari, P. Makoni, and H. Madsen, "Impact of *Sargochromis codringtonii* (Teleostei: Cichlidae) on pulmonate snails in irrigation ponds in Zimbabwe," *African Journal of Aquatic Science*, vol. 32, no. 2, pp. 197–200, 2007.

[15] M. Chimbari, "Public health and water borne diseases impacts of Kariba Dam," Working Paper for the World Commission on Dams. Kariba Dam Case Study, Zambia and Zimbabwe, 2000.

[16] M. J. Chimbari, P. D. Ndhlovu, B. Ndlela, S. Shamu, S. K. Chandiwana, and W. Mapira, "Schistosomiasis control measures for small irrigation schemes in Zimbabwe. Evaluation of Mushandike Irrigation Scheme," Report 321 (1) Blair Research Laboratory, 2001.

[17] C. J. Shiff, V. V. Clarke, A. C. Evans, and G. Barnish, "Molluscicide for the control of schistosomiasis in irrigation schemes. A study in Southern Rhodesia," *Bulletin of the World Health Organization*, vol. 48, no. 3, pp. 299–307, 1973.

[18] A. C. Evans, "Control of schistosomiasis in large irrigation schemes by use of niclosamide. A ten-year study in Zimbabwe," *The American Journal of Tropical Medicine and Hygiene*, vol. 32, no. 5, pp. 1029–1039, 1983.

[19] A. J. Draper and P. Bolton, "Design note for schistosomiasis control—mushandike irrigation scheme, Zimbabwe," HR Wallingord Report OD/TN 20, 1986.

[20] M. Chimbari, B. Ndlela, Z. Nyati, A. Thomson, S. K. Chandiwana, and P. Bolton, "Bilharzia in a small irrigation community: an assessment of water and toilet usage," *Central African Journal of Medicine*, vol. 38, no. 12, pp. 451–458, 1992.

[21] M. J. Chimbari, S. K. Chandiwana, B. Ndlela et al., "Schistosomiasis control measures for small irrigation schemes in Zimbabwe. Final report on monitoring at Mushandike Irrigation Scheme," HR Wallingford Report, OD 128, 1993.

[22] S. K. Chandiwana, P. Taylor, and D. Matanhire, "Community control of schistosomiasis in Zimbabwe," *Central African Journal of Medicine*, vol. 37, no. 3, pp. 69–77, 1991.

[23] M. J. Chimbari, S. Mukaratirwa, and J. Mutaurwa, "Impact of participatory health and hygiene education on transmission of schistosomiasis in Goromonzi district," United Nations Cheldren's Fund, Harare Internal Report.

[24] C. Suzuki, T. Mizota, T. Awazawa, T. Yamamoto, B. Makunike, and Y. Rakue, "Effects of a school-based education programme for schistosomiasis control," *Southeast Asian Journal of Tropical Medicine Public Health*, vol. 36, no. 6, pp. 1388–1393, 2005.

[25] J. Ndamba and S. K. Chandiwana, "The geographical variation in the molluscicidal potency of *Phytolacca dodecandra*—a plant molluscicide," *Social Science & Medicine*, vol. 28, no. 12, pp. 1249–1253, 1989.

[26] J. Ndamba, S. K. Chandiwana, and N. Makaza, "The use of *Phytolacca dodecandra* berries in the control of trematode-transmitting snails in Zimbabwe," *Acta Tropica*, vol. 46, no. 5-6, pp. 303–309, 1989.

[27] J. Ndamba, E. Lemmich, and P. Molgaard, "Investigation of the diurnal, ontogenetic and seasonal variation in the molluscicidal saponin content of *Phytolacca dodecandra* aqueous berry extracts," *Phytochemistry*, vol. 35, no. 1, pp. 95–99, 1994.

[28] P. R. Gwatirisa, J. Ndamba, and N. Z. Nyazema, "The impact of health education on the knowledge, attitudes and practices of a rural community with regards to schistosomiasis control using a plant molluscicide, *Phytolacca dodecandra*," *Central African Journal of Medicine*, vol. 45, no. 4, pp. 94–97, 1999.

[29] P. Mølgaard, A. Chihaka, E. Lemmich et al., "Biodegradability of the molluscicidal saponins of *Phytolacca dodecandra*," *Regulatory Toxicology and Pharmacology*, vol. 32, no. 3, pp. 248–255, 2000.

[30] A. Ndekha, E. H. Hansen, P. Mølgaard, G. Woelk, and P. Furu, "Community participation as an interactive learning process: experiences from a schistosomiasis control project in Zimbabwe," *Acta Tropica*, vol. 85, no. 3, pp. 325–338, 2003.

[31] M. J. Chimbari and C. J. Shiff, "A laboratory assessment of the potential molluscicidal potency of Jatropha curcas aqueous extracts," *African Journal of Aquatic Science*, vol. 33, no. 3, pp. 269–273, 2008.

[32] B. Ndlela and M. J. Chimbari, "A preliminary assessment of the potential of the Muschovy duck (*Cairina maschata*) as a biocontrol agent of schistosomiasis intermediate host snails," *Central African Journal of Medicine*, vol. 46, no. 10, pp. 271–275, 2000.

[33] S. S. Chiotha, K. R. McKaye, and J. R. Stauffer Jr., "Use of indigenous fishes to control schistosome snail vectors in Malaŵi, Africa," *Biological Control*, vol. 1, no. 4, pp. 316–319, 1991.

[34] N. Moyo, *The biology of Sargochromis codringtoni in Lake Kariba*, Ph.D. thesis, University of Zimbabwe, 1995.

[35] M. J. Chimbari, H. Madsen, and J. Ndamba, "Laboratory experiments on snail predation by *Sargochromis codringtoni*, a candidate for biological control of the snails that transmit schistosomiasis," *Annals of Tropical Medicine and Parasitology*, vol. 91, no. 1, pp. 95–102, 1997.

[36] M. J. Chimbari, J. Ndamba, and H. Madsen, "Food selection behaviour of potential biological agents to control intermediate host snails of schistosomiasis: *Sargochromis codringtoni* and Tilapia rendalli," *Acta Tropica*, vol. 61, no. 3, pp. 191–199, 1996.

[37] M. J. Chimbari, H. Madsen, and J. Ndamba, "Simulated field trials to evaluate the effect of *Sargochromis codringtoni* and Tilapia rendalli on snails in the presence and absence of aquatic plants," *Journal of Applied Ecology*, vol. 34, no. 4, pp. 871–877, 1997.

[38] J. Brodersen, M. J. Chimbari, and H. Madsen, "An enclosure study to evaluate the effect of *Sargochromis codringtoni* on snail populations in an irrigation canal," *African Zoology*, vol. 37, no. 2, pp. 255–258, 2002.

[39] P. Makoni, M. J. Chimbari, and H. Madsen, "Predator avoidance in *Bulinus globosus* (morelet, 1866) and *B. tropicus* (Krauss, 1848) (Gastropoda: Planorbidae) exposed to a snail predator, *Sargochromis codringtonii* (Boulenger, 1908) (Pisces: Cichlidae)," *Journal of Molluscan Studies*, vol. 70, pp. 353–358, 2004.

[40] P. Makoni, M. J. Chimbari, and H. Madsen, "Interactions between fish and snails in a Zimbabwe pond, with particular reference to *Sargochromis codringtonii* (Pisces: Cichlidae)," *African Journal of Aquatic Science*, vol. 30, no. 1, pp. 45–48, 2005.

[41] J. Brodersen, M. J. Chimbari, and H. Madsen, "Laboratory experiments on snail size selections by cichlid snail predator, *Sargochromis codringtonii*," *Journal of Molluscan Studies*, vol. 68, pp. 194–196, 2003.

[42] B. Ndlela and H. Madsen, "Laboratory and quasi-field studies on interspecific competition between *Bulinus globosus* and *B. tropicus* (Gastropoda: Planorbidae)," *African Journal of Aquatic Science*, vol. 26, no. 1, pp. 17–21, 2000.

[43] B. Ndlela, M. J. Chimbari, and H. Madsen, "Interactions between Bulinus globosus and B. tropicus (Gastropoda: Planorbidae) in a pond experiment in Zimbabwe," *African Journal of Aquatic Science*, vol. 32, no. 1, pp. 13–16, 2007.

[44] P. Sonsino, "Discovery of the life history of bilharzias haematobiu (Cobbold)," *The Lancet*, vol. 142, no. 3654, pp. 621–622, 1893.

[45] A. Fenwick, D. Rollinson, and V. Southgate, "Implementation of human schistosomiasis control: challenges and prospects," *Advances in Parasitology*, vol. 61, pp. 567–622, 2006.

[46] P. Morgan, *Rural Water Supplies and Sanitation: A text from Zimbabwe's Blair Research Laboratory*, Macmillan Education, London, UK, 1990.

[47] J. Aagaard-Hansen, J. R. Mwanga, and B. Bruun, "Social science perspectives on schistosomiasis control in Africa: past trends and future directions," *Parasitology*, vol. 136, no. 13, pp. 1747–1758, 2009.

19

The Curative and Prophylactic Effects of Xylopic Acid on *Plasmodium berghei* Infection in Mice

J. N. Boampong,[1] E. O. Ameyaw,[1] B. Aboagye,[1] K. Asare,[1] S. Kyei,[2] J. H. Donfack,[3] and E. Woode[4]

[1] *Department of Biomedical and Forensic Sciences, University of Cape Coast, Cape Coast, Ghana*
[2] *Department of Optometry, School of Physical Sciences, University of Cape Coast, Cape Coast, Ghana*
[3] *Department of Biomedical Sciences, Faculty of Sciences, University of Dschang, P.O. Box 067, Dschang, Cameroon*
[4] *Department of Pharmacology, Faculty of Pharmacy and Pharmaceutical Sciences, College of Health Sciences, KNUST, Kumasi, Ghana*

Correspondence should be addressed to J. N. Boampong; jonboamus@yahoo.com

Academic Editor: Wej Choochote

Efforts have been intensified to search for more effective antimalarial agents because of the observed failure of some artemisinin-based combination therapy (ACT) treatments of malaria in Ghana. Xylopic acid, a pure compound isolated from the fruits of the *Xylopia aethiopica,* was investigated to establish its attributable prophylactic, curative antimalarial, and antipyretic properties. The antimalarial properties were determined by employing xylopic acid (10–100 mg/kg) in ICR mice infected with *Plasmodium berghei*. Xylopic acid exerted significant ($P < 0.05$) effects on *P. berghei* infection similar to artemether/lumefantrine, the standard drug. Furthermore, it significantly ($P < 0.05$) reduced the lipopolysaccharide- (LPS-) induced fever in Sprague-Dawley rats similar to prednisolone. Xylopic acid therefore possesses prophylactic and curative antimalarial as well as antipyretic properties which makes it an ideal antimalarial agent.

1. Introduction

Malaria, caused by *Plasmodium* parasite, is a leading poverty-associated disease that undermines the development of countries. The numbers of disease cases and deaths was 225 million and 781 000 respectively in 2009 [1]. Children under five years and pregnant women (vulnerable groups) succumb to the devastating effects of the disease making the disease a major global infectious disease. Chemotherapy has ultimately been the central tool for management of malaria, and combination of drug regimens has become the practice of choice because of their increased therapeutic efficacy over monotherapy and other benefits which include decreased cytotoxicity and delay or prevention of the development of drug resistance [2]. *Plasmodium falciparum* (Pf), the most lethal malaria pathogen, has developed resistance to some antimalarials [3]. This makes it imperative to search for newer, more effective antimalarial agents. Plants have served as reliable sources of drugs especially antimalarials [4]. The fruits of *Xylopia aethiopica* are used traditionally for the treatment of malaria but the active principle(s) responsible for the observed antimalarial effect of the extract is still not known [5, 6]. Xylopic acid, a kaurene diterpene, occurs as the major constituent in the fruits of *Xylopia aethiopica* and is reported to possess analgesic properties [7]. Xylopic acid, unlike kaurenoic acid, has no cytotoxic effect against human cancer cells [8], making the compound a safe one for the treatment of diseases where selective toxicity towards the parasite is highly needed. In the light of the above, xylopic acid was evaluated for its antimalarial and antipyretic properties. The structure of the xylopic acid is shown below (Figure 1).

2. Materials and Methods

2.1. Extraction and Purification of Xylopic Acid (15β-Acetoxy-(−)-kaur-16-en-19-oic Acid). The extraction process was carried out as described elsewhere [7]. Briefly, 0.36 kg of the

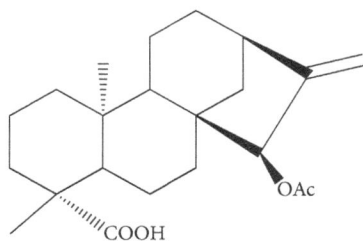

FIGURE 1: Chemical structure of xylopic acid.

fruit of *Xylopia aethiopica* was pulverized and placed in cylindrical jars. This was soaked with 5 L of petroleum ether (40–60°C) and allowed to stand for three days. The petroleum ether was drained and concentrated using rotary evaporator at a temperature of 50°C. Ethyl acetate was added to the concentrate to facilitate the crystallization of xylopic acid. Crystals (xylopic acid) formed after the concentrate had been allowed to stand for three days and were washed with petroleum ether at 40–60°C repeatedly until all unwanted materials had been removed. Crude xylopic was purified in 96% ethanol. The yield of the xylopic acid was 1.41%. The purity of the isolated xylopic acid was 95%.

2.2. Chemicals and Test Agents. The lipopolysaccharide (LPS), ethanol, petroleum ether, and ethyl acetate used for the extraction were purchased from Sigma-Aldrich Inc., St. Louis, MO, USA. Artemether/Lumefantrine (A-L) was obtained from Ajanta Pharma Ltd., Maharashtra, India, sulfadoxine/pyrimethamine was obtained from Maxheal Laboratories Pvt. Ltd. Gujarat, India, and prednisolone from (Anhui Medical Co. Ltd).

2.3. Animals. Male ICR mice (25–30 g) and Sprague-Dawley (150–200 g) rats of both sexes were housed in the animal facility of the Department of Biomedical and Forensic Sciences, University of Cape Coast (UCC). The animals were housed in groups of five in stainless steel cages (34 × 47 × 18 cm) with soft wood shavings as bedding, fed with normal commercial pellet diet (AGRICCARE, Kumasi), given water *ad libitum*, and maintained under laboratory conditions. All procedures and techniques used in these studies were in accordance with the National Institute of Health Guidelines for the Care and Use of Laboratory Animals [9]. All protocols used were approved by the Departmental Ethics Committee.

2.3.1. Source of Rodent Parasite (Plasmodium berghei NK65). The rodent parasite was obtained from Noguchi Memorial Institute for Medical Research, University of Ghana, Legon, Ghana, and maintained alive in mice by continuous intraperitoneal passage in mice after every 6 days [10]. The reinfected mice were kept in the animal house of the Department of Biomedical and Forensic Sciences.

2.3.2. Inoculation of Parasite. Total inoculum concentration of 60×10^6 of *P. berghei* parasitized erythrocytes per mL was prepared. This was carried out by determining the parasite

density of the *Plasmodium berghei*-infected mice. The blood obtained from the infected mice was diluted appropriately with EDTA-phosphate buffer saline (PBS) and subsequently washed with PBS. Each mouse was intraperitoneally inoculated on day 0 with 0.2 mL of infected erythrocytes containing 1×10^6 *P. berghei* parasitized red blood cells.

2.4. Effect of Xylopic Acid on Established Plasmodium berghei Infection. To evaluate the curative antimalarial properties of xylopic acid on established *Plasmodium berghei* infection, thirty male mice were each inoculated with 1×10^6 *P. berghei* on the first day [11]. The mice were assigned to five groups ($n = 6$). Seventy-two hours later, the animals were treated once daily with three doses of xylopic acid (10, 30, and 100 mg/kg *p.o.*) (groups 1–3), 4 mg/kg *p.o.* of artemether/lumefantrine (A-L) (standard drug: group 4), and 10 mL/kg *p.o.* normal saline (group 5) for 5 days. To determine the daily parasitaemia level, about three drops of blood were collected from the tail of each mouse and smeared onto a microscope slide to make a thin film. The thick film was prepared from two drops of blood obtained from the tail of the mice. The smears were fixed in absolute ethanol and stained with 10% Geimsa stain, and examined microscopically (×100 magnification). The parasitaemia was determined by counting infected erythrocytes in hundred fields, divided by the total erythrocytes in the hundred fields then multiplied by hundred. On the twelfth day (D 13), two animals from each treatment group were sacrificed, and the liver were taken for histopathological assay. The tissue was embedded in paraffin; 8 μm sections were cut on a microtome (Bright 5040, Bright instrument company Ltd., England) and processed for routine haematoxylin-eosin staining. Slides of tissue sections were observed using trinocular clinical light microscope with a digital camera (Olympus CX1, Japan) connected to a computer. Micrographs of the tissue were generated using the ×10 objective lens for further analysis. The mean survival time of the mice in each treatment group was determined over a period of 30 days.

2.5. Prophylactic Activity of Xylopic Acid on P. berghei Infection. Xylopic acid was further assayed for its prophylactic activity against *P. berghei* infection using the method described by Peters [12]. The mice were randomly assigned to five groups ($n = 6$) and pretreated orally with 10, 30, and 100 mg/kg/day of xylopic acid, 1.2 mg/kg/day sulfadoxine/pyrimethamine (SP, the reference drug), and 10 mL/kg/day of normal saline. The treatment was continued for 3 consecutive days. On the fourth day, all mice were infected with 1×10^6 *P. berghei*, and seventy-two hours later, blood smears were prepared from the tail. The parasite density and % chemosuppression for all the treatment groups were determined.

2.6. Lipopolysaccharide-Induced Fever. The method of Santos and Rao [13] was used with slight modification for the assessment of the antipyretic activity of xylopic acid. Rats were fasted overnight prior to induction of fever, and water was given *ad libitum*. Rectal temperature was measured using

TABLE 1: Summary of the effect of XA and A-L on established *Plasmodium berghei* infection in mice.

| Parameters | Control (vehicle) | Xylopic acid (mg/kg) | | | A-L (mg/kg) |
		10	30	100	4
Survival days	12 ± 0.4	25 ± 0.3	19 ± 1.3	29 ± 0.8	28 ± 0.2

Values are expressed as mean $\pm$ S.E.M. ($n = 6$).

a lubricated ECT-1 digital thermometer (Estar Electronic And Instrument Co., Ltd., Zhejiang, China) inserted 3 mm deep into the rectum of the rats. Fever was induced by injecting intramuscularly 1 mg/kg of LPS into the right thigh of each rat. Rectal temperature was measured again, and animals that showed an increase in temperature of 0.5°C and more were selected for the study. The animals with fever were put into five groups ($n = 6$) and were treated with three doses of xylopic acid (10–100 mg/kg), 30 mg/kg of prednisolone, or 1 mL/kg of normal saline solution (the control), orally, two hours after LPS-induced fever. Rectal temperature was measured after 1 h of the treatments.

2.7. Statistical Analysis. GraphPad Prism for Windows version 4.03 (GraphPad Software, San Diego, CA, USA) was used for all statistical analyses, and $P < 0.05$ was considered statistically significant. All data were expressed as mean $\pm$ SEM (duplicate measurement). The time-course curves were subjected to two-way (treatment $\times$ time) repeated measures analysis of variance (ANOVA) with Bonferroni's *post hoc* test. The column graphs were subjected to one-way analysis of variance (ANOVA) with Tukey's *post hoc* test.

3. Results

3.1. Curative Activities Of Xylopic Acid and A-L on P. berghei Infection in Mice. Xylopic acid and A-L reduced the parasitaemia significantly from the first day of treatment to the final day. Again, all of the treatments resulted in relatively increased the survival time of the mice compared to the control (Table 1). Xylopic acid significantly ($P < 0.0001$) reduced the level of parasitaemia from day one after treatment and achieved the highest effect on the last day (Figure 2). The % parasitaemia decreases by the 10 mg/kg xylopic acid were 6.7%, 13.9%, 54.7%, 70%, and 85.4 from day 1 to day 5 after treatment, respectively. The 30 mg/kg xylopic acid also produced % parasitaemia reductions of 5.2%, 46.1%, 53%, 86.9%, and 87.6% from day 1 to day 5 after treatment, respectively. Similarly, the % parasitaemia decreases by the 100 mg/kg xylopic acid were 7.4%, 46.1%, 84.8%, 92.8%, and 99.6% from day 1 to day 5 after treatment, respectively, (Figure 2).

The standard antimalarial drug A-L (4 mg/kg) produced a % of parasitaemia reductions of 1.1%, 35.2%, 63.6%, 91.7%, and 99.6% from day 1 to day 5 after treatment, respectively, (Figure 2).

The highest dose of xylopic acid (100 mg/kg) and 4 mg/kg A-L produced a maximum % 99.6% chemosuppression on the last day. The % chemosuppression of A-L was, however, 1.1 times greater than the maximum % chemosuppression

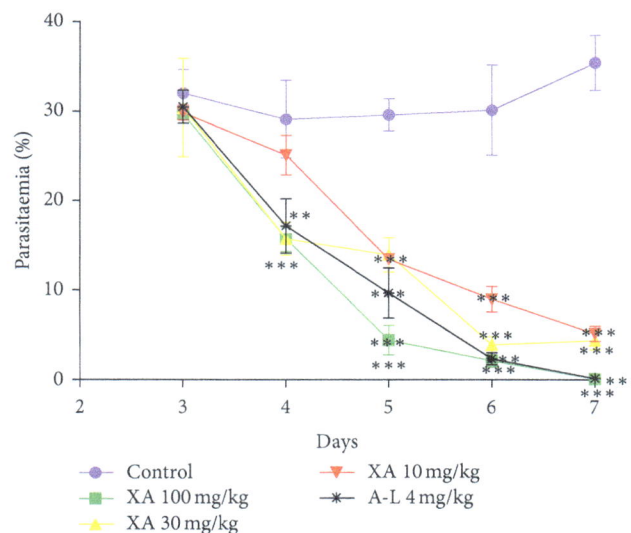

FIGURE 2: Curative effect of xylopic acid and A-L on the time-course curve of *Plasmodium berghei* infection in mice. Data is presented as mean $\pm$ SEM. *** $P < 0.001$, ** $P < 0.01$ compared to vehicle-treated group (two-way ANOVA followed by Bonferroni's *post hoc* test).

produced by 30 mg/kg dose of xylopic acid and 1.2 times greater than the maximum % chemosuppression produced by the lowest dose of xylopic acid. The % chemosuppression of A-L was not statistically significant compared to the % chemosuppression produced by the various doses of xylopic acid (Table 1).

Histopathological assessments of the hepatocytes reveal high levels of Kupffer cells in all the treated groups of animals except the mice treated with 100 mg/kg of xylopic acid. Parasites were observed in the lumen of the blood vessels of liver sections of the normal saline and middle dose of xylopic acid-treated groups but were barely seen in the lumen of A-L and the lowest and highest doses of xylopic acid-treated mice (Figure 3).

3.2. Prophylactic Activities of Xylopic Acid and SP on P. berghei Infection in Mice. Xylopic acid exhibited significant ($P < 0.05$) prophylactic activity against *Plasmodium berghei in vivo* at all of the three doses tested (Figure 4), seen as reduction in parasite count compared to the vehicle-treated group. The % chemosuppressive effect seen at the highest dose employed was 59.4%.

3.3. Lipopolysaccharide-Induced Pyrexia. Xylopic acid (30 and 100 mg kg^{-1}) reduced significantly ($P < 0.05$) lipopolysaccharide-induced fever in rats (Figure 5). Prednisolone

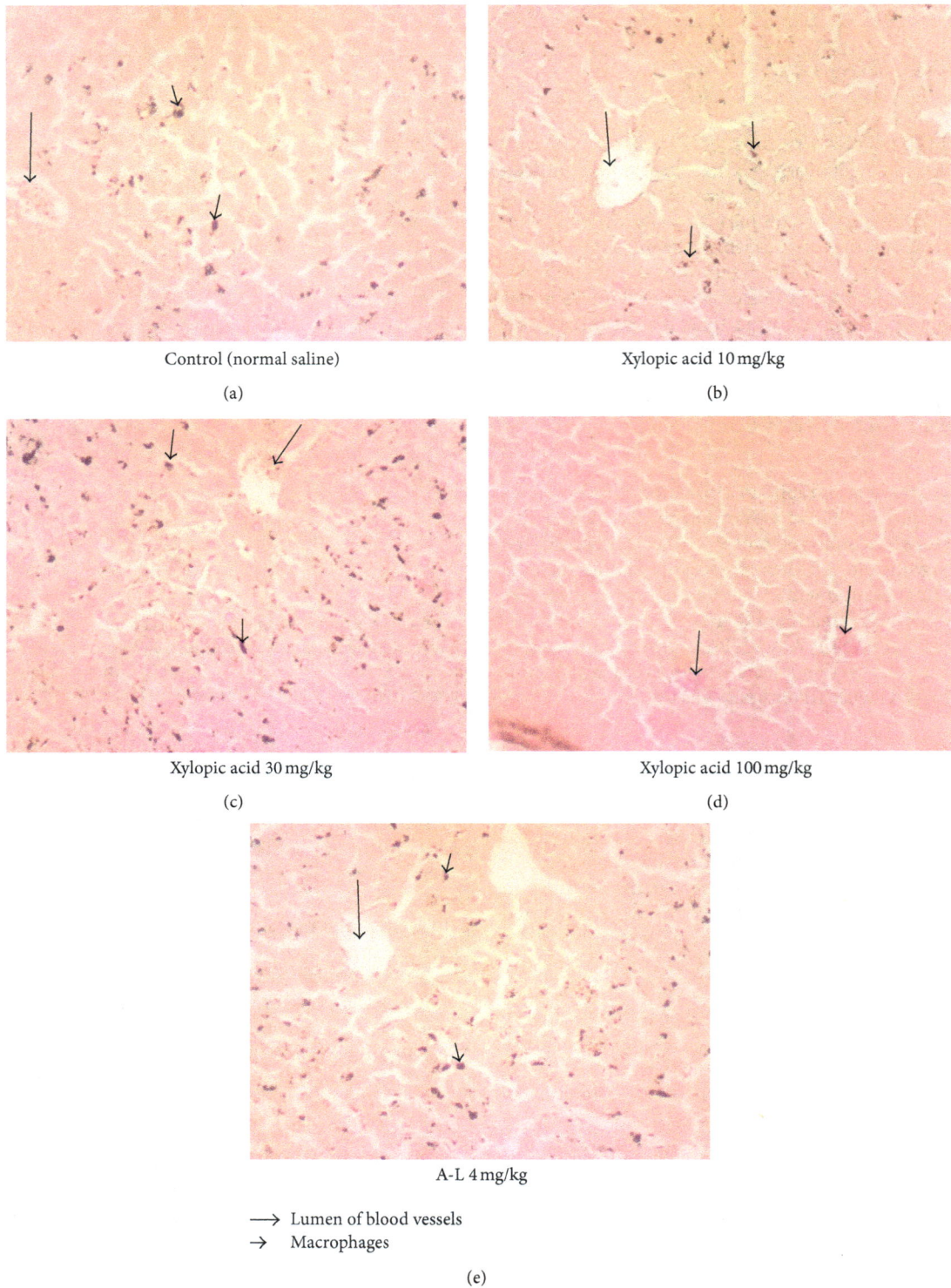

Control (normal saline)

(a)

Xylopic acid 10 mg/kg

(b)

Xylopic acid 30 mg/kg

(c)

Xylopic acid 100 mg/kg

(d)

A-L 4 mg/kg

$\longrightarrow$ Lumen of blood vessels
$\rightarrow$ Macrophages

(e)

FIGURE 3: Photomicrographs of liver cells of animals treated with xylopic acid (10–100 mg/kg), artemether/lumefantrine (A-L) 4 mg/kg, and normal saline showing some Kupffer cells in the hepatocytes and *P. berghei*-infected red cells in the lumen of blood vessels.

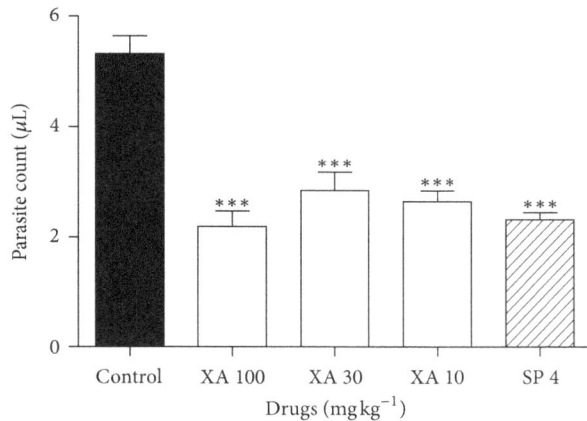

FIGURE 4: Prophylactic effects of xylopic acid and SP on *Plasmodium berghei* infection in mice. ***$P < 0.001$, compared to vehicle-treated group (One-way ANOVA followed by Tukey's *post hoc* test).

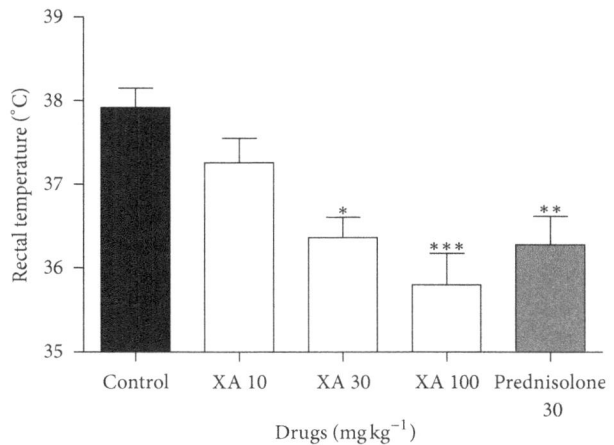

FIGURE 5: Antipyretic effects of xylopic acid (10–100 mg/kg) and prednisolone (10 mg/kg) on LPS-induced pyrexia in rats. Data plotted are means ± SEM. *$P \leq 0.05$, **$P \leq 0.01$, ***$P \leq 0.001$; the level of significance of rectal temperature reduction compared to the normal saline-treated group (One-way ANOVA followed by Tukey's *post hoc* test).

used as positive control also significantly reduced ($P < 0.05$) lipopolysaccharide-induced fever in rats (Figure 5).

4. Discussion

Malaria infections are complicated syndromes involving many inflammatory responses which may enhance cell-to-cell interaction (cytoadherence), cell stimulation involving malaria-derived antigens/toxins and host-derived factors such as cytokines. Moderate amounts of cytokines are though good for the host causes fever [14]. Clinically, it is crucial to reverse the effects of both toxins and cytokines to prevent further complications of malaria. This makes xylopic acid an ideal agent for malaria treatment because it exerted curative and prophylactic properties on *P. berghei*-induced malaria in mice as well as antipyretic activities in rats. The inflammatory condition of malaria is charcterised by free radical generation, activation of phospholipase activity resulting in generation of eicosanoids such as prostaglandins and other cytokines (TNF, IFN-γ, and IL-1β). These inflammatory mediators as well as parasite sequestration are responsible for the disease. It has been suggested that the cytokines upregulate the expression of adhesion molecules such as ICAM-1 that is involved in the binding of the parasitized red blood cells to the vascular endothelium [15]. The curative antiplasmodial properties of xylopic acid may be due to the inhibition of the production and/or release of these inflammatory mediators associated with malaria. Indeed, xylopic acid has analgesic properties [7] and preliminary data in our laboratory also indicate that xylopic acid possesses potent anti-inflammatory properties. In addition, the curative effect may be attributed to its direct cytotoxic effect on the parasites in a mechanism similar to the A-L combination. A-L is an oral fixed dose combination of artemether (20 mg) and lumefantrine (120 mg). Artemether exerts its antimalaria properties by interference with parasite transport proteins, disruption of parasite mitochondrial function, inhibition of angiogenesis, and modulation of host immune function [16]. Lumefantrine, an aryl-amino alcohol,

prevents detoxification of heam, resulting in parasite death from the toxic heam and free radicals [17, 18]. It is worth noting that xylopic acid completely eradicated parasites from the blood of the mice similar to the standard A-L. Xylopic acid again at the highest dose to a greater extent destroyed the parasites circulating in the lumen of the blood vessels of the liver [19]. The macrophages present in the liver sections of the control and low doses of xylopic acid-treated mice could be due to inflammatory processes induced by the circulating parasite [19]. The absence of macrophages in the liver sections of the highest dose of xylopic acid could be attributed to the complete elimination of the parasites from circulation [19]. Xylopic acid and A-L both prolonged the survival times of the mice and this could be attributed to the high parasitaemia clearance (reduced parasite burden) observed for these drugs [18].

Xylopic acid showed comparable efficacy to SP in the prophylaxis assay. This indicates the nonselectivity of xylopic acid on the stages of malaria parasite. It is not clear how xylopic acid exerts prophylactic activity on *P. berghei* infection but it may be inhibiting the multiplication of the parasites as well as direct cytotoxic effect on the parasites [16]. It may modulate the membrane properties of the erythrocytes preventing parasite invasion [15]. SP used in this study exerts prophylactic activities via the inhibition of dihydropteroate synthetase and dihydrofolate reductase enzymes of the parasites [20]. Generally, prophylactic antimalarial drugs work by disrupting the initial development of malaria parasites in the liver (causal activity). They may act by suppressing the emergent asexual blood stages of the parasite (suppressive activity) or by preventing the relapses induced by the latent liver forms (hypnozoites) [21]. Xylopic acid can therefore be used for malaria prophylaxis as well as a curative agent such as atovaquone/proguanil (Malarone), a drug approved in USA for malaria treatment and prophylaxis [22]. Although the rodent model presents with some limitations, it has successfully been validated through the identification of several conventional antimalarials including the currently used antimalarials, halofantrine, and the artemisinin derivatives [23].

Proinflammatory mediators, IL-2, and PGE_2, are among the important mediators of LPS-induced pyrexia. The antipyretic activity of xylopic acid in this model may be attributed to its negative effect on cytokines. This partly may explain the antimalarial effect of xylopic acid.

5. Conclusion

Xylopic acid possesses curative and prophylactic properties on *P. berghei*-induced malaria in ICR mice as well as antipyretic properties. It is therefore an ideal antimalarial drug candidate.

Conflict of Interests

The authors have no conflict of interests with the trademarks stated in this study. There is no financial gain or any other benefits from the cited trademarks.

Acknowledgments

The authors express their sincere gratitude to Mr. Amoaning and Miss Nancy Darkoa Darko for their help. The authors give special thanks to Mr. Amonoo of the Animal House of UCC.

References

[1] WHO, *World Malaria Report*, World Health Organization, Geneva, Switzerland, 2010.

[2] K. Mishra, A. P. Dash, B. K. Swain, and N. Dey, "Anti-malarial activities of *Andrographis paniculata* and *Hedyotis corymbosa* extracts and their combination with curcumin," *Malaria Journal*, vol. 8, no. 1, article 26, 2009.

[3] M. Randrianarivelojosia, V. T. Rasidimanana, H. Rabarison et al., "Plants traditionally prescribed to treat tazo (malaria) in the eastern region of Madagascar," *Malaria Journal*, vol. 2, no. 1, p. 25, 2003.

[4] L. K. Basco, S. Mitaku, A.-L. Skaltsounis et al., "In vitro activities of furoquinoline and acridone alkaloids against *Plasmodium falciparum*," *Antimicrobial Agents and Chemotherapy*, vol. 38, no. 5, pp. 1169–1171, 1994.

[5] M. M. Suleiman, M. Mamman, Y. O. Aliu, and J. O. Ajanusi, "Anthelmintic activity of the crude methanol extract of *Xylopia aethiopica* against *Nippostrongylus brasiliensis* in rats," *Veterinarski Arhiv*, vol. 75, no. 6, pp. 487–495, 2005.

[6] L. N. Tatsadjieu, J. J. Essia Ngang, M. B. Ngassoum, and F.-X. Etoa, "Antibacterial and antifungal activity of *Xylopia aethiopica*, *Monodora myristica*, *Zanthoxylum xanthoxyloides* and *Zanthoxylum leprieurii* from Cameroon," *Fitoterapia*, vol. 74, no. 5, pp. 469–472, 2003.

[7] E. Woode, E. O. Ameyaw, E. Boakye-Gyasi, and W. K. M. Abotsi, "Analgesic effects of an ethanol extract of the fruits

[8] B. C. Cavalcanti, D. P. Bezerra, H. I. F. Magalhães et al., "Kauren-19-oic acid induces DNA damage followed by apoptosis in human leukemia cells," *Journal of Applied Toxicology*, vol. 29, no. 7, pp. 560–568, 2009.

[9] National Institute of Health Guidelines for the Care and Use of Laboratory Animals and National Institutes of Health, Office of Science and Health Reports, Guide for care and use of laboratory animals 83-23, Office of Science and Health Reports, Department of Health and Human Services, Bethesda, Md, USA, 1996.

[10] A. Ishih, T. Suzuki, T. Hasegawa, S. Kachi, H. Wang, and M. Terada, "*In vivo* evaluation of combination effects of chloroquine with cepharanthin or minocycline hydrochloride against blood-induced choloquine-resistant *Plasmodium berghei* NK65 infections," *Tropical Medicine and Health*, vol. 32, pp. 15–19, 2004.

[11] A. H. Al-Adhroey, Z. M. Nor, H. M. Al-Mekhlafi, and R. Mahmud, "Ethnobotanical study on some Malaysian anti-malarial plants: a community based survey," *Journal of Ethnopharmacology*, vol. 132, no. 1, pp. 362–364, 2010.

[12] W. Peters, "Drug resistance in *Plasmodium berghei* Vincke and Lips, 1948. III. Multiple drug resistance," *Experimental Parasitology*, vol. 17, no. 1, pp. 97–102, 1965.

[13] F. A. Santos and V. S. N. Rao, "A study of the anti-pyretic effect of quinine, an alkaloid effective against cerebral malaria, on fever induced by bacterial endotoxin and yeast in rats," *Journal of Pharmacy and Pharmacology*, vol. 50, no. 2, pp. 225–229, 1998.

[14] N. Depinay, J. F. Franetich, A. C. Grüner et al., "Inhibitory effect of TNF-α on malaria pre-erythrocytic stage development: influence of host hepatocyte/parasite combinations," *PLoS One*, vol. 6, no. 3, Article ID e17464, 2011.

[15] D. S. Hansen, "Inflammatory responses associated with the induction of cerebral malaria: lessons from experimental murine models," *PLoS Pathogens*, vol. 8, no. 2, Article ID e1003045, 2012.

[16] J. Golenser, J. H. Waknine, M. Krugliak, N. H. Hunt, and G. E. Grau, "Current perspectives on the mechanism of action of artemisinins," *International Journal for Parasitology*, vol. 36, no. 14, pp. 1427–1441, 2006.

[17] P. I. German and F. T. Aweeka, "Clinical pharmacology of artemisinin-based combination therapies," *Clinical Pharmacokinetics*, vol. 47, no. 2, pp. 91–102, 2008.

[18] G. Kokwaro, L. Mwai, and A. Nzila, "Artemether/lumefantrine in the treatment of uncomplicated *falciparum* malaria," *Expert Opinion on Pharmacotherapy*, vol. 8, no. 1, pp. 75–94, 2007.

[19] A. Haque, S. E. Best, F. H. Amante et al., "High parasite burdens cause liver damage in mice following Plasmodium berghei ANKA infection independently of CD8[+] T cell-mediated immune pathology," *Infection and Immunity*, vol. 79, no. 5, pp. 1882–1888, 2011.

[20] I. Petersen, R. Eastman, and M. Lanzer, "Drug-resistant malaria: molecular mechanisms and implications for public health," *FEBS Letters*, vol. 585, no. 11, pp. 1551–1562, 2011.

[21] D. R. Hill, J. K. Baird, M. E. Parise, L. S. Lewis, E. T. Ryan, and A. J. Magill, "Primaquine: report from CDC expert meeting on malaria chemoprophylaxis I," *The American Journal of Tropical Medicine and Hygiene*, vol. 75, no. 3, pp. 402–415, 2006.

[22] G. S. Dow, A. J. Magill, and C. Ohrt, "Clinical development of new prophylactic antimalarial drugs after the 5th Amendment

of *Xylopia aethiopica* (Dunal) A. Rich, (Annonaceae) and the major constituent, xylopic acid in murine models," *Journal of Pharmacy and BioAllied Sciences*, vol. 4, no. 4, pp. 291–301, 2012.

to the Declaration of Helsinki," *Therapeutics and Clinical Risk Management*, vol. 4, no. 4, pp. 803–819, 2008.

[23] J. F. Ryley and W. Peters, "The antimalarial activity of some quinolone esters," *Annals of Tropical Medicine and Parasitology*, vol. 64, no. 2, pp. 209–222, 1970.

20

B-Cell Response during Protozoan Parasite Infections

María C. Amezcua Vesely, Daniela A. Bermejo, Carolina L. Montes,
Eva V. Acosta-Rodríguez, and Adriana Gruppi

Centro de Investigaciones en Bioquímica Clínica e Inmunología (CIBICI-CONICET), Departamento de Bioquímica Clínica,
Facultad de Ciencias Químicas, Universidad Nacional de Córdoba, Haya de la Torre y Medina Allende, Ciudad Universitaria,
5000 Córdoba, Argentina

Correspondence should be addressed to Adriana Gruppi, agruppi@fcq.unc.edu.ar

Academic Editor: Mauricio M. Rodrigues

In this review, we discuss how protozoan parasites alter immature and mature B cell compartment. B1 and marginal zone (MZ) B cells, considered innate like B cells, are activated during protozoan parasite infections, and they generate short lived plasma cells providing a prompt antibody source. In addition, protozoan infections induce massive B cell response with polyclonal activation that leads to hypergammaglobulnemia with serum antibodies specific for the parasites and self and/or non related antigens. To protect themselves, the parasites have evolved unique ways to evade B cell immune responses inducing apoptosis of MZ and conventional mature B cells. As a consequence of the parasite induced-apoptosis, the early IgM response and an already establish humoral immunity are affected during the protozoan parasite infection. Moreover, some trypanosomatides trigger bone marrow immature B cell apoptosis, influencing the generation of new mature B cells. Simultaneously with their ability to release antibodies, B cells produce cytokines/quemokines that influence the characteristic of cellular immune response and consequently the progression of parasite infections.

1. B Cells Can Play Protective and Pathogenic Roles in Protozoan Infections

Host resistance in protozoan infections is dependent on both innate and acquired cell-mediated immune responses. In addition, several studies have implicated B cells and antibodies (Abs) in host survival and protozoan parasite clearance [1–3]. B cells can function as Ab-producing cells but they can also modulate immune responses through critical Ab-independent mechanisms that include secretion of cytokines and chemokines as well as antigen presentation [4–6]. Furthermore, B cells can directly modulate dendritic cells and T-cell subsets, and, consequently, they can influence adaptive immunity and the progression of the infection [7]. Accordingly, in protozoan infections B cells may play a protective and a pathological role. In malaria and trypanosome infections, Abs appear to play a famajor role in immunity. In *Trypanosoma cruzi* and *T. brucei gambiense* infections, Ab-dependent cytotoxic reactions against the parasite have been

reported [8]. Several studies demonstrated that Abs are responsible for the survival of susceptible animals in the initial phase of *T. cruzi* infection and for the maintenance of low levels of parasitemia in the chronic phase [9, 10]. Although Abs were shown to be responsible for clearing the African trypanosomes from the blood of infected animals, recent evidence suggests that the survival time of infected mice does not necessarily correlate with the ability of the animal to produce trypanosome-specific antibody. In general, the parasite-specific immune response mounted during protozoan infections is insufficient to completely eradicate the pathogen, allowing chronic infection.

B cells do not only play protective roles in protozoan infections. In fact, they are required for the development of Th2 cell response and, consequently, for the susceptibility to infection with *Leishmania major* [11]. BALB/c uMT mice infected with *L. major* LV39 mount a Th1 response and present restricted lesion development and contained parasite replication. Adoptive transfer of B cells from BALB/c mice in

B-cell-deficient BALB/c uMT mice before infection restores susceptibility to *L. major* LV39 and Th2 cell development in resistant mice.

2. B-Cell Development

Given the role for B cells in conditioning the progression of protozoan infections, it is important to understand the kinetics and regulation of the whole B-cell cycle from the development to the differentiation into mature and memory B cells and plasma cells. The humoral immune response has been shown to be two branched providing an innate-like response (involving B1 and marginal zone (MZ) B cells) and an adaptive immune response (involving conventional B2 cells). In the adult, B cells are generated in the bone marrow (BM) and migrate to the periphery at the transitional B-cell stage, when they are still short lived and functionally immature [12, 13]. Conventional B2-cell development occurs via a series of BM stromal cell-facilitated processes that begin within the hematopoietic stem cell pool and proceed in hierarchical steps of lineage commitment [14]. B lymphopoiesis yields several developmental stages of pre-pro-B, pro-B, pre-B, and, eventually, immature B cells, which show a high expression of the IgM form of the antigen receptor and low or no expression of the IgD maturation marker. To complete their development, immature B cells migrate through the periphery; however, only 10% reach the spleen as transitional B cells of the T1 type [15]. In the spleen, transitional B cells develop into conventional and MZ B cells [16]. B1 cells are efficiently generated in fetal life and during the first few weeks after birth. The fetal liver is an efficient source of B1 cells [17]; however, it is not the only one as a recent study identified a B1-cell precursor in adult BM [18]. Interestingly, protozoan parasite infections can affect the different compartments of B-cell development (summarized in Figure 1), influencing the generation of new mature B cells or their survival and, consequently, the cellular immune response.

3. Protozoan Parasite Infections Affect BM B-Cell Development

BM is the main hematopoietic organ of an adult organism and is able to provide cells of immune systems rapidly in cases of infection. Immature B-cell reduction in BM during an infection would limit the Ab source cells and favour parasite replication and chronicity, so the identification of the mechanisms ruling B-cell depletion represents an important challenge in biomedical research.

We have reported that *T. cruzi* infection induces a marked loss of immature B cells in the BM and also compromises recently emigrated B cells in the periphery [19]. The depletion of immature BM B cells was associated with an increased rate of apoptosis, and we established that *T. cruzi* trypomastigotes failed to directly induce immature B-cell apoptosis. We proved that this cell death process occurs in a Fas/FasL-independent fashion but depends on the presence of CD11b$^+$ myeloid cells that secrete a product of the cyclooxygenase pathway that depletes immature B cells [19].

In addition, BM is compromised in other protozoan parasite infections. In fact, infections with *Neospora caninum* [20] and *T. brucei* [21] also cause a general decrease in BM cells. Recently, the *T. brucei* infection upshot on B lymphopoiesis has been examined using a C57BL/6 mouse *T. brucei* AnTat 1.1E infection model [22]. Using this model, Bockstal et al. [23] observed that the number of hematopoietic stem cells was minimally affected, but BM B lymphopoiesis was severely affected in *T. brucei*-infected mice, starting with the common lymphoid progenitor fraction. The pre-pro-B-cell population showed a 50% reduction by day 20 after infection, while the subsequent B-cell maturation stage, that is, the pro-B, pre-B, and immature B-cell populations reached more than 95% depletion by day 10 after infection and failed to recover throughout the further course of infection. In *T. brucei* infection, mice do not present increased apoptosis of BM B-cell precursors nor alteration in the expression of B-cell-development-specific transcription factors like Icaros, PU.1, EBF and E2A and the IL-7. However, *T. brucei*-infected mice show a reduction in BM CXCL12 levels [23], indicating that during early *T. brucei* infections, B-cell precursors prematurely migrate out of the BM as a result of the initiation of inflammation. Similarly, CXCL12 decreased production by BM cells was determined in *Plasmodium chabaudi* infection [24]. Furthermore, the significant reduction in CXCL12 expression in the BM of 10 days *P. chaubaudi*-infected mice correlates with a reduction in B-cell precursor cells. At days 20 and 30 of infection, a significant recovery in CXCL12 expression in BM is detected, coinciding with a slow recovery of B lymphopoiesis.

4. B1 and MZ B-Cell Response in Protozoan Parasite Infections

Among the mature B cells, MZ and B1 B cells appear to be evolutionarily selected and maintained to facilitate prompt Ab responses. Due to this, they provide a bridge between the innate and the adaptive arms of the antipathogen immune response. B1 cells, distinguished from B2 cells by their phenotype (B220^{low} CD5$^{+/-}$ CD11b$^+$) and anatomic location and functional properties, are the dominant population of B cells in the pleural and peritoneal cavities, but represent only a small fraction of splenic B cells [25]. B1 cells produce most of the natural serum IgM and much of the gut IgA and express a BCR repertoire that is enriched for highly polyspecific receptors with low affinities to a broad range of antigens [26]. Despite the fact that B1 cells are very efficient in the control of bacterial and viral infections, they apparently do not play a role in the control of protozoan parasite replication. Indeed, BALB/c Xid mice carrying an X-linked mutation (that prevents B1 cell development) infected with *T. cruzi* display poor B-cell responses to the infection, accompanied by low levels of specific and nonspecific immunoglobulins in the serum [27]. Surprisingly, Xid mice infected with *T. cruzi* were able to control parasitemia and did not show the wasting syndrome observed in wild-type mice. In addition, they developed almost no pathology early in the chronic phase. The resistance of these mice to experimental Chagas disease was associated with the absence of

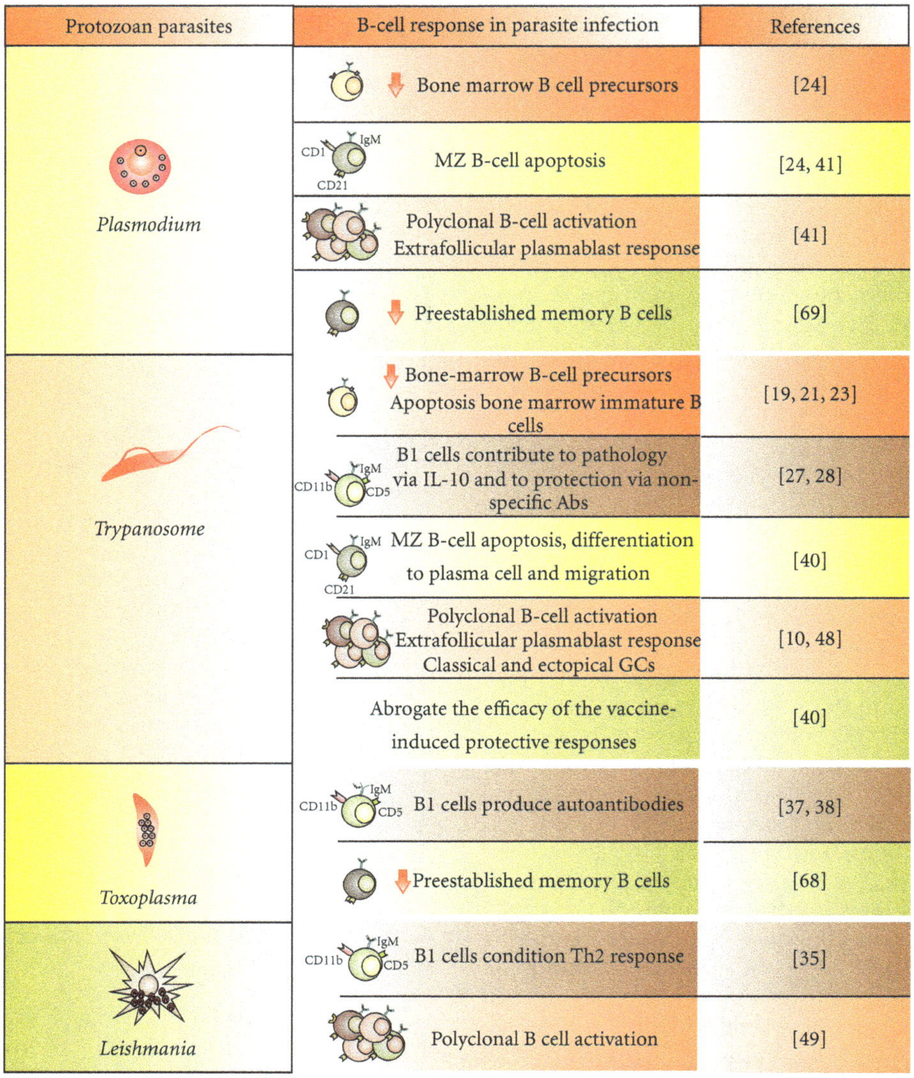

Protozoan parasites	B-cell response in parasite infection	References
Plasmodium	↓ Bone marrow B cell precursors	[24]
	MZ B-cell apoptosis	[24, 41]
	Polyclonal B-cell activation Extrafollicular plasmablast response	[41]
	↓ Preestablished memory B cells	[69]
Trypanosome	↓ Bone-marrow B-cell precursors Apoptosis bone marrow immature B cells	[19, 21, 23]
	B1 cells contribute to pathology via IL-10 and to protection via non-specific Abs	[27, 28]
	MZ B-cell apoptosis, differentiation to plasma cell and migration	[40]
	Polyclonal B-cell activation Extrafollicular plasmablast response Classical and ectopical GCs	[10, 48]
	Abrogate the efficacy of the vaccine-induced protective responses	[40]
Toxoplasma	B1 cells produce autoantibodies	[37, 38]
	↓ Preestablished memory B cells	[68]
Leishmania	B1 cells condition Th2 response	[35]
	Polyclonal B cell activation	[49]

FIGURE 1: Protozoan parasites affect the different B-cell compartments. MZ: marginal zone B cells, GCs: germinal centers.

IL-10-secreting B1 cells and high levels of IFN-gamma [28]. These results suggested that B1 cells play a pathological rather than protective role in Chagas' disease. Additionally, in *T. cruzi* infection, we observed a disappearance of peritoneal B1 cells, due to an enhanced differentiation into a particular type of plasma cells, the "Mott-like cells" [29]. Nevertheless, the specific role of these cells in the experimental Chagas disease has not been elucidated yet; their association with autoimmune manifestations in CD22-deficient mice [30] and lupus [31, 32] suggests that these cells may be involved in the autoimmune responses observed in *T. cruzi* infection. We and others have reported that the peritoneal B-cell response observed in *T. cruzi* infection is almost not specific for the invading pathogen [29, 33]. However, these "sticky" antibodies could unspecifically bind to parasites providing protection.

B1 cells are not only implicated on Ab secretion; in fact, they may modulate T-cell response. In this sense, O'Garra et al. [34] have reported that B1 cells secrete large amounts of IL-10 and, consequently, can contribute to the susceptibility of BALB/c mice to *L. major* infection by skewing the T-helper cell network towards a Th2 phenotype. In this way, it has been observed that *L. major* infection of B-cell-defective BALB/c Xid mice induces a less severe disease compared to wild-type control mice [35]. Another report indicates that the behavior of *L. major*-infected Xid mice can be explained more in relation to the high endogenous IFN-gamma production than to the lack of B1 cells. Indeed, B1-cell-depleted irradiated mice showed similar or even worse disease progression compared to control BALB/c mice [36].

B1 cells would also be implicated in the pathogenesis of toxoplasmosis through the production of Abs against the heat shock protein 70 of *T. gondii* that also recognize mice HSP70 [37]. These Abs seem to have a pathogenic role in toxoplasmosis as their injection in *T. gondii*-infected mice clearly increases the number of parasites in mice brain [38]. Moreover, IL-10 produced by B1 cells could, in turn, favor *T. gondii* survival. Then, fine tune regulation of the exacerbated

Th1 response by IL-10 is important during *T. gondii* infection.

MZ B cells are also considered innate-like cells that can be induced to differentiate into short-lived plasma cells in the absence of BCR ligation. Splenic MZ B cells can be distinguished from the other splenic B cells by CD24^{high}, IgM^{high}, IgD^{high}, CD23$^-$ expression, as well as by their higher expression of CD21. It is known that these B cells mediate humoral immune responses against blood-borne type 2 T-independent antigens [39] but their role in parasite infection has been scarcely studied. Induction of a T-independent anti-trypanosome IgM response has been shown to be a crucial factor in *T. brucei* parasite elimination [1]. Even when increased splenic cellularity occurs after *T. brucei* infection, a significant reduction of splenic IgM+ MZ B-cell numbers takes place right after the first week of infection [40]. The infection-associated disappearance of the MZ B cells from the spleen could be explained by two independent mechanisms, namely, cell differentiation and/or cell death. Supporting the first possibility is the observation that the rapid disappearance of MZ B cells coincided with the temporary accumulation of IgM+ plasma cells. The analysis of MZ B cells that remain in the spleen in the days following the clearance of the first peak of parasitemia revealed that these cells upregulated Annexin V expression. In addition, these cells exhibit caspase 3 gene expression as well as the conversion of procaspase 3 into the cleaved 12 kD and 17 kD caspase 3 activation products suggesting the induction of trypanosomiasis-associated apoptosis in the splenic MZ B-cell population [40]. As in African trypanosomes, *P. chabaudi chabaudi* infection also caused a severe depletion of MZ B cells in the spleen [41] and this loss is mainly the result of the highly increased rate of apoptosis [24]. As MZ B cells serve as an important source for T-cell independently generated IgM+ plasma cells during early stages of infection, apoptosis induction of MZ B cells can be used by parasites as strategy to avoid early IgM protective response and, consequently, prolong their survival.

5. Protozoan Infections Induce Massive B-Cell Response with Polyclonal Activation of Splenic B Cells

Whereas the BM of some protozoan-infected mice suffers from a strong B-lineage-cell depletion, the spleens show a marked cellular hyperplasia as a consequence of an intense B-cell response. A detailed analysis of splenic B-cell response was performed in experimental Chagas disease and malaria [10, 41]. An extrafollicular Ab response, in mice infected with *T. cruzi*, is evident a few days after infection and reached a peak after 18 days of infection. This extended kinetics of the extrafollicular response could be characteristic of infections caused by blood circulating protozoan parasites since in *P. chabaudi chabaudi* infection extrafollicular plasmablasts are visible from day 4, and by day 10 they are unconventionally sited in the periarteriolar region of the white pulp. In this region, in both *T. cruzi* and *Plasmodium* infection, extrafollicular plasmablasts form clusters occupying part of the area

normally filled by T cells. The kinetics of the appearance of GCs during *T. cruzi* and *Plasmodium* infection are similar to those observed after immunization with classical haptenated proteins, where GCs are visible within the 8 days of immunization [42]. In addition, we detected functional (Ab producing) GCs in atypical sites. The GCs in the spleens of *T. cruzi*-infected mice persisted for at least 32 days resembling the kinetics of the response seen in *P. chabaudi* [41] and *L. amazonensis* [43] infections. We observed that even though *T. cruzi* infection induces early, persistent, and massive extrafollicular and follicular plasmablast responses together with classical and ectopic GCs, infected mice have a delayed parasite-specific Ab response. A key finding of our study [10] is that, while an important amount of Abs is rapidly secreted during infection, antigen specific antibodies were not detected until the third week of infection.

The consequence of the massive extrafollicular and follicular B-cell response is the polyclonal B-cell activation that leads to hypergammaglobulinemia with serum Abs specific for the parasite and self- and/or nonrelated Ags [44–46]. In leishmaniasis, hypergammaglobulinemia was described in both *L. major*-susceptible and -resistant mouse strains. *T. congolense* infection also results in a strong production of non-parasite-specific Abs characterized by the predominance of IgG2a- and IgG2b isotypes [47]. All the mouse strains infected with *T. congolense* present a marked increase in splenic B cells resulting in a nonspecific polyclonal activation of lymphocytes that affects primarily B cells. In strains of *T. congolense* mice which survived longest, that is, C57Bl/6J and AKR/A, the increase in splenic B cells is less marked.

Different roles are proposed for polyclonal B-cell activation, which can be crucial for early host defense by contributing with Abs specific for a spectrum of conserved structures present in the pathogens. Additionally, polyclonal activation can be a mechanism triggered by microorganisms to escape the host-specific immune response by diluting pathogen-specific Abs while increasing irrelevant antibodies. Accordingly, recently it has been reported that C57Bl/6 mice, resistant to *T. cruzi* infection, had improved parasite-specific humoral responses that were associated with decreased polyclonal B-cell activation. In the context of parasite infection, Bryan et al. [48] study shows that Th2 cytokine responses were associated with amplified polyclonal B-cell activation and diminished specific humoral immunity. This report demonstrate, that polyclonal B-cell activation during acute experimental Chagas disease is not a generalized response and suggests that the nature of humoral immunity during *T. cruzi* infection contributes to host susceptibility. In leishmaniasis visceral, at early times after infection, there is a marked B-cell expansion in the draining lymph nodes of the site of the infection, which persists throughout infection. As early as day 7 after infection, polyclonal antibodies (TNP, OVA, chromatin) were observed in infected mice and the levels appeared comparable to the specific antileishmania response. Although B-cell-deficient JhD BALB/c mice are relatively resistant to infection, neither B-cell-derived IL-10 nor B-cell antigen presentation appears to be primarily responsible for the elevated parasitemia. Interestingly, passive transfer and reconstitution of JhD BALB/c with secretory immunoglobulins (IgM or IgG; specific or nonspecific immune complexes) results in increased susceptibility to *L. infantum* infection [49].

Another potential deleterious role for polyclonal activation is that it could potentially turn on anti-self-responses and lead to autoimmune manifestations during chronic infections. IgG autoantibodies to brain antigens are increased in *P. falciparum*-infected patients and correlate with disease severity in African children [50]. Autoreactive Abs against endocardium and nerves can be detected in mice and humans infected with *T. cruzi* [51, 52] and are thought to be responsible for much of the Chagas' disease pathological damage. Recently, we reported that BAFF-BAFF-R signaling in *T. cruzi* infection partially controls polyclonal B-cell response but not parasite-specific class-switched primary effectors B cells. BAFF (TNF superfamily B lymphocyte stimulator), a crucial factor for the survival of peripheral B cells [53–55] associated to the development of autoimmune disorders [56], is produced early and persists throughout the infection with *T. cruzi*. By BAFF blockade we observed that this cytokine mediates the mature B-cell response and the production of non-parasite-specific IgM and IgG and influences the development of antinuclear IgG [57].

In addition, polyclonal B-cell activation can be responsible for maintenance of memory B-cell responses because of the continuous, unrestricted stimulation of memory B cells whose Ab production may be sustained in the absence of the antigens binding-specific BCR [58].

6. B Cells Influence the Characteristic of Cellular Immune Response Because They Act as APC and Cytokine/Chemokine Producers

Besides being the precursors of the Ab-secreting cells, B cells are committed to do other immune functions such as Ag presentation to T cells or cytokine/chemokine production. It has been widely studied that CD8$^+$ CTL are important for protective *T. cruzi* immunity [59, 60] but generally they are not induced by soluble protein vaccines. However, a mechanism known as cross-priming has been described whereby certain professional APC can induce CD8$^+$ T-cell responses after the uptake of exogenous Ag [61]. Hoft et al. [62] have demonstrated that the APC functions of B cells may be important for the induction of optimal vaccine-induced responses in mice immunized with a mix of CpG and *T. cruzi* transsialidase, an enzyme involved in parasite infectivity. They also demonstrate that mice deficient in B cells (uMT mice) fails to induce protective immunity when they were immunized with CpG and *T. cruzi* transsialidase. This failure of uMT mice to be protected was associated with the absence of *T. cruzi* transsialidase-specific CD8$^+$ T cell response, suggesting that B cells may be important for the cross-priming of CD8$^+$ CTL. In addition, it has been reported that *T. cruzi*-infected B-cell-deficient mice have reduced numbers of CD8$^+$ splenic T cells and impaired generation of central or effector splenic memory T cells [63]. *T. gondii*-infected C57BL/6 mice develop a robust and

uncontrolled Th1 response, and it has been reported that *T. gondii*-primed B cells, but not naive B cells, were able to increase IFN gamma production by splenic T cells *in vivo*. The mechanisms involved may be linked to the presence of membrane-bound TNF on B-cell surface [64].

The fine tune regulation of migratory cells during infection is an important event in which chemokines and their receptors play a leading role. Different reports have demonstrated that the chemokine receptor CCR5 plays a role in systemic protection and cardiac inflammation during *T. cruzi* infection [65, 66]. CCR5$^+$ cells migrate to both mucosal and systemic sites in response to the chemokines CCL3 (MIP-1a), CCL4 (MIP-1b), and CCL5 (RANTES). In line with these reports, Sullivan et al. [67] showed that neutralization of CCL5 in CCR5$^{-/-}$ *T. cruzi*-immune mice results in decreased levels of *T. cruzi*-specific B cell responses and decreased mucosal protection in these mice. They also showed that CCL5 produced by B cells acts in an autocrine manner to increase B cell proliferation and total IgM secretion.

7. Established Memory B Cell Response Is Affected by Protozoan Parasite Infections

A hallmark of adaptive immunity is the ability to generate humoral immunological memory by which memory B cells could respond more rapidly and robustly to re-exposure to a new infection. Interestingly, it has been reported that *T. brucei* infection is capable of abrogating the efficacy of the vaccine-induced protective responses against nonrelated pathogens such as *B. pertussis* [40]. In the same line, Strickland and Sayles [68] showed that *T. gondii* infected mice which were immunized with SRBC had a depression not only in the primary, but also in the secondary humoral immune response, since they showed less IgM and IgG splenic Ab-secreting cells than non-infected control mice. All the data discussed in the present review indicate that protozoan parasites not only affect the development of the cells involved in Ab production [69] but also affect an already established humoral response against other pathogens. Then, the identification of mechanisms able to improve B cell response and, consequently, parasite control will also be beneficial to avoid the deterioration of a memory response to other pathogens.

Conflict of Interests

The authors have declared that no competing interests exist.

Acknowledgments

This work received financial support from CONICET, ANPCyT, and SECYT-UNC to A. Gruppi. We would like to thank to the National Institute of Vaccine (CNPq-Brazil) for paying the charges of this publication.

References

[1] S. Magez, A. Schwegmann, R. Atkinson et al., "The role of B-cells and IgM antibodies in parasitemia, anemia, and VSG

switching in Trypanosoma brucei-infected mice," *PLoS Pathogens*, vol. 4, no. 8, Article ID e1000122, 2008.

[2] T. von der Weid, N. Honarvar, and J. Langhorne, "Gene-targeted mice lacking B cells are unable to eliminate a blood stage malaria infection," *Journal of Immunology*, vol. 156, no. 7, pp. 2510–2516, 1996.

[3] S. Kumar and R. L. Tarleton, "The relative contribution of antibody production and CD8+ T cell function to immune control of Trypanosoma cruzi," *Parasite Immunology*, vol. 20, no. 5, pp. 207–216, 1998.

[4] F. E. Lund, "Cytokine-producing B lymphocytes-key regulators of immunity," *Current Opinion in Immunology*, vol. 20, no. 3, pp. 332–338, 2008.

[5] C. D. Myers, "Role of B cell antigen processing and presentation in the humoral immune response," *FASEB Journal*, vol. 5, no. 11, pp. 2547–2553, 1991.

[6] G. J. Silverman and D. A. Carson, "Roles of B cells in rheumatoid arthritis," *Arthritis Research and Therapy*, vol. 5, no. 4, pp. S1–S6, 2003.

[7] P. Youinou, "B cell conducts the lymphocyte orchestra," *Journal of Autoimmunity*, vol. 28, no. 2-3, pp. 143–151, 2007.

[8] I. A. Abrahamsohn and W. D. Silva, "Antibody dependent cell-mediated cytotoxcity against Trypanosoma cruzi," *Parasitology*, vol. 75, no. 3, pp. 317–323, 1977.

[9] L. F. Umekita, H. A. Takehara, and I. Mota, "Role of the antibody Fc in the immune clearance of Trypanosoma cruzi," *Immunology Letters*, vol. 17, no. 1, pp. 85–89, 1988.

[10] D. A. Bermejo, M. C. Amezcua Vesely, M. Khan et al., "Trypanosoma cruzi infection induces a massive extrafollicular and follicular splenic B-cell response which is a high source of non-parasite-specific antibodies," *Immunology*, vol. 132, no. 1, pp. 123–133, 2011.

[11] C. Ronet, H. Voigt, H. Himmelrich et al., "Leishmania major-specific B cells are necessary for Th2 cell development and susceptibility to L. major LV39 in BALB/c mice," *Journal of Immunology*, vol. 180, no. 7, pp. 4825–4835, 2008.

[12] R. Carsetti, "The development of B cells in the bone marrow is controlled by the balance between cell-autonomous mechanisms and signals from the microenvironment," *Journal of Experimental Medicine*, vol. 191, no. 1, pp. 5–8, 2000.

[13] M. Qing, D. Jones, and T. A. Springer, "The chemokine receptor CXCR4 is required for the retention of B lineage and granulocytic precursors within the bone marrow microenvironment," *Immunity*, vol. 10, no. 4, pp. 463–471, 1999.

[14] T. Nagasawa, "Microenvironmental niches in the bone marrow required for B-cell development," *Nature Reviews Immunology*, vol. 6, no. 2, pp. 107–116, 2006.

[15] J. B. Chung, M. Silverman, and J. G. Monroe, "Transitional B cells: step by step towards immune competence," *Trends in Immunology*, vol. 24, no. 6, pp. 342–348, 2003.

[16] T. T. Su, B. Guo, B. Wei, J. Braun, and D. J. Rawlings, "Signaling in transitional type 2 B cells is critical for peripheral B-cell development," *Immunological Reviews*, vol. 197, pp. 161–178, 2004.

[17] L. A. Herzenberg, "B-1 cells: the lineage question revisited," *Immunological Reviews*, vol. 175, pp. 9–22, 2000.

[18] E. Montecino-Rodriguez, H. Leathers, and K. Dorshkind, "Identification of a B-1 B cell-specified progenitor," *Nature Immunology*, vol. 7, no. 3, pp. 293–301, 2006.

[19] E. Zuniga, E. Acosta-Rodriguez, M. C. Merino, C. Montes, and A. Gruppi, "Depletion of immature B cells during Trypanosoma cruzi infection: involvement of myeloid cells and the cyclooxygenase pathway," *European Journal of Immunology*, vol. 35, no. 6, pp. 1849–1858, 2005.

[20] L. Teixeira, A. Marques, C. S. Meireles et al., "Characterization of the B-cell immune response elicited in BALB/c mice challenged with Neospora caninum tachyzoites," *Immunology*, vol. 116, no. 1, pp. 38–52, 2005.

[21] C. E. Clayton, M. E. Selkirk, and C. A. Corsini, "Murine trypanosomiasis: cellular proliferation and functional depletion in the blood, peritoneum, and spleen related to changes in bone marrow stem cells," *Infection and Immunity*, vol. 28, no. 3, pp. 824–831, 1980.

[22] N. Van Meirvenne, E. Magnus, and P. Büscher, "Evaluation of variant specific trypanolysis tests for serodiagnosis of human infections with Trypanosoma brucei gambiense," *Acta Tropica*, vol. 60, no. 3, pp. 189–199, 1996.

[23] V. Bockstal, P. Guirnalda, G. Caljon et al., "T. brucei infection reduces B lymphopoiesis in bone marrow and truncates compensatory splenic lymphopoiesis through transitional B-cell apoptosis," *PLoS Pathogens*, vol. 7, no. 6, Article ID e1002089, 2011.

[24] V. Bockstal, N. Geurts, and S. Magez, "Acute disruption of bone marrow B lymphopoiesis and apoptosis of transitional and marginal zone B cells in the spleen following a blood-stage plasmodium chabaudi infection in mice," *Journal of Parasitology Research*, vol. 2011, Article ID 534697, 11 pages, 2011.

[25] K. Hayakawa and R. R. Hardy, "Development and function of B-1 cells," *Current Opinion in Immunology*, vol. 12, no. 3, pp. 346–354, 2000.

[26] N. Baumgarth, J. W. Tung, and L. A. Herzenberg, "Inherent specificities in natural antibodies: a key to immune defense against pathogen invasion," *Springer Seminars in Immunopathology*, vol. 26, no. 4, pp. 347–362, 2005.

[27] P. Minoprio, A. Coutinho, S. Spinella, and M. Hontebeyrie-Joskowicz, "Xid immunodeficiency imparts increased parasite clearance and resistance to pathology in experimental Chagas' disease," *International Immunology*, vol. 3, no. 5, pp. 427–433, 1991.

[28] P. Minoprio, M. C. El Cheikh, E. Murphy et al., "Xid-associated resistance to experimental Chagas' disease is IFN-γ dependent," *Journal of Immunology*, vol. 151, no. 8, pp. 4200–4208, 1993.

[29] M. C. Merino, C. L. Montes, E. V. Acosta-Rodriguez, D. A. Bermejo, M. C. Amezcua-Vesely, and A. Gruppi, "Peritoneum from Trypanosoma cruzi-infected mice is a homing site of Syndecan-1neg plasma cells which mainly provide non-parasite-specific antibodies," *International Immunology*, vol. 22, no. 5, pp. 399–410, 2010.

[30] J. Jellusova, U. Wellmann, K. Amann, T. H. Winkler, and L. Nitschke, "CD22 x Siglec-G double-deficient mice have massively increased B1 cell numbers and develop systemic autoimmunity," *Journal of Immunology*, vol. 184, no. 7, pp. 3618–3627, 2010.

[31] Z. Xu, E. J. Butfiloski, E. S. Sobel, and L. Morel, "Mechanisms of peritoneal B-1a cells accumulation induced by murine lupus susceptibility locus Sle2," *Journal of Immunology*, vol. 173, no. 10, pp. 6050–6058, 2004.

[32] T. Ito, S. Ishikawa, T. Sato et al., "Defective B1 cell homing to the peritoneal cavity and preferential recruitment of B1 cells in the target organs in a murine model for systemic lupus erythematosus," *Journal of Immunology*, vol. 172, no. 6, pp. 3628–3634, 2004.

[33] B. Reina-San-Martin, A. Cosson, and P. Minoprio, "Lymphocyte polyclonal activation: a pitfall for vaccine design against infectious agents," *Parasitology Today*, vol. 16, no. 2, pp. 62–67, 2000.

[34] A. O'Garra, R. Chang, N. Go, R. Hastings, G. Haughton, and M. Howard, "Ly-1 B (B-1) cells are the main source of B cell-derived interleukin 10," *European Journal of Immunology*, vol. 22, no. 3, pp. 711–717, 1992.

[35] A. Hoerauf, W. Solbach, M. Lohoff, and M. Rollinghoff, "The Xid defect determines an improved clinical course of murine leishmaniasis in susceptible mice," *International Immunology*, vol. 6, no. 8, pp. 1117–1124, 1994.

[36] B. Babai, H. Louzir, P. A. Cazenave, and K. Dellagi, "Depletion of peritoneal CD5+ B cells has no effect on the course of Leishmania major infection in susceptible and resistant mice," *Clinical and Experimental Immunology*, vol. 117, no. 1, pp. 123–129, 1999.

[37] M. Chen, F. Aosai, H. S. Mun, K. Norose, H. Hata, and A. Yano, "Anti-HSP70 autoantibody formation by B-1 cells in Toxoplasma gondii-infected mice," *Infection and Immunity*, vol. 68, no. 9, pp. 4893–4899, 2000.

[38] M. Chen, F. Aosai, K. Norose, H. S. Mun, and A. Yano, "The role of anti-HSP70 autoantibody-forming VH1-JH1 B-1 cells in Toxoplasma gondii-infected mice," *International Immunology*, vol. 15, no. 1, pp. 39–47, 2003.

[39] F. Martin, A. M. Oliver, and J. F. Kearney, "Marginal zone and B1 B cells unite in the early response against T-independent blood-borne particulate antigens," *Immunity*, vol. 14, no. 5, pp. 617–629, 2001.

[40] M. Radwanska, P. Guirnalda, C. De Trez, B. Ryffel, S. Black, and S. Magez, "Trypanosomiasis-induced B cell apoptosis results in loss of protective anti-parasite antibody responses and abolishment of vaccine-induced memory responses," *PLoS Pathogens*, vol. 4, no. 5, Article ID e1000078, 2008.

[41] A. H. Achtman, M. Khan, I. C. M. MacLennan, and J. Langhorne, "Plasmodium chabaudi chabaudi infection in mice induces strong B cell responses and striking but temporary changes in splenic cell distribution," *Journal of Immunology*, vol. 171, no. 1, pp. 317–324, 2003.

[42] J. Jacob, R. Kassir, and G. Kelsoe, "In situ studies of the primary immune response to (4-hydroxy-3- nitrophenyl)acetyl. I. The architecture and dynamics of responding cell populations," *Journal of Experimental Medicine*, vol. 173, no. 5, pp. 1165–1175, 1991.

[43] A. L. Abreu-Silva, K. S. Calabrese, S. M. N. Cupolilo, F. O. Cardoso, C. S. F. Souza, and S. C. Gonçalves Da Costa, "Histopathological studies of visceralized Leishmania (Leishmania) amazonensis in mice experimentally infected," *Veterinary Parasitology*, vol. 121, no. 3-4, pp. 179–187, 2004.

[44] C. Daniel-Ribeiro, J. de Oliveira-Ferreira, D. M. Banic, and B. Galvao-Castro, "Can malaria-associated polyclonal B-lymphocyte activation interfere with the development of anti-sporozoite specific immunity?" *Transactions of the Royal Society of Tropical Medicine and Hygiene*, vol. 83, no. 3, pp. 289–292, 1989.

[45] R. E. Sacco, M. Hagen, J. E. Donelson, and R. G. Lynch, "B lymphocytes of mice display an aberrant activation phenotype and are cell cycle arrested in G0/G(1A) during acute infection with Trypanosoma brucei," *Journal of Immunology*, vol. 153, no. 4, pp. 1714–1723, 1994.

[46] P. Minoprio, "Parasite polyclonal activators: new targets for vaccination approaches?" *International Journal for Parasitology*, vol. 31, no. 5-6, pp. 588–591, 2001.

[47] W. I. Morrison, G. E. Roelants, K. S. Mayor-Withey, and M. Murray, "Susceptibility of inbred strains of mice to Trypanosoma congolense: correlation with changes in spleen lymphocyte populations," *Clinical and Experimental Immunology*, vol. 32, no. 1, pp. 25–40, 1978.

[48] M. A. Bryan, S. E. Guyach, and K. A. Norris, "Specific humoral immunity versus polyclonal B Cell activation in trypanosoma cruzi infection of susceptible and resistant mice," *PLoS Neglected Tropical Diseases*, vol. 4, no. 7, article e733, 2010.

[49] E. Deak, A. Jayakumar, K. W. Cho et al., "Murine visceral leishmaniasis: IgM and polyclonal B-cell activation lead to disease exacerbation," *European Journal of Immunology*, vol. 40, no. 5, pp. 1355–1368, 2010.

[50] V. Guiyedi, Y. Chanseaud, C. Fesel et al., "Self-reactivities to the non-erythroid alpha spectrin correlate with cerebral malaria in gabonese children," *PLoS One*, vol. 2, no. 4, article e389, 2007.

[51] J. M. Peralta, P. Ginefra, J. C. Dias, J. M. Magalhaes, and A. Szarfman, "Autoantibodies and chronic Chagas's heart disease," *Transactions of the Royal Society of Tropical Medicine and Hygiene*, vol. 75, no. 4, pp. 568–569, 1981.

[52] A. Szarfman, A. Luquetti, and A. Rassi, "Tissue-reacting immunoglobulins in patients with different clinical forms of Chagas' disease," *The American Journal of Tropical Medicine and Hygiene*, vol. 30, no. 1, pp. 43–46, 1981.

[53] P. A. Moore, O. Belvedere, A. Orr et al., "BLyS: member of the tumor necrosis factor family and B lymphocyte stimulator," *Science*, vol. 285, no. 5425, pp. 260–263, 1999.

[54] P. Schneider, F. Mackay, V. Steiner et al., "BAFF, a novel ligand of the tumor necrosis factor family, stimulates B cell growth," *Journal of Experimental Medicine*, vol. 189, no. 11, pp. 1747–1756, 1999.

[55] M. Batten, J. Groom, T. G. Cachero et al., "BAFF mediates survival of peripheral immature B lymphocytes," *Journal of Experimental Medicine*, vol. 192, no. 10, pp. 1453–1466, 2000.

[56] F. MacKay and P. Schneider, "Cracking the BAFF code," *Nature Reviews Immunology*, vol. 9, no. 7, pp. 491–502, 2009.

[57] D. A. Bermejo, M. C. Amezcua-Vesely, C. L. Montes et al., "BAFF mediates splenic B cell response and antibody production in experimental chagas disease," *PLoS Neglected Tropical Diseases*, vol. 4, no. 5, article e679, 2010.

[58] N. L. Bernasconi, E. Traggiai, and A. Lanzavecchia, "Maintenance of serological memory by polyclonal activation of human memory B cells," *Science*, vol. 298, no. 5601, pp. 2199–2202, 2002.

[59] R. L. Tarleton, "Depletion of CD8+ T cells increases susceptibility and reverses vaccine-induced immunity in mice infected with Trypanosoma cruzi," *Journal of Immunology*, vol. 144, no. 2, pp. 717–724, 1990.

[60] R. L. Tarleton, B. H. Koller, A. Latour, and M. Postan, "Susceptibility of β2-microglobulin-deficient mice to Trypanosoma cruzi infection," *Nature*, vol. 356, no. 6367, pp. 338–340, 1992.

[61] C. Kurts, B. W. Robinson, and P. A. Knolle, "Cross-priming in health and disease," *Nature Reviews Immunology*, vol. 10, no. 6, pp. 403–414, 2010.

[62] D. F. Hoft, C. S. Eickhoff, O. K. Giddings, J. R. C. Vasconcelos, and M. M. Rodrigues, "Trans-sialidase recombinant protein mixed with CpG motif-containing oligodeoxynucleotide induces protective mucosal and systemic trypanosoma cruzi immunity involving CD8+ CTL and B cell-mediated cross-priming," *Journal of Immunology*, vol. 179, no. 10, pp. 6889–6900, 2007.

[63] F. Cardillo, E. Postol, J. Nihei, L. S. Aroeira, A. Nomizo, and J. Mengel, "B cells modulate T cells so as to favour T helper type 1 and CD8 + T-cell responses in the acute phase of Trypanosoma cruzi infection," *Immunology*, vol. 122, no. 4, pp. 584–595, 2007.

[64] L. C. Menard, L. A. Minns, S. Darche et al., "B cells amplify IFN-γ production by T cells via a TNF-α-mediated

mechanism," *Journal of Immunology*, vol. 179, no. 7, pp. 4857–4866, 2007.

[65] J. L. Hardison, R. A. Wrightsman, P. M. Carpenter, W. A. Kuziel, T. E. Lane, and J. E. Manning, "The CC chemokine receptor 5 is important in control of parasite replication and acute cardiac inflammation following infection with Trypanosoma cruzi," *Infection and Immunity*, vol. 74, no. 1, pp. 135–143, 2006.

[66] F. S. Machado, N. S. Koyama, V. Carregaro et al., "CCR5 plays a critical role in the development of myocarditis and host protection in mice infected with Trypanosoma cruzi," *Journal of Infectious Diseases*, vol. 191, no. 4, pp. 627–636, 2005.

[67] N. L. Sullivan, C. S. Eickhoff, X. Zhang, O. K. Giddings, T. E. Lane, and D. F. Hoft, "Importance of the CCR5-CCL5 axis for mucosal Trypanosoma cruzi protection and B cell activation," *Journal of Immunology*, vol. 187, no. 3, pp. 1358–1368, 2011.

[68] G. T. Strickland and P. C. Sayles, "Depressed antibody responses to a thymus dependent antigen in toxoplasmosis," *Infection and Immunity*, vol. 15, no. 1, pp. 184–190, 1977.

[69] G. E. Weiss, P. D. Crompton, S. Li et al., "Atypical memory B cells are greatly expanded in individuals living in a malaria-endemic area," *Journal of Immunology*, vol. 183, no. 3, pp. 2176–2182, 2009.

21

Morphology and Developmental Rate of the Blow Fly, *Hemipyrellia ligurriens* (Diptera: Calliphoridae): Forensic Entomology Applications

Nophawan Bunchu,[1,2] **Chinnapat Thaipakdee,**[1] **Apichat Vitta,**[1,2] **Sangob Sanit,**[3] **Kom Sukontason,**[3] **and Kabkaew L. Sukontason**[3]

[1] *Department of Microbiology and Parasitology, Faculty of Medical Science, Naresuan University, Phitsanulok 65000, Thailand*
[2] *Centre of Excellence in Medical Biotechnology, Faculty of Medical Science, Naresuan University, Phitsanulok 65000, Thailand*
[3] *Department of Parasitology, Faculty of Medicine, Chiang Mai University, Chiang Mai 50200, Thailand*

Correspondence should be addressed to Nophawan Bunchu, bunchu_n@hotmail.com

Academic Editor: D. D. Chadee

Hemipyrellia ligurriens (Diptera: Calliphoridae) is a forensically important blow fly species presented in many countries. In this study, we determined the morphology of all stages and the developmental rate of *H. ligurriens* reared under natural ambient conditions in Phitsanulok province, northern Thailand. Morphological features of all stages based on observing under a light microscope were described and demonstrated in order to use for identification purpose. Moreover, development time in each stage was given. The developmental time of *H. ligurriens* to complete metamorphosis; from egg, larva, pupa to adult, took 270.71 h for 1 cycle of development. The results from this study may be useful not only for application in forensic investigation, but also for study in its biology in the future.

1. Introduction

Specimens of blow flies (Diptera: Calliphoridae), especially fly larvae found in corpses and/or at death scenes, can be used as entomological evidence in forensic investigations, that is, estimating the postmortem interval (PMI) and determining toxic substances, antemortem trauma, and whether relocation of remains had occurred [1], as already documented [2–4]. Precise morphological identification of insect specimens is one of very important factors from an applied point of view because they provide relevant evidences for forensic investigations [1].

Of blow fly species, *Hemipyrellia ligurriens* (Wiedemann) is a forensically important blow fly species, as reported previously from cases in Thailand [4] and Malaysia [3]. This fly species is distributed widely, covering Korea, Taiwan, Laos, Singapore, Papua New Guinea, Australia, India, China, The Philippines, Sri Lanka, Malaysia, Indonesia, and Thailand [5–7]. Besides its forensic importance, *H. ligurriens* can be

a nuisance in markets and gardens, and adults also are mechanical vectors of pathogens, due to their attraction to human excreta near human-occupied environments [7]. Previously, morphology of some immature stages (egg, 3rd-instar larvae, and puparium) of *H. ligurriens* has only been studied by observing under a scanning electron microscope and a light microscope [8–13]. For that reason, the available information of *H. ligurriens* is incomplete for identifying all immature evidences which can be found in the death scenes. Moreover, developmental rate of this species which is important data for estimating PMI has not been found in the cited literatures. Moreover, local population-specific developmental data are very important for estimating larval age to determine PMI [14]. To increase the accuracy and precision when applying this information in forensic investigations, the study of all its immature stages and developmental rate is of great interest because the morphology of each one could provide specific features that will become important for a proper identification, and growth data in particular condition could provide essential data for PMI estimation. Therefore, this study aimed to investigate the distinctive characteristics of all its immature stages by observing under the light microscope, to give some important details for identification and to determine its developmental rate, particularly in condition in Phitsanulok province, northern Thailand.

2. Materials and Methods

2.1. Maintenance of H. ligurriens in the Laboratory. The colony of *H. ligurriens* used in this study was obtained originally by collecting adult flies with a sweeping net in areas of Sao Hin Village, Muang Phitsanulok district, Phitsanulok province, Thailand (16°44′18N; 100°13′44E). All adults of *H. ligurriens* were collected in the field and identified based on their morphology, using the taxonomic key of Tumrasvin et al. [5], before being reared further in the laboratory. Flies were reared under natural ambient condition in the open-system rearing room, according to the method of Sukontason et al. [8]. Briefly, adults were maintained in a rearing cage (30 × 30 × 30 cm) and fed with 2 kinds of food: (i) a mixture of 10% (w/v) sugar solution and 1.5% (v/v) multivitamin syrup solution (SEVEN SEA, England) and (ii) fresh pork liver. Fresh pork liver was provided as larval food and oviposition site. The presence of eggs on the pork liver was observed daily. When eggs were found, the pork liver having fly eggs were transferred gently into a larva-rearing box by using forceps, and then, 2 or 3 pieces (≈50 g/day) of fresh pork liver were added in the box as food for the larvae. Some pieces of fresh pork liver were added daily until the larvae in the box stopped feeding and moving or they became the prepupal stage (late 3rd instar). All pieces of pork liver which remained in the rearing box were removed from the box. Only pupae were kept in the rearing box and sealed tightly until emergence of adult was found. After that, the box was placed and opened in a rearing cage in order to release adults from the box to live in the rearing cage. New fly generation was reared continuously as mentioned above.

2.2. Morphology

Egg. In this study, egg specimens of *H. ligurriens* were obtained from the laboratory colony to investigate distinct features by using the potassium permanganate staining technique, according to previous description of Sukontason et al. [9]. The main characteristics, such as morphology of median area surrounding the micropyle, and chorionic sculpturing, were observed and photographed under a light microscope (Olympus, Japan), which was connected to a digital camera (Samsung S700, Korea). Furthermore, the widths and lengths of 30 fly eggs were also measured under the calibrated light microscope. Mean and standard deviation of width and length were analyzed by using the Excel program (Microsoft office Enterprise 2007).

Larva. This study determined the morphology and developmental rate in all larval stage (1st, 2nd, and 3rd instars).

Morphology of all instars was investigated by using the hydroxide clearing method. Briefly, 30 larvae of each instar were obtained from the laboratory colony and sacrificed by placing them into a beaker containing hot water (80°C) for 30 sec to prevent them from shrinking [15]. After that, they were preserved in a small glass bottle containing 70% ethanol. The preserved larvae were dissected individually at two sites to obtain three body portions by using a sharp blade under a stereo microscope (Olympus, Japan), according to the method described by Sukontason et al. [8]. The first cut was positioned across the middle of the second thoracic segment for viewing the internal cephalopharyngeal skeleton and external anterior spiracle. The second cut was positioned across the 11th body segment in order to observe the characteristics of the posterior spiracle. As a clearing method, each pair of anterior and posterior parts was left in a glass plate containing 10% (w/v) potassium hydroxide solution for 1 day (for 1st instar) or 2 days (for 2nd and 3rd instars). Subsequently, the specimens were washed twice with distilled water and neutralized by placing them on a glass plate containing a mixture of 35% ethanol and 1% glacial acetic acid for 30 min. After that, the specimens were dehydrated serially in 50%, 70%, 80%, 95%, and absolute ethanol (RCI LABSCAN, Thailand) for 30 min per alcohol concentration. Dehydrated specimens were transferred onto a glass plate containing xylene (PROLAB, France) and left for 1 min before mounting on a glass slide with 2-3 drops of mounting medium (Permount). A cover slip was placed over each specimen. The permanent slides were left in room temperature for 2 days before observing under a light microscope. The cephalopharyngeal skeleton and posterior spiracle of each instar were photographed under the light microscope, which was connected to the digital camera. In addition, the body length and width of all instars were examined. Thirty of the first instar larvae were measured by using a calibrated microscope, while the second and third instars (*n* = 30 larvae/ each instar) were measured with vernier calipers under a dissecting microscope. Mean and standard deviation of their body width and length were analyzed by using the Excel program.

Pupa. In this stage, morphology of the anterior and posterior parts and coloration change were studied. The anterior and posterior parts of puparia were determined by using the potassium hydroxide clearing technique, previously described by Sukontason et al. [10]. The anterior parts of puparia were the remnants of the contracted head to the fourth segment of puparia after the flies had emerged. They were recruited from the rearing box that flies had already emerged. For each posterior part, the caudal segment of puparium was cut with a sharp blade under the stereo microscope and then transferred by using forceps into the same glass plate as that used for the anterior part. After that, they were soaked in 1% (v/v) dish-washing detergent to remove surface artifacts and/or pork liver tissues. As a clearing method, the anterior and posterior parts were left in a test tube containing 10% (w/v) potassium hydroxide solution, and the test tube was transferred into a water bath set at a temperature of 80°C for 1 h. Then, the specimen process was performed following that described for the larva stage. The important features of anterior and posterior parts were observed and photographed under the light microscope, connected to the digital camera. Number of papillae in each posterior spiracle of 30 anterior parts was counted and calculated for its range. In addition, thirty puparia were measured for their width and length by using vernier calipers. Mean and standard deviation of their width and length were analyzed by using the Excel program. For observation of coloration change, only one puparium which represented the color of the puparia in the rearing box was selected and photographed at 0, 3, 6, 9, 12, 15, 18, 21, and 24 h by using the digital camera. Before taking photographs, puparium was washed in distilled water to remove surface artifacts.

Adult. The young adults, 5–7 days old, were obtained from the rearing caged by using a test tube. They were sacrificed by placing them in a refrigerator ($-4°C$) for 2 h. After that, they were gently pinned with good anatomical arrangement. The important characteristics for identification were examined and photographed under the stereo microscope, connected to the digital camera. In addition, thirty adults were measured for body width and length by using vernier calipers. In this study, body width means the width of the 2nd thoracic segment, and body length is the entire length of the body, ranging from the middle compound eye to the last segment of abdomen. Mean and standard deviation of their body width and length were analyzed by using the Excel program.

2.3. Developmental Rate Assessment. For assessment of the developmental rate, the experiments were performed in the open-system rearing room under the natural ambient temperature of Phitsanulok province, northern Thailand. Temperature and relative humidity were recorded daily by using a thermometer and hygrometer (Thermo-Hygro TM870, China). The ranges (Mean ± SD) of recorded temperature and relative humidity used to determine the development rate of this fly were 26.7 ± 0.61°C and 74 ± 3%, respectively. For each experiment, the developmental rate began at finding of newly hatched larvae (or the 1st instar); the prepupal stage signified the endpoint. The newly hatched larvae were recognized as 0 h old larvae. Developmental rate assessment in this study was studied by separating the newly hatched larvae from the same adult fly colony into 3 groups. Each group consisted of 150–200 newly hatched larvae. They were transferred gently from the rearing box to the new one by using a wet paint brush (number 4). Fresh pork liver (≈ 50 g) was provided daily as a food source for the larvae in each rearing box. The five largest larvae were removed from the rearing box every 3 h. They were sacrificed by placing in hot water ($\approx 80°C$) for 30 seconds to prevent larval shrinkage [15] and were preserved in a small glass bottle containing 70% ethanol. All preserved larvae were measured body lengths by using a calibrated microscope or vernier calipers under a dissecting microscope, depending on their sizes. The relation of larval body lengths and developmental

time was analyzed using the Excel program. In addition, life spans in other stages were recorded and analyzed by using the Excel program.

3. Results

3.1. Morphology

Egg. The egg of *H. ligurriens* was elongated and tapered at both anterior and posterior ends (Figure 1(a)). It measured 1.44 ± 0.11 mm in length and 0.47 ± 0.04 mm in width (Table 1). It took time in this stage for 10.3 ± 0.30 h before moulting to be larval stage (Table 1). The unstained egg was creamy white; while those after staining with 1% potassium permanganate were light brown in color. The median area was located dorsally, positioning a Y-shape that extended from the anterior end to almost the posterior end (Figure 1(b)). The hatching line was upright, thus displaying dark thickening along the median area (Figure 1(a)). The chorionic sculpture appeared as a hexagonal pattern, with its reticular boundary slightly elevating like a net (Figure 1(c)).

Larva. Under observation with the stereo microscope, all larval instars of *H. ligurriens* displayed typical muscoid-shaped vermiform larva that was pointed in anterior and blunt in posterior. The caudal segment had a single pair of posterior spiracles. The first instar was relatively small, measuring 2.62 ± 0.70 mm in length and 0.75 ± 0.35 mm in width, with the developmental time at this stage for 12 ± 0.1 h (Table 1). The cephalopharyngeal skeleton was not well developed (Figure 2(a)), while the posterior spiracle had 2 spiracular slits merging ventrally (Figure 2(e)). The second instar was 6.24 ± 1.67 mm in length and 1.37 ± 0.19 mm in width, with 12 ± 3 h duration at this stage (Table 1). The cephalopharyngeal skeleton was almost complete (Figure 2(b)), while the posterior spiracle had 2 separated spiracular slits with weakly pigmented incomplete peritreme (Figure 2(f)). The size of the 3rd instar was the largest, measuring 12.18 ± 1.31 mm in length and 2.32 ± 0.19 mm in width. The developmental time was also the longest at this stage, being 84 ± 3 h (Table 1). The cephalopharyngeal skeletons of the early and the late 3rd instar were similar (Figures 2(c) and 2(d), resp.). The posterior spiracles of the early (Figure 2(g)) and late 3rd instar (Figure 2(h)) were similar, having 3 separated spiracular slits and complete peritreme with an interslit projection, except for the latter showing a highly pigmented peritreme (Figure 2(h)). The distinctive features of all three instars were summarized in Table 2.

Pupa. The puparium of *H. ligurriens* was of typical coarctate form (Figure 3), which measured 6.82 ± 0.27 mm in length and 2.76 ± 0.11 mm in width (Table 1). Observed color changes revealed that the early puparium was creamy white in color (0 h), with the posterior end still truncated (Figure 3(a)). However, the color of the puparium changed gradually to light yellow brown within 3 h after pupariation (Figure 3(b)) and then to brown after around 15 h

Morphology and Developmental Rate of the Blow Fly, Hemipyrellia ligurriens (Diptera: Calliphoridae): Forensic Entomology Applications

189

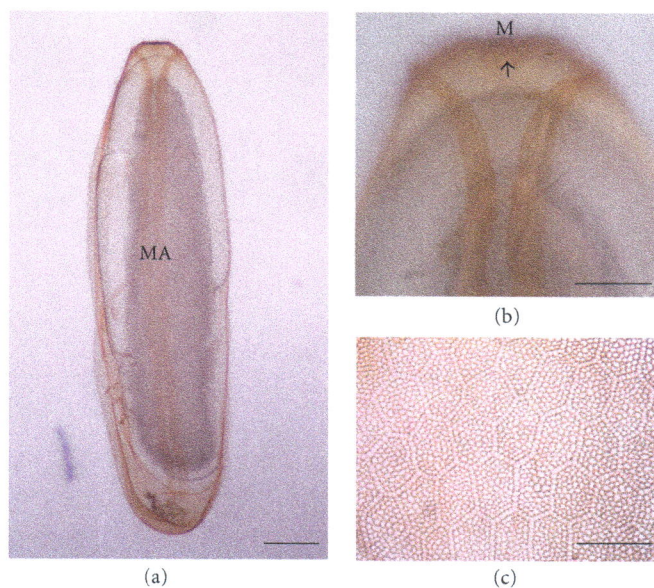

FIGURE 1: Egg of *Hemipyrellia ligurriens* stained with 1% potassium permanganate. (a) Whole egg; MA: median area. (b) Anterior end of egg showing bifurcated plastron and micropyle; M: micropyle. (c) External chorionic sculpture showing hexagonal pattern, with its reticular boundary slightly elevating like a net. Bars = 100 μm for all figures.

TABLE 1: Size and life span (Mean ± SD) of each developmental stage of *Hemipyrellia ligurriens* under natural ambient conditions (26.7 ± 0.61°C, 74 ± 3% RH) of Phitsanulok province, northern Thailand.

Stage	Length (Mean ± SD)	Width (Mean ± SD)	Duration (Mean ± SD)
Egg	1.44 ± 0.11 mm	0.47 ± 0.04 mm	10.3 ± 0.30 hrs
1st instar	2.62 ± 0.70 mm	0.75 ± 0.35 mm	12.0 ± 0.10 hrs
2nd instar	6.24 ± 1.67 mm	1.37 ± 0.19 mm	12.0 ± 3.00 hrs
3rd instar	12.18 ± 1.31 mm	2.32 ± 0.19 mm	84.0 ± 3.00 hrs
Pupa	6.82 ± 0.27 mm	2.76 ± 0.11 mm	152.5 ± 30.70 hrs
Adult	12.23 ± 0.61 mm	2.85 ± 0.25 mm	270.7 ± 30.90 hrs

(Figure 3(f)). Color of puparium changed to dark brown after about 18 h of observation (Figure 3(g)). At 24 h, the puparium showed the darkest brown in color (Figure 3(i)). The duration for the entire pupal stage was 152.5 ± 30.7 h (Table 1).

Observation of the anterior and posterior parts of puparia appeared as a pale yellow-brown color after being treated with 10% potassium hydroxide. The anterior plate, pressed with a cover slip, was trapezoid-shaped with 2 anterior spiracles located at both top ends (Figure 4(a)). Each anterior spiracle consisted of 5 to 7 papillae (Figure 4(b)). The spine observed in the third segment displayed rows of single-pointed tips (Figure 4(c)). Each posterior spiracle appeared three highly pigmented dark brown spiracular slits, with highly pigmented button and weakly pigmented peritreme (Figure 4(d)).

Adult. The adult *H. ligurriens* was 12.23 ± 0.61 mm in length, 2.85 ± 0.25 mm in width (Table 1), and metallic copper green in appearance, with grayish white pollinose on the anterior part of the thorax (Figures 5(a) and 5(b)). Head of the female was dichoptic, but that of male was subholoptic (Figures 5(c) and 5(d)). Grayish facial pollinosity was found in both sexes (Figures 5(c) and 5(d)), and the entire third antennal segment was dark brown or orange ventrally (Figures 5(c) and 5(d)). Squama was whitish (Figure 5(e)). Gena was covered with black hairs (Figure 5(f)). Stem vein was without setulae (Figure 5(g)). Two postsutural acrostichal setae were found on the thorax (Figure 5(h)). Supraconvexity was found pilose hair (Figure 5(i)). All above characteristics were important for identification of this fly adult.

3.2. Developmental Rate Assessment. Assessments of the developmental rate of *H. ligurriens* were determined during the investigation period, based on natural ambient condition of Phitsanulok province. The growth curve of larval stage showed sigmoid form (Figure 6). Data on the developmental rate of rearing revealed that developmental time from newly hatched larvae to beginning of pupariation grows rapidly in a total period of 108 h. Larvae reached their maximum median length at ≈42 h, and pupariation occurred at 108 h. The life spans of all stages were also summarized in Table 1.

4. Discussion

Data related to important insect characteristics for fast and reliable identification and development in the laboratory are crucial prerequisites for appropriate application in forensic investigations. *H. ligurriens* is a forensically important blow fly species that has been recorded in forensic cases in

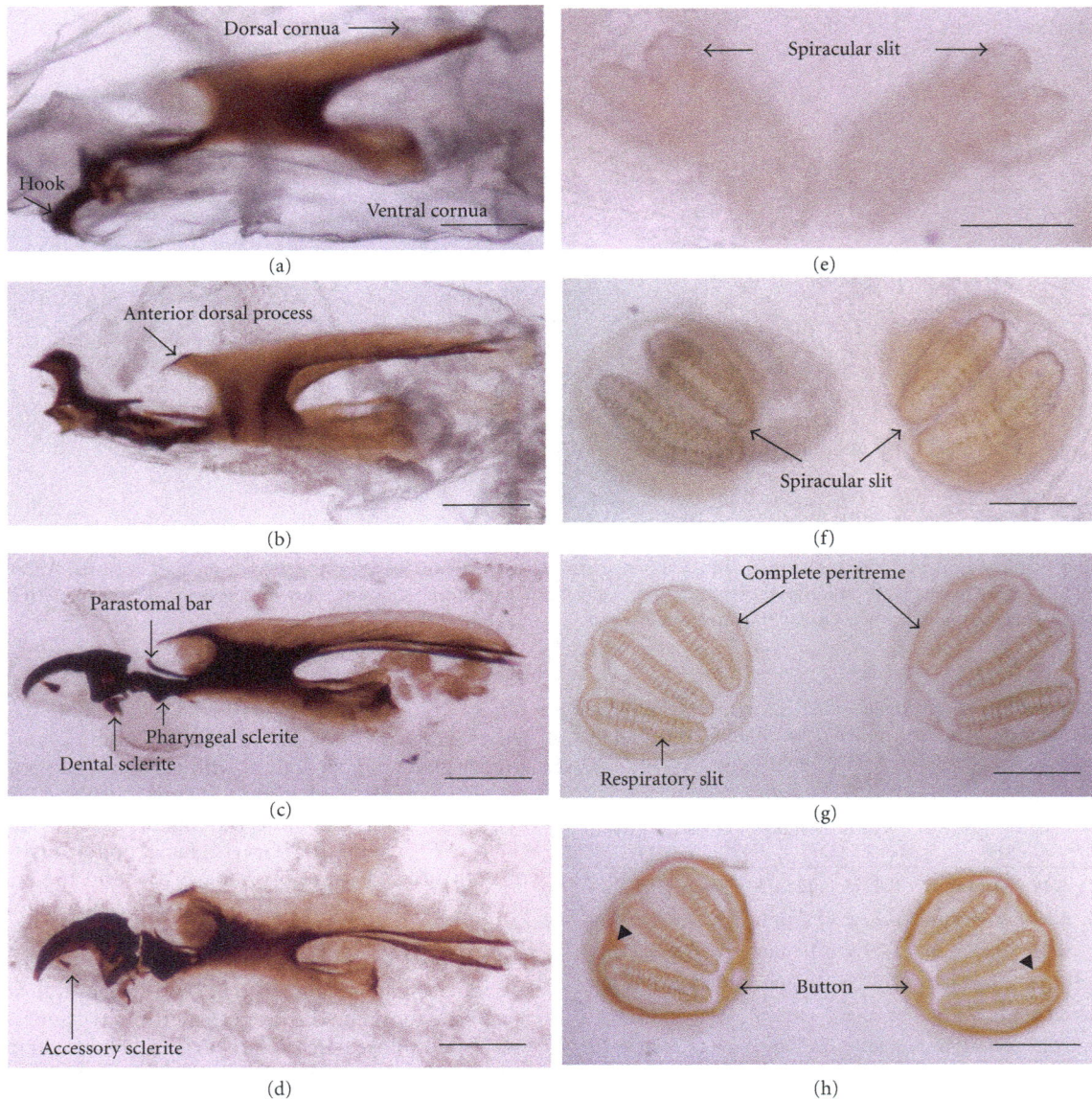

FIGURE 2: Cephalopharyngeal skeletons and posterior spiracles of *Hemipyrellia ligurriens* larvae. Cephalopharyngeal skeletons of (a) 1st instar, (b) 2nd instar, (c) early 3rd instar,. and (d) late 3rd instar. Posterior spiracles of (e) 1st instar, (f) 2nd instar, (g) early 3rd instar, and (h) late 3rd instar. Bars = 100 μm for all figures.

TABLE 2: Comparison of distinctive features of the 1st, 2nd, and 3rd instar larvae of *Hemipyrellia ligurriens*.

Character	1st instar	2nd instar	3rd instar
Cephalopharyngeal skeletal:			
accessory sclerite of cephalopharyngeal skeletal	Absent	Absent	Present
Posterior spiracle:			
button of posterior spiracle	Absent	Absent	Present
peritreme of posterior spiracle	Very weakly pigmented; incomplete peritreme	Weakly pigmented; incomplete peritreme without an interslit projection	Highly pigmented; complete peritreme with an interslit projection
Number of spiracular slit	2 slits merging ventrally	2 separated slits	3 separated slits

Morphology and Developmental Rate of the Blow Fly, Hemipyrellia ligurriens (Diptera: Calliphoridae): Forensic Entomology Applications

191

FIGURE 3: Color changes in puparia of *Hemipyrellia ligurriens* up to 24 h after pupariation. Puparium (a)–(i) at 0, 3, 6, 9, 12, 15, 18, 21, and 24 h, respectively. Bars = 100 μm for all figures.

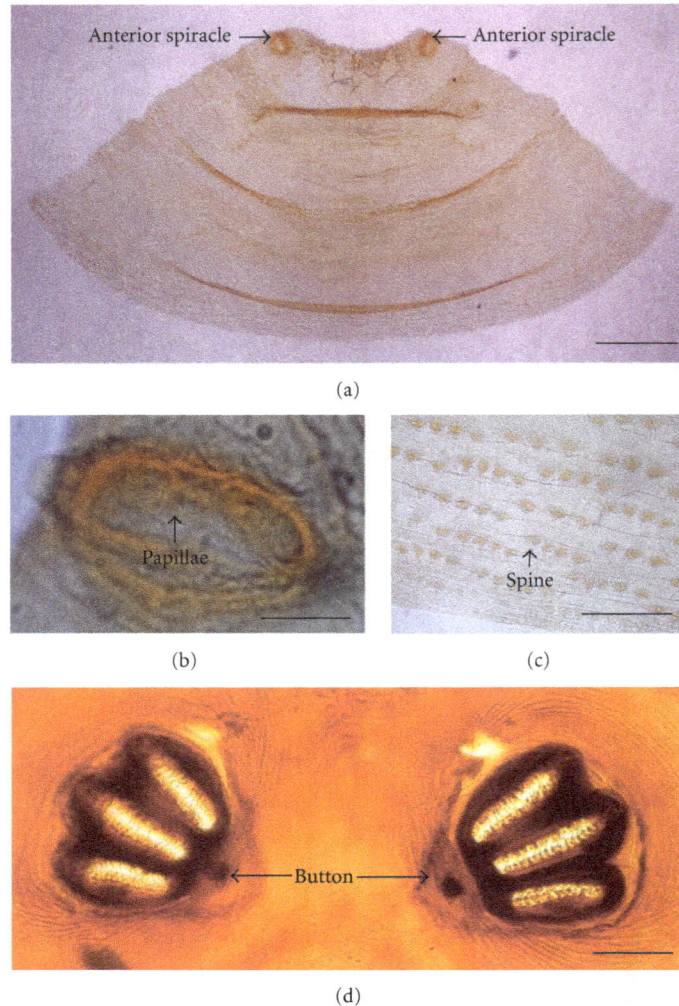

FIGURE 4: Puparium of *Hemipyrellia ligurriens* after treating with 10% KOH. (a) Anterior plate with trapezoid shape. (b) Anterior spiracle with 6 papillae. (c) Spine pattern at the 3rd segment showing single point tip. (d) Posterior spiracles. Bars = 100 μm for all figures.

Malaysia and Thailand [3, 4]. This study focused on the important characteristics for identifying all stages of *H. ligurriens*, based on observation under the light microscope and developmental time of each stage in natural ambient condition with average temperatures and relative humidity, which were 26.7 $\pm$ 0.61°C and 74 $\pm$ 3%, respectively. Although morphology at some stages of *H. ligurriens* has been studied previously using the scanning electron microscope and the light microscope [8–13], this study provided more detailed information on distinctive features and new clues for identification by using simple techniques in order to increase accuracy and precision of forensic application in the future. Moreover, this study provided developmental data of *H. ligurriens*, which can be used in estimating the PMI of corpses on which this fly species has colonized.

Chorionic sculpture as well as width and length of median area were reported previously as an important characteristic in identifying the fly egg [9, 13]. Morphology of the *H. ligurriens* egg, observed in this study by staining with 1% potassium permanganate and observing under the

light microscope, was similar to that in previous work, which investigated by using the scanning electron microscope [11]. Therefore, results from this study confirmed the effectiveness of staining method, initiated by Sukontason et al. [9], for identification purpose of fly eggs. The stained eggs could be observed clearly under the light microscope. When comparing data with previous studies, the size of the *H. ligurriens* egg in this study was larger than that of other blow flies, such as *Lucilia cuprina*, *Ceylonomyia* (= *Chrysomya*) *nigripes*, and *Aldrichina grahami*; however, its size was not different from the sizes of *Chrysomya megacephala* and *Achoetandrus* (= *Chrysomya*) *rufifacies* [9]. Nevertheless, size of fly egg cannot be used as the primary characteristics for identification, according to information from Erzinclioglu [16], who reported that the size of fly egg depended on the dietary levels. Therefore, identification of fly egg should use characteristics of the width of plastron, morphology of plastron area surrounding the micropyle as the main criteria and use size as a supplemental feature for egg identification [9].

Morphology and Developmental Rate of the Blow Fly, Hemipyrellia ligurriens (Diptera: Calliphoridae): Forensic Entomology Applications

193

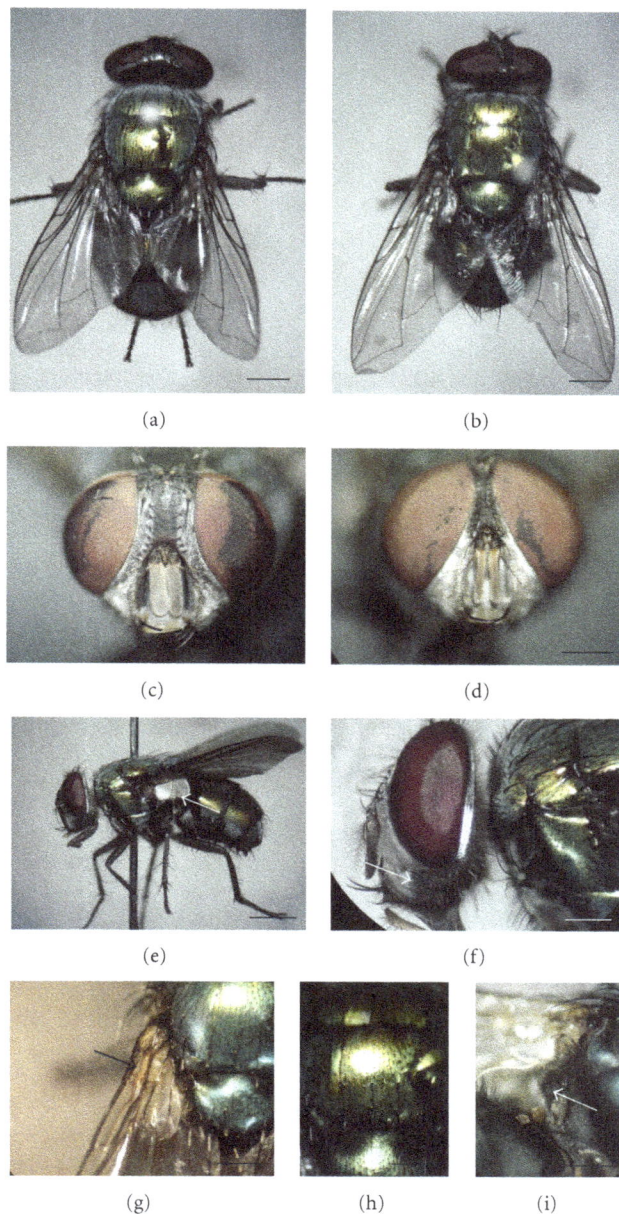

FIGURE 5: Adults and important characteristics for identification of *Hemipyrellialigurrien*. (a) Dorsal view of the female. (b) Dorsal view of the male. (c) Frons of the female showing dichoptic eyes. (d) Frons of the male showing subholoptic eyes. (e) Lateral view of the adult showing whitish squamae (arrow). (f) Higher magnification of the head showing black hairs on the gena (arrow). (g) Wing showing the stem vein without setulae (arrow). (h) Dorsal thorax. (i) Pilose hair on the supraconvexity (arrow). Bars = 200 μm for all figures.

Based on our references, this study was the first study, showing morphology of cephalopharyngeal skeleton and posterior spiracle in all instar larvae by observing under the light microscope. Cephalopharyngeal skeletons and posterior spiracles of larvae were seen clearly as being different in each instar. The results of this study agreed with the previous reports [11–13]. In addition, this study showed degree of pigmentation of peritreme in each instar. It may be a new clue for differentiating age of larvae, especially in between early and late instar. With each developmental instar definitely being different as well as the degree of pigmentation, the data

from this study may be useful in identifying the larval stages and age of larvae of this blow fly in detail as mentioned above.

Study on morphology of pupal stage provided the results which were similar to the previous reports [10]. The results from observation in coloration changes by time of *H. ligurriens* puparia were firstly reported in this study and may be another new supportive evidence to determine at least the approximate age of puparia for increasing accuracy of PMI value.

This study demonstrated some photographs of distinctive features for using in identification of adult *H. ligurriens*.

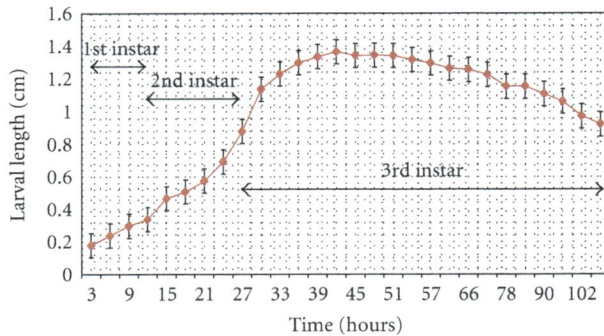

FIGURE 6: Relation between developmental time and the larval length of *Hemipyrellia ligurriens* under natural ambient conditions (26.7 ± 0.61°C; 74 ± 3% RH) of Phitsanulok province, northern Thailand.

These photographs of peculiar characteristics from this study may be useful for people who are not familiar with the terminology in the taxonomic keys. Current taxonomic key for identification of blow fly species in Thailand was given by Kurahashi and Bunchu [17]. By the naked eyes of a nonentomologist, adult *H. ligurriens* seems similar to *A. rufifacies* in appearance. Therefore, the chance of misidentification may occur in both species, especially in Thailand because *A. rufifacies* was the second most predominant blow fly species of Thailand, and coincidence of both species in some provinces in this country was found [18].

This study was the first report that provided developmental time of *H. ligurriens* in all stages. The results showed that the developmental time of *H. ligurriens* under natural conditions in this study (an average temperature of 26.7 ± 0.61°C and relative humidity of 74 ± 3%) was faster than that of *C. megacephala* and *A. rufifacies*, which were studied at an average temperature of 27.4°C [19]. Furthermore, the developmental time of each stage of *H. ligurriens* was different from that of *C. megacephala* and *A. rufifacies*. The developmental rate of each species was specified, although they grew in the same or related conditions [18]. Furthermore, variation of developmental times within a blow fly species was found in geographically different populations [14]. The local population-specific developmental data are needed for estimating larval age to determine PMI. Therefore, the data from this study are very important for further application, particularly in Phitsanulok province.

Data, obtained from both the morphological characteristics and developmental time, fulfill the previous information and provide new supportive evidences for identification of this blow fly species and application in forensic investigation, particularly when *H. ligurriens* is present in human cadavers. Moreover, they may be useful as baseline data in its biology study in the future.

Acknowledgments

This study was supported mainly by a grant provided to N. Bunchu from the Faculty of Medical Science, Naresuan University, and another from the Thailand Research Fund (MRG5280194). The authors also thank the Centre of Excellence in Medical Biotechnology, Faculty of Medical Science, Naresuan University, and the Division of Research Administration, Naresuan University, for support and defraying the publication cost.

References

[1] E. P. Catts and M. L. Goff, "Forensic entomology in criminal investigations," *Annual Review of Entomology*, vol. 37, no. 1, pp. 253–272, 1992.

[2] H. L. Lee, "Recovery of forensically important insect larvae from human cadavers in Malaysia (1993–1996)," *The Malaysian Journal of Pathology*, vol. 18, no. 2, pp. 125–127, 1996.

[3] H. L. Lee, M. Krishnasamy, A. G. Abdullah, and J. Jeffery, "Review of forensically important entomological specimens in the period of 1972–2002," *Tropical Biomedicine*, vol. 21, no. 2, pp. 69–75.

[4] K. Sukontason, P. Narongchai, C. Kanchai et al., "Forensic entomology cases in Thailand: a review of cases from 2000 to 2006," *Parasitology Research*, vol. 101, no. 5, pp. 1417–1423, 2007.

[5] W. Tumrasvin, H. Kurahashi, and R. Kano, "Studies on medically important flies in Thailand. VII. Report on 42 species of calliphorid flies, including the taxonomic keys Diptera: Calliphoridae)," *The Bulletin of the Tokyo Medical and Dental University*, vol. 26, no. 4, pp. 243–272, 1979.

[6] H. Kurahashi and L. Chowanadisai, "Blow flies (Insecta: Diptera: Calliphoridae) from Indochina," *Species Diversity*, vol. 6, pp. 185–242, 2001.

[7] H. Kurahashi, N. Benjaphong, and B. Omar, "Blow flies (Insecta: Diptera: Calliphoridae) of Malaysia and Singapore," *Raffles Bulletin of Zoology*, vol. 5, pp. 1–88, 1997.

[8] K. Sukontason, R. Methanitikorn, K. L. Sukontason, S. Piangjai, and J. K. Olson, "Clearing technique to examine the cephalopharyngeal skeletons of blow fly larvae," *Journal of Vector Ecology*, vol. 29, no. 1, pp. 192–195, 2004.

[9] K. Sukontason, K. L. Sukontason, S. Piangjai et al., "Identification of forensically important fly eggs using a potassium permanganate staining technique," *Micron*, vol. 35, no. 5, pp. 391–395, 2004.

[10] K. L. Sukontason, R. Ngern-Klun, D. Sripakdee, and K. Sukontason, "Identifying fly puparia by clearing technique: application to forensic entomology," *Parasitology Research*, vol. 101, no. 5, pp. 1407–1416, 2007.

[11] K. L. Sukontason, P. Sribanditmongkol, T. Chaiwong, R. C. Vogtsberger, S. Piangjai, and K. Sukontason, "Morphology of immature stages of *Hemipyrellia ligurriens* (Wiedemann) (Diptera: Calliphoridae) for use in forensic entomology applications," *Parasitology Research*, vol. 103, no. 4, pp. 877–887, 2008.

[12] K. Sukontason, P. Sribanditmongkol, R. Ngoen-Klan, T. Klong-Klaew, K. Moophayak, and K. L. Sukontason, "Differentiation between *Lucilia cuprina* and *Hemipyrellia ligurriens* (Diptera: Calliphoridae) larvae for use in forensic entomology applications," *Parasitology Research*, vol. 106, no. 3, pp. 641–646, 2010.

[13] M. S. Ahmad Firdaus, M. A. Marwi, R. A. Syamsa, R. M. Zuha, Z. Ikhwan, and B. Omar, "Morphological descriptions of second and third instar larvae of *Hypopygiopsis violacea* Macquart (Diptera: Calliphoridae), a forensically important

Morphology and Developmental Rate of the Blow Fly, Hemipyrellia ligurriens (Diptera: Calliphoridae): Forensic Entomology Applications

195

fly in Malaysia," *Tropical Biomedicine*, vol. 27, no. 1, pp. 134–137, 2010.

[14] M. B. Gallagher, S. Sandhu, and R. Kimsey, "Variation in developmental time for geographically distinct populations of the common green bottle fly, *Lucilia sericata* (Meigen)," *Journal of Forensic Sciences*, vol. 55, no. 2, pp. 438–442, 2010.

[15] Z. J. O. Adams and M. J. R. Hall, "Methods used for the killing and preservation of blowfly larvae, and their effect on post-mortem larval length," *Forensic Science International*, vol. 138, no. 1–3, pp. 50–61, 2003.

[16] Y. Z. Erzinclioglu, "The value of chorionic structure and size in the diagnosis of blowfly eggs," *Medical and Veterinary Entomology*, vol. 3, no. 3, pp. 281–285, 1989.

[17] H. Kurahashi and N. Bunchu, "The blow flies recorded from Thailand, with the description of a new species of *Isomyia* Walker (Diptera, Calliphoridae)," *Japanese Journal of Systematic Entomology*, vol. 17, no. 2, pp. 217–278, 2011.

[18] N. Bunchu, "Blow fly (Diptera: Calliphoridae) in Thailand: distribution, morphological identification and medical importance appraisals," *International Journal of Parasitology Research*, vol. 4, no. 1, pp. 57–64, 2012.

[19] K. Sukontason, S. Piangjai, S. Siriwattanarungsee, and K. L. Sukontason, "Morphology and developmental rate of blowflies *Chrysomya megacephala* and *Chrysomya rufifacies* in Thailand: application in forensic entomology," *Parasitology Research*, vol. 102, no. 6, pp. 1207–1216, 2008.

22

Alimentary Canal of the Adult Blow Fly, *Chrysomya megacephala* (F.) (Diptera: Calliphoridae)—Part I: Ultrastructure of Salivary Glands

Worachote Boonsriwong,[1] Kabkaew L. Sukontason,[2] Tarinee Chaiwong,[3] Urai Chaisri,[4] Roy C. Vogtsberger,[5] and Kom Sukontason[2]

[1] *Faculty of Allied Health Science, Burapha University, Chonburi 20131, Thailand*
[2] *Department of Parasitology, Faculty of Medicine, Chiang Mai University, Chiang Mai 50200, Thailand*
[3] *College of Medicine and Public Health, Ubon Ratchathani University, Ubon Ratchathani 34190, Thailand*
[4] *Department of Tropical Pathology, Faculty of Tropical Medicine, Mahidol University, Bangkok 10400, Thailand*
[5] *Department of Biology, Midwestern State University, Wichita Falls, TX 76308, USA*

Correspondence should be addressed to Kom Sukontason, ksukonta@med.cmu.ac.th

Academic Editor: Wej Choochote

The salivary gland ultrastructure of the adult male blow fly, *Chrysomya megacephala* (F.) (Diptera: Calliphoridae), was investigated at the ultrastructural level using light microscopy (LM), scanning electron microscopy (SEM), and transmission electron microscopy (TEM). The salivary glands are paired structures composed of a single median deferent duct bifurcated into two long, narrow efferent ducts connected to the coiled tubular glands. The SEM image of the gland surface revealed that the basal lamina is relatively smooth in general, but the whole surface appeared as a trace of rough swollen insertion by intense tracheal ramification. Ultrastructurally, the salivary gland is enclosed within the basal lamina, and interdigitation cytoplasmic extensions were apparent between the adjacent gland cells. The basement membrane appeared infoldings that is similar to the complex of the labyrinth channel. The cytoplasm characteristic of the gland revealed high activity, based on the abundance of noticeable secretory granules, either singly or in an aggregated reservoir. In addition, mitochondria were found to intersperse among rich parallel of arrays rough endoplasmic reticulum. Thick cuticle, which was well-delineated and electron dense, apically lined the gland compartments, with discontinuity of the double-layer cuticle revealing a trace of secretion discharged into the lumen. Gross anatomy of the adult salivary gland was markedly different from that of the third instar of the same species, and structural dissimilarity is discussed briefly.

1. Introduction

Chrysomya megacephala (F.), or the Oriental latrine fly, is a medically important blow fly species. Its adults are not only annoying to humans and animals, but they also act as a potential mechanical disseminator of pathogens that may cause diseases [1, 2]. In some Southeast Asian countries, adult flies cause damage in fermented fish when females oviposit on this product, resulting in infestation of fly larvae [3]. Furthermore, myiasis produced by the larvae of this fly has been reported increasingly in human cases [4–6]. Geographically, *C. megacephala* is distributed widely over continents worldwide, extending from Oriental Asia, Australasia, Africa, Europe, the Mediterranean to North and South America [7–9]. In northern Thailand, systematic surveys revealed that *C. megacephala* is the most common species collected in many habitats, ranging from urban human to rural and forest environments, from which the number of *C. megacephala* collected was more than that for the house fly, *Musca domestica* [10].

Based on the close association of *C. megacephala* with humans and/or animals, which may be either disadvantageous or desirable from a forensic entomology viewpoint, diverse biological knowledge pertaining to this fly is essential

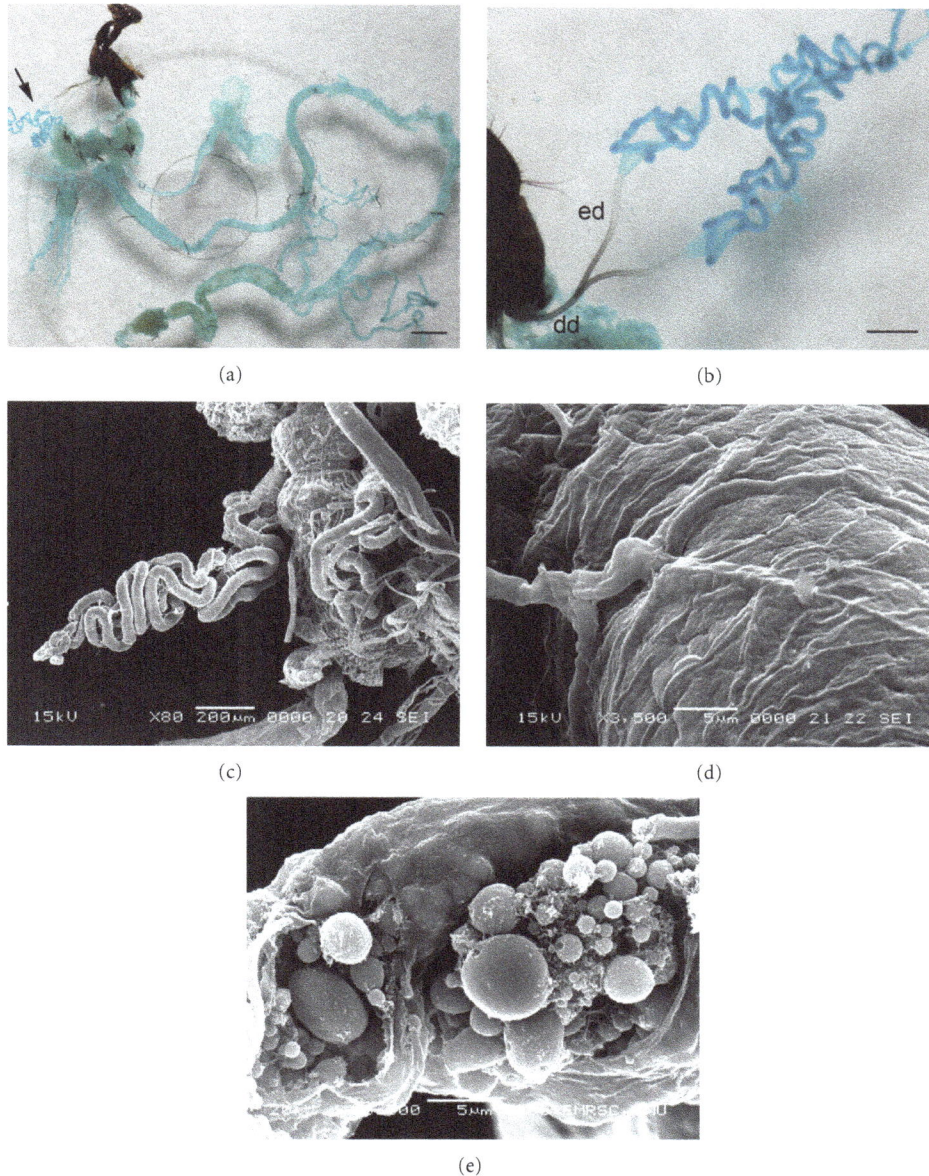

FIGURE 1: Micrographs of the salivary gland in adult male *C. megacephala*. (a) Light micrograph of the whole excised alimentary canal showing salivary glands (arrow) situated in the foregut region. Bar = 10 μm. (b) Light micrograph indicating a translucent single median deferent duct (dd) bifurcated into two long, narrow efferent ducts (ed) connected to the coiled tubular glands. Bar = 5 μm. (c–e; SEM micrographs) (c) Coil salivary gland at one side. (d) Intense tracheal ramification onto the gland surface. (e) Ruptured gland exhibiting numerous rounded vesicles that are probably secretory granules.

in order to manage it. Morphological information on both the gross and ultrastructural level is not exempt in this regard. With respect to the various internal systems of flies, the alimentary one is accountable for successful feeding and it dislodges unused food materials; therefore, ultrastructure of the alimentary canal has been investigated intensively in several species of insects and flies. Examples of this were recorded in the salivary glands of *Calliphora erythrocephala* (Diptera: Calliphoridae) [11], *Dermatobia hominis* (Diptera: Oestridae) [12], *Heliothrips haemorrhoidalis* (Thysanoptera: Thripidae) [13], *Atta sexdens rubropilosa* (Hymenoptera:

Formicidae) [14], *Cimex hemipterus* [15], *Triatoma infestans* [16], *Mahanarva fimbriolata* [17], and *Ixodes ricinus* (Acari: Ixodidae) [18], or the midgut of *D. hominis* [19], *Ceroplastes japonicus* (Hemiptera: Coccidae) [20], and *Belgica Antarctica* (Diptera: Chironomidae) [21]. Ultrastructure of the salivary glands should be researched increasingly, based on verification of salivary gland hypertrophy viruses (SGHVs), which are entomopathogenic and induce salivary gland hypertrophy in dipteran hosts [22]. Recently, two hytrosaviruses, MdSGHV and GpSGHV, were found to induce distinct cytopathology in the salivary glands of *M. domestica* and

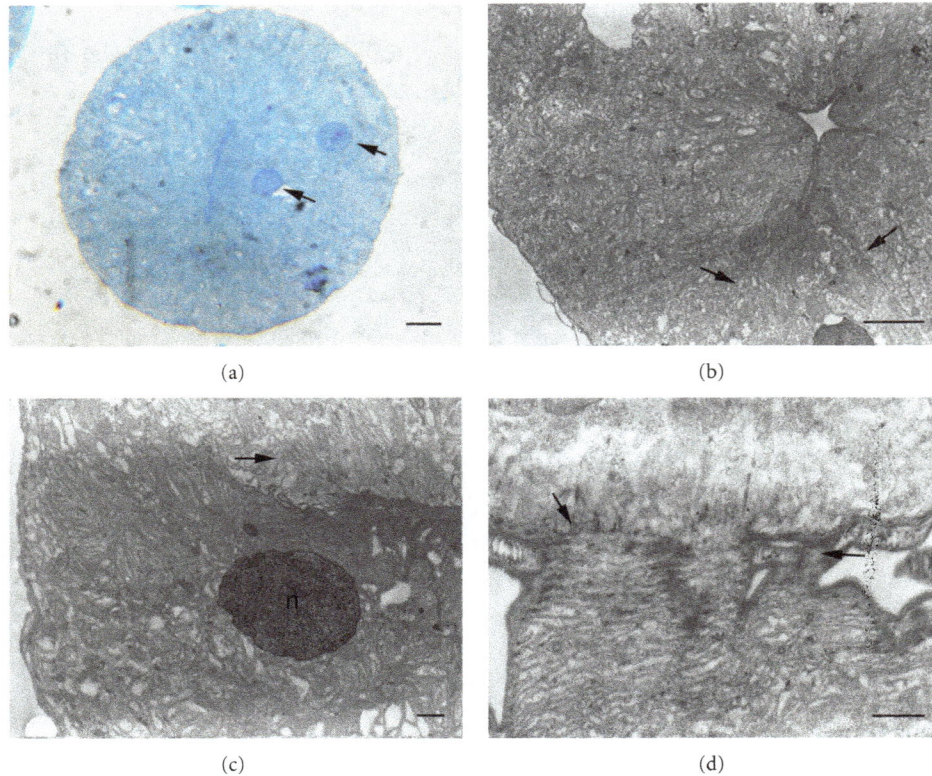

FIGURE 2: Micrographs of the salivary gland in adult male *C. megacephala*. (a) Light micrograph of the thick section showing the narrow central gland lumen. Arrows indicate a large nucleus. Bar = 5 μm. (b–d; TEM micrographs) (b) Compartment of the gland context (arrows). Bar = 5 μm. (c) Baso-lateral region of the gland cell, exhibiting infoldings of the basement membrane (arrow). A large nucleus (n) is observed. Bar = 1 μm. (d) Interdigitation of cytoplasmic extensions between the adjacent gland cells (arrows). Bar = 1 μm.

the tsetse fly, *Glossina pallidipes*, respectively [23], causing pathological symptoms after infection. Thereby, MdSGHV may have potential in a management strategy for house fly populations [24]. As for *C. megacephala*, the alimentary canal ultrastructure of the third instar was documented [25], but information pertaining to the salivary gland in its adult is lacking. Therefore, this study aimed to investigate the salivary gland of adult male *C. megacephala* at the ultrastructural level using light microscopy (LM), scanning electron microscopy (SEM), and transmission electron microscopy (TEM) to provide relevant baseline information. Special attention has been placed on both gross anatomy and cellular structure of this organ.

2. Materials and Methods

2.1. Rearing of C. megacephala . Flies were collected from local marketplaces in Chiang Mai, northern Thailand, and subsequently reared. This fly colony was maintained at ambient temperature (18–27°C), with a natural light/dark photoperiod, in a cabinet in the rearing room of the Department of Parasitology, Faculty of Medicine, Chiang Mai University. Adults were reared on two kinds of food: (i) a mixture of 10% (w/v) multivitamin syrup solution and

(ii) fresh pork liver (used as both a food source and oviposition site), while larvae were provided with fresh pork liver.

2.2. Dissection of the Salivary Gland. Dissection of 7-day-old males was performed in a phosphate buffer solution (pH = 7.4) using two fine forceps under a binocular dissecting microscope (Olympus, Japan). Photographs were taken using an Olympus C4040Z digital camera (Olympus, Japan). The protocol of dissection has been described previously [26].

2.3. Scanning Electron Microscopy (SEM). The dissected salivary gland was processed for the SEM study, with the procedure being recorded [25]. The micrographs were viewed with a JEOL JSM-5910LV scanning electron microscope (JEOL, Japan).

2.4. Transmission Electron Microscopy (TEM). Preparation of the salivary gland specimens was processed, with the protocol being the same as that for the third instar [25]. Ultrathin sections (90 nm) were cut and stained with uranyl acetate and lead citrate before being observed in a Hitachi

Alimentary Canal of the Adult Blow Fly, Chrysomya megacephala (F.) (Diptera: Calliphoridae)—Part I:
Ultrastructure of Salivary Glands

199

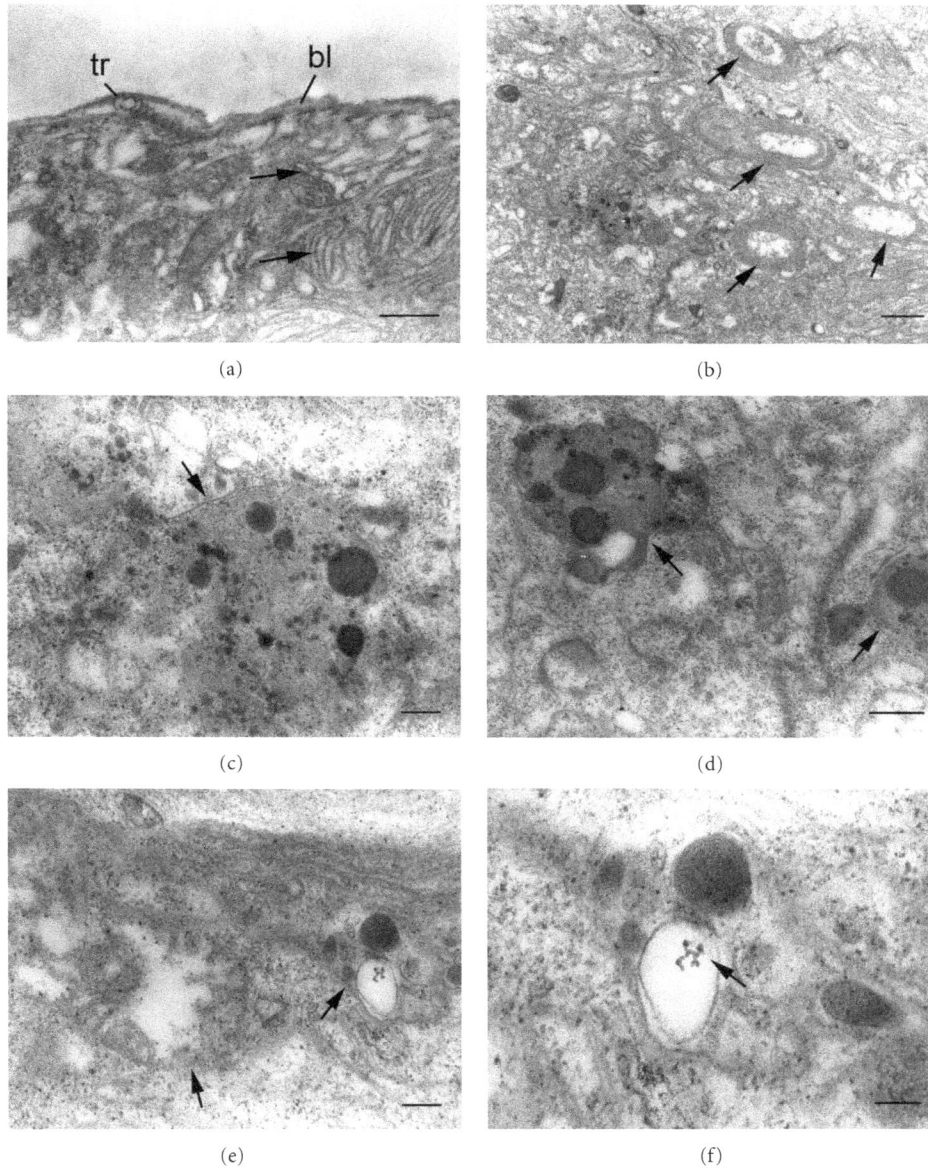

(a)

(b)

(c)

(d)

(e)

(f)

FIGURE 3: TEM micrographs of the salivary gland in adult male *C. megacephala*. (a) Salivary gland cells highlighting the basal lamina (bl), and infoldings of the basement membrane (arrows). Evidence of small tracheole (tr) insertion. Bar = 1 μm. (b) The gland cell cytoplasm showing aggregation of the oval reservoir of secretory materials (arrows). Bar = 1 μm. (c) Secretory reservoirs containing electron dense secretory granules (arrow). Bar = 2 μm. (d) Aggregation of secretory reservoirs (arrows). Bar = 0.5 μm. (e) Variable secretory reservoir (arrows). Bar = 0.2 μm. (f) Secretory reservoir demonstrating globulin-like material (arrow). Bar = 1 μm.

H700 transmission electron microscope (Japan) operated at 100 kV.

3. Results

Observation of the whole excised alimentary canal of male *C. megacephala* under LM demonstrated that the salivary glands were a paired structure situated in the foregut region (Figure 1(a)). The gland comprised a translucent single median deferent duct was inserted into the mouthpart at the junction between the rostrum and haustellum (Figures 1(a) and 1(b)). Two long and narrow efferent ducts, bifurcated from the deferent duct, were connected to the long narrow coiled tubular glands (Figures 1(b) and 1(c)). The SEM image of the gland surface revealed that the basal lamina was relatively smooth in general, but the whole surface appeared as a trace of rough swollen insertion by intense tracheal ramification (Figure 1(d)). A ruptured gland in another SEM image exhibited numerous rounded vesicles inside the gland, which were probably secretory granules of variable sizes (Figure 1(d)).

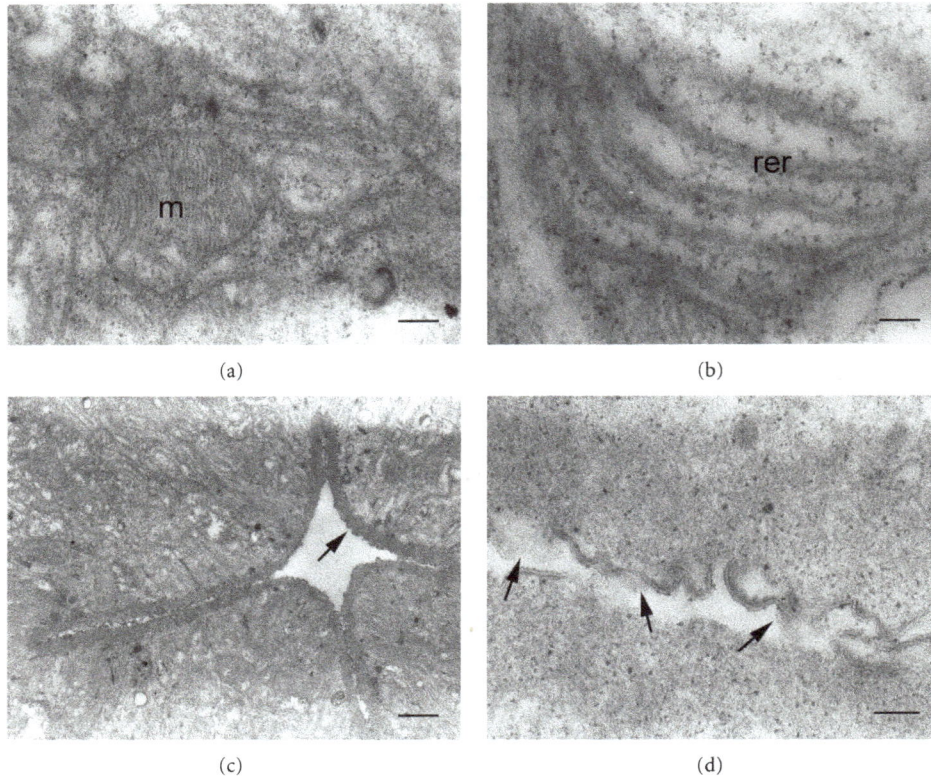

(a)   (b)

(c)   (d)

FIGURE 4: TEM micrographs of the salivary gland in adult male *C. megacephala*. (a) Salivary gland cell presenting mitochondria (m). Bar = 0.2 μm. (b) Parallel arrays rough endoplasmic reticulum (rer). Bar = 0.1 μm. (c) Gland lumen indicating an electron dense thick cuticle (arrow). Bar = 1 μm. (d) High magnification of a cuticle exhibiting trace of secretory discharge into the lumen (arrows). Bar = 0.1 μm.

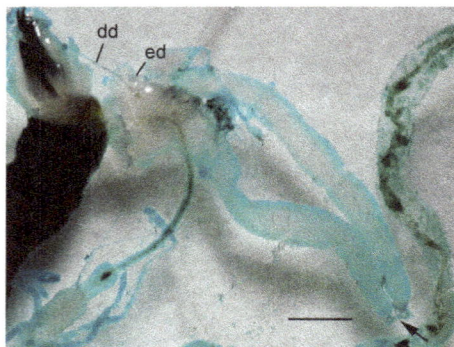

FIGURE 5: Light micrograph of the salivary gland in third instar *C. megacephala* revealing a translucent single median deferent duct (dd) bifurcated into two long, narrow efferent ducts (ed) terminally connected to the large tubular glands (arrow). Bar = 10 μm.

The salivary gland revealed a rounded structure in the thick section that was enclosed within the basal lamina. The narrow gland lumen was situated centrally, with an apparently large nucleus (Figure 2(a)). The gland structure was more noticeable under TEM images, illustrating the

compartment of the gland context (Figures 2(b) and 2(c)). Observation under higher magnification illustrated that the basolateral region of the gland cell exhibited infoldings of the basement membrane (Figure 2(c)). Images also suggested interdigitation cytoplasmic extensions between adjacent gland cells (Figure 2(d)).

Ultrastructurally, the outermost part of the gland was outlined with thin basal lamina, inserted with small tracheoles (Figure 3(a)). The basement membrane was very close to the basal lamina and appeared as infoldings similar to the complex of the labyrinth channel (Figure 3(a)). The gland cell cytoplasm was rich with aggregation of variable electron dense secretory granules (Figures 3(b)–3(f)), some of which had the oval structure of secretory materials wrapped in several thin layers (Figure 3(b)). A cluster of reservoirs containing electron dense secretory granules was prominent (Figures 3(c)–3(f)), while globulin-like material, similar to protein synthesis, was observed in the reservoirs (Figure 3(f)). In addition, the gland cell cytoplasm presented mitochondria (Figure 4(a)) interspersed among rich parallel arrays rough endoplasmic reticulum (Figure 4(b)). When highlighting the gland lumen, infoldings of the basement membrane in each gland compartment tended to orient in the gland lumen. The well-delineated electron density of the thick cuticle apically lined the gland compartments (Figure 4(c)). Close-up investigation of the cuticle facing the lumen exhibited a double-layer of discontinuity, which most likely represented a trace of secretion discharged from the gland into the lumen space (Figure 4(d)).

4. Discussion

This study presented morphology of the salivary gland in male *C. megacephala* at the ultrastructural level. Gross morphology of the salivary gland in adult male *C. megacephala* appears as long narrow coiled tubular glands, which are similar to those in females of the same species (data not shown). However, these glands are different in appearance from those already demonstrated in the third instar, which comprise a large long tube (Figure 5) [25]. This information suggests that much anatomical transformation occurs during metamorphosis from larva to adult. Ultrastructural observation of the salivary gland in adult *C. megacephala* presented a thick cuticle layer lined along the apical border of the gland; however, no cuticle was apparent in the third instar [25]. This phenomenon is similar to that found in *Ceratitis capitata* (Diptera: Tephritidae) [27], of which Riparbelli et al. [27] suggested that the larval glands are histolyzed during metamorphosis and the adult glands form imaginal cells situated at the junction between the gland and duct. This transformation reflects adaptation to difference in feeding behavior between larvae and adults, for example, type of food source, food nutrient, or physiological process, as previously described in *D. hominis* [19].

The gross anatomy of salivary gland observed in *C. megacephala* was greatly similar to the illustrated micrograph in adult *M. domestica* [22]. Interestingly, the results under TEM observation revealed doubt in musculature in the outermost part of the salivary gland in male *C. megacephala*. This phenomenon was related to the absence described in *M. domestica* [23] or in the salivary duct of the female scorpion fly, *Panorpa obtusa* [28]. The membrane infoldings of salivary gland cells was observed as analogous to the complex of the labyrinth channel, in accordance with investigations in *C. capitata* [27], *C. hemipterous* [15], or even mouse [29]. Similarly, the rough endoplasmic reticulum, which is often arranged in parallel arrays observed in male *C. megacephala,* was related to the stack parallel rough endoplasmic reticulum described in *P. obtusa* [28] or the mouse examined using high-resolution scanning electron microscopy [29]. On the gland surface of *C. megacephala*, a trace of intense tracheal ramification was observed by SEM in this study, which implied that a high level of high oxygen was supplied from the respiratory system, similar to descriptions of the salivary gland in *D. hominis* larvae [12]. An abundance of mitochondria, rough endoplasmic reticulum, and high oxygen supplies, therefore represents metabolic activity of the salivary gland in *C. megacephala*.

With respect to the fine secretory products of *C. megacephala*, results obtained from this study exhibited

ultrastructurally heterogenous profiles, with either electron-dense or electron lucid content, or material related to protein synthesis. A large number of reservoirs containing such secretory materials, including possibly enzymes, were predominant. This characteristic was in line with presence of the salivary reservoir in male *P. obtusa*, of which Liu and Hua [28] suggested that the salivary reservoir is used mainly for temporary accumulation and storage of salivary products secreted from salivary cells in the secretory region.

The result examined under TEM clearly demonstrated that salivary secretion was conveyed from the salivary gland cell membrane into the gland lumen (see Figure 4(d)), indicating a particular transportation port of the secretory secretion from the salivary cells. The inclusive data observed demonstrated not only intense salivary secretion, but also variable secretory products synthesized by the salivary gland of male *C. megacephala*. Further investigation into research pertaining to the molecular approach, when identifying these secretory products in adult *C. megacephala*, is therefore merited.

Acknowledgments

The authors appreciate the support given by the Thailand Research Fund (RMU4980007 to K. L. Sukontason) and Royal Golden Jubilee Ph.D. Program (PHD/0114/2547 to K. Sukontason and W. Boonsriwong). they also thank the Faculty of Medicine and Chiang Mai University for subsidizing publication costs. This paper was presented as a poster presentation at the Microscopy Conference held in Kiel, Germany, from August 28 to September 2, 2011.

References

[1] B. Greenberg, *Flies and Disease. Vol. II. Biological and Disease Transmission*, Princeton University Press, Princeton, NJ, USA, 1973.

[2] K. L. Sukontason, M. Bunchoo, B. Khantawa, S. Piangjai, Y. Rongsriyam, and K. Sukontason, "Comparison between *Musca domestica* and *Chrysomya megacephala* as carriers of bacteria in northern Thailand," *Southeast Asian Journal of Tropical Medicine and Public Health*, vol. 38, no. 1, pp. 38–44, 2007.

[3] H. J. de Boer, C. Vongsombath, and J. Kafer, "A fly in the ointment: evaluation of traditional use of plants to repel and kill blowfly larvae in fermented fish," *PLoS One*, vol. 6, no. 12, Article ID e29521, 2011.

[4] S. P. W. Kumarasinghe, N. D. Karunaweera, and R. L. Ihalamulla, "A study of cutaneous myiasis in Sri Lanka," *International Journal of Dermatology*, vol. 39, no. 9, pp. 689–694, 2000.

[5] K. L. Sukontason, P. Narongchai, D. Sripakdee et al., "First report of human myiasis caused by *Chrysomya megacephala* and *Chrysomya rufifacies* (Diptera: Calliphoridae) in Thailand, and its implication in forensic entomology," *Journal of Medical Entomology*, vol. 42, no. 4, pp. 702–704, 2005.

[6] A. C. P. Ferraz, B. Proenea, B. Q. Gadelha et al., "First record of human myiasis caused by association of the species *Chrysomya megacephala* (Diptera: Calliphoridae), *Sarcophaga (Liopygia) ruficornis* (Diptera: Sarcophagidae), and *Musca domestica* (Diptera: Muscidae)," *Journal of Medical Entomology*, vol. 47, no. 3, pp. 487–490, 2010.

[7] H. Kurahashi and F. R. Magpayo, "Blow flies (Insecta: Diptera: Calliphoridae) of the Philippines," *Raffles Bulletin of Zoology*, vol. 48, supplement 9, pp. 1–78, 2000.

[8] A. Martínez-Sánchez, M. A. Marcos-García, and S. Rojo, "First collection of *Chrysomya megacephala* (Fabr.) in Europe (Diptera: Calliphoridae)," *Pan-Pacific Entomologist*, vol. 77, no. 4, pp. 240–243, 2001.

[9] M. S. Olea, M. J. D. Juri, and N. Centeno, "First report of *Chrysomya megacephala* (Diptera: Calliphoridae) in North-western Argentina," *Florida Entomologist*, vol. 94, no. 2, pp. 345–346, 2011.

[10] R. Ngoen-klan, K. Moophayak, T. Klong-klaew et al., "Do climatic and physical factors affect populations of the blow fly *Chrysomya megacephala* and house fly *Musca domestica*?" *Parasitology Research*, vol. 109, pp. 1279–1292, 2011.

[11] M. J. Berridge, B. J. Gupta, J. L. Oschman, and B. J. Wall, "Salivary gland development in the blowfly, *Calliphora erythrocephala*," *Journal of Morphology*, vol. 149, no. 4, pp. 459–482, 1976.

[12] L. G. Evangelista and A. C. R. Leite, "Salivary glands of second and third instars of *Dermatobia hominis* (Diptera: Oestridae)," *Journal of Medical Entomology*, vol. 44, no. 3, pp. 398–404, 2007.

[13] G. del Bene, V. Cavallo, P. Lupetti, and R. Dallai, "Fine structure of the salivary glands of *Heliothrips haemorrhoidalis* (Bouche) (Thysanoptera: Thripidae)," *International Journal of Insect Morphology and Embryology*, vol. 28, no. 4, pp. 301–308, 1999.

[14] J. B. do Amaral and G. M. Machado-Santelli, "Salivary system in leaf-cutting ants (*Atta sexdens rubropilosa* Forel, 1908) castes: a confocal study," *Micron*, vol. 39, no. 8, pp. 1222–1227, 2008.

[15] J. E. Serrão, M. I. Castrillon, J. R. Dos Santos-Mallet, J. C. Zanuncio, and T. C. M. Gonçalves, "Ultrastructure of the salivary glands in *Cimex hemipterus* (Hemiptera: Cimicidae)," *Journal of Medical Entomology*, vol. 45, no. 6, pp. 991–999, 2008.

[16] M. M. Reis, R. M. S. Meirelles, and M. J. Soares, "Fine structure of the salivary glands of *Triatoma infestans* (Hemiptera: Reduviidae)," *Tissue and Cell*, vol. 35, no. 5, pp. 393–400, 2003.

[17] P. H. Nunes and M. I. Camargo-Mathias, "Ultrastructural study of the salivary glands of the sugarcane spittlebug *Mahanarva fimbriolata* (Stal, 1854) (Euhemiptera: Cercopidae)," *Micron*, vol. 37, no. 1, pp. 57–66, 2006.

[18] M. Vancova, K. Zacharovova, L. Grubhoffer, and J. Nebesarova, "Ultrastructure and lectin characterization of granular salivary cells from *Ixodes ricinus* females," *Journal of Parasitology*, vol. 92, no. 3, pp. 431–440, 2006.

[19] L. G. Evangelista and A. C. R. Leite, "Optical and ultrastructural studies of midgut and salivary glands of first instar of *Dermatobia hominis* (Diptera: Oestridae)," *Journal of Medical Entomology*, vol. 42, no. 3, pp. 218–223, 2005.

[20] Y. Xie, W. Liu, Y. Zhang, Q. Xiong, J. Xue, and X. Zhang, "Morphological and ultrastructural characterization of the alimentary canal in Japanese wax scale (*Ceroplastes japonicus* Green)," *Micron*, vol. 42, pp. 898–904, 2011.

[21] J. B. Nardi, L. A. Miller, C. M. Bee, R. E. Lee Jr, and D. L. Denlinger, "The larval alimentary canal of the Antarctic insect, *Belgica antarctica*," *Arthropod Structure and Development*, vol. 38, no. 5, pp. 377–389, 2009.

[22] A. M. M. Abd-Alla, F. Cousserans, A. G. Parker et al., "Genome analysis of a *Glossina pallidipes* salivary gland hypertrophy virus reveals a novel, large, double-stranded circular DNA virus," *Journal of Virology*, vol. 82, no. 9, pp. 4595–4611, 2008.

[23] V. U. Lietze, A. M. M. Abd-Alla, and D. G. Boucias, "Two hytrosaviruses, MdSGHV and GpSGHV, induce distinct cytopathologies in their respective host insects," *Journal of Invertebrate Pathology*, vol. 107, no. 2, pp. 161–163, 2011.

[24] C. J. Geden, T. Steenberg, V. U. Lietze, and D. G. Boucias, "Salivary gland hypertrophy virus of house flies in Denmark: prevalence, host range, and comparison with a Florida isolate," *Journal of Vector Ecology*, vol. 36, no. 2, pp. 231–238, 2011.

[25] W. Boonsriwong, K. Sukontason, J. K. Olson et al., "Fine structure of the alimentary canal of the larval blow fly *Chrysomya megacephala* (Diptera: Calliphoridae)," *Parasitology Research*, vol. 100, no. 3, pp. 561–574, 2007.

[26] W. Boonsriwong, K. Sukontason, R. C. Vogtsberger, and K. L. Sukontason, "Alimentary canal of the blow fly *Chrysomya megacephala* (F.) (Diptera: Calliphoridae): an emphasis on dissection and morphometry," *Journal of Vector Ecology*, vol. 36, no. 1, pp. 2–10, 2011.

[27] M. G. Riparbelli, G. Callaini, and R. Dallai, "Cytoskeleton of larval and adult salivary glands of the dipteran *Ceratitis capitata*. Implication of microfilaments and microtubules in saliva discharge," *Bolletino di Zoologia*, vol. 61, no. 1, pp. 9–17, 1994.

[28] S. Liu and B. Hua, "Histology and ultrastructure of the salivary glands and salivary pumps in the scorpionfly *Panorpa obtusa* (Mecoptera: Panorpidae)," *Acta Zoologica*, vol. 91, no. 4, pp. 457–465, 2010.

[29] L. C. Picoli, F. J. Dias, J. P. M. Issa et al., "Ultrastructure of submandibular salivary glands of mouse: TEM and HRSEM observations," *Microscopy Research and Technique*, vol. 74, pp. 1154–1160, 2011.

Permissions

The contributors of this book come from diverse backgrounds, making this book a truly international effort. This book will bring forth new frontiers with its revolutionizing research information and detailed analysis of the nascent developments around the world.

We would like to thank all the contributing authors for lending their expertise to make the book truly unique. They have played a crucial role in the development of this book. Without their invaluable contributions this book wouldn't have been possible. They have made vital efforts to compile up to date information on the varied aspects of this subject to make this book a valuable addition to the collection of many professionals and students.

This book was conceptualized with the vision of imparting up-to-date information and advanced data in this field. To ensure the same, a matchless editorial board was set up. Every individual on the board went through rigorous rounds of assessment to prove their worth. After which they invested a large part of their time researching and compiling the most relevant data for our readers. Conferences and sessions were held from time to time between the editorial board and the contributing authors to present the data in the most comprehensible form. The editorial team has worked tirelessly to provide valuable and valid information to help people across the globe.

Every chapter published in this book has been scrutinized by our experts. Their significance has been extensively debated. The topics covered herein carry significant findings which will fuel the growth of the discipline. They may even be implemented as practical applications or may be referred to as a beginning point for another development. Chapters in this book were first published by Hindawi Publishing Corporation; hereby published with permission under the Creative Commons Attribution License or equivalent.

The editorial board has been involved in producing this book since its inception. They have spent rigorous hours researching and exploring the diverse topics which have resulted in the successful publishing of this book. They have passed on their knowledge of decades through this book. To expedite this challenging task, the publisher supported the team at every step. A small team of assistant editors was also appointed to further simplify the editing procedure and attain best results for the readers.

Our editorial team has been hand-picked from every corner of the world. Their multi-ethnicity adds dynamic inputs to the discussions which result in innovative outcomes. These outcomes are then further discussed with the researchers and contributors who give their valuable feedback and opinion regarding the same. The feedback is then collaborated with the researches and they are edited in a comprehensive manner to aid the understanding of the subject.

Apart from the editorial board, the designing team has also invested a significant amount of their time in understanding the subject and creating the most relevant covers. They scrutinized every image to scout for the most suitable representation of the subject and create an appropriate cover for the book.

The publishing team has been involved in this book since its early stages. They were actively engaged in every process, be it collecting the data, connecting with the contributors or procuring relevant information. The team has been an ardent support to the editorial, designing and production team. Their endless efforts to recruit the best for this project, has resulted in the accomplishment of this book. They are a veteran in the field of academics and their pool of knowledge is as vast as their experience in printing. Their expertise and guidance has proved useful at every step. Their uncompromising quality standards have made this book an exceptional effort. Their encouragement from time to time has been an inspiration for everyone.

The publisher and the editorial board hope that this book will prove to be a valuable piece of knowledge for researchers, students, practitioners and scholars across the globe.

List of Contributors

Liam Reilly
Ashworth Laboratories, Institute for Immunology and Infection Research, School of Biological Sciences, The University of Edinburgh, King's Buildings, Edinburgh EH9 3JT, UK
Institute of Tropical Medicine Antwerp, Nationalestraat 155, 2000 Antwerp, Belgium

Norman Nausch and Francisca Mutapi
Ashworth Laboratories, Institute for Immunology and Infection Research, School of Biological Sciences, The University of Edinburgh, King's Buildings, Edinburgh EH9 3JT, UK

Nicholas Midzi
Schistosomiasis Section, National Institute of Health Research, Box CY 570, Causeway, Harare, Zimbabwe

Takafira Mduluza
Biochemistry Department, University of Zimbabwe, P.O. Box MP 167, Mount Pleasant, Harare, Zimbabwe
Harvard School of Public Health, Botswana Havard Aids Institute, P. Bag 320, Gaborone, Botswana

James P. Whitcomb, Mary De Agostino, Mark Ballentine, Jun Fu and Mary Ann McDowell
Eck Institute for Global Health, Department of Biological Sciences, University of Notre Dame, Notre Dame, IN 46556, USA

Martin Tenniswood and Jo Ellen Welsh
Cancer Research Center, University at Albany, 1 Discovery Drive, Rensselaer, NY 12144, USA

Margherita Cantorna
Department of Veterinary and Biomedical Sciences, Pennsylvania State University, University Park, PA 16802, USA

Nadine Rujeni, David W. Taylor and Francisca Mutapi
Institute of Immunology and Infection Research, Centre for Immunity, Infection, and Evolution, School of Biological Sciences, University of Edinburgh, Ashworth Laboratories, King's Buildings, West Mains Rd, Edinburgh EH9 3JT, UK

Luis H. Franco and Dario S. Zamboni
Department of Cell Biology, School of Medicine of Ribeirao Preto, University of Sao Paulo, FMRP/USP, 14049-900, Ribeirao Preto, SP, Brazil

Maria Fatima Horta, Barbara Pinheiro Mendes, Eric Henrique Roma, Juan Pereira Macedo and Luciana Souza Oliveira
Departamento de Bioquimica e Imunologia, Instituto de Ciencias Biologicas, Universidade Federal de Minas Gerais, 31270-901 Belo Horizonte, MG, Brazil

Stephen M. Beverley
Department of Molecular Microbiology, Washington University School of Medicine, 660 S. Euclid Avenue, St. Louis, MO 63110, USA

Fatima Soares Motta Noronha and Myrian Morato Duarte
Departamento de Microbiologia, Instituto de Ciencias Biologicas, Universidade Federal de Minas Gerais, 31270-901 Belo Horizonte, MG, Brazil

Leda Quercia Vieira
Nucleo de Pesquisas em Ciencias Biologicas (NUPEB), Instituto de Ciencias Biologicas e Exatas, Universidade Federal de Ouro Preto, Morro do Cruzeiro, 35400-000 Ouro Preto, MG, Brazil

R. A. Khan
Department of Biology, Memorial University of Newfoundland, St. John's, NL, Canada

Rita Medina Costa, Karina Pires de Sousa, Jorge Atouguia, Luis Távora Tavira and Marcelo Sousa Silva
Unidade de Ensino e Investigacao de Clinica Tropical, Centre for Malaria and Tropical Diseases, Instituto de Higiene e Medicina Tropical, Universidade Nova de Lisboa, Rua da Junqueira 100, 1349-008 Lisbon, Portugal

Filipa Santana Ferreira, Joana da Graça Matias Gomes, Rúben Miguel Lopes Rodrigues, Jorge Luís Marques da Silva Atouguia and Sónia Chavarria Alves Ferreira Centeno-Lima
Unidade de Clinica Tropical e Centro de Malaria e Doencas Tropicais-LA, Instituto de Higiene e Medicina Tropical, Universidade Nova de Lisboa, Rua da Junqueira 100, 1349-008 Lisboa, Portugal

Rita Alexandre dos Santos Soares de Bellegarde Machado Sá da Bandeira
Departamento de Pediatria Medica, Hospital Dona Estefania, Rua Jacinta Marto, 1169-045 Lisboa, Portugal

Cláudia Alexandra Cecílio de Sampaio Ferreira Constantino
Departamento de Pediatria Medica, Hospital Dona Estefania, Rua Jacinta Marto, 1169-045 Lisboa, Portugal
Servico de Pediatria Instituto Portugues de Oncologia de Lisboa Francisco Gentil, Rua Professor Lima Bastos, 1099-023 Lisboa, Portugal

Ana Maria Teixeira Duarte Cancela da Fonseca
Unidade de Clinica Tropical e Centro de Malaria e Doencas Tropicais-LA, Instituto de Higiene e Medicina Tropical, Universidade Nova de Lisboa, Rua da Junqueira 100, 1349-008 Lisboa, Portugal
Graduated Program in Areas of Basic and Applied Biology, Instituto de Ciencias Biomedicas Abel Salazar, University of Porto, Rua Dr. Roberto Frias, s/n, 4200-465 Porto, Portugal

Elisama Azevedo and Ana Karina Castro Lima
Laboratorio de Imunologia e Bioquimica de Protozoarios, Departamento de Microbiologia, Imunologia e Parasitologia, FCM, UERJ, Avenida Professor Manuel de Abreu 444 5 andar, Vila Isabel, 20550-170 Rio de Janeiro, RJ, Brazil
Programa de Pos-Graduacao em Microbiologia Medica, Faculdade de Ciencias Medicas, UERJ, 20550-170 Rio de Janerio, RJ, Brazil

Patricia Maria Lourenco Dutra
Programa de Pos-Graduacao em Microbiologia Medica, Faculdade de Ciencias Medicas, UERJ, 20550-170 Rio de Janerio, RJ, Brazil

Veronica P. Salerno and Leandro Teixeira Oliveira
Departamento Biociencias, Escola de Educacao Fisica e Desportos, Universidade Federal do Rio de Janeiro, 21941-599 Rio de Janerio, RJ, Brazil

Rodrigo Terra
Laboratorio de Imunologia e Bioquimica de Protozoarios, Departamento de Microbiologia, Imunologia e Parasitologia, FCM, UERJ, Avenida Professor Manuel de Abreu 444 5 andar, Vila Isabel, 20550-170 Rio de Janeiro, RJ, Brazil
Programa de Pos-Graduacao em Biodinamica do Movimento, EEFD, UFRJ, 21941-599 Rio de Janerio, RJ, Brazil

Roger Wumba, Zanga Josue and Sala Jean
Departement de Medecine Tropicale, Maladies Infectieuses et Parasitaires, Cliniques Universitaires de Kinshasa, Universite de Kinshasa, 747 Kinshasa XI, Democratic Republic of Congo

Menotti Jean
Service de Parasitologie-Mycologie, Hopital Saint-Louis, Assistance Publique-Hopitaux de Paris et Faculte de Medecine Lariboisiere-Saint-Louis, Universite Paris VII, 75010 Paris, France

Longo-Mbenza Benjamin
Faculty of Health Sciences, Walter Sisulu University, Private Bag XI, Mthatha, Eastern Cape 5117, South Africa

Mandina Madone and Kintoki Fabien
Department of Internal Medicine, University of Kinshasa, 783 Kinshasa XI, Democratic Republic of Congo

Thellier Marc
National Center for Malaria Research, AP-HP, CHU Pitie Salpetriere, 75013 Paris, France
AP-HP, Groupe hospitalier Pitie-Salpetriere, Service de Parasitologie-Mycologie, Universite Pierre et Marie Curie, 75013 Paris, France

Kendjo Eric
National Center for Malaria Research, AP-HP, CHU Pitie Salpetriere, 75013 Paris, France

Guillo-Olczyk AC
AP-HP, Groupe hospitalier Pitie-Salpetriere, Service de Parasitologie-Mycologie, Universite Pierre et Marie Curie, 75013 Paris, France

Issa N. Lyimo, Kija R. Nghabi, Monica W. Mpingwa, Ally A. Daraja, Dickson D. Mwasheshe, Nuru S. Nchimbi and Ladslaus L. Mnyone
Biomedical and Environmental Thematic Group, Ifakara Health Institute, P.O. Box 53, Off Mlabani, Ifakara, Morogoro, Tanzania

Dickson W. Lwetoijera
Biomedical and Environmental Thematic Group, Ifakara Health Institute, P.O. Box 53, Off Mlabani, Ifakara, Morogoro, Tanzania
Vector Group, Liverpool School of Tropical Medicine, Liverpool L3 5QA, UK

Rebecca M. Minneman, Monique M. Hennink, Andrea Nicholls, Sahar S. Salek, Francisco S. Palomeque, Amina Khawja and Juan S. Leon
Rollins School of Public Health, Emory University, Atlanta, GA 30322, USA

Lauren C. Albor and Chester C. Pennock
Emory College, Emory University, Atlanta, GA 30322, USA

Hetron Mweemba Munangandu
Section of Aquatic Medicine and Nutrition, Department of Basic Sciences and Aquatic Medicine, Norwegian School of Veterinary Sciences, Ullevalsveien 72, P.O. Box 8146 Dep, 0033, Oslo, Norway

Victor M. Siamudaala
Kavango Zambezi Transfrontier Conservation Area Secretariat, Kasane 821, Gaborone, Botswana

Musso Munyeme
Department of Disease Control, School of Veterinary Medicine, University of Zambia, P.O. Box 32379, Lusaka 10101, Zambia

King Shimumbo Nalubamba
Department of Clinical Studies, School of Veterinary Medicine, University of Zambia, P.O. Box 32379, Lusaka 10101, Zambia

Maria Cecilia F. Almeida, Givaneide S. Lima, Robson P. de Souza, Regis A. Campos and Joanemile P. Figueiredo
Servico de Imunologia, Universidade Federal da Bahia (UFBA), 40110-160 Salvador, BA, Brazil

Luciana S. Cardoso
Servico de Imunologia, Universidade Federal da Bahia (UFBA), 40110-160 Salvador, BA, Brazil
Departamento de Ciencias da Vida, Universidade do Estado da Bahia UNEB, 41.150-000 Salvador, BA, Brazil
Instituto Nacional de Ciencias e Tecnologia em Doencas Tropicais (INCT-DT/CNPq-MCT), Brazil

Alvaro A. Cruz
ProAR, Nucleo de Excelencia em Asma, Universidade Federal da Bahia 40110-160 Salvador, BA, Brazil

Ricardo R. Oliveira
Servico de Imunologia, Universidade Federal da Bahia (UFBA), 40110-160 Salvador, BA, Brazil
Instituto Nacional de Ciencias e Tecnologia em Doencas Tropicais (INCT-DT/CNPq-MCT), Brazil
Faculdade de Farmacia, UFBA 40110-160 Salvador, BA, Brazil

Edgar M. Carvalho and Maria Ilma Araujo
Servico de Imunologia, Universidade Federal da Bahia (UFBA), 40110-160 Salvador, BA, Brazil
Instituto Nacional de Ciencias e Tecnologia em Doencas Tropicais (INCT-DT/CNPq-MCT), Brazil
Escola Bahiana de Medicina e Sadude Publica, EBMSP, Salvador, Bahia, Brazil

Malinee Anantaphruti, Urusa Thaenkham, Teera Kusolsuk, Wanna Maipanich, Surapol Saguankiat, Somjit Pubampen and Orawan Phuphisut
Department of Helminthology, Faculty of Tropical Medicine, Mahidol University, 420/6 Ratchawithi Road, Bangkok 10400,Thailand

Thales A. Barcante
Departamento de Medicina Veterinaria, Pontificia Universidade Catolica de Minas Gerais (PUC Minas), 30535-901 Belo Horizonte, MG, Brazil
Departamento de Medicina Veterinaria, Preventiva, Universidade Federal de Lavras, 37200-000 Lavras, MG, Brazil

Joziana M. P. Barcante
Departamento de Medicina Veterinaria, Preventiva, Universidade Federal de Lavras, 37200-000 Lavras, MG, Brazil

Ricardo T. Fujiwara and Walter S. Lima
Departamento de Parasitologia, Universidade Federal de Minas Gerais, 31270-901 Belo Horizonte, MG, Brazil

Carmen Aranzamendi and Elena Pinelli
Centre for Infectious Disease Control Netherlands, National Institute for Public Health and the Environment (RIVM), P.O. Box 1, 3720 BA Bilthoven, The Netherlands

Ljiljana Sofronic-Milosavljevic
Institute for the Application of Nuclear Energy (INEP), University of Belgrade, Banatska 31b, 11080 Belgrade, Serbia

Moses J. Chimbari
Okavango Research Institute, University of Botswana, P. Bag 285, Maun, Botswana

J. N. Boampong, E. O. Ameyaw, B. Aboagye and K. Asare
Department of Biomedical and Forensic Sciences, University of Cape Coast, Cape Coast, Ghana

S. Kyei
Department of Optometry, School of Physical Sciences, University of Cape Coast, Cape Coast, Ghana

J. H. Donfack
Department of Biomedical Sciences, Faculty of Sciences, University of Dschang, P.O. Box 067, Dschang, Cameroon

E. Woode
Department of Pharmacology, Faculty of Pharmacy and Pharmaceutical Sciences, College ofHealth Sciences, KNUST, Kumasi, Ghana

Maria C. Amezcua Vesely, Daniela A. Bermejo, Carolina L. Montes, Eva V. Acosta-Rodriguez and Adriana Gruppi
Centro de Investigaciones en Bioquimica Clinica e Inmunologia (CIBICI-CONICET), Departamento de Bioquimica Clinica, Facultad de Ciencias Quimicas, Universidad Nacional de Cordoba, Haya de la Torre y Medina Allende, Ciudad Universitaria, 5000 Cordoba, Argentina

Nophawan Bunchu and Apichat Vitta
Department of Microbiology and Parasitology, Faculty of Medical Science, Naresuan University, Phitsanulok 65000, Thailand
Centre of Excellence in Medical Biotechnology, Faculty of Medical Science, Naresuan University, Phitsanulok 65000, Thailand

Chinnapat Thaipakdee
Department of Microbiology and Parasitology, Faculty of Medical Science, Naresuan University, Phitsanulok 65000, Thailand

Sangob Sanit, Kom Sukontason and Kabkaew L. Sukontason
Department of Parasitology, Faculty of Medicine, Chiang Mai University, Chiang Mai 50200, Thailand

Worachote Boonsriwong
Faculty of Allied Health Science, Burapha University, Chonburi 20131, Thailand

Kabkaew L. Sukontason and Kom Sukontason
Department of Parasitology, Faculty of Medicine, Chiang Mai University, Chiang Mai 50200, Thailand

Tarinee Chaiwong
College of Medicine and Public Health, Ubon Ratchathani University, Ubon Ratchathani 34190, Thailand

Urai Chaisri
Department of Tropical Pathology, Faculty of Tropical Medicine, Mahidol University, Bangkok 10400, Thailand

Roy C. Vogtsberger
Department of Biology, Midwestern State University, Wichita Falls, TX 76308, USA